D0886990

PIGMENT OF THE IMAGINATION

PIGMENT OF
THE IMAGINATION
A History of Phytochrome Research

LINDA C. SAGE
St. Louis, Missouri

ACADEMIC PRESS, INC.
Harcourt Brace Jovanovich, Publishers
San Diego   New York   Boston
London   Sydney   Tokyo   Toronto

581·192
S 12 p

Cover photograph: *Catalpa bignonioides.* See Figure 7.7, page 97 (From Borthwick, 1957).

This material is based upon work supported by the U.S. Department of Agriculture, Agricultural Research Service, under agreement 59-32U4-8-25. Any opinions, findings, conclusions or recommendations expressed in this publication are those of the author and do not necessarily reflect the view of the U.S. Department of Agriculture.

This book is printed on acid-free paper. ∞

Copyright © 1992 by ACADEMIC PRESS, INC.

All Rights Reserved.

No part of this publication may be reproduced or transmitted in any form or by any means, electronic or mechanical, including photocopy, recording, or any information storage and retrieval system, without permission in writing from the publisher.

Academic Press, Inc.
1250 Sixth Avenue, San Diego, California 92101

United Kingdom Edition published by
Academic Press Limited
24–28 Oval Road, London NW1 7DX

Library of Congress Cataloging-in-Publication Data

Sage, Linda C.
    Pigment of the imagination : a history of phytochrome research /
Linda C. Sage.
        p.   cm.
    Includes index.
    ISBN 0-12-614445-1
    1. Phytochrome.   2. Phytochrome--Research--History.   I. Title.
QK898.P67S24   1992
581.19'218--dc20   91-38325
                                                              CIP

PRINTED IN THE UNITED STATES OF AMERICA
92 93 94 95 96 97   BB   9 8 7 6 5 4 3 2 1

To Joe Varner,
who knew this story should be told.

University Libraries
Carnegie Mellon University
Pittsburgh, Pennsylvania 15213

Contents

7
The 1950s

8
High-Energy Responses

9
Detection

10
Partial Purification

11
Properties of Phytochrome

12
In Vivo Spectrophotometric Studies

13
Seed Germination and Flowering

14
Photoperiodic Timing

15
Mougeotia

16
Mode of Action

17
The Hartmann Model

PART II _____

18
Large Phytochrome

19
Gene Regulation Hypothesis

20
Dark Transformations

21
Localization

22
Properties of Large Phytochrome

23
Membrane Hypothesis

24
High-Irradiance Responses

25
Very Low Fluence Response; A Dimer Model

32
Gene Regulation

33
Autoregulation

34
Phytochrome and Flowering

Preface

Of the many intricate and beautiful control mechanisms living organisms have evolved to optimize their survival in a variable and changing environment, none is more elegant than the phytochrome system of plants.

Warren L. Butler

Two and a half years ago, I began a journey through the United States, Europe, and Japan to interview scientists who study phytochrome, a bright-blue plant pigment. Between these travels, I journeyed through time, sifting through the letters, personnel records, and publications of those who deduced the pigment's existence. An unlikely trio of government workers led the way: "A laid-back, shy, introspective seed researcher, a very sharp, I-can-do-it-all, politically astute smoothie, and an oddball nonconformist genius from the south trained in the physical chemistry of soils."

The work revealed that a photoreversible pigment controls seed germination, emergence from the soil, neighbor detection, and daylength perception. Thus weed seeds sprout after the barest glimpse of light, seedlings grow sturdy and green when they poke through soil, shade-avoiding plants overtop their neighbors, and members of the same species flower at the same time of year. For plants, which depend on light as a source of energy, also obtain information from light. Unable to better their location, they monitor light quality second by second and adjust their development accordingly. They also distinguish between light and dark—day and night and growth below and above soil. This constant wariness involves at least three pigment systems that collectively scan the entire spectrum. But only the pigment phytochrome has been extracted from plants.

---

1. Arthur W. Galston, personal communication.

The account spans 80 years, documenting an entire field of scientific research. It focuses on the development of new ideas and on the researchers who, through flashes of inspiration, endless labor, and judicious collaboration, uncovered the sophistication of the enigmatic phytochrome system. Events in Part I are arranged in chronological order as U. S. Department of Agriculture researchers discover the "photoperiodic pigment" and extract it from seedlings. Topics in Part II are also in chronological order, but each chapter covers the last 20 years. Thus Chapter 23—Membrane Hypothesis—is placed in the mid-1970s, when interest in this topic peaked.

Footnotes make the text accessible to students and scientists in other fields. Quotations phrased in the present tense that lack citations were obtained by personal communication. In the interests of readability, biographical information and chapters on phytochrome intermediates and phytochrome ubiquitination were removed in revising this book prior to publication. The unabridged manuscript is deposited in the National Agricultural Library, Beltsville, Maryland.

I would like to thank those quoted on the following pages for sharing their expertise. I am also indebted to the staff of Washington University Biology Library and the Missouri Botanical Garden Library.

PART I

CHAPTER 1

Daylength and Flowering

Two men were lunching at a government research farm, on the site of the present Pentagon. They had eaten together for years, but this day their conversation veered in a new direction. The younger, shorter man, 38-year-old Harry Ardell Allard (see Figs. 1-2 and 4-6) had a keen, naturalist's eye and wanted to share his latest observation. The previous fall, he had taken tobacco stumps into a greenhouse to protect them from frost. By Christmas, they were flowering, even though the plants from which they were cut had grown only leaves all summer. By April, they stopped flowering promptly, though little in the greenhouse had changed. So there seemed to be "a definite and sharp seasonal incidence, a flowering impulse invariably associated with late autumn and winter conditions" (Allard, 1931).

Wightman Wells Garner (Fig. 1-1), soon to be 43, brought his sharp, analytical mind to bear on the problem. When temperature and quantity of light as flower-promoting factors were ruled out, one possibility remained. "In the course of the casual conversation, we thought of length of day," Allard recalled (Allard, 1931). "At this stage, the problem appeared to assume a somewhat hopeless trend" (Murneek, 1948).

Hopeless or not, the two men studied the effects of daylength on flowering. Combining simple experiments with clever deductions, they established a field of research that would lead to better understanding—and control—of every stage of plant development.[1]

---

1. When work began in July 1918, the effects of daylength on living organisms were largely unsuspected. It was known in Japan, however, that extending days with artificial light hastened the onset in early spring of twittering in the Japanese bush warbler (Senkado-Sancho, 1799). Also, several workers had noted early blooming when certain plants were moved to the continuous summer days of northern Scandinavia (Schübeler, 1879) or when days were extended with artificial light (Bailey, 1891, 1892, 1893; Rane, 1894; Corbett, 1899).

3

FIGURE 1-1   Wightman Wells Garner. Courtesy of
Hunt Institute for Botanical Documentation, Carnegie
Mellon University, Pittsburgh, Pennsylvania.

JULIEN TOURNOIS

Two European researchers had already made a connection between
daylength and time of flowering. In 1910, a doctoral student at the Labora-
toire de Botanique de l'École Normale Supérieure in Paris, Julien Tournois, began to study sexuality in hops (*Humulus japonicus and Humulus
lupulus*), which normally bear male and female flowers on different plants.
Hoping to get a head start in 1911, he sowed seeds in February. To his
surprise, the plants burst into flower in April, even though they were only 6
inches tall (Tournois, 1912). The hop is normally a straggly, twining plant
whose main stem grows a meter long before blooming.

When plants sown in January 1912 bloomed in March, Tournois began to
think that daylength might affect time of reproduction. So he wondered if
shortened days would hasten flowering during the normal growing season.
In April 1912 and 1913, he sowed Japanese hop seeds in well-ventilated
garden frames. The plants on normal daily cycles grew tall and produced
nine pairs of leaves before flowering modestly on the main stem at the
beginning of July. Those covered with a dark screen for all but 6 h daily
grew just taller than the ankle and developed only three or four pairs of
leaves before starting to flower vigorously on June 20. "Chez le Houblon
japonais," Tournois decided, "la diminution de l'éclairement quotidien
pendant la période normale de végétation provoque des floraisons
progénétiques"[2] (Tournois, 1914).

2. In the Japanese hop, a decrease in daylength during the normal growth period provokes
some floral reproduction.

Obtaining similar results with hemp, he concluded that young plants flower when exposed to short days from germination onward. And he deduced that night length, rather than the brevity of the day, was the determining factor (Tournois, 1914).

Tournois received his doctorate in June 1914 but was then called to the front. On November 27, at Mourmelon-le-Grand, he was struck in the head by a bullet (Desroche, 1915).

GEORG ALBRECHT KLEBS

In Heidelberg, meanwhile, the distinguished plant physiologist Georg Albrecht Klebs was gathering evidence that plant development is influenced by the environment rather than being solely "the expression of the inscrutable inner nature of plants" (Klebs, 1910). From an extensive series of experiments with an alpine plant, the house leek (*Sempervivum funkii*), he concluded that plants must reach a certain stage of maturity before they can flower and that nutritional status determines whether flowering will then occur. Thus house leeks were able to bloom in the third summer of life but remained vegetative when cultivated early that year on a warm, moist bed rich with manure. Two or 3 weeks of darkness in April also prevented sexual reproduction. "The experiments prove," said Klebs, "that the transformation of a plant, ripe to flower, into a vegetative one is possible on the one hand by means of an increase of temperature and manure (that is to say, of inorganic salts), and, on the other hand, by means of a decrease of carbon assimilation" (Klebs, 1910).

Klebs had been unable to alter the season of blooming, however. For 10 years, microscopic flower buds had appeared with amazing regularity between the end of April and beginning of May; flowering then ceased in June. Low temperatures only slightly delayed flowering.

During the winter of 1912, Klebs exposed house leeks to a few days of continuous electric illumination. The plants flowered 3 weeks later, under conditions where they otherwise would have remained vegetative.

Klebs decided that flowering might be similarly triggered in nature. As days lengthened after the spring equinox, they must eventually become long enough to promote blooming. And because even house leeks with ample stored food normally failed to flower in winter, the additional hours of light seemed not to act simply by increasing the carbon–nitrogen ratio: "Das Licht wirkt wohl nicht als ernährender Faktor, sondern mehr katalytisch"[3] (Klebs, 1913).

---

3. Light operates perhaps not as a nutritive factor, but, on the contrary, more catalytically.

During 1914 and 1915, Klebs discovered that floral induction required a certain number of hours of light per day, which varied according to temperature. Because there was no simple relationship between light quantity and the extent or speed of blooming, he deduced that it is "nicht allein auf die Lichtmenge ankommt, sondern daß die Unterbrechung durch die Dunkelheit einen gewissen hemmenden Einfluß ausübt"[4] (Klebs, 1918).

To make a convincing case, Klebs needed more data. But he died in 1918, just before his 61st birthday.

GARNER AND ALLARD

Initially unaware of these studies, Garner and Allard were working for the U.S. Department of Agriculture (USDA) at Arlington Experiment Farm in Virginia, where Allard was determining "why the Maryland Mammoth Tobacco failed to flower in season like other commercial varieties, and why it usually failed to flower in the field, so that seed could not be secured before frost killed the plants" (Allard, 1931). Maryland Mammoth was a giant strain of tobacco that appeared spontaneously in a field of Maryland Narrowleaf tobacco in 1906. It had enormous commerical potential because it could grow to twice the size of a person and yield as many as a hundred leaves.

Allard found that "dry soils, pot-bound conditions or stunting due to any other soil conditions" (Allard, 1931) did not affect time of flowering. But he noticed that the change from vegetative to reproductive growth seemed to be tied to a seasonal change. First, there was the observation with the stumps of field plants. Second, he and Garner supposed that Maryland Mammoth would eventually flower at a certain age, so they sowed seed in the greenhouse in fall to get an early start. But the young plants began to flower along with the stumps (see Borthwick, 1946). Both bloomed all winter, growing no taller than shoulder height. Then in April, "I found that strictly sterile vegetative shoots developed among the winter flowering stems" (Allard, 1931). This synchronous onset and cessation of flowering despite enormous differences in age and size suggested that reproduction was triggered by an external, seasonal factor rather than the condition of the plants.

There were clues that soybean might also be cued by the calendar because some varieties bloomed around a fixed date, no matter when they were planted in the field (Mooers, 1908). Studying the oil content of soybeans (Garner et al., 1914), the tobacco workers had tried to investigate

---

4. It is not only the amount of light but, on the contrary, the interruption by darkness that has a negative influence.

temperature effects. Rather than refrigerate the plants during the summer, they heated a greenhouse during the winter. The experiments never got off the ground, however, because the soybeans began to bloom—with flowers that failed to open—before they had a chance to make normal growth (Garner and Allard, 1920).

Garner and Allard were intrigued. "We mentioned various aspects of seasonal conditions," Allard recalled, "including temperature and light relations. . . ." (Allard, 1931).

It was generally assumed that temperature was the major environmental influence on plant development. But a temperature drop could not have prompted the soybeans to flower, for the greenhouse was heated in winter. Water supply or soil fertility had no effect on date of flowering either. So perhaps the change was triggered by another factor that changes with the seasons: light.

Quantity of light is one variable: in winter, both the number of hours of sunlight and the brightness of the sun's rays lessen. But during the soybean studies, Garner and Allard had noticed that plants shaded with cheesecloth bloomed at the same time as unshaded plants, even though they became straggly. The flowering time of Maryland Mammoth was also unaltered by shading.

And so the discussion turned to daylength. "It hardly seemed likely," Garner wrote, "that the other primary factor controlling the radiation received by the plant—namely, the length of the daily maximum amount of exposure—could be responsible for the effects in question" (Garner and Allard, 1920).

1918 EXPERIMENTS

To test this idea, Allard "devised a simple small box-like or primitive dog-house . . . furnished with a door and ventilation" (Allard, 1931). This was dark inside, thanks to a tightly fitting door and felt draped over the flat roof. At 4:00 P.M. on July 10, 1918—a cool, dry, partly cloudy day—Allard carried one box of Peking soybean plants and three pots of Maryland Mammoth into the shack. At 9:00 A.M. the following day he opened the door and removed the plants, leaving them in daylight until 4:00 P.M. This backbreaking procedure was repeated at the same time every morning and afternoon, day after day, weekends included, giving the test plants 7 h daylight and 17 h darkness in each 24-h period. In July in Arlington, the natural daylength was just more than 14 h.

Under this regime, the tobacco plants switched from making leaves—as those under natural daylength continued to do—to making flowers and then seeds. The soybeans, which were blossoming when they first entered the dark chamber, sported many full-grown pods within a couple of weeks, whereas none of the control plants' pods were more than half-grown. A month later, the leaves of the darkened soybeans yellowed and fell, and some of the seed pods were fully ripe. The leaves and pods of the control plants were still green.

"The results of the experiments," Garner and Allard concluded, "make it plain that of the various factors of the environment which affect plant life the length of the day is unique in its action on sexual reproduction" (Garner and Allard, 1920). They soon realized that daylength is, in fact, a logical factor to guide the development of living organisms. Unlike temperature, rainfall, or light intensity, daylength is "the only consistently rhythmic feature of the external environment" (Garner and Allard, 1923). Moreover, days shorten in the fall and lengthen in the spring before the weather changes drastically; thus a response to daylength allows plants to anticipate seasonal changes. And it enables plants of the same species to synchronize their reproduction.

Garner introduced the term *photoperiodism* for the response of organisms to relative length of night and day. The word was concocted by USDA scientist O. F. Cook, much to Allard's dismay. "The more correct word, which I myself suggested, since Garner was determined to make use of a big-sounding term," Allard said, "was the word haemero-nyctotropism, or day and night response. . . . In my contentions on this point I pointed out that nothing but a light source itself could be said to possess photoperiods or light periods. . . . A firefly or a light house may have its photoperiods, or a flashing electric sign, but not plants which cannot generate light" (Allard, 1931).

The two men believed that photoperiodism might also account for the seasonal reproduction of certain algae and the timing of bird migration. Shortly after, the first example of the many photoperiodic responses of animals was documented by S. J. Marcovitch, an entomologist at the Tennessee Agricultural Experiment Station. In 1922, after reading Garner and Allard's paper, Marcovitch demonstrated that the appearance of sexual forms of the strawberry louse in late fall is regulated by photoperiod and not by temperature, as was previously believed (Marcovitch, 1923, 1924). Reports of photoperiodic responses of birds soon followed (Rowan, 1926).

LONG-DAY AND SHORT-DAY PLANTS

By the end of 1918, Garner and Allard had submitted an account of their work to the *Journal of Agricultural Research* (Kellerman, 1919). This paper did not appear in its original form, but an enlarged version was published more than a year later (Garner and Allard, 1920).

Extending the work required more sophisticated facilities, and in the spring of 1919, the researchers acquired three larger dark houses. The inside walls were painted black, every tiny crevice was covered, and the doors were made snug and tight. And heavy plants no longer had to be hauled around; wooden trucks holding buckets or trays of plants were pushed in and out of each house on rails (Fig. 1-2).

That spring, summer, and fall, four varieties of soybean, three strains of tobacco, beans from Peru and Bolivia, and radish, carrot, lettuce, cabbage, wild aster, climbing hempweed, ragweed, hibiscus, wild violet, and early goldenrod plants were moved in and out of the darkhouses on various schedules, some beginning at the crack of dawn. Rows of soybeans were also sown at regular intervals outdoors.

Garner and Allard compared the times to flowering for the plants on shortened days with those for similar plants under natural daylengths. The results amply confirmed the findings of their 1918 experiments. Biloxi soybean plants on 7-h days flowered 26 days after germination, whereas those that remained outdoors took 110 days. Maryland Mammoth, aster, ragweed, and the tropical beans also bloomed more rapidly under short days (SD) than under natural Virginia daylengths.

Not all species behaved this way, however. Climbing hempweed, radish, and hibiscus bloomed only during the long summer days. Radish plants did flower in the winter of 1919, however, when the day was extended from 4:30 P.M. to 12:30 A.M. with incandescent greenhouse lamps. Iris, cosmos, spinach, and radish also flowered under the artificially long days, (LD) but not under the SD of winter, whereas the reverse was true for Maryland Mammoth tobacco (Fig. 1-3), Peruvian beans, beggarticks, and three varieties of soybean.

The Maryland Mammoth, radish, and soybean data were particularly satisfying because they were obtained under conditions opposite to those used the previous summer. Thus Maryland Mammoth and soybean flowered after SD but not LD, whereas radish flowered after LD but not SD. Clearly the duration rather than total quantity of light determined time of flowering.

Realizing that the plants fell into at least two groups, Garner and Allard coined the terms *short-day plants* (SDP) and *long-day plants* (LDP). Later they defined these terms more precisely (Garner and Allard, 1933). SDP, such as Maryland Mammoth, are those that flower and fruit (or flower and fruit sooner) through a wide range of daylengths up to but not more than a critical number of hours. LDP, such as radish, are those that flower and fruit through a wide range of daylengths above a critical length. Thus LDP

FIGURE 1-2 Harry Allard in front of the darkhouses at Arlington Experiment Farm, Virginia. Courtesy of USDA.

FIGURE 1-3   Maryland Mammoth tobacco plants of Garner and Allard. The one on the *right* grew in a greenhouse with supplementary electric light (long days); the one on the *left* grew in a greenhouse under natural short days. Courtesy of USDA.

identify the lengthening days of spring and tend to flower in the summer, whereas SDP bloom in the spring or fall. Many LDP also have a rosette habit, because stem elongation occurs with flowering, whereas SDP elongate their stems even when conditions do not permit flowering.

Two more groups were later added to the classification (Garner and Allard, 1923). One was intermediate plants, which flower when days are neither too long nor too short; the other was indeterminate plants (also called *day-neutral plants*), whose reproduction is indifferent to daylength, provided there is adequate light for photosynthesis. Early goldenrod, which ranges over the vast differences in daylength between Canada and North Carolina, is an indeterminate plant. Other plants are indeterminate over one temperature range but photoperiodic over another (Thompson, 1939). Many are photoperiodic when young but indeterminate later.[5]

5. Other photoperiodic responses have since been observed. Short–long-day plants, such as *Poa pratensis* (Kentucky blue grass) (Peterson and Loomis, 1949), require SD for floral induction but LD for floral development; they therefore flower in late spring or early summer. Long–short-day plants, which flower in late summer or early fall, have the reverse requirement (Resende, 1952; Sachs, 1956). Ambiphotoperiodic plants flower after LD or SD but not after intermediate daylengths.

"It is evident," the researchers pointed out, "that a plant can not persist in a given region or extend its range in any direction unless it finds conditions not only favorable for vegetative activity but also for some form of successful reproduction" (Garner and Allard, 1920). Allard later learned (Allard, 1944) that a British biologist, Arthur Henfrey, had suggested in 1852 that the distribution of certain plants depends on daylength or the amount of direct sunlight rather than on temperature or other climatic conditions (Henfrey, 1852).

Garner and Allard's first photoperiodism paper finally appeared on March 20, 1920, in the *Journal of Agricultural Research,* though its publication was bitterly opposed by one of the journal's reviewers (see Borthwick, 1956). The *New York Times* carried an article about the joint work on February 10, 1921, whereas some journalists interviewed Garner about "his" great discovery. "Time and again," Allard complained, "these reports extolled, alone, the supreme efforts of Dr. Garner in bringing this great discovery of plant behavior to the scientific world" (Allard, 1931).

In 1930, the American Society of Plant Physiologists awarded Garner the second annual Stephen Hales Prize for his development of the concept of photoperiodism. "There has been a tendency on the part of my colleague, Dr. Garner, to seek to center all the credit of the entire work on himself," Allard argued (Allard, 1931).

Garner and Allard maintained outwardly cordial relations despite this tension. Publishing jointly until 1941, they inaugurated a new era of plant research that would inspire, frustrate, and intrigue many future generations of researchers. As well as studying the photoperiodic responses of an increasing number of species, they showed that other aspects of development, including branching, root growth, leaf senescence, reddening of poinsettia bracts, tuberization, and senescence of annuals, are also affected by daylength.

BENEFITS

Computing how much time Garner and Allard had spent on daylength studies during a 10-year period, the USDA put the cost of discovering photoperiodism at $10,000 (USDA, 1906–1946). This modest investment in basic research brought billions of dollars of benefits to farmers, horticulturists, and plant breeders.

Farmers had often been frustrated by the failure of new crops despite fertile soil and clement weather. After Garner and Allard discovered that many economically important plants have critical daylengths for flowering and fruiting, testing new crop varieties for their photoperiodic requirements became standard practice. Although altering daylength over acres of farmland was impractical, crops could now be chosen for their suitability to different latitudes. They were then introduced only where the growing season would provide the necessary length of day, saving farmers much expense and disappointment.

In 1947, Garner and Allard's discovery rescued farmers in Trinidad. Early that year, an oil company erected a 25-foot-high pipe on the edge of a swamp padi field: waste natural gas ran up the pipe and burned at the top as a flare. To the farmers' dismay, the padi around the flare failed to flower in October, even 300 feet away where the light was very dim. When government scientists discovered flowers on plants shaded by a tree, they deduced that light from the flare had upset the photoperiodic response of padi to shortening days. The company put out the flare, the padi flowered in November, and the farmers reaped a late harvest (Benson and Murray, 1949).

Horticulturists also applied Garner and Allard's principles, using black cloth to shorten days or supplementary lighting to lengthen them (e.g., Laurie, 1930). Calceolaria, China aster, coreopsis, dahlia, feverfew, nasturtium, pansy, scabious, and Shasta daisy were some of the plants that flowered earlier and more profusely under extended days.

Today's major horticultural crop, chrysanthemum, was simply a fall-blooming garden plant in the 1920s. But Allard advanced flowering from mid-October to mid-July by darkhousing plants during the latter part of each day, starting in May (Allard, 1928). In the 1930s, horticulturists began to cover chrysanthemums with black sateen cloth (e.g., Post, 1931). Then they obtained blooms in all seasons by selecting responsive cultivars, manipulating temperature (Post, 1939), and altering daylength. To keep cuttings vegetative until flowering was desired, they extended days with artificial light (Poesch, 1935).

Plant breeders manipulated daylength to obtain three generations of plants per year—two in the greenhouse and one outdoors. They could also obtain seeds of new varieties that failed to reproduce outdoors by growing plants under shortened days in the greenhouse. Moreover, varieties of plants that normally flowered at different times could be crossed if flowering times could be synchronized.

The plant that inspired Garner and Allard, Maryland Mammoth, was also "established agriculturally solely through the utilization of length of day response. . . . As this variety of tobacco was found to be a short day plant, the problem of securing seed is satisfactorily met by growing the seed crop in Southern Florida during the winter, for under that daylight period the Mammoth variety does not continue its vigorous giant growth but flowers and fruits in practically the same manner as other varieties of tobacco" (Kellerman, 1926).

REFERENCES

Allard, H. A. (1928). Chrysanthemum flowering season varied according to daily exposure to light. *In* "1928 Yearbook of Agriculture," pp. 194–195. U.S. Department of Agriculture, Washington, D.C.

Allard, H. A. (1931). Letter to Paul De Kruif. Harry A. Allard papers #2977, Southern Historical Collection, University of North Carolina Library, Chapel Hill.

Allard, H. A. (1944). An interesting reference to length of day as affecting plants. *Science* 99,263.

Bailey, L. H. (1891). Some preliminary studies of the influence of the electric arc light upon greenhouse plants. *Cornell U. Agric. Exp. Stn. Bull.* 30,83–122.

Bailey, L. H. (1892). Second report on electrohorticulture. *Cornell U. Agric. Exp. Stn. Bull.* 42,133–146.

Bailey, L. H. (1893). Greenhouse notes for 1892–93. I. Third report upon electrohorticulture. *Cornell U. Agric. Exp. Stn. Bull.* 55,147–157.

Benson, E. G., and Murray, D. E. (1949, December). Light effects. *Caribbean Commission Monthly Information Bulletin.*

Borthwick, H. A. (1946). Daylength and flowering. *In* "1943–47 Yearbook of Agriculture," pp. 273–283. U.S. Department of Agriculture, Washington, D.C.

Borthwick, H. A. (1956). Light in relation to growth and development of plants. Presidential address, August 27, 1956, annual meeting. American Society of Plant Physiologists, University of Connecticut, Storrs. Unpublished.

Corbett, L. C. (1899). A study of the effect of incandescent gas light on plant growth. *W.V. Univ. Agric. Exp. Stn. Bull.* 62,77–110.

Garner, W. W. (1914). Tobacco Culture. *Farmers Bull.* #571. U.S. Department of Agriculture, Washington, D.C.

Garner, W. W., and Allard, H. A. (1920). Effect of the relative length of day and night and other factors of the environment on growth and reproduction in plants. *J. Agric. Res.* 18,553–606.

Garner, W. W., and Allard, H. A. (1923). Further studies in photoperiodism, the response of the plant to relative length of day and night. *J. Agric. Res.* 23,871–920.

Garner, W. W., and Allard, H. A. (1933). Comparative responses of long-day and short-day plants to relative length of day and night. *Plant Physiol.* 8,347–356.

Garner, W. W., Allard, H. A., and Foubert, C. L. (1914). Oil content of seed as affected by the nutrition of the plant. *J. Agric. Res.* 3,227–249.

Henfrey, A. (1852). "The Vegetation of Europe, Its Condition and Cause." J. van Voorst, London.

Kellerman, K. F. (1919). Memorandum to Mr. Fitts. Wightman Wells Garner, USDA personnel records. Jan 2 entry. Unpublished.

Kellerman, K. F. (1926). A review of the discovery of photoperiodism: The influence of the length of daily light periods upon the growth of plants. *Q. Rev. Biol.* 1,87–94.

Klebs, G. (1910). Alterations in the development and forms of plants as a result of environment. *Proc. R. Soc. Lond. B* 82,547–558.

Klebs, G. (1913). Ueber das Verhältnis der Aussenwelt zur Entwicklung der Pflanzen. *Sitzungsber. Heidelb. Akad. Wiss. Math. Naturwiss. Kl. Ser. B* 5,1–47.

Klebs, G. (1918). Über die Blütenbildung von *Sempervivum. Flora (Jena)* 111/112,128–151.

Laurie, A. (1930). Photoperiodism—Practical application to greenhouse culture. *Proc. Am. Soc. Hort. Sci.* 27,319–322.

Marcovitch, S. J. (1923). Plant lice and light exposure. *Science* 58,537–538.

Marcovitch, S. J. (1924). The migration of the Aphididae and the appearance of the sexual forms as affected by the relative length of daily light exposure. *J. Agric. Res.* 27,513–522.

Mooers, C. A. (1908). The soy bean. A comparison with the cowpea. *Tenn. U. Agric. Exp. Stn. Bull.* 82,75–104.

Murneek, A. E. (1948). History of research in photoperiodism. *In* "Vernalization and Photoperiodism" (A. E. Murneek and R. O. Whyte, eds.), pp. 39–61. Chronica Botanica Co., Waltham, Mass.

Peterson, M. L., and Loomis, W. E. (1949). Effects of photoperiod and temperature on growth and flowering of Kentucky bluegrass. *Plant Physiol.* 24,31–43.

Poesch, G. H. (1935). Supplementary illumination from mazda, mercury and neon lamps on some greenhouse plants. *Proc. Am. Soc. Hort. Sci.* 33,637–638.

Post, K. (1931). Reducing the day length of chrysanthemum for the production of early blooms by the use of black sateen cloth. *Proc. Am. Soc. Hort. Sci.* 28,382–388.

Post, K. (1939). The relationship of temperature to flower bud formation in chrysanthemums. *Proc. Am. Soc. Hort. Sci.* 37,1003–1006.

Rane, F. W. (1894). Electrohorticulture with the incandescent lamp. *W.V. Univ. Agric. Exp. Stn. Bull.* 37, 27 pp.

Resende, F. (1952). "Long–short" day plants. *Port. Acta Biol. Ser. A* 3,318–322.

Rowan, W. (1926). On photoperiodism, reproductive periodicity, and the annual migration of birds and certain fishes. *Proc. Boston Soc. Nat. Hist.* 38,147–189.

Sachs, R. M. (1956). Floral initiation in *Cestrum nocturnum*. I. A long–short day plant. *Plant Physiol.* 31,185–192.

Schübeler, (1879). See: The effects of uninterrupted sunlight on plants. *Nature* 21 (1880) 311–312.

Senkado-Sancho (1799). "Shocho Shiyo Momochidori" ("Rearing Methods of Various Birds"). Edo, Tokyo.

Thompson, H. C. (1939). Temperature in relation to vegetative and reproductive development in plants. *Proc. Am. Soc. Hort. Sci.* 37,672–679.

Tournois, J. (1912). Influence de la lumiére sur la floraison du Houblon japonais et du Chanvre. *C. R. Hebd. Seances Acad. Sci.* 155,297–300.

Tournois, J. (1914). Études sur la sexualité du Houblon. *Ann. Sci. Nat. Bot. Biol. Veg.*, Ser. IX, 19,49–191.

USDA (1906–1946). Harry Ardell Allard. Personnel records. Unpublished.

CHAPTER **2** _____

The 1930s

Garner and Allard presented such persuasive evidence for photoperiodism that workers across the world were attracted to the field. With a comprehensive theory to explain seasonal development, details could now be sketched in.

SITE OF INDUCTION

In 1933 and 1934, James Knott at Cornell University exposed spinach plants in a greenhouse to light for 15 h daily, while covering others with a black cloth after 10 h. Yet others were exposed to light for 10 h, but the growing points of their stems poked through washer-lined holes in the cloth to be illuminated for an additional 5 h. The remaining plants were on 15-h days, but their growing points were covered with weighted thimbles for 5 h daily.

The plants on 15-h days, including those whose growing points received only 10 h of light daily, produced seedstalks—spinach is an LDP. Those on 10-h days, including those whose buds received 15 h of daily illumination, remained vegetative. So the daylength at the growing points did not determine time of flowering, even though flowers, when they formed, arose at those locations. "Though the response of the plant may be localized in the bud," Knott concluded, "the leaves appear to function in some way to hasten the reproductive response to the appropriate photoperiod. . . . Accordingly, the part played by the foliage of spinach in hastening the response to a photoperiod favorable to reproductive growth may be in the production of some substance, or stimulus, that is transported to the growing point" (Knott, 1934).

More direct studies, with leaves themselves, were performed in the USSR by B.S. Moshkov (Moshkov, 1936, 1937, 1939a, 1939b, 1941). And at the Institute of Plant Physiology and Biochemistry, Academy of Sciences in Moscow, Mikhail Khristoforovich Chailakhyan stripped successively larger numbers of leaves from potted plants of Saratov millet, an SDP, and Krimsky 016 barley, an LDP, and kept the plants under different daylengths. Both species became less able to respond to photoperiod as the number of leaves declined. Defoliated plants also flowered[1] much later than plants with leaves and were insensitive to the length of the photoperiod.

Chrysanthemum stripped of all but one leaf remained vegetative under normal daylength but flowered after 6 weeks if the leaf was covered for all but 10 h each day. So Chailakhyan concluded in 1935 that "the processes induced by changes in the length of day light and leading to the reproductive development of plants (flowering and fruiting) occur within the leaf tissues. The formative processes occurring in the zones of growth (growing points) are secondary changes dependent upon the functional activity of the leaf" (Chailakhyan, 1936a).

FLORIGEN

This implied—as Knott had suggested—that leaves perceive daylength and pass a stimulus to growing points. Half a century earlier, the German plant physiologist Julius von Sachs had proposed that flowers develop in response to a specific flower-forming substance transported from leaves (Sachs, 1880, 1882). Because he was unable to induce flowering experimentally, this suggestion had met strong opposition.

The hypothesis cleanly explained Chailakhyan's data. It also seemed more plausible than 50 years earlier, because Frits Went had discovered in the Netherlands that a specific substance controls growth (Went, 1929). "Just as, in the growth processes, the regulatory function is fulfilled by the hormone of growth," Chailakhyan wrote, "so, in the processes of development a hormone of flowering, or a flowering hormone fulfills this function. The flowering of plants is beginning as the result of the formation in the leaves of a sufficient quantity of the flower hormone and of its transferring into the growing point" (Chailakhyan, 1936b).

1. Chailakhyan, like Garner and Allard, documented floral induction by the appearance of visible flower buds. Klebs, however, had monitored a much earlier parameter (Klebs, 1910, 1918), observing that growing points switch from making cells that form new stem and leaves (leaf primordia) to making those that form flower parts (flower primordia), a change that can be seen only with a microscope. Olive Norah Purvis, at Imperial College, London University, urged that "in assigning a plant to its photoperiodic category, the *time of formation of flower primordia* should be considered rather than the time of emergence of the inflorescence" (Purvis, 1934). She had observed that induction of flower primordia in winter cereals and emergence of flowers (ears) were favored by different conditions.

Working with *Perilla nankinensis,* an ornamental plant that failed to flower under Moscow daylengths, Chailakhyan grafted scions onto decapitated stems. He then exposed the scions to natural daylengths and the stocks to 10-h days. The stocks produced fruit shoots and then clusters of small pink blossoms. The scions also bloomed—but only if the flowers on the stocks were removed. "Together with nutrient substances there is a movement of blossom hormone from the leaves of the stock into the growing points of the engrafted top, where they cause formation of flowers," Chailakhyan (1936c) deduced.

He also grafted scions of *Helianthus tuberosus,* which does not bloom in Moscow, to stocks of *Helianthus annuus* (sunflower), which does. Thirty days later, the scions were blooming, whereas those grafted onto their own stocks were still vegetative. "The blossom hormone is not specific in its action for separate plant species," Chailakhyan said, "but has the same nature in different plants." He named the putative hormone "florigen, meaning 'blossom-former,' which expresses the basic function of this substance in the vegetable organism" (Chailakhyan, 1936c).

INHIBITOR OF FLOWERING

At the Kaiser Wilhelm Institute for Biology in Berlin, Georg Melchers and Anton Lang extended Chailakhyan's approach with grafts between plants of different genera and photoperiodic classes. Leaves of Maryland Mammoth tobacco, an SDP, could clearly control the flowering of henbane (*Hyoscyamus niger*), an LDP (Melchers and Lang, 1941a).

Stripped of all its leaves, *Hyoscyamus* could also be induced by SD or even total darkness[2] (Melchers and Lang, 1941b). But grafting back a single leaf prevented flowering on SD. Thus *Hyoscyamus* leaves appeared to produce an inhibitor of floral induction during long nights. Because plants flowered more quickly on LD with one leaf than when defoliated, the leaves appeared capable of releasing a positive stimulus also (Lang and Melchers, 1943).

TIMING MECHANISM

Amid the search for flowering hormones, a young German botanist named Erwin Bünning (Fig. 2-1) drew attention to a much-neglected aspect of photoperiodism: time measurement. Bünning arrived at the Institute for the Physical Basis of Medicine in Frankfurt in 1928 with another postdoctorate, Kurt Stern, to investigate the effects of atmospheric ions on plants.

---

2. A storage organ allows *Hyoscyamus* to survive defoliation.

FIGURE 2-1   Erwin Bünning. Courtesy of Hunt Institute for Botanical Documentation, Carnegie Mellon University, Pittsburgh, Pennsylvania.

Looking for a response to study, Bünning and Stern chose the movements of bean leaves, which expose their surfaces to the sun by day and fold vertically at night, a behavior known as *nyctinasty*. It had been known for a hundred years that *Mimosa* leaves continue to open and close for several days after being placed in constant darkness (de Mairan, 1729). Building on subsequent studies (de Candolle, 1832; Darwin and Darwin, 1880), Wilhelm Pfeffer collected convincing evidence for the operation of an internal clock that regulated such movements (Pfeffer, 1915).

Rose Stoppel, a researcher from Hamburg, helped Bünning and Stern set up their experiments, and she watered the plants each morning after turning on a flashlight covered with dark red paper. After observing, in 1920, that bean leaves in her lab always reached maximal night position between 3:00 and 4:00 A.M. she reasoned that an environmental factor must reset the clock each day. Stern and Bünning wondered if an atmospheric ion constituted the unknown factor, but enrichment or depletion of ions from the air failed to perturb the rhythm.

After Stoppel returned to Hamburg, the two young men moved the plants into Stern's potato cellar to obtain a more constant temperature. Because the house was far from the lab, they visited the cellar every afternoon. The plants responded by shifting their maximal night position to between 10:00 A.M. and noon. But without daily exposure to the flashlight, the leaves moved with a periodicity of 25.4 h. Thus the plants had an internal clock whose rhythm failed to coincide with the normal daily cycles of day and night but could be entrained to a 24-h cycle by daily exposure to a dim red light (Bünning and Stern, 1930).

After moving to Jena in 1930, Bünning proposed a function for this clock. During a visit to Köningsberg in 1934, "the great professor Kurt Noack . . . mentioned that discoveries were being made that were so remarkable one simply couldn't believe in them. Such a one, for example, was photoperiodism. He, as a specialist in the field of photosynthesis, must certainly know that it makes no difference whatsoever what program is followed in giving the plant the necessary quantities of light. Then, as I was riding back on the train, the idea came to me—aha, for the plant it does make a difference at which time light is applied, if not exactly in photosynthesis, then for its development" (Bünning, 1970). At the 1935 annual meeting of the German Botanical Society, Bünning suggested that endogenous rhythms constitute the timing mechanism of photoperiodism. This hypothesis appeared in the literature a year later (Bünning, 1936).

Bünning was forced to leave Germany in 1935 because of his opposition to Nazism. After the war, at the University of Tübingen, he outlined his ideas in, *Die physiologische Uhr* (Bünning, 1958). He proposed that the environment entrains a plant's natural internal rhythm to 24-h cycles, each consisting of a 12-h light-loving or photophile phase, in which light promotes flowering, followed by a 12-h dark-loving or scotophile phase, in which light inhibits flowering. To explain the difference between SDP and LDP, he originally suggested that the photophile phase begins at dawn in the former but is delayed by 12 h in the latter.

REFERENCES

Bünning, E. (1936). Die endonome Tagesrhythmik als Grundlage der photoperiodischen Reaktion. *Ber. Dtsch. Bot. Ges.* 54,590–607.
Bünning, E. (1958). "Die physiologische Uhr." Springer, Berlin, Göttingen, Heidelberg.
Bünning, E. (1970). Potato cellars, trains, and dreams: Discovering the biological clock. Letter to Frank Salisbury, June 2, 1970. *In* Salisbury, F. B., and Ross, C. W. (1978). "Plant Physiology," 2nd ed., p. 312. Wadsworth, Belmont, California.
Bünning, E., and Stern, K. (1930). Über die tagesperiodischen Bewegungen der Primärblätter von *Phaseolus multiflorus*. II. Die Bewegungen bei Thermokonstanz. *Ber. Dtsch Bot. Ges.* 48,227–252.
Chailakhyan, M. Kh. (1936a). On the mechanism of the photoperiodic reaction. *C. R. (Dokl.) Acad. Sci. URSS* 10 (1936 part I), 89–93.
Chailakhyan, M. Kh. (1936b). On the hormonal theory of plant development. *C. R. (Dokl.) Acad. Sci. URSS* 12 (1936 part III), 443–447.
Chailakhyan, M. Kh. (1936c). New facts in support of the hormonal theory of plant development. *C. R. (Dokl.) Acad. Sci.* 13, (1936 part IV), 79–83.
Darwin, C., and Darwin, F. (1880). "The Power of Movement in Plants." John Murray, London.
de Candolle, A. P. (1832). "Physiologie Végétale." Béchet jeune. Paris.
de Mairan, M. (1729). Observation botanique. *In "Histoire de l'Academie Royale des Sciences, Paris,"* p. 35. Pierre Mortier, Amsterdam.
Klebs, G. (1910). Alterations in the development and forms of plants as a result of environment. *Proc. R. Soc. Lond. B* 82,547–558
Klebs, G. (1918). Über die Blütenbildung von *Sempervivum*. *Flora (Jena)* 111–112, 128–151.

Knott, J. E. (1934). Effect of localized photoperiod on spinach. *Proc. Am. Soc. Hort. Sci.* 31,152–154.

Lang, A. (1980). Some recollections and reflections. *Annu. Rev. Plant Physiol.* 31,1–28.

Lang, A., and Melchers, G. (1943). Die photoperiodische Reaktion von *Hyoscyamus niger*. *Planta* 33,653–702.

Maximov, N. A. (1929). Experimentelle Änderungen der Lange der Vegetations-periode bei den Pflanzen. *Biol. Zentralb.* 49,513–543.

Melchers, G., and Lang, A. (1941a). Weitere Untersuchungen zur Frage der Blühhormone. *Biol. Zentralb.* 61,16–39.

Melchers, G., and Lang, A. (1941b). Über den hemmenden Einfluß der Blätter in der photoperiodischen Reaktion der Pflanzen. *Naturwissenschaften 29,82–83.*

Melchers, G., and Lang, A. (1948). Versuche zur Auslösung der Blütenbildung an zweijähr-igen *Hyoscyamus-niger*-Pflanzen durch Verbindung mit einjährigen ohne Gewebever-wachsung. *Z. Naturforsch. B* 3,105–107.

Moshkov, B. S. (1936). Role of leaves in photoperiodic reaction of plants. *Bull. Appl. Bot. Genet. Plant Breed. A* 17,25–30.

Moshkov, B. S. (1937). Photoperiodism and a hypothesis as to hormones of flowering. *C. R. (Dokl.) Acad. Sci. URSS 15,211–214.*

Moshkov, B. S. (1939a). Minimum intervals of darkness and light to induce flowering in short day plants. *C. R. (Dokl.) Acad. Sci. URSS* 22,456–459.

Moshkov, B. S. (1939b). Transfer of photoperiodic reaction from leaves to growing points. *C. R. (Dokl.) Acad. Sci. URSS* 24,489–491.

Moshkov, B. S. (1941). On the photoperiodic after-effect. *C. R. (Dokl.) Acad. Sci. URSS* 31,699–701.

Pfeffer, W. (1915). Beiträge zur Kenntnis der Entstehung der Schlafbewcgungen. *Abh. Math. Phys. Kl. Sächs. Akad. Wiss.* 34,1–154.

Purvis, O. N. (1934). An analysis of the influence of temperature during germination on the subsequent development of certain winter cereals and its relation to the effect of length of day. *Ann. Bot.* 48,919–955.

Sachs, J. (1880, 1882). Stoff und Form der Pflanzenorgane. *Arb. Bot. Inst. Würzburg* 3,452–488; 4,689–713.

Went, F. W. (1929). Wuchsstoff und Wachstum. *Recl. Trav. Bot. Neerl.* 25,1–116.

CHAPTER 3

Spectral Studies

Nineteenth-century workers had discovered that light promotes the germination of certain seeds (Caspary, 1860; Stebler, 1881a,b,c,d) and that some colors stimulate germination while others inhibit (Cieslar, 1883). Adolph Cieslar had noted that light-sensitive seeds tend to be small and therefore to have little stored food. They germinate only when they are sufficiently near the soil surface to receive a light signal.

In the early 20th century, Wilhelm Kinzel tested 964 species of seeds and found that, under certain experimental conditions, white light (WL) favored the germination of 672, inhibited that of 258, and had no effect on the remaining 35 (Kinzel, 1913–1926). But there was no consensus among researchers, even concerning seeds of the same species (see Crocker, 1938).

In 1920, Eben H. Toole (Fig. 3-1) took charge of research in the USDA Seed Testing Laboratory, which ensured that seed for sale in the United States met legal standards of viability. In 1934, Toole offered a temporary job to his friend Lewis H. Flint, asking him to determine which parts of the spectrum could most effectively break the dormancy[1] of light-requiring lettuce seed.

SEED GERMINATION

Flint soon discovered that a mere 4-s exposure to sunlight prompted imbibed Arlington Fancy lettuce seed to sprout. And a 1-min irradiation with 64 footcandles[2] of incandescent light made more than half the seeds

---

1. Dormancy is defined here as a block in development that delays germination until environmental conditions are suitable for the growth of an emerging seedling.

2. A footcandle is the illuminance on a surface of 1 square foot in area placed 1 foot from a uniformly distributed flux of 1 lumen or from a uniform point source of one candela, which, until 1948, was the international candle.

FIGURE 3-1   Eben H. Toole about 1956. Courtesy of USDA.

germinate, although none began to develop without a light treatment. "The light-sensitivity of prepared photographic film is now familiar as to be rated a commonplace," Flint declared, "but that an exposure of a few seconds may mean the difference between no germination and complete germination in moist lettuce seed has only recently been appreciated through the studies here reported" (Flint, 1934).

Using commercial glass filters, he observed that the longer wavelengths—yellow, orange, and red—promoted lettuce seed germination, whereas shorter wavelengths—violet, blue, and green—did not. And when blue light (B) followed red (R), the stimulatory effect of R was abolished. So wavelengths emerging from some of the filters not only failed to promote germination—they also inhibited.

Strong B induced light sensitivity in imbibed, light-insensitive lettuce seeds. Such seeds germinated under R or became light-insensitive again after R exposure and drying. So Flint concluded that "the reaction involved in the above procedure is thus reversible" (Flint, 1934).

Using a Mazda bulb without filters, he showed that the photoreaction that promoted germination obeyed the reciprocity law formulated by Bunsen and Roscoe in 1850.[3] Thus brief irradiations at higher energy levels caused the same percentage of seeds to germinate as longer irradiations at lower energy levels (Table 3-1).

3. The law states that a response involving a single photoreceptor should depend on total energy (fluence) absorbed and not also on the rate at which that energy is delivered (fluence rate). Thus a long, dim irradiation should be as effective as, but not more effective than, a short, bright pulse delivering the same number of photons.

TABLE 3-1

Germination of Arlington Fancy Lettuce Seed
Imbibed in Water at 20°C for 2 h Before Light
Treatment[a]

Illumination in footcandles	Exposure time	Percentage germination
64	1	53.5
32	2	52.8
16	4	54.5
8	8	58.1
4	16	47.5
2	32	58.2

[a] From Flint, 1934.

More sophisticated studies were needed, and Toole knew where to go for help. On the Mall, just along the street from the USDA, was the Smithsonian Institution (Fig. 3-2), where investigations of solar radiation had been in progress since Secretary Samuel Pierpont Langley had established an Astrophysical Observatory in 1890. Taking advantage of this expertise and a large collection of optical apparatus, Secretary Charles Greeley Abbot established, in 1929, a unit within the Astrophysical Observatory named the Division of Radiation and Organisms. It was to conduct "(1) direct investigations upon living organisms, and (2) fundamental molecular structure and photochemical investigations related to the biological problems" (Brackett, 1930). Work began in 1931, after wine cellars in the western basement of the Smithsonian Castle were converted to labs. This was an ideal site for photobiological research because many of the rooms lacked windows and the temperature underground was fairly constant.

In 1932, a plant physiologist from the University of Maryland, Earl Steinford Johnston, became head of the division. Studying oat coleoptiles (the protective sheaths of monocotyledonous seedlings), he repeated the pioneering work of A. H. Blaauw, who, as a student of F.A.F.C. Went (Frits Went's father) at the University of Utrecht, had determined which parts of the spectrum induce phototropism, the bending of plants toward light (Blaauw, 1909). Like Blaauw, Johnston discovered that the shorter wavelengths of visible light evoked the phototropic response most effectively, but he distinguished two peaks in the B region (Johnston, 1934). When Flint observed that shorter wavelengths seemed to inhibit seed germination, it appeared that the two processes might share a common mechanism. Consequently, Johnston asked Edward D. McAlister, a phys-

FIGURE 3-2   The Smithsonian Castle. From Smithsonian Institution Archives RU95.

icist who had joined the Division of Radiation and Organisms in 1930, to collaborate with Flint.

McAlister, an expert in the measurement of infrared radiation, made a crucial observation about Flint's previous experiments: A major difference between the filters that had promoted germination and those that had inhibited it was that the latter transmitted more infrared and less ultraviolet (UV) radiation. So he enlarged Flint's study to include radiation bands outside the range of the visible spectrum.

To obtain purer bands of light, McAlister used a spectrograph, which physicist Fred S. Brackett had built for Johnston. This was a large light-proof box containing a streetlight bulb (Fig. 3-3). The bulb cast a beam through lenses and a slit into a huge quartz prism. The prism split the beam into a rainbow of colors, which bounced down from a mirror onto precisely arranged biological specimens. Thus rows of seeds or seedlings could sit in an area about a foot long, and each row or group of rows would be bathed with a different wavelength.

The two men placed imbibed Arlington Fancy lettuce seeds on the spectrograph in divided brass boxes. They irradiated some seeds with monochromatic light for 24 h and exposed others to an amount of R (from two additional lamps wrapped in red cellophane) that had caused half the seeds to germinate in Flint's previous experiments. Yet others received

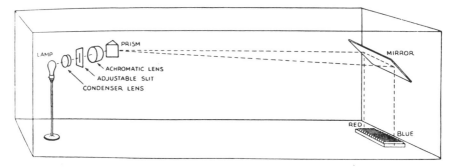

FIGURE 3-3   The Smithsonian spectrograph. (From Flint and McAlister, 1935. Courtesy of the Smithsonian Institution.)

both R and monochromatic light. The R treatment provided a baseline from which any stimulation or inhibition of germination could be seen.

Ultraviolet had no effect on percentage germination. But far-red (FR) rays, between red and infrared, were strongly inhibitory. Taking into account that the value of a quantum of energy decreases as wavelength increases, Flint and McAlister related percentage inhibition to equal amounts of energy at different wavelengths. The resulting graph was an action spectrum—a plot of the effectiveness of different wavelengths of light in provoking or inhibiting a biological response. It revealed that light in the violet-blue-green region was indeed inhibitory to germination but that the most marked effect was seen at about 760 nm.[4]

Sunlight contains large amounts of infrared radiation, the authors pointed out, and "in consequence, it would appear that under natural outdoor conditions . . . radiation of a wavelength in the long red exerts a relatively powerful inhibitory influence upon the germination of dormant lettuce seed, although this influence is ordinarily more than counteracted by the promoting influence in the yellow-orange-red region" (Flint and McAlister, 1935).

In Germany, meanwhile, Dieter Meischke was determining the spectral sensitivities of both light-promoted and light-inhibited seeds, with the aid of glass filters. The wavelengths 550 nm and 600–650 nm were most effective, whereas 435–490 nm and 750 nm were least effective. Light filtered through *Paulownia imperialis* leaves also inhibited the germination

4. Sunlight is a continuous series of wavelengths, with lower and upper limits at about 285 nm and 2,400 nm (a nanometer is a millionth of a millimeter). Wavelengths between about 380 and 770 nm constitute light or visible radiation because the human eye responds to that region of the spectrum. The invisible rays less than 380 nm are UV radiation, whereas wavelengths greater than 770 nm are heat rays or infrared radiation. (Source: Illuminating Engineering Society of North America).

of light-promoted seeds, and the addition of a heat filter reduced this inhibition. This suggested that leaves transmit infrared rays that might inhibit germination of shaded, light-sensitive seeds (Meischke, 1936).

In 1936, Flint published an action spectrum for germination of Arlington Fancy lettuce seed (Fig. 3-4A). Its most striking feature was a sharp swing at 700 nm from promotion to complete inhibition. There was little detail in

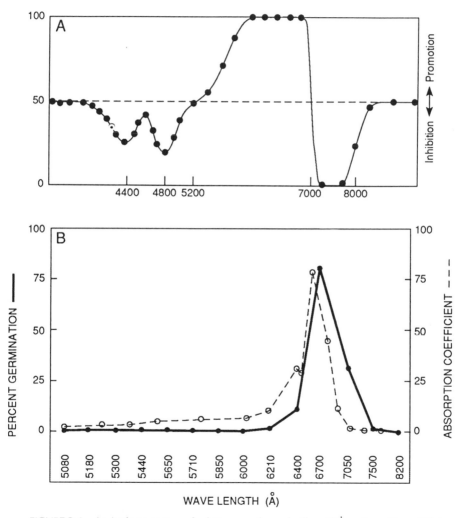

FIGURE 3-4   A: Action spectrum for lettuce seed germination. 10Å = 1nm. (From Flint, 1936.) B: Partial action spectrum for germination of seeds inhibited by blue light (——); absorption of chlorophyll in ether (----). (From Flint and McAlister, 1937. Courtesy of the Smithsonian Institution.)

the red region, however (Flint, 1936). Subsequently Flint and McAlister reprinted this figure, plus the results of a more detailed study (Fig. 3-4B). By overlaying monochromatic light with inhibitory B, they ensured—"by the well-known system of trial and error" (Flint and McAlister, (1937)— that 100% germination occurred in only one compartment of the brass boxes. This had been in the 670-nm band of the spectrum, indicating that this wavelength promoted germination most effectively.

Because light can trigger a biological response only if it is absorbed by a pigment in the responding organism, Flint and McAlister suggested that chlorophyll, the green pigment that harnesses light energy during photosynthesis, might be the photoreceptor in seed germination (Flint and McAlister, 1937). There was some resemblance between the absorption of R by chlorophyll and the action spectrum for this part of the spectrum (Fig. 3-4B).

By the time the 1937 paper was published, Flint's temporary position had ended, and he had moved to the Boyce Thompson Institute in Yonkers, New York, where pioneering studies on plant growth in controlled environments had begun in the mid 1920s. In 1939, Toole moved to the USDA Plant Industry Station in Beltsville, Maryland. Thus Flint and McAlister's data languished in the literature. "I remember being very impressed by their paper," recalls plant physiologist Kenneth Thimann, "but it didn't seem to be in the forefront of the main lines of research at that time. It was an oddity, an interesting phenomenon, but not much to the point."

MESOCOTYL ELONGATION

There were studies, meanwhile, on the effects of monochromatic light on other stages of plant development. At the Smithsonian Institution, Johnston had continued to work with oat seedlings, adding studies of photomorphogenesis (the effects of light on plant development) to those of phototropism. Comparing the growth of seedlings exposed to darkness or continuous monochromatic light, he found that light inhibited the elongation of the first internode (mesocotyl) but promoted that of the coleoptile, which lengthened most in R (640–660 nm) and least in B (480–500 nm) and FR (~ 760 nm). "The similarity of this curve to that found by Flint and McAlister representing the germination of light-sensitive lettuce seed," wrote Johnston, " . . . is suggestive of a common physiological process" (Johnston, 1937).

In 1938, a newly appointed biochemist, Robert L. Weintraub, took up the seedling work. In 1942, Weintraub and McAlister published an action

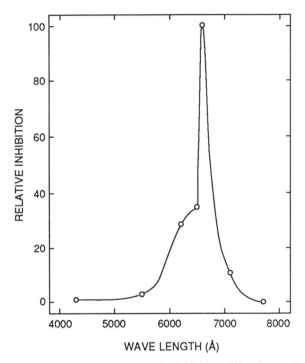

FIGURE 3-5 Quantitative action spectrum for inhibition of first internode elongation of oat seedlings. (From Weintraub and McAlister, 1942. Courtesy of the Smithsonian Institution.)

spectrum for first internode inhibition of oat seedlings under continuous illumination (Weintraub and McAlister, 1942), obtaining different energy levels at each wavelength by inserting layers of colored cellophane between the seedlings and the light source. The resulting data allowed them to plot the reciprocal of the irradiance required to produce a given (40% inhibition) response and thus to derive an action spectrum that truly compared the effectiveness of different parts of the spectrum[5] (Weintraub and McAlister, 1942). The curve revealed that R was the most effective wave band, for again there was a sharp peak at 660 nm (Fig. 3-5). Weintraub and Leonard Price extended this work to 12 members of the grass family (Weintraub and Price, 1947).

   Frits Went noted during a 1946 visit that "the rooms are chock-full of the most elaborate equipment: galvanometers, carbon dioxide meters, lighting

5. Such a comparison is impossible when plants are exposed to a single irradiance because in some regions of the spectrum that level of energy might be insufficient to trigger a response, whereas in others it might be many times the required amount.

equipment, transformers, etc., but they are painfully devoid of plants. In one place Weintraub pointed out that in three cans corn seedlings were growing but that was all. They have no greenhouses to start their experimental plants and they mostly use seedlings" (Went, 1946).

FLORAL INDUCTION

The Smithsonian facilities were thus not suited to photoperiodism research, which at that time required a constant supply of mature plants. Workers elsewhere were looking at the effects of colored light on flowering, however. After Klebs reported that house leeks bloomed in a red but not blue glass house (Klebs, 1910), many researchers confirmed the inductive effect of R but disagreed over the effectiveness of B (Rasumov, 1933; Withrow and Benedict, 1936; Withrow and Withrow, 1940).

At the Institute of Plant Physiology in Moscow, Victor M. Katunskij filtered light through colored solutions to control flowering in several SDP and LDP. He concluded that "the maximum photoperiodic effect . . . is yielded by the red rays. The minimum or zero-effect is yielded by the green rays. Blue light in the majority of test plants resulted in a distinct photoperiodic effect, completely analogical with that caused by red light, but in a less intensive degree than the latter. It is thus evident that the difference in the effect of blue and red rays in respect of photoperiodic reaction is purely of a quantitative nature" (Katunskij, 1937).

Studying the SDP *Perilla ocymoides,* A. F. Kleshnin at the K. A. Timiriazev Institute of Plant Physiology delayed the onset of budding by extending the day to 24 h for several weeks with different energy levels at various wavelengths. "It is only the amount of energy absorbed by some acceptor of the leaf that is important. . . ." he explained. "For every region of the spectrum there is a definite threshold of radiation intensity below which this region is perceived as darkness in the photoperiodical process; this threshold is different for different plants" (Kleshnin, 1943).

Quantitative action spectra were thus needed for floral induction. But they had to be obtained with pure wave bands of light, because contamination of B with FR had confounded many of the previous studies.

CRUCIFERAE

Inadequate experimental design and impure light sources did not entirely explain why some plants responded to B and others did not: there seemed to be genuine interspecific differences. Thus G. L. Funke, who conducted an extensive series of experiments in Ghent between 1936 and his sudden death in 1946, noticed that plants whose flowering was hastened by a photoperiod extension of B or WL but not R were all members of the Cruciferae (Funke, 1948).

This cruciferous quirk was confirmed with better light sources at the Agricultural University (Landbouwhogeschool) of Wageningen by Evert C. Wassink and colleagues. A photosynthesis researcher from the University of Utrecht, Wassink (see Fig. 21-4) moved to Wageningen in 1947 to direct the Laboratory for Plant Physiological Research. A lead-lined cellar in the underlying glacial moraine provided lightproof conditions and constant temperature.

Wanting to irradiate mature plants with pure bands of monochromatic light without having to purchase prohibitively expensive optical glass for filters, Wassink combed local glass companies with a hand spectroscope, selecting panes that transmitted only narrow bands of light. He obtained additional spectral bands from newly developed colored fluorescent sources (Wassink and van der Scheer, 1950). His group confirmed that flowering in the crucifers *Brassica rapa* var. (Wassink et al., 1950) and *Sinapis alba* (Stolwijk, 1952) was especially sensitive to B and discovered that mature plants respond to the "infrared" rays of the spectrum (see Chapter 5).

REFERENCES

Blaauw, A. H. (1909). Die Perzeption des Lichtes. *Recl. Trav. Bot. Neerl.* 5,209–372.

Brackett, F. S. (1930). *Smithson. Inst. Annu. Rep.,* pp. 115,117.

Caspary, R. (1860). *Bulliarda aquatica* D. C. *Schriften königlichen Physikalisch-Ökonomischen Gesellschaft Königsberg* 1,66–91.

Cieslar, A. (1883). Untersuchungen über den Einfluss des Lichtes die Keimung der Samen. *Forsuch. Geb. Agrik. Physik.* 6,270–295.

Crocker, W. (1938). Effect of the visible spectrum upon the germination of seeds and fruits. *In* "Biological Effects of Radiation" (B. M. Duggar, ed.), vol. II, pp. 791–827. McGraw-Hill, New York, London.

Flint, L. H. (1934). Light in relation to dormancy and germination in lettuce seeds. *Science* 80,38–40.

Flint, L. H. (1936). The action of radiation of specific wave-lengths in relation to the germination of light-sensitive lettuce seed. *Proc. Int. Seed. Test. Assoc.* 8,1–4.

Flint, L. H., and McAlister, E. D. (1935). Wavelengths of radiation in the visible spectrum inhibiting the germination of light-sensitive lettuce seed. *Smithson. Misc. Collect.* 94(5),1–11.

Flint, L. H., and McAlister, E. D. (1937). Wavelengths of radiation in the visible spectrum promoting the germination of light-sensitive lettuce seed. *Smithson. Misc. Collect.* 96(2),1–8.

Funke, G. L. (1948). The photoperiodicity of flowering under short day with supplemental light of different wavelengths. *Lotsya* 1,79–82.

Johnston, E. S. (1934). Phototropic sensitivity in relation to wavelength. *Smithson. Misc. Collect.* 92(11),1–17.

Johnston, E. S. (1937). Growth of *Avena* coleoptile and first internode in different wavelength bands of the visible spectrum. *Smithson. Misc. Collect.* 96(6),1–19.

Katunskij, V. M. (1937). Dependency of photoperiodic reactions of plants on the spectral composition of light. *C. R. (Dokl.) Acad. Sci. URSS* 15,509–512.

Kinzel, W. (1913–1926). "Frost und Licht als beeinflussende Kräfte bei der Samenkeimung." Ulmer, Stuttgart.

Klebs, G. (1910). Alterations in the development and forms of plants as a result of environment. *Proc. R. Soc. Lond. B* 82,547–558.

Kleshnin, A. F. (1943). On the role of spectral composition of light in photoperiodic reaction. *C. R. (Dokl.) Acad. Sci. URSS* 40,208–211.

Meischke, D. (1936). Über den Einfluss der Strahlung auf Licht-und Dunkelkeimer. *Jahrb. Wiss. Bot.* 83,359–405.

Rasumov, V. I. (1933). The significance of the quality of light in photoperiodical response. *Bull. App. Bot. Genet. Plant Breed. Ser. III* 3,217–251.

Stebler, F. G. (1881a). Über das Einfluss des Lichtes auf die Keimung ölhaltiger Samen. *Bot. Zeit.* 39,470–471.

Stebler, F. G. (1881b). Über den Einfluß des Lichtes auf die Keimung. *Vierteljahrsschr. Naturforsch. Ges. Zurich* 26,102–104.

Stebler, F. G. (1881c). Licht and Keimung. *Fühling's Landwirtsch. Ztg.* 502.

Stebler, F. G. (1881d). Über den Einwirkung des Licht auf die Keimung. *Bot. Zentralb.* 2,157–158.

Stolwijk, J. A. J. (1952). Photoperiodic and formative effects of various wavelength regions in *Cosmos bipinnatus, Spinacia oleracea. Sinapis alba,* and *Pisum sativum.* I. *Proc. K. Ned. Akad. Wet. Ser. C.* 55,489–502.

Wassink, E. C., and van der Scheer, C. (1950). On the study of the effects of light of various spectral regions on plant growth and development. *Proc. K. Ned. Akad. Wet. Ser. C* 53,1064–1072.

Wassink, E. C., Sluysmans, C. M. J., and Stolwijk, J. A. J. (1950). On some photoperiodic and formative effects of coloured light in *Brassica rapa,* f. *oleifera,* subf. *annua. Proc. K. Ned. Akad. Wet. Ser. C* 53,1466–1475.

Went, F. W. (1946). Diary entry. April 16. Unpublished.

Weintraub, R. L., and McAlister, E. D. (1942). Developmental physiology of the grass seedling. I. Inhibition of the mesocotyl of *Avena sativa* by continuous exposure to light of low intensities. *Smithson. Misc. Collect.* 101(17),1–10.

Weintraub, R. L., and Price, L. (1947). Developmental physiology of the grass seedling. II. Inhibition of mesocotyl elongation in various grasses by red and by violet light. *Smithson. Misc. Collect.* 106(21),1–15.

Withrow, R. B., and Benedict, H. M. (1936). Photoperiodic responses of certain greenhouse annuals as influenced by intensity and wavelength of artificial light used to lengthen the daylight period. *Plant Physiol.* 11,225–249.

Withrow, R. B., and Withrow, A. P. (1940). The effect of various wavelengths of supplementary radiation on the photoperiodic response of certain plants. *Plant Physiol.* 15,609–624.

CHAPTER 4

Action Spectra for Floral Induction

In 1933, the USDA Bureau of Plant Industry decided to move much of its research program from Arlington Experiment Farm to Beltsville, Maryland, where it planned to establish one large federal research center. The War Department regained jurisdiction over the Arlington site in 1940, returning it to Arlington Military Reservation. The Beltsville site, about 11 miles from Washington, D.C.,[1] had been established for animal research in 1910 (Wiser and Rasmussen, 1966).

A building later named the West Building was completed at Beltsville in the spring of 1936 (Fig. 4-1). A small winery in its basement and first floor included tiled laboratories, underground wine cellars, and controlled-temperature fermentation rooms. But not a single grape was fermented because, despite the repeal of the Prohibition Amendment at the end of 1933, experimental winemaking proved too controversial for Congress.

On June 29, 1935, the Bankhead-Jones Act had authorized funding for "research into laws and principles underlying basic problems of agriculture in its broadest aspects" (*United States at Large,* 1935–1936). So the USDA used part of the winery space to create a unit named Basic Studies of Plant Growth and Development, "bearing on the relation of hormones to plant growth" (USDA, 1936–1960).

EZRA JACOB KRAUS

The Department invited Ezra Jacob Kraus, head of the Hull Botanical Laboratory at the University of Chicago, to be a collaborator without compensation "in order that the outstanding work being done at the University of Chicago and the California Institute of Technology could be

---

1. The USDA Beltsville Agricultural Research Center now lies on the Washington Beltway.

33

FIGURE 4-1   West Building, 006, Beltsville, Maryland. Courtesy of USDA.

specifically coordinated with the work'' (USDA, 1936–1960). Four scientists were hired to work under his direction: Paul C. Marth and John W. Mitchell in the field of plant hormones and growth regulators, and Harry Alfred Borthwick and Marion Wesley Parker in the field of photoperiodism.

In March 1938, Kraus was appointed principal plant physiologist in the Division of Fruit and Vegetable Crops and Diseases. His duties were to organize a national research program on the developmental responses of plants to environmental conditions. He was unable to take up this task, however, because the medical exam required of new employees revealed a condition that ''although uncomplicated, renders him unacceptable for permanent retention in the Service'' (USDA, 1936–1960). He therefore served the USDA as a temporary employee until 1943, spending several weeks each year at Beltsville. In 1948, he moved from Chicago to Oregon State College in Corvallis. He died in 1960 at age 74.

HAMNER AND BONNER

After Kraus failed to obtain a permanent position with the USDA, he offered 27-year-old James Bonner (Fig. 4-2) a faculty position at the Uni-

FIGURE 4-2 James Bonner, 1952, with cocklebur plants.

versity of Chicago. Unsure about leaving the California Institute of Technology, where he had done his graduate work and part of his undergraduate studies, Bonner took a paid leave of absence. In the summer of 1938, he arrived at the Hull Botanical Laboratory, where he collaborated with assistant professor Karl C. Hamner.

Bonner and Hamner decided to "do experiments to find out how photoperiodism works. We thought if we could get florigen, that would really be something."

They chose to work with cocklebur, *Xanthium pensylvanicum,* obtaining seeds from the inbred population at Chicago Midway airport. This hardy weed would flower after just one short photoperiod.

All attempts to identify a flower-promoting substance failed, although Hamner and Bonner dissected about 4,500 cuttings treated with 1,150 preparations. The latter included yeast extract, several vitamins, many organic acids, and 246 different extracts from the leaves of plants in bloom (Hamner and Bonner, 1938). Bonner returned to Caltech, deciding not to accept the University of Chicago position.

THE LIGHT BREAK

Despite the failure of their initial project, Hamner and Bonner made a crucial contribution to photoperiodism research during their summer collaboration. Both Klebs and Tournois had suggested that events during the night might influence flowering, and in 1936, British plant physiologist Vernon Herbert Blackman stated that "in the case of short-day plants, there is some reason for believing that it is the corollary of the period of illumination, the *period of darkness,* to which attention should be directed" (Blackman, 1936).

This was indicated in Garner and Allard's experiments with abnormal light–dark cycles (Garner and Allard, 1931), although the photoperiodism pioneers failed to grasp the significance of their findings. In 1926, Garner reported the "a reduction of 65 per cent of the total light energy received by the plants by darkening during the middle of the day was less effective in hastening flowering in certain varieties of soybean than was a reduction of $2\frac{1}{2}$ per cent of total light energy by excluding the early morning daylight. In this case the mid-day darkening period was 6 h while the early morning darkening period averaged about 40 minutes"(Garner, 1929).

Hamner and Bonner were unaware of Blackman's suggestion or this aspect of Garner and Allard's studies. But they wanted to test the assumption that the length of the light period rather than the length of the dark period was critical. Trucking cocklebur plants in and out of supplementary lighting, they exposed some to 12-h cycles (4 h WL/8 h darkness) and others to 48-h cycles (16 h WL/32 h darkness). The critical photoperiod for cocklebur was between 15.5 and 15.75 h, leaving 8.5–8.25 h darkness in a 24-h cycle. Although the dark period was twice as long as the photoperiod in both abnormal cycles, the dark period was shorter than 8.5 h and the light period shorter than 15.5 h in the 12-h cycle, whereas both exceeded these lengths in the 48-h cycle. When the plants on 48-h cycles formed floral primordia and then abundant flowers, while those on the 12-h cycles remained vegetative, Bonner and Hamner concluded that the length of the dark period, rather than the length of the light period, was critical.

They also replaced half a 16-h photoperiod with darkness during 48-h cycles to create a single 40-h dark period; plants on this schedule flowered as vigorously at suitable night temperatures as if they had received the full 16-h photoperiod plus 32 h darkness. But plants on 24-h cycles exposed to one 17-h cycle of 9 h light and 8 h darkness remained vegetative (Table 4-1). Thus the dark period and not the light period had to exceed a critical length for floral induction.

TABLE 4-1

Effect on Floral Induction in *Xanthium* of Exposure to One Long Dark Period and/or One Short Photoperiod[a]

Photoperiod	Dark period	Visible flowers and fruit
S (9 h)	S (8 h)	−
S (8 h)	L (40 h)	+
L (16 h)	L (32 h)	+

[a] Data from Hamner and Bonner, 1938. S, short; L, long.

The collaborators reasoned that if flowering is induced during the night, it might be prevented by a brief light exposure. Amazingly, just 1 min of light half way through the night, given under normally inductive conditions, caused cocklebur to remain completely vegetative (Hamner and Bonner, 1938). They assumed that such a "light break" destroyed the flowering hormone that normally accumulated during the dark period.

Hearing of this discovery from the Chicago workers, Borthwick and Parker demonstrated the light-break effect in soybean (see Emsweller et al., 1941). Three of their colleagues then applied the principle to the culture of chrysanthemums, hoping to develop a cheaper and more reliable method of delaying blooming than supplementary lighting, which had to extend daylengths to 15–17 h. Although the brief light break that inhibited soybean and cocklebur was ineffective, 60 min of light in the middle of the night delayed flowering by 2–3 months, without reducing the quality or quantity of blooms (Stuart, 1943). Application of this technique enabled growers to produce high-quality blooms even during January and February (Stuart, 1943). Most varieties required yet longer light interruptions, and 2–4 h became standard (Cathey, 1969), although briefer light breaks were effective with many other greenhouse crops.

HARRY ALFRED BORTHWICK

In 1938, meanwhile, Basic Studies of Plant Growth and Development at Beltsville lacked a full-time head. This problem was solved by splitting the unit into a Growth Regulator Project, headed by Mitchell, and a Photoperiod Project, headed by Borthwick, which was to focus on photoperiodism and dormancy. "Photoperiodism . . . was practically synonymous with control of flowering at that time and dormancy was something entirely unrelated," said Borthwick. "Because of the keen interest at that time in photoperiodic control of flowering, it is perhaps not surprising that dormancy was ignored by us for about 15 years" (Borthwick, 1956).

Borthwick (Fig. 4-3), cousin of future U.S. vice-president Hubert Horatio Humphrey, was 38 years old when he moved to Beltsville. He had entered the University of Minnesota School of Agriculture in 1917 and majored in botany at Stanford University in 1921. He received master's (1924) and doctoral (1930) degrees from Stanford while working as a research assistant at the Agriculture School of the University of California at Davis. He spent the next 6 years as assistant professor of Botany at Davis, collaborating with L. T. Emsweller on several projects. After Emsweller left Davis for Beltsville in 1935 to head research in floriculture, he recommended Borthwick for the position of morphologist.

FIGURE 4-3   Harry Alfred Borthwick. Courtesy of USDA.

MARION WESLEY PARKER

Parker (Fig. 4-4), the other Photoperiod Project scientist, was nearly 10 years younger than Borthwick. He held a botany degree (1928) from Hampden-Sydney College in Virginia and master's (1930) and doctoral degrees (1932) in plant physiology from the University of Maryland in

FIGURE 4-4 Marion Wesley Parker (*left*) with Borthwick at the Plant Industry Station at Beltsville, Maryland, about 1940. Courtesy of USDA.

College Park. A plant nutritionist, Parker's duties were "to determine the cause of the effect of length of day on plants; to determine internal changes in plant tissues as result of varying the length of day; and to conduct photosynthetic studies on such plants and measure their respiration and transpiration under various environmental conditions" (USDA, 1936–1966).

Internal changes included alterations in the carbon–nitrogen ratio (see Chapter 1), because Kraus was renowned for his doctoral study, conducted jointly at the University of Chicago with Henry Reist Kraybill, on the influence of carbon–nitrogen balance on fruiting in tomato (Kraus and Kraybill, 1918).

CARBON-ARC ROOMS

Two controlled environment rooms in the wine cellar of the West Building became operative in 1937. Carbon-arc lighting, which Mitchell had selected for plant growth at the Hull Botanical laboratory (Mitchell, 1935), provided the brightest artificial light available. One hospital sunlamp was placed in each 18-ft-long, 9-ft-wide, 10-ft-high room. As a large current arced between one or more pairs of carbon rods, the carbon ignited, burning with a very hot flame.

The carbon arc was not a trouble-free source. Each lamp had two sets of carbons that burned alternately and had to be changed at 8:00 A.M. and

FIGURE 4-5   Borthwick examines Biloxi soybean plants grown under carbon-arc plus incandescent lighting. Plants were arranged on semicircular benches in three concentric circles, because each carbon-arc lamp was equipped with a reflector that directed light onto a circular area, 9-ft in diameter. Height of the circles varied to ensure that all plants received same amount of light. Courtesy of USDA.

4:00 P.M. daily (Went, 1946). Moreover, the fumes from the burning carbon had to be vented through a pipe that pulled in outdoor air, passed it across the lamp, and carried the exhaust out of the basement (Fig. 4-5). The window glass casing had to be inspected daily for heat cracks so that plants would not be exposed to deadly UV rays. Despite these drawbacks, the carbon-arc lamps served the Beltsville researchers for 26 years (Kasperbauer et al., 1963), until it became impossible to obtain the necessary carbon holders and circuit breakers.

Soybean plants grown under these lamps alone tended to fall over during the fourth week, when growth was most rapid. Sturdier plants were obtained from 1938 on by encircling each lamp with eight 200-W incandescent filament bulbs. Because the carbon arc emitted mainly B, whereas the incandescent lamps emitted R and FR, the combination provided energy over a large portion of the spectrum.

FIGURE 4-6   Allard (*right*) pays a visit to Borthwick in 1960. Courtesy of USDA.

EARLY PAPERS

Borthwick and Parker chose their experimental plant after consulting with Allard (Fig. 4-6), who recommended "a short-day one. Long-day plants, he said, were much less precise in their daylength responses and, therefore, less suitable for our purpose. . . . Among the many he mentioned, he recommended the Biloxi soybean because of ready availability of seed of known heredity, earliness of flowering, ease of culture, and freedom from disease" (Borthwick, 1972).

The studies began with 1,500 young Biloxi soybean (*Glycine max* var. *Biloxi*) plants exposed to a variable number of SD. By checking anatomical development daily, the researchers determined that two short photoperiods were sufficient for the differentiation of flower primordia in this variety. The microscopic beginnings of flower buds were first visible 3 days after the end of a 2-day treatment and were very evident after 10 days, appearing first in the axil of the fourth leaf primordium from the tip of the main stem, from undifferentiated tissue. Because the speed at which buds developed into flowers depended both on the length and number of inductive photoperiods (Borthwick and Parker, 1938a), time to flowering did not indicate the effectiveness of light treatments. So it was necessary to dissect sample plants before a batch was treated to discover whether treatments initiated flowering or merely hastened the development of flower primordia that were already present.

These years were productive despite a severe shortage of funds for day-to-day expenses. "Borthwick didn't have his own budget," says Albert Piringer, a horticulturist who joined the group later. "At the end of the year, other people who had been budgeted would give him surplus money, and he would buy supplies. He literally was doing research with crumbs from the table."

The studies revealed that, in soybean, the number of flower primordia induced by SD increases with the age of the plant, reaching a maximum at 6 weeks (Borthwick and Parker, 1938b) and that every expanded leaf is capable of generating the flowering stimulus, although the youngest is most effective (Borthwick and Parker, 1940). The researchers also showed that a minimum irradiance during the photoperiod is necessary for floral induction (Borthwick and Parker, 1938c; Parker and Borthwick, 1940). Other workers, including Hamner (1940, 1944), soon confirmed this observation. Hamner also discovered that Biloxi soybean fails to flower under normally inductive nights if photoperiods exceed an upper limit (20 h) (Hamner, 1940).

Low temperatures inhibited floral induction (Parker and Borthwick, 1939b), and although growing points and leaf stalks were involved in this response (Borthwick et al., 1941a), events in the leaves were the most temperature-sensitive component (Parker and Borthwick, 1943). Because changes in the carbon–nitrogen ratio were insufficient to account for the temperature effects, they appeared not to be a major factor in floral induction (Parker and Borthwick, 1939a, b).

The group studied other varieties of soybean, finding early ones unsuitable for photoperiodism research because they formed floral primordia regardless of daylength, though LD delayed or even prevented the subsequent development of visible buds (Borthwick and Parker, 1939). By grafting leaves of varieties that could flower and fruit under continuous light to whole Biloxi plants, they found that the flower-initiating stimulus could be transferred between the leaf and the plant. Thus the autonomous induction that occurs in day-neutral plants appeared to occur in the same organ—the leaf—as in photoperiodic plants and to involve a similar transmissible stimulus (Heinze et al., 1942).

There were investigations into the effects of temperature and photoperiod on a LDP, barley (Borthwick et al., 1941b), and then several applied studies. The group determined the temperature and daylength requirements of the Russian dandelion (*Taraxacum kok-saghyz*), which was being tested as a potential source of rubber (Borthwick et al., 1943). They also discovered how plant breeders could manipulate the photoperiod and use nitrogenous fertilizers to avoid a dormant period, thereby producing successive generations of onions in the shortest time (Scully et al., 1945a). And they determined how interactions of photoperiod and nitrogen supply might affect soybean yield (Scully et al., 1945b).

The outcome was that, by the mid 1940s, the Beltsville group had extended the applications of photoperiodism research and knew a great deal about a suitable experimental system, the Biloxi soybean. But neither they nor anyone else could begin to explain how daylength controls flowering. "It began to dawn on them," says R. Jack Downs, "that running around in all directions wasn't the way to get the answer. So they had to make a choice. They knew that if you give inductive photoperiods, plants ultimately flower. So you can study this by starting at one end or the other. . . . Starting at the end where you get changes in morphology is pretty late—an awful lot of things have gone on by then. So it seemed more reasonable to start with the first step, which was the absorption of light. If light is absorbed, it is obviously absorbed by a pigment of some kind. So what is this pigment? One way to find out was to run an action spectrum."

The researchers sought advice from a brilliant physical chemist, Sterling Brown Hendricks, in the nearby Soils Building. "We soon began an informal cooperation," Hendricks wrote later, "that would endure for the next twenty-five years" (Hendricks, 1976).

STERLING BROWN HENDRICKS

Hendricks (Fig. 4-7), who was 42 when the cooperative work began in 1944, had graduated in 1922 from the University of Arkansas at Fayetteville with a bachelor's degree in chemical engineering. In 1926, he received a doctorate from Caltech with a major in chemistry and minors in mathematical physics and physics. After joining the Fixed Nitrogen Laboratory at Beltsville in 1928, he "made monumental contributions to mineralogy and the study of soils" (Butler and Wadleigh, 1987). His interests also included the structural aspects of organic and inorganic chemistry, chemistry and physics of crystal structure, insecticides, phase rule, X-ray diffraction of solids, and electron diffraction of crystals and gas molecules (Pauling, 1982).

When Hendricks learned of Borthwick's need for help, he agreed to get involved. The decision was clinched when Borthwick described the effects of a light break and they realized it might be possible to apply monochromatic light briefly during this inhibitory period. "The importance of a dark period interruption in connection with action-spectrum measurements for flowering," said Borthwick," was that it provided a means by which a very

FIGURE 4-7   Sterling Brown Hendricks. Courtesy of Frank B. Salisbury.

small amount of light energy (compared to that needed to meet daily photosynthetic requirements) would produce a very striking photomorphogenic response. It minimized confounding of photosynthetic and photoperiodic effects and, of great importance, permitted application of photomorphogenically effective irradiance before the physiological consequences of such an irradiance became involved'' (Borthwick, 1972).

THE SPECTROGRAPH

Before the work could begin, monochromatic light sources were needed. Wanting purer wave bands than could then be obtained with most filters, Hendricks built a spectrograph (Fig. 4-8). Because parts and money were scarce, he begged and borrowed components, keeping the cost to $50. The pieces were hauled down the awkward spiral staircase of the West Building into the winery's windowless racking room. About 18 m long, the room provided sufficient space for a 14-m beam of light which ensured adequate separation between different wavelengths.

Only a powerful lamp would provide bright enough light at that distance, so a 10,000-W direct-current carbon-arc light "was cadged from a Baltimore movie theater, with memories of pulchritude" (Hendricks, 1970). A carbon in this lamp lasted for about an hour and had to be checked every few minutes to avoid changes in energy output. "The DC motor-generator was scrounged from elsewhere in the USDA, and the large series resistance was one discarded by the Georgetown streetcar system upon changing from DC to AC operation" (Hendricks, 1970).

FIGURE 4-8 Beltsville spectrograph, side view. (From Parker et al., 1946.)

The beam from the arc lamp was focused through a slit and then bounced off a concave mirror into two 16-cm-high flint glass prisms, which split it into light of different colors (Fig. 4-9). The prisms came from Hendricks' laboratory, where they had been used in pioneering studies of the hydrogen bond. Hendricks originally borrowed them from the Smithsonian Institution, where Langley had used them to study solar radiation.

The spectrum emerging from the prisms illuminated an area between 1.5 to 2 m long and 10 cm wide with very pure bands of different wavelengths.

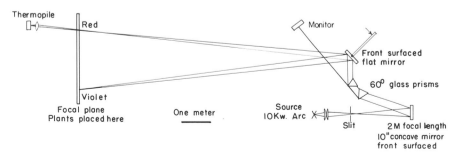

FIGURE 4-9 Optical path of Beltsville spectrograph. (From Parker et al., 1946.)

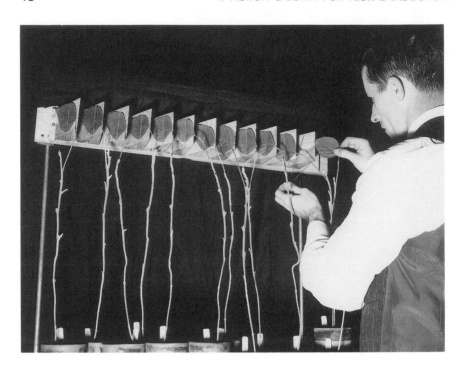

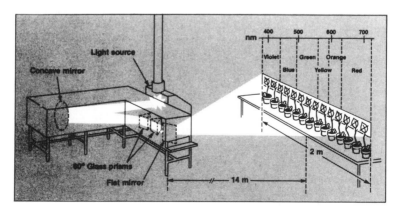

FIGURE 4-10   (*Top*) Method of holding leaflets in the image plane of spectrograph (From Hendricks and Borthwick, 1954); (*Bottom*) schematic drawing of spectrograph in action. Courtesy of USDA.

This was much larger than the spectrum at the Smithsonian, so it could accomodate leaves of mature plants as well as seeds and seedlings. Because Borthwick and Parker had already determined that the youngest fully expanded leaf was the most sensitive to light, they stripped Biloxi soybeans of all but the central leaflet of this leaf. The leaflet was temporarily secured to a predetermined position on a board for a brief irradiation, along with leaflets of plants positioned in other parts of the spectrum (Fig. 4-10).

Preliminary studies revealed that it was just as valid to vary the energy at each wavelength by altering the time of irradiation as by altering irradiance, which was less practical. Thus the light-break response obeyed the Law of Reciprocity. The researchers decided to apply the light break between 1 h before and 2 h after the middle of the dark period because the energy required to prevent floral initiation remained constant during this interval. The middle of the plant's night was set to coincide with the first hour of the working day, allowing the researchers time to warm up the spectrograph before treatments.

The soybean plants were grown under noninductive (16-h) days and then exposed to inductive 10-h days. In the middle of the 14-h dark period, test plants were irradiated on the spectrograph for various times, varying from 1 to (in the blue-violet region) 50 min. To avoid unintentional light breaks, leaflets were arranged on the spectrograph with the aid of a green safelight. Because the area between the growth rooms and the spectrograph was dark, anyone who entered had to grope through a doorway covered with two layers of black cloth and then use a dim flashlight fitted with a gelatin filter.

SDP ACTION SPECTRA

About a year elapsed between the beginning of the collaboration and the publication of an action spectrum for floral initiation in Biloxi soybean. The paper appeared in *Science* on August 10, 1945, the day after the United States dropped a plutonium bomb on Nagasaki. Opposed to nuclear weapons, Hendricks had refused to join the Manhattan Project.

To construct the curve, the researchers drew a line on graph paper from left to right for each amount of energy applied as a light break. On the lines they marked the total number of flower primordia initiated in four plants at each wavelength (Fig. 4-11). Then they superimposed freehand curves connecting the energies required for equal degrees of floral induction—total suppression (solid line) or various numbers of flower primordia (dotted lines).

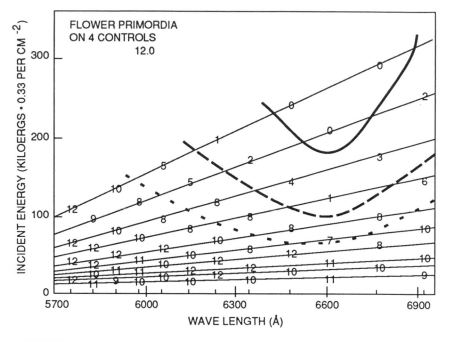

FIGURE 4-11   Action spectrum in yellow-red region for floral initiation in Biloxi soybean. Cross lines indicate energies used during light breaks. Figures on these lines are the total number of flower primordia on four plants. Solid curve shows the minimal energy required to prevent the development of the primordia. (From Parker et al., 1945.)

Combining the curves from different parts of the spectrum, they obtained an action spectrum for floral initiation that ran from violet (380 nm) to FR (720 nm). The curve had two dips, representing the most inhibitory spectral regions (Fig. 4-12). Thus "floral initiation can be suppressed by interruption of the dark period with light of sufficient energy from any region of the visible spectrum, but there are two regions of maximum efficiency, one in the yellow, orange and red and the other in the violet near 4,000 Å" (Parker et al., 1945).

Because R light breaks were most inhibitory, the photoreceptor obviously absorbed the R part of the spectrum most efficiently; therefore it was green or blue—a complementary color to red. When the most effective wavelengths for photosynthesis were marked on the action spectrum (Fig. 4-12), the researchers found "striking similarities to the curve for photosynthetic utilization of carbon dioxide. . . . It is likely that the action spectrum is due to a porphyrin-like material, which is probably chlorophyll" (Parker et al., 1945).

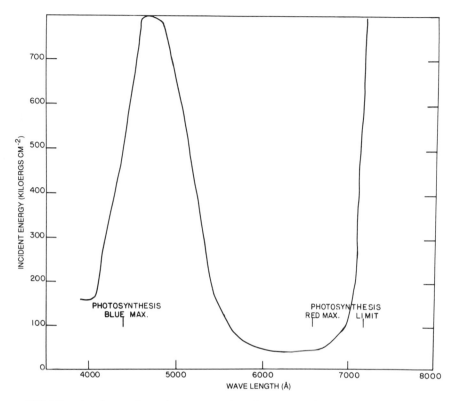

FIGURE 4-12 Composite action spectrum based on 21 experiments. Any point on curve shows amount of energy that must be applied to a Biloxi soybean leaflet during the middle of dark period to prevent floral initiation. Points of maximal utilization of carbon dioxide in photosynthesis and limits of this reaction are included for reference. (From Parker et al., 1945.)

A more extensive account of the work appeared a year later, along with an action spectrum (Fig. 4-13) for another SDP, cocklebur (*Xanthium saccharatum*).[2] Similarities between the two curves indicated "that the same pigment or pigments are involved in the photoperiodic response of the two plants" (Parker et al., 1946).

2. Downs explains that "*X. pensylvanicum* and *X. saccharatum* were described as new species in the same publication. When they were later combined into a single species, the rules of nomenclature required that the one described first—*pensylvanicum*—must be the name of the combination. Somehow, a taxonomist at Beltsville told Borthwick that *saccharatum* should be used. Later we changed to *pensylvanicum*. When Wallroth described *pensylvanicum* he either misspelled it or, more likely, a typographical error occurred. Once published, the name must be retained as written."

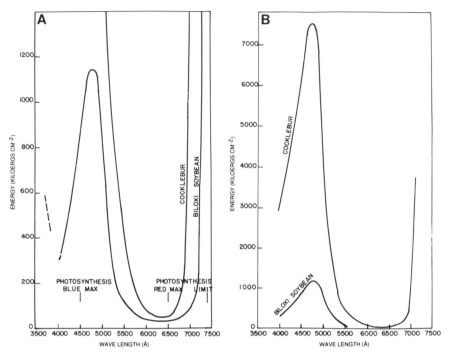

FIGURE 4-13   Composite action spectra for suppression of floral initiation in soybean and cocklebur. (A) Points on soybean curve give energy required at middle of 14-h dark period to prevent floral initiation; (B) Points on cocklebur curve give energy required at middle of 12-h dark period to prevent floral initiation. (From Parker et al., 1946.)

LDP ACTION SPECTRA

The next step was to see if floral initiation in LDP had a similar action spectrum, and Borthwick and Parker chose a barley they had already studied (Borthwick et al., 1941b), a noncommercial variety named Wintex. Modifying the spectrograph to accomodate tall plants, the group irradiated barley in the middle of dark periods to determine the minimum amount of energy at each wavelength that would trigger the formation of spikelet primordia when the plants were kept on SD. At the same time, they measured stem elongation, which in many LDP precedes the appearance of visible flowers.

The results (Borthwick et al., 1947, 1948) revealed that the action spectrum for floral induction in barley was very similar to that for the prevention of floral induction in soybean and cocklebur (Fig. 4-14). In all three plants, the most effective region of the spectrum lay between 600 and

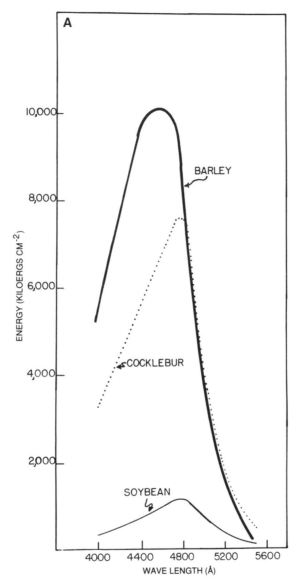

FIGURE 4-14   Composite action spectra for floral initiation in Wintex barley and suppression of floral initiation in soybean and cocklebur in (A) the blue-violet region of the spectrum and (B) the red region. Action spectrum for SDP gives energy required to suppress floral initiation when applied as a light break. Action spectrum for barley gives energy required to initiate spike development and stem elongation when applied in middle of 12.5-h dark period. (From Borthwick et al., 1948.) (*Figure continues*)

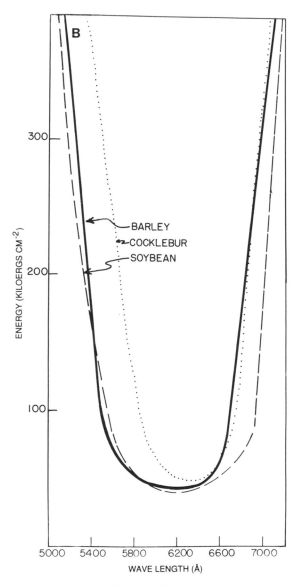

FIGURE 4-14   Continued

660 nm, where a similarly small amount of energy was required in each case. The least effective region was always near 480 nm, whereas small changes of wavelength produced the most rapid changes in effectiveness between 500 and 560 nm. A sharp cutoff at 720 nm was a further similarity.

The spectral sensitivity and energy requirements for stem elongation were very similar to those for floral induction or inhibition, suggesting that, in LDP, "the one process was linked to and dependent upon the other" (Borthwick et al, 1948).

"These points of similarity in the response curves for the three plants," the group said, "indicate that absorption of effective radiation is due to the same type of pigment both for long-day and short-day plants" (Borthwick et al., 1948). An action spectrum for inhibition of floral induction in a second LDP, *Hyoscyamus niger,* supported this deduction (Parker et al., 1950). Both plants were grown on near-inductive light–dark cycles, and light breaks were applied during nine successive dark periods.

The concept of a common basic mechanism was not new; there was suggestive evidence from the earlier spectral work (Withrow and Withrow, 1940). But the Beltsville group's action spectra provided the first convincing evidence because they were both quantitative and specific for the light-break response.

REFERENCES

Blackman, V. H. (1936). Light and temperature and the reproduction of plants. *Nature* 137,931–934.

Bonner, J., and Bonner, D. (1948). Note on induction of flowering in *Xanthium*. *Bot. Gaz.* 110,154–156.

Borthwick, H. A. (1956). Presidential address, August 27, 1956, annual meeting, American Society of Plant Physiologists, University of Connecticut, Storrs. Unpublished.

Borthwick, H. A. (1972). History of phytochrome. *In* "Phytochrome" (K. Mitrakos and W. Shropshire, Jr., eds.), pp. 4–23. Academic Press. London, New York.

Borthwick, H. A., and Parker, M. W. (1938a). Influence of photoperiods upon the differentiation of meristems and the blossoming of Biloxi soy beans. *Bot. Gaz.* 99,825–839.

Borthwick, H. A., and Parker, M. W. (1938b). Effectiveness of photoperiodic treatments of plants of different age. *Bot. Gaz.* 100,245–249.

Borthwick, H. A., and Parker, M. W. (1938c). Photoperiodic perception in Biloxi soybeans. *Bot. Gaz.* 100,374–387.

Borthwick, H. A., and Parker, M. W. (1939). Photoperiodic responses of several varieties of soybeans. *Bot. Gaz.* 101,341–365.

Borthwick, H. A., and Parker, M. W. (1940). Floral initiation in Biloxi soybeans as influenced by age and position of leaf receiving photoperiodic treatment. *Bot. Gaz.* 101,806–817.

Borthwick, H. A., Parker, M. W., and Heinze, P. H. (1941a). Influence of localized low temperature on Biloxi soybean during photoperiodic induction. *Bot. Gaz.* 102,792–800.

Borthwick, H. A., Parker, M. W., and Heinze, P. H. (1941b). Effect of photoperiod and temperature on development of barley. *Bot. Gaz.* 103,326–341.

Borthwick, H. A., Parker, M. W., and Scully, N. J. (1943). Effects of photoperiod and temperature on growth and development of kok-saghyz. *Bot. Gaz.* 105,100–107.

Borthwick, H. A., Hendricks, S. B., and Parker, M. W. (1947). Action spectrum for the control of flowering in winter barley, a long-day plant. *Am. J. Bot.* 34,598.

Borthwick, H. A., Hendricks, S. B., and Parker, M. W. (1948). Action spectrum for photo-periodic control of floral initiation of a long-day plant, Wintex barley (*Hordeum vulgare*). *Bot. Gaz.* 110,103–118.

Butler, W. L., and Wadleigh, C. H. (1987). Sterling Brown Hendricks, April 13, 1902–January 4, 1981. *Biogr. Mem. Natl. Acad. Sci. USA* 56,181–212.

Cathey, H. M. (1969). *Chrysanthemum morifolium* (Ramat.) Hemsl. In "The Induction of Flowering" (L. T. Evans, ed.), pp. 268–290. Cornell University Press, Ithaca, New York.

Emsweller, S. L., Stuart, N. W., and Byrnes, J. W. (1941). Using a short interval of light during night to delay blooming of chrysanthemums. *Proc. Am. Soc. Hort. Sci.* 39,391–392.

Gardner, F. E. (1946). Letter to M. W. Parker, July 19, Unpublished.

Garner, W. W. (1929). Effect of length of day on growth and development of plants. *Proc. Int. Congr. Plant Sci. 4th.* pp. 1050–1055.

Garner, W. W., and Allard, H. A. (1931). Effect of abnormally long and short alternations of light and darkness on growth and development of plants. *J. Agric. Res.* 42,629–651.

Hamner, K. C. (1940). Interrelation of light and darkness in photoperiodic induction. *Bot. Gaz.* 101,658–687.

Hamner, K. C. (1944). Photoperiodism in plants. *Annu. Rev. Biochem.* 13,575–590.

Hamner, K. C., and Bonner, J. (1938). Photoperiodism in relation to hormones as factors in floral initiation and development. *Bot. Gaz.* 100,388–431.

Heinze, P. H., Parker, M. W., and Borthwick, H. A. (1942). Floral initiation in Biloxi soybean as influenced by grafting. *Bot. Gaz.* 103,518–530.

Hendricks, S. B. (1970). The passing scene. Quotations reproduced, with permission, from the *Annual Review of Plant Physiology,* Vol. 21, pp. 1–10. © 1970 by Annual Reviews Inc.

Hendricks, S. B. (1976). Harry Alfred Borthwick 1898–1974. *Biogr. Mem. Natl. Acad. Sci. USA* 48,105–122.

Hendricks, S. B., and Borthwick, H. A. (1954). Photoperiodism in plants. *Proc. Int. Congr. Photobiol. 1st,* pp. 22–35.

Kasperbauer, M. J., Borthwick, H. A., and Hendricks, S. B. (1963). Inhibition of flowering of *Chenopodium rubrum* by prolonged far-red radiation. *Bot. Gaz.* 124,444–451.

Kraus, E. J., and Kraybill, H. R. (1918). Vegetation and reproduction with special reference to the tomato. *Oreg. Agric. Exp. Stn. Bull.* 149,1–90.

Loehwing, W. F. (1948). The developmental physiology of seed plants. *Science* 107,529–533.

Mitchell, J. W. (1935). A method of measuring respiration and carbon fixation of plants under controlled environmental conditions. *Bot. Gaz.* 97,376–387.

Parker, M. W., and Borthwick, H. A. (1939a). Effect of photoperiod on development and metabolism of the Biloxi soybean. *Bot. Gaz.* 100,651–689.

Parker, M. W., and Borthwick, H. A. (1939b). Effect of variation in temperature during photoperiodic induction upon initiation of flower primordia in Biloxi soybean. *Bot. Gaz.* 101,145–167.

Parker, M. W., and Borthwick, H. A. (1940). Floral initiation in Biloxi soybeans as influenced by photosynthetic activity during the induction period. *Bot. Gaz.* 102,256–268.

Parker, M. W., and Borthwick, H. A. (1943). Influence of temperature on photoperiodic reactions in leaf blades of Biloxi soybean. *Bot. Gaz.* 104,612–619.

Parker, M. W., Hendricks, S. B., Borthwick, H. A., and Scully, N. J. (1945). Action spectrum for the photoperiodic control of floral initiation in Biloxi soybean. *Science* 102,152–155. Copyright 1945 by the American Association for the Advancement of Science.

Parker, M. W., Hendricks, S. B., Borthwick, H. A., and Scully, N. J. (1946). Action spectrum for the photoperiodic control of floral intiation of short-day plants. *Bot. Gaz.* 108,1–26.

Parker, M. W., Hendricks, S. B., and Borthwick, H. A. (1950). Action spectrum for the photoperiodic control of floral initiation of the long-day plant *Hyoscyamus niger. Bot. Gaz.* 111,242–252.

Scully, N. J., Parker, M. W., and Borthwick, H. A. (1945a). Interaction of nitrogen nutrition and photoperiod as expressed in bulbing and flower-stalk development of onion. *Bot. Gaz.* 107,52–61.

Scully, N. J., Parker, M. W., and Borthwick, H. A. (1945b). Relationship of photoperiod and nitrogen nutrition to initiation of flower primordia in soybean varieties. *Bot. Gaz.* 107,218–231.

Stuart, N. W. (1943). Controlling time of blooming of chrysanthemums by the use of lights. *Proc. Am. Soc. Hort. Sci.* 42,605–606.

United States at Large, 74th Congress, 1935–1936, Vol. 49, Part I, Chapter 338, Stat. 436.

USDA (1936–1960). Ezra Jacob Kraus. Personnel records. Unpublished.

USDA (1936–1966). Marion Wesley Parker. Personnel records. Unpublished.

Went, F. W. (1946). Diary entry, March 21, 1946. Unpublished.

Wiser, V., and Rasmussen, W. D. (1966). Background for plenty: A national center for agricultural research. *Maryland Historical Magazine* 61,283–304.

Withrow, R. B., and Withrow, A. P. (1940). The effect of various wavebands of supplementary radiation on the photoperiodic response of certain plants. *Plant Physiol.* 15,609–624.

CHAPTER **5** _____

Action Spectra for Elongation Growth

In 1946, Frits Went (Fig. 5-1) and Henry O. Eversole, builder of the first air-conditioned greenhouse in the United States, set out to view plant growth facilities in the East while planning a "phytotron" at Caltech.[1] On March 21, they toured the Beltsville facilities, where they saw the spectrograph. That evening Went raised the possibility of using the device "for a few experiments with peas which I always had wanted to do. Perhaps after our return to New York we will perform these experiments together" (Went, 1946).

PEA LEAF EXTENSION

A few years earlier, Went had noticed that pea seedlings at the front of darkroom shelves were squatter and had larger leaves than those farther back. Intrigued that the minute amount of energy from his orange-red safelight could so drastically affect development, he began to investigate the effects of light of different wavelengths.

Using mercury vapor or neon lamps plus glass filters, he found that, although seedling development was affected most markedly by R and yellow light, green light affected leaf growth least, whereas B affected stem growth least. The phototropic response, however, only occurred in B or green light. "So the first steps (light absorption)," Went concluded, "for each of the three processes, phototropism, leaf growth and stem elongation, are different" (Went, 1941).

The spectrograph offered purer bands of light, however. "Friday morning we went to the lab, and there Borthwick, Parker, Scully and Hendricks had everything arranged in the most excellent manner," Went noted. "As

1. The $407,000 Earhart Plant Research Laboratory, in which practically any climatic condition could be duplicated, was dedicated in 1949.

FIGURE 5-1   Frits W. Went and tomato plants, 1945, in the first trial model of the Caltech Phytotron. Courtesy of California Institute of Technology Archives, Pasadena.

I had expected to be back in Beltsville on the previous day, when the illuminations with the spectrum were to start, they had given the first illumination before I was there. Since we had talked the experiment over in great detail, it could not have been done better than if I had been there'' (Went, 1946).

The peas were planted in long, redwood boxes, which were irradiated every day for 60, 20 , 6, 2 or $\frac{1}{4}$ min. ''For the changing of boxes, control of the light, etc., usually two or three of us work together,'' Went said. ''During the last four days the illuminations took 7 h in a stretch, but now we are through most of the work. Today it took us from 8:00 to 16:45 with only a short time off for lunch to get 550 peas measured. I did all the measuring, Marion Parker recorded, and Scully cut the peas off, placed them in order and pressed the ones I had measured. Harry Borthwick and Sterling Hendricks took care of the illuminations of the other series'' (Went, 1946).

The work was finished the following Saturday, and Went and Eversole left the next day. Two years later, after the Beltsville group had varied the irradiance rather than the duration of the light treatments, the collaborators submitted a joint paper (Parker et al., 1949). This did not include an

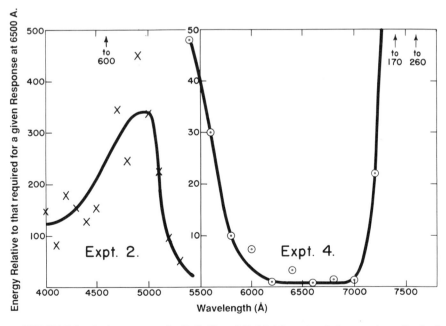

FIGURE 5-2 Action spectra in B (*left*) and R (*right*) parts of the spectrum for leaf lengthening in etiolated pea seedlings. (From Parker, et al., 1949.)

action spectrum for inhibition of stem growth because there was too much variability among the controls. But the experiments confirmed that R inhibits growth far more than other parts of the spectrum. The action spectrum for leaf elongation had many similarities to those for the light-break effect (Fig. 5-2), indicating "that the two processes are controlled by the same mechanism which is probably also effective in control of other growth phenomena" (Parker et al., 1949).

This was a remarkable conclusion, for it implicated a single mechanism in two broad classes of responses: photoperiodic ones such as flowering, in which a timing mechanism was involved, and nonphotoperiodic responses such as leaf expansion, which were not linked to time measurement.

Arthur W. Galston, who visited Beltsville in 1949 to derive an action spectrum for flavin-sensitized oxidation of indoleacetic acid, remembers that "the paper was seminal because it convinced everyone that there was a common photoreceptor. Also, it really shook things up because dark rooms all over the country had to be refurbished. Before then, plants had been grown in red safelights. But now we suddenly learned that red light was doing things to plants we don't really know about. The red filters were thrown away and people tried to get narrow-pass green filters that would grow plants properly."

IDENTITY OF THE PIGMENT

The pea action spectrum prompted an addendum to a *Chronica Botanica* review. "A fundamental similarity in controlling mechanism is thus established between floral initiation of long- and short-day plants, of stem elongation in long-day plants, of leaf elongation, and of internode shortening in etiolated pea seedlings," the Beltsville workers asserted. "Naturally one wonders what other phenomena of growth are controlled through this same mechanism and one looks more deeply for the nature of the mechanism itself than merely effect upon a meristem" (Borthwick et al., 1948).

Looking more deeply, they reconsidered the identity of the photoperiodic pigment, as they continued to call the substance controlling photomorphogenesis. In 1946, they had suggested chlorophyll (see Chapter 4), but in 1948, Richard H. Goodwin and Olga v. H. Owens of Connecticut College in New London proposed that chlorophyll's immediate precursor, protochlorophyll, might trigger inhibition of first internode elongation in oat seedlings. Both protochlorophyll and chlorophyll are cyclic tetrapyrroles, having four pyrrole units joined in a circle. A pyrrole is a ring of one nitrogen and four carbon atoms.

Extending the studies of Weintraub and McAlister (see Chapter 3) but irradiating only briefly each day with bands of filtered light, Goodwin and Owens had derived the 400–630-nm portion of the action spectrum for this response. With peaks at 577 and 623 nm, the curve bore a striking resemblance in the yellow and R regions to the absorption spectrum of protochlorophyll, although it lacked a peak in the B, where protochlorophyll absorbs strongly (Goodwin and Owens, 1948). The Connecticut researchers stressed that the only pigment in etiolated oat seedlings known to absorb R was protochlorophyll (Goodwin and Owens, 1951).

The Smithsonian workers had also favored protochlorophyll initially (Johnston, 1940). But "further study of spectral effectiveness of radiation for growth inhibition of oat mesocotyl had indicated that the maximum response occurs at [660 nm]," Johnston reported. "It is highly suggestive that both chlorophyll *a* and a pigment as yet unidentified which has been found in dark-grown oat seedlings exhibit an absorption band at this position" (Johnston, 1941).

Attempts to isolate the unknown pigment (Johnston, 1942) were halted in 1943, when the Division of Radiation and Organisms began to study the deterioration of fabric, cardboard, and electrical insulation by UV and molds, at the request of the Naval Research Laboratory. But even without the disruption of World War II, it is unlikely that much progress would have been made, for there was as yet no assay for the pigment.

By the end of the decade, the Beltsville group was arguing against a cyclic tetrapyrrole (Parker and Borthwick, 1950a), as Went had done much earlier when discussing the pigment controlling leaf growth (Went, 1941). Went pointed out that a single light exposure too short to trigger chlorophyll formation promoted the elongation of leaves (Trumpf, 1921) and that the absorption characteristics of the two processes were very different. "I think we have enough data to definitely prove that another pigment is involved in the light absorption which affects leaf growth in etiolated peas," he wrote in his diary (unpublished) on leaving Beltsville. "Then it becomes very much worth while to extract such a pigment."

Parker visited Went at Caltech several years later with this goal in mind. But again, the pigment was not to be found.

PHYCOCYANIN?

Many years would pass before the pigment would be in hand, but Hendricks was able to guess its identity. In the summer of 1949, "Sterling came back from the library one day and said, 'It has to be *C*-phycocyanin or something very close,' " recalls Galston.

C-phycocyanin is a member of the phycobilin family of red (phycoerythrins) or blue-green (phycocyanins) pigments. In the late 1940s, the phycobilins were little known, but one of Hendricks' acquaintances, Francis Haxo of Stanford University's Hopkins Marine Station in Pacific Groves, proposed that they were the chief photosynthetic pigments in red algae (Haxo and Blinks, 1950). Like the chlorophylls, phycobilins are tetrapyrroles, but their four pyrroles lie in an open chain rather than a closed circle (see Fig. 27-7). It is now known that opening up the circle has little effect on the absorption of R but drastically decreases absorption in the B part of the spectrum.

Using data published 20 years earlier (Svedberg and Katsurai, 1929), Hendricks constructed an absorption curve for C-phycocyanin. When he placed this alongside the action spectra for floral initiation and leaf growth (Fig. 5-3), the similarities were striking (Parker et al., 1950). That phycocyanin had never been found in seed plants was not a serious problem because the photoperiodic pigment was obviously present at very low concentrations. Nevertheless, "the pronouncement was a dud," Hendricks later noted, "for few physiologists trusted anything found out by physiological methods—the demand was for a bottle full of the stuff" (Hendricks, 1970).

ALBINO SEEDLINGS

Before dismissing cyclic tetrapyrroles, the Beltsville workers asked whether their action spectra were distorted by screening pigments, a possibility suggested by Goodwin and Owens. Using brief, daily irradiations, they derived an action spectrum for inhibition of second internode elongation in albino barley seedlings. This curve resembled those obtained previously, suggesting that screening was not a serious factor (Borthwick et al., 1951)—and showing beyond doubt that chlorophyll itself was not the photoreceptor.

Attempts to detect the photoperiodic pigment in albino seedlings were unsuccessful. When clippings of dark-grown leaves were placed in a spectrophotometer—which measures transmission of monochromatic light—the tissue absorbed less than 0.1% of the R that passed through it. Calculations suggesting that "not more than one part in one hundred thousand of the incident light is utilized even in the region of maximum effectiveness" were "discouraging for direct detection" (Borthwick et al., 1951).

FAR-RED RADIATION

At Purdue University Agricultural Experiment Station, Robert Bruce Withrow (Fig. 5-4A) and his wife, Alice Philips Withrow (Fig. 5-4B), were also investigating the spectral sensitivity of various stages of plant development (see Chapter 3).

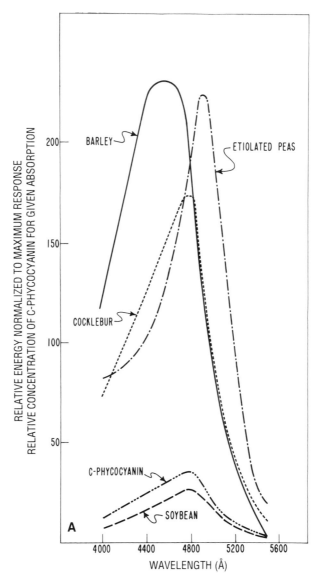

FIGURE 5-3 Composite action spectra for control of floral initiation and leaf growth, together with curve showing relative concentrations of C-phycocyanin for a given absorption. (From Parker et al., 1950.)

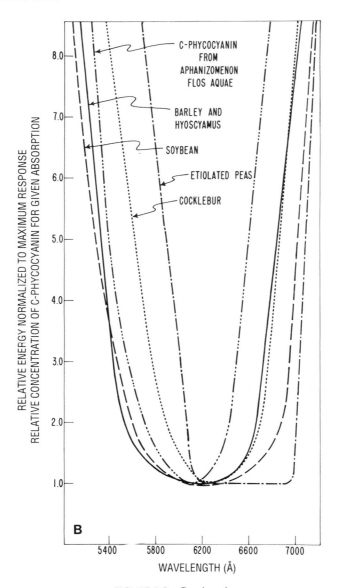

FIGURE 5-3 Continued

FIGURE 5-4　(*Left*) Robert B. Withrow, 1955; (*Right*) Alice P. Withrow, 1960.

The Withrows compared China aster, spinach, soybean, and tomato plants grown under mercury, incandescent, and fluorescent lamps, which had recently become available (Naylor and Gerner, 1940). Only the incandescent and fluorescent sources allowed satisfactory growth, but the former produced plants with "relatively long internodes, long petioles, and weak stems as compared with the growth obtained under solar radiation" (Withrow and Withrow, 1947). Unlike fluorescent lamps, incandescent lamps emit FR as well as R (Fig. 5-5).

In the fall of 1948, the Withrows moved to the Smithsonian Institution, where Robert became chief of the Division of Radiation and Organisms in place of Johnston, who had died the previous December. Withrow promptly reorganized the Smithsonian labs, overseeing installation of more modern lighting. Five rooms were converted to constant-environment rooms, one of which made use of a new type of fluorescent lamp, the 8-ft slimline. Because of their shape, these lamps could be packed together to give ample light for plant growth—2,500 footcandles 2 ft below the unit. Behind the fluorescent lamps, a bank of 60-W incandescent lamps provided supplementary radiation in the longer wavelengths.

A fluorescent-lamp room was installed in the Beltsville West Building (Fig. 5-6) after several years of comparative studies with different light sources (Parker and Borthwick, 1950b). In the course of this work, Parker and Borthwick noted that soybean plants illuminated solely with carbon rods that produced more R than usual were less sturdy and contained less carbohydrate than those grown under the usual combination of carbon-arc and incandescent lighting. Thus "the incandescent-filament lamps contrib-

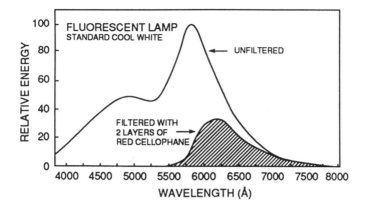

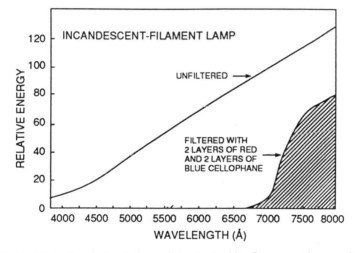

FIGURE 5-5 Spectral emissions of standard white fluorescent lamp and incandescent filament lamp. (From Downs et al., 1958.)

uted something more to the production of dry weight than might have been expected from photosynthesis due to the additional red radiation'' (Parker and Borthwick, 1949). Withrow found this manuscript ''very interesting as it again indicates some controlling mechanism due to the longer wavelengths. I would like to see sometime if we cannot get our fingers on to some measurable factor here as I believe it would bring to light a very interesting phenomenon'' (Withrow, 1949).[2]

2. Withrow also ''changed intensity to irradiance in every case but one. The term *intensity* applies to the source and is a measure of its radiant power'' (Withrow, 1949).

FIGURE 5-6   New fluorescent room 1, West Building basement, Beltsville. This was one of earliest plant growth chambers to use Power-Groove, 1,500 ma, fluorescent lamps. Courtesy of USDA.

Wassink and colleagues issued several reports on the effects of B and long, invisible wavelengths on the development of light-grown plants (e.g., Wassink et al., 1950). They observed striking stem and leaf elongation in lettuce seedlings receiving 10 h WL daily followed by 8 h violet, B, or infrared (Fig. 5-7). The effect was not produced by supplementary light from other spectral regions nor by supplementary WL. Nor was it seen when darkness immediately followed a daily WL photoperiod (Wassink et al., 1951).

Doctoral candidate Johannes A. J. Stolwijk found that the B response was mostly due to a very small amount of "infrared" emitted by the B source (Stolwijk, 1954a, b.) This was confirmed subsequently (Wassink et al., 1957; de Lint, 1961). Thus the Dutch workers determined that some parts of the spectrum that later proved to be FR promoted stem and leaf elongation, whereas researchers at the Smithsonian and Beltsville found that R inhibits stem extension in seedlings.

Observing macroscopic flower bud development in *Brassica,* Wassink suspected that different parts of the spectrum might have opposite effects on a physiological response. Plants receiving violet, B, or infrared sup-

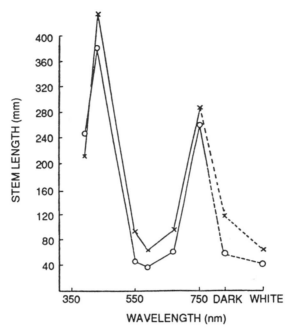

FIGURE 5-7   Stem length in *Lactuca sativa* as dependent on wavelength of supplementary radiation in a 10 + 8 = h experiment. After 40 days (o) and 67 days (x) of treatment. Measurements made in June 1950. (From Wassink et al., 1951.)

plementary radiation developed large flower buds from primordia, even—in the case of B—when 4 h of supplementary radiation followed a WL period of only 5 h. Plants exposed to 12-h cycles of WL developed primordia but not large buds, even though they received much more B than the B-irradiated plants after hour 5 of each cycle. Wassink conjectured "that other parts of the spectrum have an antagonistic effect" (Wassink et al., 1951). His group also observed that internodes of *Cosmos bipinnatus* elongated in violet, B, and infrared but not in WL or in the dark. "This might be explained by assuming that besides the part of the spectrum (violet, blue, and infrared) which actively promotes elongation, another part (green, yellow, and red) inhibits it in some way. Of course these processes need not be directly related" (Wassink et al., 1951).

When Hendricks visited Wageningen in 1951, "Wassink asked Hendricks if he had ever seen a publication on the effect of infrared on plants," recalls Carel J. P. Spruit, who joined the Wageningen group that year to study photosynthesis. "Hendricks said that Flint and McAlister had found some effects on the germination of lettuce seeds [see Chapter 3]. Wassink suggested that Hendricks should improve on this work."

REFERENCES

Borthwick, H. A., Parker, M. W., and Hendricks, S. B. (1948). Wave length dependence and the nature of photoperiodism. *In* "Vernalization and Photoperiodism, a Symposium" (A. E. Murneek and R. O. Whyte, eds.), pp. 71–78. Chronica Botanica, Waltham, Massachusetts.

Borthwick, H. A., Hendricks, S. B., and Parker, M. W. (1951). Action spectrum for inhibition of stem growth in dark-grown seedlings of albino and nonalbino barley (*Hordeum vulgare*). *Bot. Gaz.* 113,95–105.

de Lint, P. J. A. L. (1961). Dependence of elongation on wavelength of supplementary irradiation. *Meded. Landbouwhogesch. Wageningen* 61 (16),1–14.

Downs, R. J., Borthwick, H. A., and Piringer, A. A. (1958). Comparison of incandescent and fluorescent lamps for lengthening photoperiods. *Proc. Am. Soc. Hortic. Sci.* 71,568–578.

Goodwin, R. H., and Owens, O. v. H. (1948). An action spectrum for inhibition of the first internode of *Avena* by light. *Bull. Torrey Bot. Club* 75,18–21.

Goodwin, R. H., and Owens, O. v. H. (1951). The effectiveness of the spectrum in *Avena* internode inhibition. *Bull. Torrey Bot. Club* 78,11–21.

Haxo, F. T., and Blinks, L. R. (1950). Photosynthetic action spectra of marine algae. *J. Gen. Physiol.* 33,389–422.

Hendricks, S. B. (1970). The passing scene. Reproduced, with permission, from the *Annual Review of Plant Physiology*, Vol. 21, pp. 1–10. © 1970 by Annual Reviews Inc.

Johnston, E. S. (1940). *Smithson. Inst. Annu. Rep.*, p.93.

Johnston, E. S. (1941). *Smithson. Inst. Annu. Rep.*, p. 114. Reprinted by permission of the *Smithsonian Institution Annual Report, 1941*, p. 114. © Smithsonian Institution.

Johnston, E. S. (1942). *Smithson. Inst. Annu. Rep.*, pp. 84–85.

Naylor, A. W., and Gerner, G. (1940). Fluorescent lamps as a source of light for growing plants. Bot. Gaz. 101,715–716.

Parker, M. W., and Borthwick, H. A. (1949). Growth and composition of Biloxi soybean grown in a controlled environment with radiation from different carbon-arc sources. *Plant Physiol.* 24,345–358.

Parker, M. W., and Borthwick, H. A. (1950a). Influence of light on plant growth. *Annu. Rev. Plant Physiol.* 1,43–58.

Parker, M. W., and Borthwick, H. A. (1950b). A modified circuit for slimline fluorescent lamps for plant growth chambers. *Plant Physiol.* 25,86–91.

Parker, M. W., Hendricks, S. B., Borthwick, H. A., and Went, F. W. (1949). Spectral sensitivities for leaf and stem growth of etiolated pea seedlings and their similarity to action spectra for photoperiodism. *Am. J. Bot.* 36,194–204.

Parker, M. W., Hendricks, S. B., and Borthwick, H. A., (1950). Action spectrum for the photoperiodic control of floral initiation of the long-day plant *Hyoscyamus niger. Bot. Gaz.* 111,242–252.

Stolwijk, J. A. J. (1954a). Some characteristics of internode elongation. *Proc. Int. Congr. Photobiol. 1st*, pp. 78–82.

Stolwijk, J. A. J. (1954b). Wave length dependence of photomorphogenesis in plants. *Meded. Landbouwhogesch. Wageningen* 54(5),181–244.

Svedberg, T., and Katsurai, T. (1929). The molecular weights of phycocyanin and of phycoerythrin from *Porphyra tenera* and of phycocyanin from *Aphanizomenon flos aquae. J. Am. Chem. Soc.* 51,3573–3583.

Trumpf, C. (1921). Über den Einfluss intermittierender Belichtung auf das Etiolement der Pflanzen. Dissertation, Hamburg.

Wassink, E. C., Sluysmans, C. M. J., and Stolwijk, J. A. J. (1950). On some photoperiodic and formative effects of coloured light in *Brassica rapa*, f. *oleifera* subf. *annua*. *Proc. K. Ned. Akad. Wet. Ser. C* 53,1466–1475.

Wassink, E. C., Stolwijk, J. A. J., and Beemster, A. B. R. (1951). Dependence of formative and photoperiodic reactions in *Brassica rapa* var., *Cosmos* and *Lactuca* on wavelength and time of irradiation. *Proc. K. Ned. Akad. Wet. Ser. C* 54,421–432.

Wassink, E. C., Bensink, J., and de Lint, P. J. A. L. (1957). Formative effects of light quality and intensity on plants. *Proc. Int. Congr. Photobiol. 2nd,* pp. 196–213.

Went, F. W. (1941). Effects of light on stem and leaf growth. *Am. J. Bot.* 28,83–95.

Went, F. W. (1946). Diary entries. Unpublished.

Withrow, R. B. (1949). Letter to Marion Parker, March 10. Unpublished.

Withrow, A. P., and Withrow, R. B. (1947). Plant growth with artificial sources of radiant energy. *Plant Physiol.* 22,494–513.

CHAPTER 6

Photoreversible Pigment

When Eben Toole (see Fig. 3-1) arrived in Beltsville in 1939, he took charge of a newly created lab, Vegetable Seed Investigations. During World War II, when the supply of seeds from Europe seemed likely to dwindle, he and his wife, Vivian Kearns Toole (Fig. 6-1) studied the production and storage of vegetable seeds. But continuing the basic research on dormancy and germination, they became eager to expand Flint and McAlister's studies (see Chapter 3).

The photoperiodism workers were initially reluctant to be distracted by a seemingly unrelated project. But in 1951, when action spectra were amassed but the pigment was still not found, they finally agreed. "Little did they know," says Vivian Toole, "that seed would reveal the way to the secret of flowering."

In preparation for the collaborative work, the Tooles tested 31 lots of various black and white lettuce seed varieties in total darkness for response to light at different temperatures. Selecting a variety named Grand Rapids because three-fourths of that lot was light-requiring at 20°C, they stored the seed in small vials inside a large jar at 10°C. "Careful storage was very important," Vivian Toole explains, "because we did not want the effects of light to be confounded with changes due to loss of viability or to an increase in the number of seeds capable of germinating in darkness."

PROMOTION OF GERMINATION

An action spectrum for promotion of germination was obtained in November 1951. With the aid of a vacuum seed counter, the researchers placed seeds on wetted blotters in plastic sandwich boxes in four separate groups of three rows. Immediately after planting, the boxes were placed in double-thickness black sateen bags at 20°C. After an imbibition period,

70

FIGURE 6-1   Vivian K. Toole, 1963. Courtesy of
E. Dale Kearns Studio.

they were arranged across the spectrum at positions that placed each
group of seeds in a different wave band. After removal from the bags in
darkness, the seeds were irradiated for a few seconds or minutes.

Whereas Flint and McAlister had irradiated seeds for 24 h at single
energy levels, the Beltsville workers used very brief exposures at a range
of energy levels, for they wanted to obtain an action spectrum that was
both quantitative and specific for germination. After irradiating the seeds,
they immediately returned the boxes to their bags and placed them back in
the germination chamber at 20°C. Two days later, they read the results—a
germinated seed was one in which the radical had protruded through the
seed coat. "We were in such a hurry to get the results that we all sat down
and counted," Vivian Toole recalls. "It was as exciting as opening Christmas packages every 24 h."

The collaborators determined how much energy at each wavelength
made half the seeds germinate, thus obtaining the first quantitative action
spectrum for light-induced germination (Fig. 6-2A). The curve had the
familiar shape, with maximum sensitivity in the R, at about 660 nm. Thus
the photoreaction that sparked germination seemed to be identical to the
one that controlled flowering and stem and leaf growth.

This was a surprise, because the Beltsville workers had assumed that a
different photomechanism was involved (Parker and Borthwick, 1950).
Moreover, similar energy levels per unit area of tissue were required (Fig.
6-2A). "One could hardly believe such an astounding result," Hendricks
wrote, "showing that the control by light of a phenomenon at the start and
termination of plant growth—the germination of the seed and the eventual

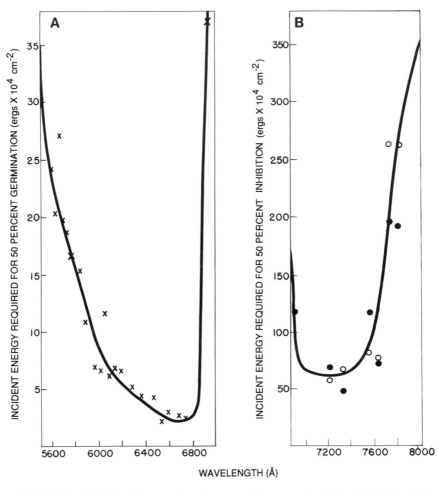

FIGURE 6-2   The Beltsville group's action spectra for promotion (A) and inhibition (B) of germination of Grand Rapids lettuce seed. (From Borthwick et al., 1954. Copyright by the University of Chicago.)

flowering of the plant—were the same not only in a qualitative sense but on an absolute basis as well" (Hendricks, 1958).

Faced with evidence of a common photoreceptor for so many different plant responses, the researchers were tempted to predict that it might operate in the animal kingdom also. "A hasty examination of the scarce but contradictory literature on the subject and a brief excursion of our own into experimentation in this field," Borthwick said, "prompted us to beat a hasty retreat to a larger limb of the tree of knowledge with the conviction that perhaps life is not quite so simple after all" (Borthwick, 1956).

INHIBITION OF GERMINATION

Deriving the germination action spectrum, the collaborators noticed that very few seeds germinated in the boxes from the region of the spectrum between 700 and 800 nm. In fact, the percentage was even lower than among the dark controls. Thus, through the serendipitous use of seeds that were not 100% light-requiring, they realized that some wavelengths actually inhibit germination, as Flint and McAlister had concluded earlier.

Irradiations to obtain an action spectrum for inhibition of germination were performed on February 18, 1952. Whereas Flint and McAlister had exposed seeds to sufficient R to promote 50% germination without further irradiation, the Beltsville group applied sufficient to induce germination in more than 95% of the viable seeds. The seeds then received a second irradiation on the spectrograph. Plotting an action spectrum for 50% inhibition of germination, the researchers discovered that wavelengths in the FR,[1] between 710 and 750 nm, were the most effective (Fig. 6-2B).

At this point, "we were turning out so many experiments," Vivian Toole recalls, "that the dishes stacked up before they were washed. The non-germinating seeds from the experiment all germinated (except for the 2–3% percent that decayed) on the lab bench in 2 days—Eben called attention to this." Thus the seeds that had been inhibited by FR were repromoted by the light in the lab.

Huddled together in Borthwick's first-floor office, the group pondered this observation. They also reexamined data from an earlier experiment. Because seeds vary in light sensitivity according to how much water they have absorbed, the researchers had determined when to place the seeds on wetted blotters. The data revealed that seeds became more sensitive to R during imbibition as they became less sensitive to FR, and vice versa (Fig. 6-3). This striking reciprocity suggested that R and FR act on the same process (Borthwick et al., 1952; Borthwick et al., 1954).

As ideas for further tests bounced back and forth, someone thought up a flip-flop experiment. What would happen if seeds were exposed to repeated alternations of R and FR?

THE FLIP-FLOP EXPERIMENT

On April 9, 1952, the collaborators exposed petri dishes of imbibed lettuce seed to Slimline fluorescent tubes wrapped in two layers of red cellophane (R) and to an incandescent filament lamp combined with a red-purple ultra filter (FR) and a water filter to remove heat rays. Using eight dishes, they

---

1. The papers describing the experiments refer to these wavelengths as "infrared." Withrow later informed Hendricks that this terminology was incorrect.

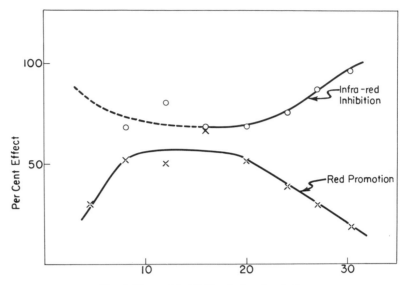

FIGURE 6-3   Variation in germination response of Grand Rapids lettuce seed with period of imbibition before a fixed irradiance in the R (lower curve) or FR (upper curve) portions of the spectrum. (From Toole et al., 1953.)

retained one as a dark control and exposed the others to alternating R and FR treatments, removing one dish to its black bag after each irradiation. Thus some dishes received R last and some FR last, and the final dish accumulated four R and three FR treatments.

The results were spectacular. In every dish that received R last, all the viable seeds (97–98%) germinated (Table 6-1). In every dish that received FR last, only about half the seeds germinated (Borthwick et al., 1952a). When the seeds were imbibed before the light treatments for 3 h, rather than 16 h, the difference was even more dramatic (Fig. 6-4). So FR not only inhibited, it reversed the very process that R triggered.[2]

The antagonistic effects of R and FR was one twist that even Hendricks had failed to predict, but he was quick to offer an explanation. He proposed that the pigment controlling germination existed in two interconvertible forms, one absorbing R and the other FR (Borthwick et al., 1952a). Thus the R-absorbing form, when exposed to R, would change to the

---

2. Later tests showed that, even with 100 alternations of R and FR, the last treatment decided the outcome (Borthwick, 1972).

TABLE 6-1

Germination of Grand Rapids Lettuce Seed After Alternating
Brief R and FR Exposures[a]

Irradiation	Germination (%)
None (dark control)	8.5
Red	98
Red + FR	54
Red + FR + Red	100
Red + FR + Red + FR	43
Red + FR + Red + FR + Red	99
Red + FR + Red + FR + Red + FR	54
Red + FR + Red + FR + Red + FR + Red	98

[a] From Borthwick et al., 1952a.

FR-absorbing form, and germination would be promoted. The FR-absorbing form, when exposed to FR, would convert back to the R-absorbing form, and germination would be inhibited. The unidentified photoreceptor was a photoreversible pigment!

THERMAL REACTIONS

The group deduced that conversion of the pigment's FR-absorbing form to its R-absorbing form occurred in darkness as well as in light, albeit much more slowly. This process—dark reversion—was evident in white lettuce seeds that germinated in light or darkness at 20°C but not at 30°C. After remaining in darkness at the higher temperature for several days, the seeds failed to germinate when the temperature was lowered to 20°C. But they germinated if they saw R at the end of the 30°C treatment. Thus at the higher temperature, the active form of the pigment must have changed thermally in darkness to the inactive form (Borthwick et al., 1952a).

The active form of the pigment was obviously able to function in darkness. After Grand Rapids seeds were fully promoted with R, half were still able to germinate if FR was given 7 h later. This amount of FR, if applied immediately after R, was sufficient to inhibit all the seeds. Thus, in this rapidly germinating species, the pigment must have completed its action in half the seeds in the 7 h between the R and FR irradiations (Borthwick et al., 1952a).

The researchers could hardly wait to publish these findings. "Since the authors are extremely anxious to submit this paper promptly for pub-

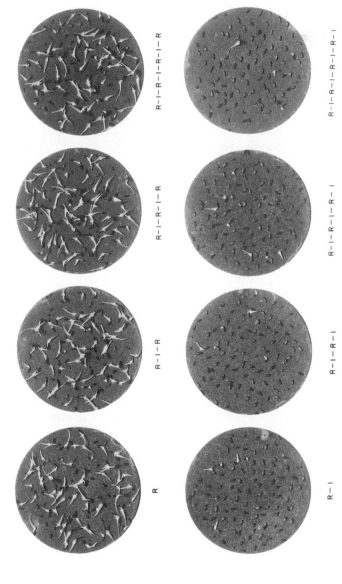

FIGURE 6-4 Germination of lettuce seed after alternating exposures to R and FR (I) radiation. (From Toole et al., 1953.)

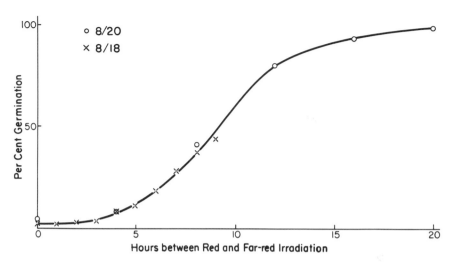

FIGURE 6-5   Rate of escape at 20°C from germination control by FR after promotion of Grand Rapids lettuce seed by R. (From Borthwick et al., 1954. Copyright by the University of Chicago.)

lication, I have read it hastily,'' a USDA editorial reviewer wrote at the end of May 1952. ''It seems to me that botanists might have trouble in understanding certain points unless the paper is elaborated. I assume, however, that it is written primarily for physicists'' (Gravatt, 1952).

A fuller account appeared 2 years later (Borthwick et al., 1954). Studying the action of the pigment in greater detail, the group had fully promoted Grand Rapids seeds with R and then exposed them to enough FR to fully inhibit germination at various times thereafter. Percentage germination increased with the interval between the two treatments (Fig. 6-5). Half the seeds germinated after about 8 h, and escape from inhibition by FR was just about complete in 15 h (Borthwick et al., 1954). Thus the FR-absorbing form of the pigment, produced by R, had triggered a sequence of events that at some point proceeded even after the pigment was converted back to its inactive form.

This observation suggested that the FR-absorbing form of the pigment was the active form. Alternatively, the R-absorbing form was an inhibitor. This seemed unlikely, Hendricks argued, because some seeds germinate if they see a minute amount of light after being buried for decades. Such an exposure would presumably convert only a small amount of pigment from one form to the other, barely altering the concentration of the R-absorbing form (the predominant form during dormancy), which therefore was unlikely to be inhibitory. The light exposure would, however, greatly change the concentration of the FR-absorbing form, suggesting that the appearance of this form provoked a physiological response (Hendricks et al., 1956).

INTERCONVERSION

Interpreting the flip-flop data, Hendricks proposed that the photoreversible pigment reacted with another substance, *RX*, when changing between its two forms (Borthwick et al., 1952a). Thus either the R-absorbing or FR-absorbing form would be pigment *X* (Fig. 6-6).

The 1954 paper reported that the photoreversible reaction was independent of temperature, for the flip-flop worked just as well when the seeds were held at 6–8°C as at 20°C. This strongly implicated a single pigment. And it suggested that either the pigment and its proposed reactant (*RX*) were in continuous contact or that no other reactant was involved and the pigment rearranged itself internally—isomerized[3]—to convert from one form to another (Borthwick et al., 1954).

But the researchers noted that imbibition in the presence of potassium nitrate changed the sensitivity of *Lepidium* seeds to R while generally changing that to FR in the opposite direction. "The coupled change in sensitivities must be operating through other molecules than the pigment in the photoreversible reaction," they argued (Toole et al., 1955; see also Hendricks and Borthwick, 1955). The most probable reactant was a hydrogen donor–acceptor:

$$P + RH_2 \underset{\text{far-red}}{\overset{\text{red}}{\rightleftarrows}} PH_2 + R$$

Because the pigment could effect such profound physiological changes but was present in tissue at very low concentrations, Hendricks proposed that its active form—PH_2 according to this scheme—functioned as an enzyme[4] (Hendricks, 1956; Hendricks, 1958). Influenced by the concept that a single enzyme in a biochemical pathway may act as a pacemaker that controls the supply of substrate for subsequent reactions (Krebs and Kornberg, 1957), he suggested that biological responses to R result from major metabolic changes that occur on photoactivation of the pigment (Hendricks, 1960a,b).

3. *Cis* and *trans* isomers have the same chemical formula but different structures. In the *cis* form, two atoms are on the same side of a double bond. In the *trans* form, they are on opposite sides of the double bond.
4. Because enzymes are proteins, this implied that the photoperiodic pigment was a protein, as did Hendricks's assertion that the pigment was a phycobilin.

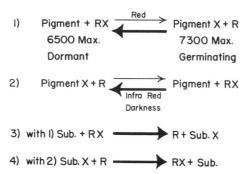

FIGURE 6-6 The Beltsville group's original scheme for the photoreversible reaction controlling seed germination. (From Borthwick et al., 1952a.)

REVERSIBILITY OF FLORAL INDUCTION

The discovery of a reversible photoreaction in seeds "caused us to wonder whether this far-red action might perhaps also be present in photoperiodism," Borthwick said. "If present it might be expected to act in opposition to red there also. That is, far red might be expected to reinduce flowering of short-day plants after they had been treated with red to inhibit flowering" (Borthwick, 1956).

The researchers had already demonstrated that a brief FR light break was unable to suppress flowering in SDP or promote it in LDP. And they had noted sharp cut-offs in the long wavelengths of their action spectra. But no one had thought to ask if FR could reverse the effects of R.

Borthwick, Hendricks, and Parker answered this question by first exposing cocklebur leaves to a white light break that normally would prevent flowering and subsequently irradiating them on the spectrograph with FR (Borthwick et al., 1952b). "Flowering of these few plants was very impressive," Borthwick later wrote, "because never before had we known a cocklebur plant to flower after its dark period was adequately interrupted with . . . light. This experiment proved that far red does, indeed, reverse the action of red in flowering and these reversible responses clearly showed that flowering and seed germination were truly controlled by the same photoreaction" (Borthwick, 1972).[5]

5. This paper dates the demonstration of R/FR-reversibility of floral induction to 1951. But lab notebooks reveal that photoreversibility was discovered in 1952, as stated in Hendricks and Borthwick, 1954. The results of the seed flip-flop were presented to the National Academy of Sciences on June 7, 1952, 2 months after Hendricks was elected to membership. The paper describing photoreversibility of floral induction was presented on September 9, 1952.

PHOTOPERIODIC TIMING

Because seed germination and floral induction appeared to be regulated by the same pigment, Hendricks and Borthwick proposed that the dark reversion they had detected in seeds might also occur in mature plants, enabling the pigment to function as a photoperiodic timer. Thus the photoperiod would place the pigment in its FR-absorbing form, which, when darkness fell, would slowly revert to its R-absorbing form, like sand trickling through an hourglass. SDP would flower only if reversion was completed before dawn, whereas LDP would fail to flower under these conditions (Borthwick et al., 1952b).

By exposing cocklebur to FR for 30 min just before the start of the dark period, the Beltsville workers decreased the critical night length from 8.5 h to less than 7.0 h. A R exposure at the beginning of the dark period lengthened it to 9.0 h (Borthwick et al., 1952b). Thus production of additional FR-absorbing pigment or removal of R-absorbing pigment did appear to affect dark timing.

When Borthwick and Hendricks left for vacations, they asked a new member of the lab to repeat and extend the experiments. "I was unable to," says Robert Jack Downs, "so upon returning, Borthwick suggested we repeat the experiment together. We made several additional tries without success, and at this point Borthwick and Hendricks became upset—embarrassed would be a better description. This was the only experiment before or since where their results could not be duplicated easily by anyone desiring to do so. But there is no question that they obtained the published results."

BEAN HOOK OPENING

At the Smithsonian meanwhile, Withrow was expanding the research program, which had flagged with Johnston's illness and death. Among several projects begun in the early 1950s was a study of the effects of light on dark-grown bean seedlings, which have a staple-shaped stem—a plumular hook—that protects the developing shoot from physical damage as it pushes through the soil. Once in sunlight, the hook opens, pulling the cotyledons above the ground (Fig. 6-7).

A dissertation 30 years earlier had revealed that short periods of dim light induce hook opening, apparently without the involvement of chlorophyll (Trumpf, 1921). Studying the response at Purdue, Withrow concluded that it was strongly promoted by wavelengths of light that produce very little photosynthetic pigment (Withrow, 1941).

FIGURE 6-7 Hook opening in hydroponically grown Black Valentine bean seedlings. Exposures 24 h apart. Courtesy of Smithsonian Institution.

Victor B. Elstad, who joined the Division of Radiation and Organisms in 1947, and William H. Klein, who arrived in 1952, removed hooks from germinated Black Valentine beans under a green safelight and placed them in moist petri dishes. During exposure to weak R for 24 h, rate of opening related logarithmically to irradiance (Withrow, 1953).

Withrow, Klein, and Elstad set out to derive an action spectrum for the response, which was much less variable than internode or coleoptile elongation. In 1953 and 1954, the group performed two series of experiments in the R part of the spectrum, looking at factors that influenced reproducibility. Amazingly, the data from the two consecutive years fell on the same straight line when angle of opening was plotted against logarithm of irradiance (Klein et al., 1956).

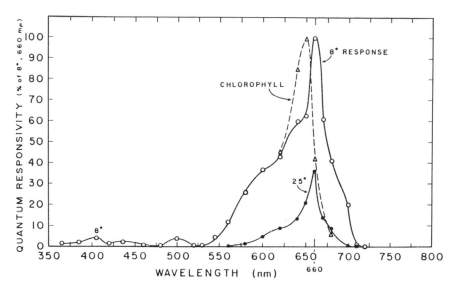

FIGURE 6-8   Action spectra for hook opening and chlorophyll synthesis in Black Valentine bean seedlings. (From Withrow et al., 1957. Reproduced with permission from the American Society of Plant Physiologists.)

Because Withrow wanted to control irradiance in different parts of the spectrum independently, he designed a series of monochromators, each of which generated a different but very narrow wave band through combined interference filters (Withrow, 1957). Arranged on a table, the monochromators delivered equal irradiances in 10 bands through a wide range of energy levels.

By the summer of 1954, the group obtained an action spectrum for hook opening at 20 h after isolated hooks were irradiated for 1,000 s. Like the Beltsville curves for floral induction, seed germination, and leaf expansion, it had a strong maximum in the R, at 660 nm (Fig. 6-8).

To see if absorption of light by chlorophyll accounted for this peak, the researchers derived an action spectrum for chlorophyll formation, extracting the photosynthetic pigment from hooks irradiated in the same manner as those used previously. The chlorophyll synthesis curve had a peak at 650 nm rather than 660 nm, suggesting that the conversion of protochlorophyll to chlorophyll was a different light reaction than the one controlling photomorphogenesis (Withrow et al., 1957). Moreover, flip-flop experiments revealed that R promotion of hook opening was reversed by FR (Withrow, 1954). Thus the unknown pigment with tell-tale photo-

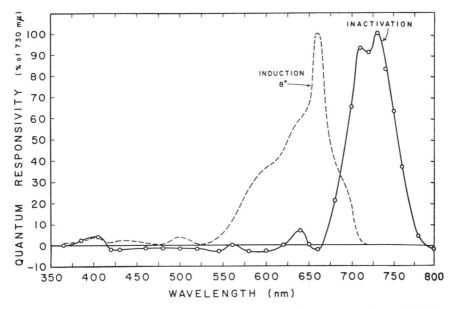

FIGURE 6-9 Action spectrum for reversal of R-promoted bean hook opening (solid line). Dashed curve is action spectrum for promotion shown in Fig. 6-8. (From Withrow et al., 1957. Reproduced with permission from the American Society of Plant Physiologists.)

reversibility controlled hook opening as well as germination and floral induction.

A reversal action spectrum was completed by the summer of 1955 (Withrow, 1955; Withrow et al., 1955); this had two peaks, at 710 nm and 730 nm (Fig. 6-9). "It seems very possible that the absorption spectra of the photoinduction and photoreversal pigments may be quite similar in shape, with the latter shifted 20 to 30 [nm] to the longer red wavelengths," the Smithsonian workers suggested (Withrow et al., 1957). They were convinced, however, that the photoinduction and photoreversal pigments were two separate substances rather than two forms of the same pigment (e.g., Klein et al., 1957).

The Wageningen workers had also found it unnecessary "to accept interconvertibility of the pigments which act antagonistically in the cases we studied" (Stolwijk, 1952). They proposed that "two pigment systems are present in constant concentrations, their ratio varying from species to species. One pigment, associated with the action of blue and infrared, has absorption peaks around 750 and 480 [nm]. The other pigment, associated with the action of red light, has an absorption peak around 650 [nm], and some absorption throughout the region 400–700 nm. Each pigment is engaged in a light reaction. This light reaction is preceded and followed by dark reactions . . . ." (Stolwijk, 1954).

GATLINBURG SYMPOSIUM

The skeptics gathered in 1957, in Gatlinburg, Tennessee, where Withrow organized a symposium on plant and animal photoperiodism. On the first page of the resulting symposial volume (Withrow, 1959), Gerald Oster of the Polytechnic Institute of Brookhaven considered "the possible nature of what Dr. French calls 'this pigment of the imagination.' " C. Stacy French was a photosynthesis researcher and director of the Carnegie Institution of Washington in Stanford. Initially, he had coined "pigments of the imagination" for the multiple forms of chlorophyll a his spectroscopic studies had uncovered. But the phrase seemed even more apt for a supposedly photoreversible pigment. Such a bizarre molecule was unknown among living organisms; thus French, like Withrow, thought "it" was really two separate pigments.

In the middle of the symposium, Withrow suffered a massive heart attack. He recovered sufficiently to edit most of the papers but was stricken again the following spring, while lecturing to a hospital group in Minneapolis. He died, at age 53, on April 8, 1958.

REFERENCES

Borthwick, H. A. (1956). Light in relation to growth and development of plants. Presidential address at the annual meeting of the American Society of Plant Physiologists, University of Connecticut, Storrs, August 27. Unpublished.

Borthwick, H. A. (1972). History of phytochrome. In "Phytochrome" (K. Mitrakos and W. Shropshire, Jr., eds.), pp. 4–23. Academic Press, London, New York.

Borthwick, H. A., Hendricks, S. B., Parker, M. W., Toole, E. H., and Toole, V. K. (1952a). A reversible photoreaction controlling seed germination. Proc. Natl. Acad. Sci. USA 38,662–666.

Borthwick, H. A., Hendricks, S. B., and Parker, M. W. (1952b). The reaction controlling floral initiation. Proc. Natl. Acad. Sci. USA 38,929–934.

Borthwick, H. A., Hendricks, S. B., Toole, E. H., and Toole, V. K. (1954). Action of light on lettuce-seed germination. Bot. Gaz. 115,205–225.

Gravatt, A. R. (1952). Letter to Dr. Magness, May 26. Unpublished.

Hendricks, S. B. (1956). Control of growth and reproduction by light and darkness. Am. Sci. 44,229–247.

Hendricks, S. B. (1958). Photoperiodism. Agron. J. 50,724–729. Quotation reprinted with permission from the American Society of Agronomy, Inc.

Hendricks, S. B. (1960a). The photoreactions controlling photoperiodism and related responses. In "Symposia on Comparative Biology. 1. Comparative Biochemistry of Photoreactive Systems" (M. B. Allen, ed.), pp. 303–321. Academic Press, New York, London.

Hendricks, S. B. (1960b). Control of plant growth by phytochrome. *In* "Comparative Effects of Radiation" (M. Burton, J. S. Kirby-Smith, and J. L. Magee, eds.), pp. 22–48. John Wiley, New York, London.

Hendricks, S. B., and Borthwick, H. A. (1954). Photoperiodism in Plants. *Proc. Int. Congr. Photobiol. 1st* pp. 23–35.

Hendricks, S. B., and Borthwick, H. A. (1955). Photoresponsive growth. *In* "Aspects of Synthesis and Order in Growth" (D. Rudnick, ed.), pp. 149–169. Princeton University Press, Princeton, New Jersey.

Hendricks, S. B., Borthwick, H. A., and Downs, R. J. (1956). Pigment conversion in the formative responses of plants to radiation. *Proc. Natl. Acad. Sci. USA* 42,19–26.

Klein, W. H., Withrow, R. B., and Elstad, V. B. (1956). Response of the hypocotyl hook of bean seedlings to radiant energy and other factors. *Plant Physiol.* 31,289–294.

Klein, W. H. Withrow, R. B., Withrow, A. P., and Elstad, V. (1957). Time course of far-red inactivation of photomorphogenesis. *Science* 125,1146–1147.

Krebs, H. A., and Kornberg, H. L. (1957). "Energy Transformations in Living Matter." Springer, Berlin, Göttingen, Heidelberg.

Parker, M. W., and Borthwick, H. A. (1950). Influence of light on plant growth. *Annu. Rev. Plant Physiol.* 1,43–58.

Stolwijk, J. A. J. (1952). Photoperiodic and formative effects of various wavelength regions in *Cosmos bipinnatus, Spinacia oleracea, Sinapis alba* and *Pisum sativum*. I. *Proc. K. Ned. Akad. Wet. Ser. C* 55,489–502.

Stolwijk, J. A. J. (1954). Wave length dependence of photomorphogenesis in plants. *Meded. Landbouwhogesch. Wageningen* 54(5),181–244.

Toole, E. H., Borthwick, H. A., Hendricks, S. B., and Toole, V. K. (1953). Physiological studies of the effects of light and temperature on seed germination. *Proc. Int. Seed Test. Assoc.* 18,267–276.

Toole, E. H., Toole, V. K., Borthwick, H. A., and Hendricks, S. B. (1955). Photocontrol of *Lepidium* seed germination. *Plant Physiol.* 30,15–21.

Trumpf, C. (1921). Über den Einfluss intermittierender Belichtung auf das Etiolement der Pflanzen. Dissertation. Hamburg.

Withrow, R. B. (1941). Response of seedlings to various wave bands of low intensity irradiation. *Plant Physiol.* 16,241–256.

Withrow, R. B. (1953). *Smithson. Inst. Annu. Rep.*, pp. 123–125.

Withrow, R. B. (1954). *Smithson. Inst. Annu. Rep.*, pp. 48–50.

Withrow, R. B. (1955). *Smithson. Inst. Annu. Rep.*, pp. 62–63.

Withrow, R. B. (1957). An interference-filter monochromator system for the irradiation of biological material. *Plant Physiol.* 32,355–360.

Withrow, R. B. (1959). "Photoperiodism in Plants and Animals. "Association for the Advancement of Science, Washington, D.C.

Withrow, R. B., Klein, W. H., and Elstad, V. (1955). The action spectrum of the reversal of the photoinduction of hook opening in the hypocotyl of bean. *Plant Physiol.* 30(suppl. ix).

Withrow, R. B., Klein, W. H., and Elstad, V. (1957). Action spectra of photomorphogenic induction and its photoinactivation. *Plant Physiol.* 32,453–462.

CHAPTER 7

The 1950s

Parker left the Photoperiod Project in 1952—"just at the moment of discovery of photoreversibility of flowering"[1]—to become a USDA administrator. He died from a heart attack on October 8, 1966, at age 58. Albert Aloysius Piringer (Fig. 7-1) and Robert Jack Downs (Fig. 7-2) were hired in 1952 to replace Parker.

TOMATO COLOR

As well as growing top-quality crop plants—and weeds—for the group's experiments, Piringer studied the effects of light on tomato color. Heinze and Borthwick had noticed that tomatoes ripened in the dark were pink whereas those exposed briefly every day to incandescent light became red-orange by synthesizing a yellow skin pigment (Heinze and Borthwick, 1952). Obtaining an action spectrum for the yellowing, Piringer obtained the usual curve; the response was also repeatedly R/FR-reversible (Piringer and Heinze, 1954). The photoperiodic pigment therefore controlled skin coloring, even though, in this day-neutral plant, it did not control flowering.

Tomatoes wrapped in foil yellowed only in portions still exposed to light; the rest of the skin stayed transparent. Thus the photoreversible pigment triggered a strictly localized response, in contrast to its action in flowering.

FLORAL INDUCTION IN XANTHIUM

The Beltsville group had previously assayed floral induction in cocklebur by counting the number of plants that were induced to flower (Parker et al.,

---

1. Borthwick, 1972, which incorrectly dates Parker's departure to 1951.

FIGURE 7-1 Albert A. Piringer, about 1960. Courtesy of USDA.

1946). But the developmental stage of floral primordia varied with photoperiod length. "It thus seemed possible to obtain a more precise measure of the photoperiodic response by differentiating several stages in the development of the reproductive primordium," Downs decided (Downs, 1954). He assigned numbers 0 through 7 to the identifiable stages (Downs, 1954, 1956). Frank B. Salisbury, a graduate student with James Bonner at Caltech, developed a similar index for cocklebur (Salisbury, 1955), but the Beltsville workers continued to use Downs' system.

FIGURE 7-2 Robert Jack Downs. Courtesy of USDA.

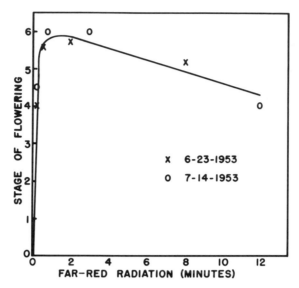

FIGURE 7-3   Repromotion of cocklebur flowering by various FR energies immediately after a saturating R exposure. (From Downs, 1956. Reproduced with permission from the American Society of Plant Physiologists.)

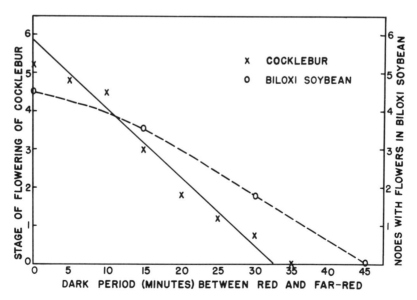

FIGURE 7-4   Effect of interval between R and FR on floral induction in cocklebur and Biloxi soybean. (From Downs, 1956. Reproduced with permission from the American Society of Plant Physiologists.)

TABLE 7-1

Effects on Floral Induction of Cocklebur and Soybean of Consecutive R and FR Light Breaks[a]

Treatment	Mean stage of floral development in cocklebur[b]	Mean no. of flowering nodes in Biloxi soybean[c]
Dark control	6.0	4.0
R	0.0	0.0
R, FR	5.6	1.6
R, FR, R	0.0	0.0
R, FR, R, FR	4.2	1.0
R, FR, R, FR, R	0.0	—
R, FR, R, FR, R, FR	2.4	0.6
R, FR, R, FR, R, FR, R	0.0	0.0
R, FR, R, FR, R, FR, R, FR	0.6	0.0

[a] From Downs, 1956. Reproduced with permission from the American Society of Plant Physiologists.

[b] Two min far-red from the sun. Values are for lots of 5 plants.

[c] Eight min of far-red from three 300-W internal reflector flood lamps. Values are for lots of 4 plants.

To avoid personal bias in scoring, "Borthwick would take the larger leaves off to nearly expose the terminal meristem and then hand it to me," Downs recalls. "I would put it under the microscope to make the final check, without knowing what the treatments had been."

The researchers had used only a few plants to demonstrate photoreversibility in *Xanthium* (see Chapter 6), because space was limited on the spectrograph. Downs repeated the work with larger numbers, using filtered sources of light. To ensure success, he applied FR for longer than before. "We ran 10 successive experiments and failed to reinduce a single plant" Borthwick later related. "When we reduced the period of far red radiation to that used in the spectrograph experiment flowering was reinduced and the spectrograph results were thus confirmed" (Borthwick, 1972). Thus short periods of FR reversed the effect of R, but prolonged FR acted like R. The R-like action of FR began after only a few minutes (Fig. 7-3) (Downs, 1956).

Downs performed the flip-flop for floral induction using short FR exposures. Although the plants always remained vegetative when R came last and formed flower primordia when FR was last, "the repromotive effect of FR diminshed with successive cycles" (Table 7-1).

Wondering if this was due to the passage of time, Downs varied the interval between R and subsequent FR treatment in the middle of a 12-h dark period. Although flowering was induced when FR immediately followed R, reversal lessened as time between treatments increased (Fig. 7-4).

When the interval reached about half an hour, floral induction ceased entirely. Thus escape time was about 30 min and, in soybean, about 45 min (Downs, 1956). It appeared that the FR-absorbing form of the pigment initiated events that in 30–45 min prevented flowering. Escape was much slower in both species when the temperature was lowered from 20°C to 5°C (Downs, 1954, 1956), which indicates, Downs says, "that certainly there is a chemical reaction." Downs subsequently demonstrated photoreversibility for the light-break response in the SDP Biloxi soybean and pigweed and the LDP Wintex barley and henbane (Downs, 1956).

DARK-GROWN BEAN SEEDLINGS

Downs also studied leaf and hypocotyl elongation in Red Kidney bean. Dark-grown seedlings exposed for 2 min to R on days 6, 7, and 8 after planting had shorter hypocotyls and much larger leaves than dark controls, and the effect was FR/R-reversible even after 8 h. Thus bean seedling development, like seed germination, was controlled by the same photosystem as flowering, but escape from photoreversibility was much less rapid.

Action spectra for hypocotyl inhibition and leaf growth had peaks near 640 nm. When the seedlings were first briefly exposed to FR, the reversal spectrum peaked at 730 nm (Downs, 1954, 1955).

LIGHT-GROWN BEAN SEEDLINGS

While this work was in progress, Piringer removed bush beans from the fluorescent/incandescent room to make space for Borthwick's specimens. The next time he tried to grow bush beans, he got pole beans instead. "What had happened," Piringer explains, "was that the fluorescent lights [high in R] and incandescent lights [R/FR mixture] were on different time clocks. Although we set them to go off at the same time, there was a difference of some seconds between them."

To investigate this end-of-day response, Downs subjected Pinto bean seedlings to cycles of 8 h fluorescent light/16 h darkness. When he replaced all or part of the dark periods with incandescent illumination, he got pole beans. A few hours of incandescent light after the fluorescent photoperiod had the same effect. But internodes did not elongate when fluorescent light extended into or replaced the dark period.

Suspecting that FR alone would be even more effective, Downs exposed some seedlings to 24 min of incandescent light on days 7 to 10 after emergence and others to 3 min of the same light filtered through red plus dark-blue cellophane. Second internodes of the FR-treated seedlings elongated three times as much as those exposed to unfiltered radiation. Until food supplies were exhausted, elongation was proportional to the number of hours in darkness after FR irradiation.

These observations suggested that the photoperiodic pigment might control stem elongation in light-grown as well as dark-grown plants. Running the definitive flip-flop test, Downs showed that second internodes were more than twice as long after terminal FR as after terminal R (Fig. 7-5). Several varieties of bean, as well as sunflower and morning glory, responded similarly. "The bean plant, when grown in fluorescent light for several hours per day," Downs explained, "is in a red-saturated condition at the beginning of darkness and is immediately responsive to far red but not to red. It becomes responsive to red, however, immediately after it receives far red. The dark-grown bean plant, on the contrary, is in a far-red saturated condition in darkness and is therefore initially responsive to red but not to far red until after it has received red" (Downs et al., 1957).

This paper was seminal, for it showed that the photoperiodic pigment operates during normal conditions of growth as well as during the fleeting,

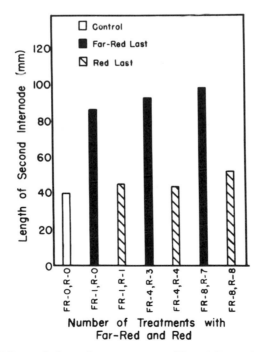

FIGURE 7-5 Effects of alternating exposures to FR and R on elongation of second internode of Pinto bean plants. (From Downs, et al., 1957. Copyright by the University of Chicago.)

etiolated phase of life between germination in the soil and emergence into sunlight. It also revealed that, in light-grown plants, the FR-absorbing form of the pigment persists and is active for at least 20 h after the end of the photoperiod. This important finding was neglected for many years.

PIGMENT CONVERSION

Hendricks realized he could use the results of Downs's experiments with light-grown Pinto beans to calculate what fraction of the photoperiodic pigment is in the active form at any time and how efficiently each form absorbs and uses radiant energy. He also used data from lettuce and *Lepidium* (peppergrass) seed experiments.[2]

Otto Warburg had performed a similar *tour de force* in Berlin in the late 1920s, after detecting an iron-containing "respiratory enzyme" (now named *cytochrome oxidase*) in yeast cells (Warburg and Negelein, 1928, 1929). Warburg reasoned that, when the respiratory enzyme was combined with carbon monoxide, the rate at which light released the poisonous gas from the complex was proportional to the energy absorbed and used. Therefore he measured how long it took for half the complex to decompose at a given light intensity and calculated the photoefficiency.

Rather than measuring decomposition times of a pigment complex, Downs had measured internode elongation rates, which, Hendricks assumed, related to the ratio of the pigment's two forms. Thus after 8-h fluorescent days, the short length of the second internode would be that attained when most of the pigment was in the FR-absorbing form. Maximal elongation after a saturating end-of-day FR irradiation would be due to conversion of the pigment to its R-absorbing form. Intermediate lengths would result from various ratios. Moreover, any given intermediate length could be obtained in two ways: by irradiating seedlings at the end of the light period with just the right amount of FR, or by irradiating with saturating FR and then the amount of R needed to get the ratio obtained in the first procedure. Just how much energy was required in each instance would depend on how efficiently each form of the pigment absorbed and used energy. These parameters could therefore be calculated once the relationship between energy level and fractional conversion of the pigment was determined (Hendricks et al., 1956).

Hendricks arbitrarily chose two internode lengths from Downs's data (Table 7-2): 40.3 mm and 65.2 mm. These resulted from FR exposures of 4

2. The group had recently obained action spectra for promotion and inhibition of *Lepidium virginicum* seed that were very similar on a relative scale to those for Grand Rapids lettuce seed (Toole et al., 1955).

TABLE 7-2

Mean Lengths of Second Internodes of Pinto Beans Resulting from Various Periods of FR or R after Saturating FR Exposures

Periods of irradiation (min)[b]	Mean internode lengths after irradiation (mm)		Periods of irradiation (min)[b]	Mean internode lengths after irradiation (mm)	
	Far-red	Red		Far-red	Red
0.00	26.3	88.0	2.00	31.7	59.6
0.25	27.0	86.6	4.00	40.3	50.3
0.50	28.7	81.3	8.00	65.2	35.0
1.00	29.6	77.5	16.00	96.0	30.7

[a] From Hendricks et al., 1956.

[b] Each radiation source had a power equivalent to about 0.25 mW/cm² at the wave length of maximum action integrated over the region of action.

and 8 min, respectively. He then determined which periods of R (after FR) had produced the same two lengths. The answer was 5.95 and 1.63 min. Turning each pair of numbers into a ratio, he obtained 2.0 in the case of FR (8 divided by 4), and 3.65 in the case of R (5.95 divided by 1.63). He then placed each ratio into a mathematical equation that related it to fractional conversion of the pigment, assuming the reaction was first-order.[3] Because the remaining unknowns, F_1 (fraction of the pigment in the FR-absorbing form that gives an internode length of 65.2 mm) and F_2 (fraction of the pigment in the FR-absorbing form that gives a length of 40.3 mm), appeared in each equation, a solution was possible. F_1 was 0.10 and F_2 was 0.316.

Repeating this procedure, Hendricks constructed a curve relating internode length to the fraction of the pigment in the FR-absorbing form over the entire range of the photoconversion (Fig. 7-6). As well as generating data for the next stage in his calculations, the graph showed that photoconversion of just a small fraction of the pigment to the FR-absorbing form caused a large increase in activity (inhibition of internode lengthening). This was further evidence that the FR-absorbing form was active. The fact that all the points fell on a smooth curve, regardless of whether the data were derived from FR or FR/R exposures, confirmed the photoreaction's first-order kinetics.

3. The rate of a first-order reaction depends only on the concentration of one reactant. Unimolecular reactions are first-order, as are reactions between one reactant and another that is present in such excess that its concentration does not change perceptibly during the reaction. Hendricks's assumption was based on the temperature independence of the photoreversible reaction (see Chapter 6).

Using this curve, Hendricks deduced that 9.1×10^{-7} einstein cm^{-2} (9.1×10^{-3} mol m^{-2}) FR converted 0.81 of FR-absorbing form to the R-absorbing form of the pigment.

Three factors determine how efficiently a pigment inside a plant makes use of radiant energy: (1) What fraction of the radiation falling on the plant's surface reaches the pigment (X); (2) how efficiently the pigment absorbs the energy it receives—this factor is the absorption coefficient (ω); (3) what fraction of the absorbed energy the excited pigment uses to change in a particular way (in this case to convert from the FR-absorbing to the R-absorbing form)—this factor is named *quantum efficiency* (Φ). Thus photoeffectiveness equals $\omega\Phi X$, which in the case of the FR-absorbing pigment involved in internode growth had a value of 0.09×10^7 (0.81 divided by 9.1×10^{-7}). The value for the R-absorbing form was 0.11×10^7.

Applying the same reasoning to the lettuce and *Lepidium* seed data, Hendricks obtained values for light absorption and use during seed germination. Although germination is an all-or-nothing process—seeds either

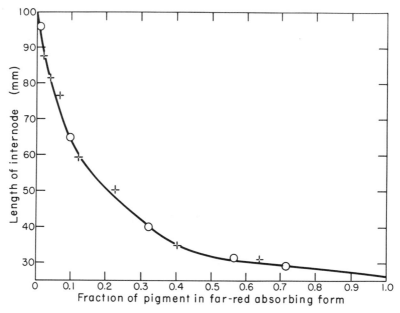

FIGURE 7-6 Internode lengths of Pinto beans as a function of pigment conversion: (o) irradiation in R part of spectrum of plants with pigment initially in R-absorbing form; (+) irradiation in FR with pigment initially in FR-absorbing form. (From Hendricks, et al., 1956.)

germinate or they do not—percentage germination in a batch of light-requiring seeds depends on the amount of R applied, until saturation is reached. So as in internode elongation, an increment in physiological response is proportional to light energy. From graphs relating percent germination of lettuce and pepper grass seeds to fractional conversion of the pigment, Hendricks derived values for $\omega\Phi X$. In the case of *Lepidium,* these were 0.067×10^7 for the R-absorbing form of the pigment and 1.9×10^7 for the FR-absorbing form. For lettuce seed, they were 5.0×10^7 and 0.11×10^7. Thus all three sets of values had the same order of magnitude, but there was some variation, as one might expect it X—the fraction of incident light reaching the pigment—differed.

In a strictly photochemical reaction, quantum efficiency has a maximum value of 1.0 because one quantum cannot activate more than one pigment molecule, assuming one chromophore per molecule. X cannot exceed 1.0, either, because not more than 100% of incident radiation can be absorbed. Assuming that both these values would in fact be smaller than 1.0, the greatest values derived for $\omega\Phi X$—5.0×10^7 for R absorption in lettuce seed and 1.9×10^7 for FR absorption in peppergrass seed—would be minimal values for ω. Dividing each of these figures by 1,000 to change the units (from $cm^2\ mol^{-1}$) and obtain more commonly used molar absorption coefficients (ϵ = the absorbance of a 1 molar solution of 1-cm path-length—formerly named *molar extinction coefficient*), Hendricks obtained values of 50,000 $dm^3\ mol^{-1}\ cm^{-1}$ for the R-absorbing form of the pigment and 19,000 $dm^3\ mol^{-1}\ cm^{-1}$ for the FR-absorbing form (Hendricks et al., 1956). Such high values were comparable with those for most dyes, for chlorophyll in the R region of the spectrum, and for a tetrapyrrole chromophore of a biliprotein such as *C*-phycocyanin. They revealed that the photoperiodic pigment was an intensely colored substance that, being invisible in albino seedlings, had to have such a low *in vivo* concentration that detection would require an exceedingly sensitive spectrophotometer.

Hendricks assumed that the relationship between incident energy and amount of pigment converted was linear, but it is actually logarithmic. So even higher molar absorption coefficients were obtained when this error was corrected. The lettuce seed data gave a minimum value for the R-absorbing form of 69,000 $dm^3\ mol^{-1}\ cm^{-1}$. Using the *Lepidium* data, the minimum value for the FR-absorbing form became 41,000 (Butler, 1982).

This exercise in logic revealed Hendricks at his best—able to extract sophisticated information about an unidentified and unextracted pigment from a few simple measurements. Thus he regarded it as one of his greatest achievements (Butler, 1982). But the paper befuddled most plant physiologists, for Hendricks communicated his ideas poorly. "The problem," recalls Downs, "was that he assumed that everybody knew what he was talking about.'' Downs's paper on light-grown bean seedlings offered a rerun "for the common herd" (Downs et al., 1957).

OVERCOMING DORMANCY IN SHRUBS AND TREES

Studies of immediate practical value continued alongside attempts to characterize and detect the photoperiodic pigment. Thus the Beltsville workers investigated the photoperiodic responses of azalea, coffee, gladioli, hemp, hydrangea, poinsettia, *Rauvolfia,*[4] strawberry, soybean, tulip, wheat, and barley. They determined that poinsettia will produce its showy red bracts just in time for Christmas if cuttings started at the beginning of October are kept in the dark for 14–16 h daily (Parker et al., 1950) and that the vegetative growth of tropical plants, such as *Rauvolfia* and coffee, is markedly affected by daylengths that exceed or fall short of those in their natural environments (Piringer et al., 1958).

Downs also studied the seasonal cessation of growth by trees, shrubs, and herbaceous perennials. In temperate climates, woody plants blossom in spring and then grow rapidly. But sometime during the summer, many prepare for winter by covering their buds with protective scales.

Downs used the Tooles' germinated pine seeds plus seedlings of some deciduous trees. Most species on 8-h days grew for about 4 weeks and then stopped, but horse chestnut formed large, sticky buds right away, whereas elm ceased growth only after about 20 weeks. Some species, such as *Weigela florida,* catalpa (Fig. 7-7), tulip poplar, birch, red maple, dogwood, and elm, grew continuously when the day exceeded a critical length. Thus horticulturists would be able to raise certain shrubs and trees more quickly by extending daylength to circumvent dormancy. Other species, such as horse chestnut, which stops growing in Beltsville in midsummer, eventually became dormant no matter what the photoperiod (Downs and Borthwick, 1956a)

Trying to break dormancy, Downs once again noted differences between species. At one extreme, catalpa failed to resume growth unless it was chilled for 2–3 weeks (Downs and Borthwick, 1956a). But *Weigela* responded promptly to the resumption of LD and would even resume growth on SD if defoliated (Downs and Borthwick, 1956b).[5]

---

4. Rauvolfia is a tropical tree whose roots produce the drug reserpine.

5. Downs and Borthwick shared the 1957 Leonard H. Vaughan Award in Horticulture (American Society of Horticultural Science) for their work with *Weigela*. Piringer and N. W. Stuart received this award in 1956 for studies of hydrangea (Piringer and Stuart, 1955).

FIGURE 7-7   *Catalpa bignonioides. Left:* 23 weeks of 8-h photoperiods. *Center:* 8 weeks of 16-h photoperiods followed by 15 weeks of 8-h photoperiods. *Right:* 23 weeks of 16-h photoperiods. (From Borthwick, 1957.)

From June 1956 to June 1957, Downs and Piringer studied the photoperiodic responses of pines, which produced the greatest juvenile stem growth on 14–16-h days. Total growth was greatest under continuous light, however, suggesting that a dark period is less important to the adult than the juvenile growth phase. Comparing daylength responses, the two men found that ponderosa pine (*Pinus ponderosa* var. *ponderosa*) from northern seed lots grew more vigorously on LD than plants from southern seed lots, whereas loblolly pine (Pinus taeda) showed quite the opposite response. The differences within both species lessened between January and July, as maximum temperature in the greenhouse rose, indicating that pine ecotypes are determined by temperature as well as photoperiod (Downs and Piringer, 1958).

Discovering supplemental incandescent lighting to be much more effective than supplemental fluorescent lighting in evoking the photoperiodic growth response of both pines and deciduous trees, Downs and Borthwick concluded that "the photochemical reaction controlling germination and flowering and growth of herbaceous plants also controls onset of dormancy and the elongation of new structures of woody plants" (Downs and Borthwick, 1956a).

BUSHY PETUNIAS

In 1958 and 1959, Piringer and Henry Marc Cathey, a research horticulturist in the Crops Research Division, determined how growers could manipulate the environment to obtain compact, branched petunias suited to bedding. "If you grew a seedling on short days, you got a plant with many branches that flowered like a dinner plate," Piringer explains. "But on long days, you got a single stem, and the plants looked like a different variety. So, if you started seedlings in Florida, where the days are short in early spring, you got very branched plants. But if you started them in New York, the days were already too long in late spring, and you got single-stemmed plants."

Manipulating daylength, temperature, and lighting, Piringer and Cathey developed a reliable method for obtaining compact, bushy petunias. They advised growers to germinate seeds and maintain the resulting seedlings on 10-h days by covering them with black cloth. Then, as soon as branching began, they were to switch to days longer than 12 h, using incandescent rather than fluorescent lamps if supplemental lighting was necessary. With this regime, compact plants were obtained even from varieties such as Ballerina (Fig. 7-8) that normally were too tall for bedding (Piringer and Cathey, 1960). This work greatly boosted the petunia's popularity and

FIGURE 7-8 *Petunia hybrida* var. Ballerina grown for 64 days. Each day the plants received 8 h sunlight plus (*left to right*) 0,1,2,4,6, or 8 h of illumination from a low-intensity incandescent lamp. (From Piringer and Cathey, 1960.)

earned Piringer and Cathey special recognition (bronze medals) from the American Society for Horticultural Science in 1960 and again in 1961. These concepts have now been applied to production of most annuals and perennials sold for yearly bedding out. Supplemental high-intensity discharge lamps are used to create the same effects on seedlings and to greatly accelerate development and improve subsequent performance.

REFERENCES

Borthwick, H. A. (1957). Light effects on tree growth and seed germination. *Ohio J. Sci.* 57,357–364.

Borthwick, H. A. (1972). History of phytochrome. *In* ''Phytochrome'' (K. Mitrakos and W. Shropshire, Jr., eds.), pp. 4–23. Academic Press, London, New York.

Butler, W. L. (1982). Sterling Hendricks: A molecular plant physiologist. *Plant Physiol.* 70,S1–S3.

Downs, R. J. (1954). Regulatory effects of light on plant growth and reproduction. Ph. D. dissertation. George Washington University.

Downs, R. J. (1955). Photoreversibility of leaf and hypocotyl elongation of dark grown Red Kidney bean seedlings. *Plant Physiol.* 30,468–473.

Downs, R. J. (1956). Photoreversibility of flower initiation. *Plant Physiol.* 31,279–284.

Downs, R. J., and Borthwick, H. A. (1956a). Effects of photoperiod on growth of trees. *Bot. Gaz.* 117,310–326.

Downs, R. J., and Borthwick, H. A. (1956b). Effect of photoperiod upon the vegetative growth of *Weigela florida* var. *variegata*. *Proc. Am. Soc. Hort. Sci.* 68,518–521.

Downs, R. J. and Piringer, A. A. (1958). Effects of photoperiod and kind of supplemental light on vegetative growth of pines. *For. Sci.* 4,185–195.

Downs, R. J. Hendricks, S. B., and Borthwick, H. A. (1957). Photoreversible control of elongation of Pinto beans and other plants under normal conditions of growth. *Bot. Gaz.* 118,199–208. Copyright by the University of Chicago.

Heinze, P. H., and Borthwick, H. A. (1952). The light-controlled production of a pigment in the skins of tomato fruit. Program American Society of Plant Physiologists, Ithaca, New York, September 7–10, p 37.

Hendricks, S. B., Borthwick, H. A., and Downs, R. J. (1956). Pigment conversion in the formative responses of plants to radiation. *Proc. Natl. Acad. Sci. USA* 42,19–26.

Parker, M. W., Hendricks, S. B., Borthwick, H. A., and Scully, N. J. (1946). Action spectra for photoperiodic control of floral initiation in short-day plants. *Bot. Gaz.* 108,1–26.

Parker, M. W., Borthwick, H. A., and Rappleye, L. (1950). Photoperiodic responses of poinsettia. *Florists Exch.,* Nov 11.

Piringer, A. A., and Cathey, H. M. (1960). Effect of photoperiod, kind of supplemental light, and temperature on the growth and flowering of petunia plants. *Proc. Am. Soc. Hort. Sci.* 76,649–660.

Piringer, A. A., and Heinze, P. H. (1954). Effect of light on the formation of a pigment in the tomato fruit cuticle. *Plant Physiol.* 29,467–472.

Piringer, A. A., and Stuart, N. W. (1955). Response of hydrangea to photoperiod. *Proc. Am. Soc. Hort. Sci.* 65,446–454.

Piringer, A. A., Downs, R. J., and Borthwick, H. A. (1958). Effect of photoperiods on *Rauvolfia*. *Am. J. Bot.* 45,323–326.

Salisbury, F. B. (1955). The dual role of auxin in flowering. *Plant Physiol.* 30,327–334.

Toole, E. H., Toole, V. K., Borthwick, H. A., and Hendricks, S. B. (1955). Photocontrol of *Lepidium* seed germination. *Plant Physiol.* 30,15–21.

Warburg, O., and Negelein, E. (1928). Über den Einfluβ der Wellenlänge auf die Verteilung des Atmungsferment. *Biochem Z.* 193,339–346.

Warburg, O., and Negelein, E. (1929). Über das Absorptionsspektrum des Atmungsferments. *Biochem Z.* 214,64–100.

CHAPTER **8** _____

High-Energy Responses

In 1957, the Photoperiodism Unit was transformed into a Pioneering Laboratory for Plant Physiology. This unit and the Pioneering Laboratory for Mineral Nutrition, headed by Hendricks, were two of five pioneering groups chartered by the Agricultural Research Service that year to promote the search for fundamental new principles in agricultural science.

H. WILLIAM SIEGELMAN

An agricultural biochemist from Post-Harvest Physiology, Harold William (Bill) Siegelman (Fig. 8-1) filled a new position in the Pioneering Laboratory for Plant Physiology. After arriving in Beltsville in 1953, Siegelman had wondered if skin color could reveal the optimal time to harvest apples, which develop apple scald if taken too early and water core if harvested too late.

The red color in apple skin is due to the pigment idaein, a member of the anthocyanin family of 15-carbon aromatic compounds that color flowers, stems, leaves, and roots red, blue, and purple. In 1949, Kenneth Thimann at Harvard University had found that anthocyanin formation was irradiance-dependent (Thimann and Edmondson, 1949). The Smithsonian workers demonstrated that FR increased anthocyanin synthesis in bean seedlings without promoting chlorophyll synthesis (Withrow, 1952; Withrow et al., 1953).

Hearing of Siegelman's idea, Hendricks agreed to collaborate. Measuring the formation of the red pigment in bright, continuous light, the researchers uncovered two distinct light-dependent phases (Fig. 8-2). The first was an induction period, during which there was no additional anthocyanin production; this lasted for 15 min in red cabbage seedlings, 2 h in turnip seedlings, and 20 h in apple skin. The second was a linear period, when synthesis was proportional to total light energy (Siegelman and Hendricks, 1956, 1957a, 1958a).

FIGURE 8-1 Harold William Siegelman, 1968. Courtesy of Brookhaven National Laboratory, Long Island, New York.

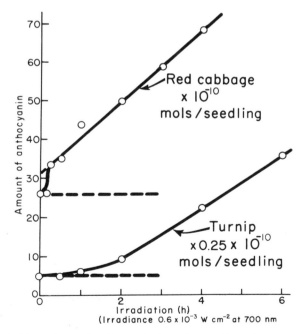

FIGURE 8-2 Dependence of anthocyanin formation on time of exposure to fluorescent light. Anthocyanin was extracted 24 h after beginning of irradiation. (From Siegelman and Hendricks, 1957b. Reproduced with permission from the American Society of Plant Physiologists.)

Action spectra for the second —high-energy—response (HER) showed peaks at 690 and 450 nm for red cabbage seedlings; 725, 620, and 450 nm for turnip seedlings (Siegelman and Hendricks, 1957b); and 650 and 600 nm for apple skin, a light-grown tissue (Fig. 8-3). In apple, the induction period had similar spectral sensitivity (Siegelman and Hendricks, 1958a). Thus both light-dependent phases seemed to involve a pigment that absorbed most strongly in the FR, R, and B.

This appeared not to be the photoperiodic pigment, which was inactivated by saturating levels of FR. Moreover, low irradiances saturated its phototransformation. In red cabbage, the R/FR-reversible reaction did appear to operate initially, resulting in the formation of a small additional amount of anthocyanin over the amount produced in darkness.

Siegelman and Hendricks speculated that the HER photoreceptor might be a flavoprotein. Recently described acyl dehydrogenase enzymes, which

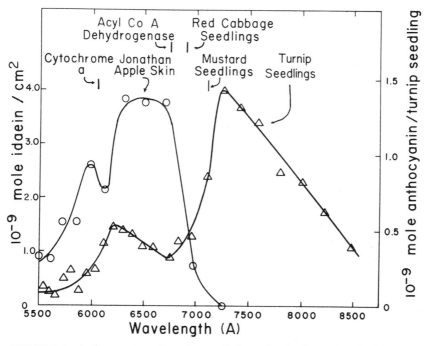

FIGURE 8-3   Action spectrum for anthocyanin formation in pieces of apple skin floating on sucrose solution contrasted with that in turnip seedlings. Wavelengths for action maxima of anthocyanin formation in red cabbage and mustard seedlings are also shown. (From Siegelman and Hendricks, 1958a. Reproduced with permission from the American Society of Plant Physiologists.)

remove hydrogen atoms from 2-carbon fragments, had the right absorption characteristics, and 2-carbon fragments are precursors of anthocyanins.

HANS MOHR

Shortly after Siegelman and Hendricks began to study anthocyanin synthesis, 26-year-old Hans Mohr (Fig. 8-4) arrived at Beltsville from Tübingen, where he had just completed a doctorate with Bünning on fern spore germination. An action spectrum revealed that R was the most effective wave band for promotion (Mohr, 1956). Thus germination in lower plants was sensitive to the same part of the spectrum as seed germination.

Mohr wanted to find a biochemical model for photomorphogenesis "that allowed biochemical or biophysical access to the causalities: How does the pigment operate? My aim was to reach the level of enzymes." Choosing anthocyanin synthesis, he selected white mustard (*Sinapis alba*), which produces large quantities of the red pigment in the light but none in the dark. As in red cabbage, two photoreactions were involved: the low-energy R/FR-reversible reaction, which was saturated after a brief irradiation, and a high-energy reaction. An HER action spectrum had peaks in the B and FR. Hypocotyl lengthening in dark-grown mustard seedlings showed similar spectral sensitivity.

Mohr concluded that "the region of action indicates that the pigment that is involved in the high-energy reaction might be a copper-

FIGURE 8-4   Hans Mohr, Beltsville, 1956.

flavoprotein'' (Mohr, 1957). But he was skeptical of Siegelman and Hendricks' hypothesis that this pigment directly participated in the pathway of anthocyanin synthesis (Mohr, 1959a, 1962). And he proposed that ''under natural conditions of radiation, for example in the open air amidst nature itself, this high energy reaction seems to be extremely important for photomorphogenesis'' (Mohr, 1962).

Returning to Tübingen, Mohr derived an action spectrum for leaf expansion, a further HER in mustard (Mohr and Lünenschloss, 1958). Moreover, a colleague had uncovered effects of B and FR on radicle elongation. This response differed from previously studied HERs because it was promoted by FR but inhibited by other wavelengths, particularly B (Kohlbecker, 1957).

BELTSVILLE PROPOSAL

Great Lakes lettuce seed was thought to be light-insensitive, but when the Beltsville workers exposed samples to FR for 24 h, the germination percentage halved. The seeds that failed to grow did germinate if they were next exposed to 5 min R, but they remained dormant if 5 min FR followed the R, although a brief FR treatment was ineffective by itself. Thus prolonged FR allowed the low-energy reversible photoreaction to subsequently control germination (Hendricks et al., 1959).

The action spectrum for the HER resembled that for promotion of anthocyanin synthesis in red cabbage seedlings, because both had peaks in the R and FR, with maximal inhibition at about 700 nm (Fig. 8-5). But an action spectrum derived just as the radicle was emerging was surprisingly different: elongation was suppressed maximally at about 610 nm and enhanced maximally at about 750 nm, with very little response elsewhere in the longer wavelengths (Fig. 8-5). Thus two physiological responses in the same material yielded two completely different curves.

At this point the researchers might have evoked a plethora of pigments—the photoreversible one for the low-energy response, a flavoprotein for the HER that inhibited germination, and another for the HER controlling radicle elongation in Great Lakes lettuce and mustard. But Hendricks conjured up a much more elegant explanation.

Combining absorption coefficients calculated from Great Lakes lettuce seed data with those derived from bean seedlings and Grand Rapids lettuce and pepper grass seeds, Hendricks constructed hypothetical absorption curves for the R-absorbing and FR-absorbing forms of the photoreversible pigment. Then he vertically aligned these curves (Fig. 8-6C) with the low-energy action spectra for Grand Rapids seed germination (Fig. 8-6A),

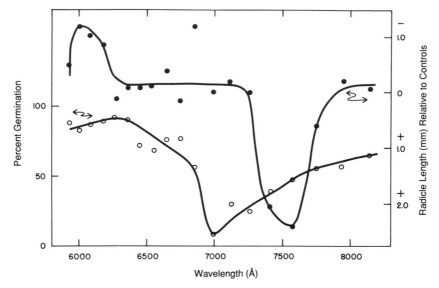

FIGURE 8-5   Action spectra (>580 nm) for germination (○) and radicle elongation (●) of *Lactuca sativa* var. Great Lakes seed. Seeds were imbibed for 12 h in the dark, irradiated for 12 h, and then returned to darkness for 24 h at 20°C. (From Hendricks et al., 1959. Copyright by the University of Chicago.)

the high-energy action spectra for Great Lakes lettuce seed germination and radicle elongation (Fig. 8-6B), and the high-energy action spectra for anthocyanin synthesis in red cabbage seedlings, turnip seedlings, and apple skin (Fig. 8-6D). Cutting through the resulting morass—which looked more like a Miró painting than a figure in a scientific paper— Hendricks proposed that the photoperiodic pigment also mediated the HER (Hendricks and Borthwick, 1959).

He reached this stunning conclusion after dividing the absorption curve into five zones:

1.   In wavelengths between 580 to 630 nm, the pigment would be largely converted to the FR-absorbing form. This zone corresponded to the peak for suppression of radicle elongation in Great Lakes lettuce and to subsidiary peaks for anthocyanin synthesis in apple skin and turnip seedlings.

2.   Between 630 to 670 nm, both pigment forms would be present, with the FR-absorbing form dominant. This zone corresponded to promotion of germination in Grand Rapids lettuce and promotion of anthocyanin synthesis in apple skin.

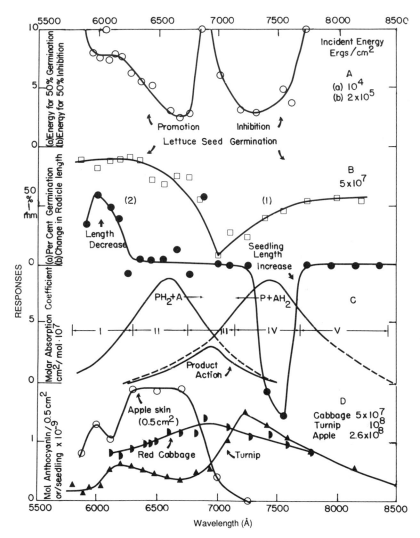

FIGURE 8-6 Action spectra in the region of 580–850 nm for control of various aspects of plant development. (From Hendricks and Borthwick, 1959.)

3.   The region between 670 to 720 nm contained the wavelength 695 nm, where the R-absorbing and FR-absorbing forms had equal absorbance (the isosbestic point). Thus the two forms of the pigment would be present in roughly equal amounts. This zone corresponded to inhibition of germination in Great Lakes lettuce and promotion of anthocyanin synthesis in red cabbage.

4. Between 720 to 780 nm, both pigment forms would be present, but the R-absorbing form would predominate. This zone corresponded to inhibition of germination of Grand Rapids lettuce, promotion of radicle elongation in Great Lakes lettuce, and promotion of anthocyanin synthesis in turnip seedlings.

5. >780 nm. The small absorbance would be due largely to the FR-absorbing form. Radiation in this region had little physiological activity.

Hendricks concluded that the photoperiodic pigment might mediate HERs, at least at wavelengths greater than 580 nm, by constantly cycling between its two forms without being converted from one to the other, thus freeing absorbed energy for other process, such as anthocyanin synthesis. He explained the differing spectral characteristics of the various HERs in terms of availability of reduced and oxidized reactant.

Reading this manuscript (Hendricks and Borthwick, 1959), Mohr found the hypothesis "very interesting and exciting. Since some time we also have had the idea that the high energy reaction must be closely related to the low energy reaction, and that perhaps the reversible red/far-red pigment system may function as a sensitizer for other reactions. . . . Today I would like to put only a few questions concerning some of our experimental results, which apparently do not agree with your theory" (Mohr, 1959b).

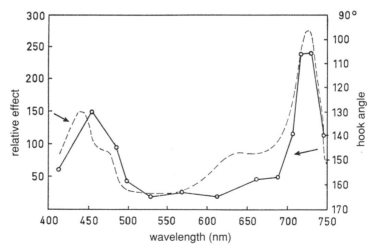

FIGURE 8-7 Action spectrum (solid line) for reopening of plumular hook in lettuce seedlings. Action spectrum of HER in mustard seedlings is indicated by dashed curve (From Mohr, 1962, after Mohr and Noblé, 1960. Copyright 1960, Springer-Verlag.)

Mohr was not convinced that the HER could involve changes in the ratio of reduced to oxidized reactant or that a reactant was even involved in the photoconversion itself. He also questioned whether the action of a R/FR-absorbing pigment could explain the B high-energy peak (Mohr, 1959b; Mohr and Wehrung, 1960). Moreover, he had derived an action spectrum for hook opening in dark-grown lettuce seedlings (Mohr and Noblé, 1960), which form a plumular hook only after exposure to R. The R/FR-reversible photosystem controlled this induction. But hook opening was promoted by high irradiances in the B and FR regions of the spectrum (Fig. 8-7), making it a typical HER. Uncovering these antagonistic responses "greatly strengthened our conception of phytochrome and high energy reaction as two entirely different reaction systems" (Mohr, 1962). In 1960, Mohr left Tübingen to become professor of biology at the University of Freiburg.

REFERENCES

Hendricks, S. B., and Borthwick, H. A. (1959). Photocontrol of plant development by the simultaneous excitations of two interconvertible pigments. *Proc. Natl. Acad. Sci. USA* 45,344–349.

Hendricks, S. B., Toole, E. H., Toole, V. K., and Borthwick, H. A. (1959). Photocontrol of plant development by the simultaneous excitations of two interconvertible pigments. III. Control of seed germination and axis elongation. *Bot. Gaz.* 121,1–8.

Kohlbecker, R. (1957). Die Abhängigkeit des Längenwachstums und der phototropischen Krümmungen von der Lichtqualität bei Keimwurzeln von *Sinapis alba*. *Z. Bot.* 45,507–524.

Mohr, H. (1956). Die Beeinflussung der Keimung von Farnsporen durch Licht und andere Faktoren. *Planta* 46,534–551.

Mohr, H. (1957). Der Einfluss Monochromatischer Strahlung auf das Langenwachstum des Hypokotyls und auf die Anthocyanbildung bei Keimlingen von *Sinapis alba* L. *Planta* 49,389–405.

Mohr, H. (1959a). The photocontrol of anthocyanin synthesis and correlated photomorphogenesis in dark-grown seedlings of *Sinapis alba* L. *Proc. Int. Bot. Congr. 9th* 2,267–268.

Mohr, H. (1959b). Letter to Sterling Hendricks. January 16. Unpublished.

Mohr, H. (1962). Primary effects of light on growth. *Annu. Rev. Plant Physiol.* 13,465–488.

Mohr, H., and Lünenschloss, A. (1958). Weitere Studien zur Photomorphogenese der Keimlinge von *Sinapis alba* (L.) *Naturwissenschaften* 45,578–579.

Mohr, H., and Noblé, A. (1960). Die Steuerung der Schliessung und Öffnung des Plumula-Hakens bei Keimlingen von *Lactuca sativa* durch sichtbare Strahlung. *Planta* 55,327–342.

Mohr, H., and Wehrung, M. (1960). Die Steuerung des Hypokotylwachstums bei den Keimlingen von *Lactuca sativa* L. durch sichtbare Strahlung. *Planta* 55,438–450.

Siegelman, H. W., and Hendricks, S. B. (1956). Two photochemically distinct controls of anthocyanin formation in *Brassica* seedlings. *Plant Physiol.* 31,(suppl. xiii).

Siegelman, H. W., and Hendricks, S. B. (1957a). Photocontrol of anthocyanin synthesis in apple hypodermis. *Plant Physiol.* 32(suppl. ix).

Siegelman, H. W., and Hendricks, S. B. (1957b). Photocontrol of anthocyanin formation in turnip and red cabbage seedlings. *Plant Physiol.* 32,393–398.

Siegelman, H. W., and Hendricks, S. B. (1958a). Photocontrol of anthocyanin synthesis in apple skin. *Plant Physiol.* 33,185–190.

Siegelman, H. W., and Hendricks, S. B. (1958b). Photocontrol of alcohol, aldehyde, and anthocyanin production in apple skin. *Plant Physiol.* 33,409–413.

Thimann, K. V., and Edmondson, Y. H. (1949). The biogenesis of the anthocyanins. I. General nutritional conditions leading to anthocyanin formation. *Arch. Biochem.* 22,33–53.

Withrow, R. B. (1952). *Smithson. Inst. Annu. Rep.*, pp. 134–135.

Withrow, R. B., Klein, W. H., Price, L., and Elstad, V. (1953). Influence of visible and near infrared radiant energy on organ development and pigment synthesis in bean and corn. *Plant Physiol.* 28,1–14.

CHAPTER 9

Detection

In 1945, the Beltsville workers derived the first quantitative action spectrum for photoperiodism. Seven years later, they assigned photoreversibility to the photoperiodic pigment. Seven years after that, in 1959, they finally detected and extracted the R/FR-reversible photoreceptor.

The initial failure to find the pigment in albino barley seedlings (see Chapter 5) had emphasized three major hurdles. First, the pigment's concentration was very low—assaying anthocyanin formation in apple skin exposed to a measured amount of radiation, Hendricks had estimated levels at 10^{-6} to 10^{-7}M (Hendricks, 1960). Second, commercial spectrophotometers, designed to assay solutions, were inadequate for light-scattering materials such as seedlings. Third, there was not yet a specific assay for the pigment. A biochemical assay was unavailable because the pigment's activity was unknown. The usual alternative—measurement of a physiological response on reintroducing the substance into biological material—was unlikely to be successful if, as Hendricks had suggested, the pigment was a protein. The discovery of photoreversibility in 1952 solved this problem, because activation by R and reversal by FR would betray the pigment's presence.

INSTRUMENTATION RESEARCH LABORATORY

The photoperiodism workers gained access to the necessary spectrophotometers by collaborating with the Instrumentation Research Laboratory of the Agricultural Marketing Service. This unit was created in 1956, and located in building 002, so its director, Karl H. Norris, soon got to know Borthwick and Hendricks.

Norris was an electronics engineer who had moved in 1950 from the University of Chicago to Beltsville, where he developed electronic egg-grading equipment (Fig. 9-1). For thousands of years, eggs had been

FIGURE 9-1   Karl H. Norris (white shirt) demonstrating egg-color machine to President
Eisenhower and Secretary Benson (Agriculture) in 1956. Courtesy of USDA.

graded through visual inspection in front of a light source, such as a candle.
"But candling takes a lot of time and is a subjective method of evaluation," says Norris. "Our goal was to develop instruments that would
measure the light transmission characteristics of eggs."

Spectrophotometers measure light transmission [and, conversely, absorbance (A) or optical density], using a lamp, lenses to focus the beam,
optical filters to select appropriate wavelengths, a compartment to hold the
sample, and some type of detector, such as a multiplier-type phototube.
Norris developed several variations on this basic theme, including an
instrument that sorted eggs by color. Sundry egg problems later, he realized that other agricultural commodities, such as fruits and vegetables,
could also be graded spectrophotometrically. This noninvasive approach
was a marked improvement over chlorophyll extraction or cutting produce
open.

The first instruments had a single light beam, and they eventually were fitted with hollow globes called *integrating spheres* (Norris, 1958), which collected light from the entire surfaces of objects. In the late 1950s, an associate agricultural engineer named Gerald S. Birth developed an instrument for grading tomatoes (Birth et al., 1957) that employed two beams of different wavelengths. This dual-monochromator spectrophotometer (Birth, 1960) was a simplified version of one developed by Britton Chance at the University of Pennsylvania (Chance, 1951). Whereas absorbance measurements at a single wavelength are affected by sample size and other variables, measurements at two wavelengths and calculation of the absorbance difference (Δ A) between the two cancels out such variables.

The two-beam instrument contained two monochromators, each of which could be set to any wavelength between 500 and 1,000 nm. A chopper blade rotating in the paths of the two beams alternately blocked one beam and the other, so that pulses of the two passed alternately through the sample. After the transmitted light was collected, the dial displayed Δ A directly.

Another member of the Instrumentation Research Laboratory was a biophysicist named Warren L. Butler (Fig. 9-2). Butler had earned a physics degree at Reed College in Oregon in 1949 and completed a doctorate in biophysics at the University of Chicago in 1955 under the guidance of Nobel Laureate James Franck. During World War II, when he was just 18, he had stepped on a land mine and lost parts of an arm and leg. Butler modified one of Norris's single-beam spectrophotometers for his photosynthesis studies. Because his samples were small, he dispensed with the integrating sphere and placed the photosensitive surface of the phototube right on top of a sample, close enough to collect most of the scattered light (Butler and Norris, 1960). This change, and the availability by then of better electronic components, allowed the instrument to detect very weak signals, such as those that would be generated by a photo-reversible pigment that was present in such minute quantities that 1 gram might be hidden in 10 million grams of tissue.

DETECTION AND EXTRACTION

Seeing Norris' spectrophotometers, Hendricks realized that detection of the photoperiodic pigment might at last be possible. So every time the Plant Physiology Lab experimented with a new plant, he ran over to building 002 with a sample. But the spectrophotometrists were unable to find the pigment in materials where it was obviously present, such as lettuce seed and leaves of *Xanthium*.

FIGURE 9-2   Warren L. Butler. Courtesy of Helga
Ninnemann.

In June 1959, Hendricks decided that tissue containing anthocyanin
might have substantial amounts of the pigment if the sought-after substance was indeed the HER photoreceptor. He therefore carried several
petri dishes of dark-grown turnip seedlings to Norris' lab. "We removed
the cotyledons," recalled Butler, "pressed them loosely into the cuvette
to a depth of about 1.5 cm and measured the absorption spectrum of the
sample after irradiation with red and far-red light. To our amazement and
delight, mixed with skepticism, we found that the difference spectrum
between the red and far-red irradiated sample was precisely that predicted
for [the photoperiodic pigment] by the physiological action spectra" (Butler, 1979).

Hendricks and Butler initially used a single-beam spectrophotometer, dimming the lamp to avoid unintentional phototransformation. Between measurements, the pigment was deliberately converted from its FR-absorbing form to its R-absorbing form with a flashlight covered with red plus blue cellophane. To drive the photoreaction in the reverse direction, the athletic Hendricks jumped up on a lab bench and held the sample under a fluorescent ceiling lamp.

Because the turnip seedlings were grown in the dark in the presence of the antibiotic chloramphenicol, they contained no chlorophyll and very little protochlorophyll to interfere with the measurements. Nevertheless, the seedlings were first irradiated with R to convert whatever protochlorophyll was present. This exposure also transformed the R-absorbing form of the pigment to the FR-absorbing form. After recording the absorption spectrum of this preparation, Butler, Hendricks, and Norris converted the pigment back to its R-absorbing form and ran a second absorption spectrum. As they jumped with excitement, the pen traced a curve that differed from the first in the crucial R (655 nm) and FR (735 nm) regions. After FR treatment, the absorbance of the seedlings decreased in the FR region, while increasing in the R region; the reverse happened after R treatment.

Flushed with success, the researchers looked for the pigment in the other dark-grown seedlings they had on hand. Finding it in all cases, they realized that the use of dark-grown rather than anthocyanin-containing tissue was the key. The pigment was especially abundant in maize (*Zea mays*) shoots, which yielded two curves that differed strikingly at 655 nm and 735 nm (upper portion of Fig. 9-3). Subtracting one curve from the other gave a difference spectrum, which clearly showed that FR removed a substance with maximal absorbance at 735 nm and generated a substance with maximal absorbance at 635 nm. R had the reverse effects (lower half of Fig. 9-3).

Siegelman immediately made a cell-free extract of the dark-grown maize seedlings by breaking open the cells and spinning off the debris in a centrifuge. The researchers placed the extract in the dual-monochromator spectrophotometer to directly measure the difference between R and FR absorption. As they had hoped, the extract exhibited the telltale photo-reversible absorbance changes (Fig. 9-4), leaving little doubt that a single pigment was responsible for the R/FR-reversible responses observed in whole plants. The pigment's ready solubility suggested that it was a cytoplasmic rather than a particulate component in the cell; it also seemed to be a protein, because heat destroyed its photoreversibility. This raised hopes for its purification and strengthened the suspicion of its enzymatic nature (Butler, 1979).

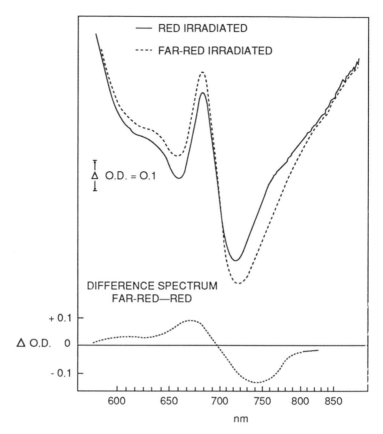

FIGURE 9-3 *Upper portion:* Absorption spectra of maize shoots after treatment with R (solid line) and FR (broken line). Peak near 672 nm resulted from conversion of protochlorophyll to chlorophyll by the R source. *Lower portion:* Curve showing difference between the two curves in upper portion. (From Butler et al., 1959.)

MONTREAL

The group announced its triumph 2 months later at the Ninth International Botanical Congress, in Montreal, Canada. Hendricks, who was a guest speaker (Hendricks, 1959), asked Butler to give a practical demonstration, using a portable interference filter photometer made by Norris. "The instrument was provided with a large circular meter (about the size of a large wall clock)," Butler explained, "with the sensitivity adjusted so that the meter would swing between 9 o'clock and 3 o'clock following irradiation of the sample with red and far-red light. Numerous practice runs

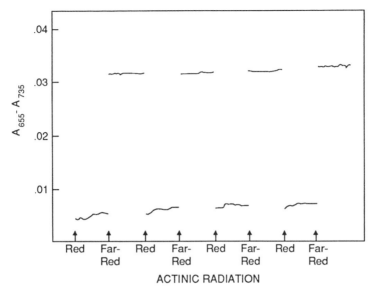

FIGURE 9-4   Absorbance changes in maize shoots alternately irradiated with R and FR (From Butler et al., 1959.)

were made in preparation for Montreal and the measurement never failed.''

Butler ''prepared the sample, irradiated it with red light and obtained the first reading which I zeroed to 9 o'clock on the meter. I then irradiated the sample with far-red light and again measured the absorbance. To my utter dismay the meter still read at 9 o'clock. I reirradiated with red and far-red light but the meter never moved. I rapidly prepared another sample only to repeat the failure'' (Butler, 1979).

Hendricks, Butler, Borthwick, Downs, and Siegelman returned forlornly to Beltsville, assuming the photometer had let them down. But the instrument worked perfectly, suggesting that the problem lay with the plants. Knowing that the pigment in dark-grown tissue was entirely in the R-absorbing form, Butler and Siegelman ''soon made the corollary observation that the far-red absorbing form . . . was not stable in the tissue but rather was destroyed in the dark. Whenever someone had opened the trunk of the car on the way to Montreal, the light transformed most of the [R-absorbing] pigment to the [FR-absorbing form], which then decayed in the dark. After several of these exposures the total amount of [pigment] in the tissue had decreased to levels we could not detect'' (Butler, 1979).

FIGURE 9-5   Borthwick demonstrating R/FR reversibility for Soviet Premier Nikita Khrushchev in November 1959. Courtesy of USDA.

The group got a second chance, when Soviet Premier Nikita Khrushchev visited the Agricultural Research Center at Beltsville the following November. In a jammed auditorium, Borthwick shone R and FR on maize seedlings, causing the needle on the huge photometer dial to swing flamboyantly back and forth (Fig. 9-5). Borthwick was relieved, Khrushchev was intrigued, and the auditorium rang with thunderous applause.

PHYTOCHROME

The historic paper describing the detection of the photoperiodic pigment appeared in December (Butler et al., 1959). It dubbed the R-absorbing form of the pigment "P_{655}" and the FR-absorbing form "P_{735}" although the subscripts were revised several times. In the spring of 1960 (USDA, 1960), the pigment was named *phytochrome* (Gk. = plant color). More convenient abbreviations were also invented—the R-absorbing form became known as *Pr* and the FR-absorbing form was named *Pfr* (Borthwick and Hendricks, 1961). "Phytochrome" made its public debut in June 1960 at a Cold Spring Harbor Symposium (Hendricks, 1961) and appeared in print shortly after (Butler et al., 1960; Borthwick and Hendricks, 1960).

Borthwick, who did not take part in the spectroscopic studies, later credited Butler with naming the pigment (Borthwick, 1972), but *phytochrome* appears to have been Siegelman's invention (Ninnemann, 1985). "After the initial detection, we were kicking around names," Siegelman says, "and to the best of my recollection I think I made the suggestion of *phytochrome,* which was a combination of *phytotron* and *Kodachrome.*"

Detected and named, the pigment was a reality and its photoreversible nature was confirmed. "The proposed action of the pigment seems almost obvious," Butler wrote later, "but, when reviewed in the context of the time at which it was made, this succinct hypothesis stands out as probably the boldest and most brilliant single stroke in the history of plant physiology" (Butler, 1982).

REFERENCES

Birth, G. S. (1960). Agricultural applications of the dual-monochromator spectrophotometer. *Agric. Eng.* 41,432–435.

Birth, G. S., Norris, K. H., and Yeatman, J. N. (1957). Non-destructive measurement of internal color of tomatoes by spectral transmission. *Food Technol.* 11,552–557.

Borthwick, H. A., and Hendricks, S. B. (1960). Photoperiodism in plants. *Science* 132,1223–1228.

Borthwick, H. A., and Hendricks, S. B. (1961). Effects of radiation on growth and development. *In* "Handbuch der Pflanzenphysiologie" (W. Ruhland, ed.), vol. XVI, pp. 299–330. Springer, Berlin-Göttingen, Heidelberg.

Butler, W. L. (1979). Remembrances of phytochrome twenty years ago. *In* "Photoreceptors and Plant Development" (J. de Greef, ed.), pp. 3–7. Antwerpen University Press, Antwerpen.

Butler, W. L. (1982). Sterling Hendricks: A molecular plant physiologist. *Plant Physiol.* 70,S1–S3.

Butler, W. L., and Norris, K. H. (1960). The spectrophotometry of dense light-scattering material. *Arch. Biochem. Biophys.* 87,31–40.

Butler, W. L., Hendricks, S. B., and Siegelman, H. W. (1960). *In vivo* and *in vitro* properties of phytochrome. *Plant Physiol.* 35 (suppl. xxxii).

Butler, W. L., Norris, K. H., Siegelman, H. W., and Hendricks, S. B. (1959). Detection, assay, and preliminary purification of the pigment controlling photoresponsive development of plants. *Proc. Natl. Acad. Sci. USA* 45,1703–1708.

Chance, B. (1951). Rapid and sensitive spectrophotometer. III. A double beam apparatus. *Rev. Sci. Instrum.* 22,634–638.

Hendricks, S. B. (1958). Photoperiodism. *Agron. J.* 50,724–729.

Hendricks, S. B. (1959). Some photo-reactions controlling plant development. *Proc. Int. Bot. Congress 9th* 2,161.

Hendricks, S. B. (1960). The photoreactions controlling photoperiodism and related responses. *In* "Symposia on Comparative Biology. 1. Comparative Biochemistry of Photoreactive Systems" (M. B. Allen, ed.), pp. 303–321. Academic Press, New York, London.

Hendricks, S. B. (1961). Rates of change of phytochrome as an essential factor determining photoperiodism in plants. *Cold Spring Harbor Symp. Quant. Biol.* 25,245–248.

Ninnemann, H. (1985). Warren L. Butler, 1925–1984. *Photochem. Photobiol.* 42,619–624.

Norris, K. H. (1958). Measuring light transmittance properties of agricultural commodities. *Agric. Eng.* 39,640–643,651.

USDA. (1960). Pigment is "phytochrome." *Agri. Res. (USDA)* 8,15.

CHAPTER 10

Partial Purification

"There would seem to be no essential barrier to finding the nature of the enzymatic action of the pigment P_{735}," the Beltsville workers predicted after the photoreversible pigment was first extracted (Butler et al., 1959). If phytochrome could be purified, its mode of action would surely soon be known. The task fell to Siegelman, advised by protein chemists at the National Institutes of Health (NIH) in nearby Bethesda, Maryland.

SIEGELMAN BEGINS

Marshalling the protein purification techniques that were then available, Siegelman took advantage of the fact that large molecules in living cells differ in overall electrical charge. One procedure (salting out) added ammonium sulfate to plant extracts, causing some proteins to precipitate while leaving many unwanted molecules still dissolved. The precipitate could then be redissolved in buffer to form a purer preparation. During ion-exchange chromatography, a preparation was poured into a glass column containing a charged resin that adsorbed some proteins more tightly than others. Buffer added to the top of the column washed the proteins off the resin at different rates, allowing them to be collected from the bottom in separate tubes.

Looking for a suitable source of phytochrome, Siegelman consulted Eben Toole, who advised him to choose big seeds that would produce large seedlings—but not oily ones that would turn into mayonnaise when ground. Siegelman considered chlorophyll-free tissues of light-grown plants, but the phytochrome in cauliflower florets (see chapter 11) could not be extracted.

By the summer of 1960, the biochemist had obtained maize preparations 20 times as concentrated as the one examined in the spectrophotometer in

1959. But the solutions were badly contaminated with brown impurities, and the sought-after pigment proved difficult to salt out.

Siegelman therefore turned to barley seedlings, which brewers routinely grow in vast quantities in the dark. In August 1961, he reported a 50-fold concentration and fourfold purification after adsorption and elution from a column of the ion-exchange resin diethylaminoethyl (DEAE)-cellulose followed by dialysis against polyethylene glycol. Centrifugation of the dialysed preparation produced a blue-green pellet that paled when exposed to R.

BRUCE A. BONNER

In the Botany Department at Yale, meanwhile, a postdoctoral fellow had also extracted phytochrome. Bruce A. Bonner, who held a Ph.D. in botany from the University of California at Berkeley, was hired by Galston to find physical evidence for the photoreversible pigment in pea. He had no specialized spectrophotometers and at first used an old Beckman in which "the knob that twisted the wavelength around was modified by a little chain—the type you have on a bathtub." He then bought an instrument that could detect phytochrome in clear solutions. Extracting the pigment blind from the stem tips of dark-grown pea seedlings and then subjecting the extracts to ammonium sulfate precipitation and DEAE-cellulose chromatography, he soon obtained stable preparations (Bonner, 1960, 1961). Thus he may have been the first person ever to see the photoreversible pigment. When Hendricks visited Yale in 1960, "I had it cleaned up enough," Bonner recalls, "so that when I showed him a cuvette with the preparation, he said, 'That's the first time I've seen a color due to phytochrome itself and not to chlorophyll and other contaminants.'" R irradiation gave the blue solution a greenish hue.

BRUSHITE

Back in Beltsville, Siegelman and his assistant, Ed Firer, were using so much ammonium sulfate that their colleagues regularly took home jars of waste to fertilize their lawns. Then Siegelman learned that brushite, a form of calcium phosphate, was being used with algal biliproteins. "One advantage was that we didn't have much money and brushite cost us only about a dollar a pound if we made it ourselves," he explains. "We used to mix up solutions of calcium chloride and sodium phosphate in each of two calibrated garbage cans. Using a couple of pumps and a big outboard motor stirrer, we would pump these two salts into a third container and precipitate out brushite.''

At first the short, fat columns flowed painfully slowly, but Hendricks—who had elucidated the crystal structure of calcium phosphate—remedied that. He knew that when calcium phosphate was imported for fertilizer from Germany before World War I, the salt was made from sodium phosphate, like Siegelman's brushite. But after the war, when the United States had its own chemical industry, potassium phosphate was used, and the resulting product contained larger sheets of crystals than when the sodium salt was the source of the phosphate. When Siegelman changed to potassium phosphate, the flow rates of his columns increased dramatically. Later he partly converted the brushite, $CaHPO_4 \cdot 2H_2O$, to hydroxylapatite, $Ca_{10}(PO_4)_6(OH)_2$. By August 1962, he had clear blue solutions from dark-grown maize and barley seedlings. But the phytochrome content was low, and the barley preparations were unstable.

OAT PHYTOCHROME

Winslow R. Briggs, associate professor of biology at Stanford University, arrived to spend a sabbatical year in the Plant Physiology Lab shortly thereafter. After writing a review article on phototropism, his principal field, he tried to detect the photoreceptor for this response, which, as Blaauw and Johnston had shown, is promoted maximally by B. Because this proved to be impossible, Siegelman diverted Briggs into phytochrome research. The outcome was a much-cited survey of the distribution of the photoreversible pigment in etiolated seedlings (Briggs and Siegelman, 1965).

Briggs concluded that, in the six species he studied, the concentration of phytochrome and its ratio to other proteins was highest in meristemmatic regions or newly differentiated tissues. These included the tips and nodes of coleoptiles, stem apices, and leaf tips. Elongating stems had very low levels, despite their high sensitivity to R.

Briggs decided that oat would best suit Siegelman's purpose because its phytochrome-rich tissue could be isolated by cutting seedlings in the mesocotyl region. So Siegelman bought 100-lb bags of large-seeded oats from a farmer's supply company in Baltimore. Using the Tooles' protocol, he germinated the seeds in the dark on wadding-lined aluminum bakery trays that fitted into proofing boxes designed for bread rolls. Because these were standard bakery items, they could be bought cheaply. He kept the boxes in a controlled temperature chamber until the oats were 5 days old

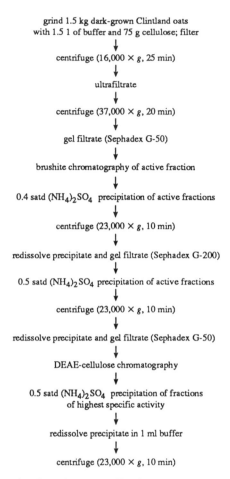

FIGURE 10-1   Steps in phytochrome purification. (From Siegelman and Firer, 1964. Reprinted with permission from *Biochemistry* vol. 3, pp. 418–423. Copyright 1964 American Chemical Society.)

and then chilled them overnight without exposing the seedlings to light. After harvesting bucketfuls of seedlings with electric hedge sheers, he ground them in a commercial babyfood grinder with buffer that neutralized the acid contents of the cells' vacuoles.

The resulting slurry was filtered through the grinder's screen into a stainless steel milk bucket and centrifuged in a continuous centrifuge hooked directly to a chromatography column. Prior to centrifugation, Siegelman had to make the preparation mildly alkaline because the phytochrome stuck to cellular debris under acid conditions—a trait that confounded early attempts to determine the pigment's intracellular location (see Chapter 21).

Siegelman performed the subsequent steps in the cold room under normal lighting. These involved repeated chromatography on Sephadex, brushite, and DEAE-cellulose columns, interspersed with ammonium sulfate precipitations (Fig. 10-1). Sephadex is a chromatography gel that separates proteins according to molecular size and shape rather than charge. NIH researchers Herbert A. Sober and Elbert A. Peterson[1] introduced Siegelman to this new material.

The phytochrome first became visible on the brushite column, where it moved as a yellow-green band. Its blue color appeared in the centrifuge pellet after the yellow-green fractions were treated with ammonium sulfate. The final product, obtained from 1.5-kg seedlings in about a week, was 1 ml of solution containing about 6% of the initial phytochrome purified 61-fold (Siegelman and Firer, 1964).

Siegelman and Firer used a portable difference meter to measure how much phytochrome was present after the various purification steps. Soon after the pigment's detection, Birth and Norris had built four such meters (Fig. 10-2), which were more rugged and sensitive than the dual-wavelength monochromator (Birth and Norris, 1963, 1965). The two earmarked for phytochrome research had built-in R and FR sources to interconvert the pigment forms plus interference filters that isolated the wavelengths needed for the difference measurements.

Agricultural Specialties, a Hyattsville company founded by one of Norris's former colleagues, marketed a commercial version, the Ratiospect R-2. This instrument could measure an optical density difference at two different wavelengths to an accuracy of ± 0.001 in materials with total optical densities up to 6 or 7. So by mid 1963, it was no longer necessary to custom-build a spectrophotometer to assay phytochrome *in vivo.*

By this time, Sober and Peterson were using an analytical technique called *disc electrophoresis,* which applies opposite electric charges to the ends of a small tube of acrylamide gel. By causing differentially charged macromolecules to separate into stainable bands, this technique revealed how many proteins were in a sample. When an NIH technician subjected Siegelman's phytochrome preparation to disc electrophoresis, she found several proteins, with phytochrome accounting for about one-third of the total. It was impossible to determine the pigment's exact size, although analytical ultracentrifugation revealed the weight-average molecular weight to be between 90,000 to 150,000 (Siegelman and Firer, 1964).

---

1. Sober and Peterson had previously appropriated DEAE-cellulose for laboratory use. This material was synthesized by USDA workers developing wash-and-wear cotton.

FIGURE 10-2   Karl H. Norris in 1962 with portable difference meter. Courtesy of USDA.

Sephadex gel filtration had given a value of about 150,000 (Siegelman and Hendricks, 1964).

Siegelman modified the purification protocol to alternately separate proteins by size (ultrafiltration and gel chromatography) and charge (brushite/hydroxylapatite chromatography and ammonium sulfate precipitation). The final preparations were now 40% pure and contained half the phytochrome from the initial seedling extracts (Siegelman and Hendricks, 1965).

REFERENCES

Birth, G. S., and Norris, K. H. (1963). Paper No. 63-330, annual meeting of the American Society of Agricultural Engineers, Miami Beach, Florida, June 1963.

Birth, G. S., and Norris, K. H. (1965). The difference meter for measuring interior quality of foods and pigments in biological tissues. Technical Bulletin #1341, USDA.

Bonner, B. A. (1960). Partial purification of the photomorphogenic pigment from pea seedlings. *Plant Physiol.* 35(suppl. xxxii).

Bonner, B. A. (1961). Properties of phytochrome from peas. *Plant Physiol.* 36(suppl. xlii).

Briggs, W. R., and Siegelman, H. W. (1965). Distribution of phytochrome in etiolated seedlings. *Plant Physiol.* 40,934–941.

Butler, W. L., Norris, K. H. Siegelman, H. W., and Hendricks, S. B. (1959). Detection, assay, and preliminary purification of the pigment controlling photoresponsive development of plants. *Proc. Natl. Acad. Sci. USA* 45,1703–1708.

Siegelman, H. W., and Firer, E. M. (1964). Purification of phytochrome from oat seedlings. *Biochemistry* 3,418–423.

Siegelman, H. W., and Hendricks, S. B. (1964). Phytochrome and its control of plant growth and development. *Adv. Enzymol.* 26,1–33.

Siegelman, H. W., and Hendricks, S. B. (1965). Purification and properties of phytochrome: A chromoprotein regulating plant growth. *Fed. Proc.* 24,863–867.

CHAPTER 11

Properties of Phytochrome

Hendricks had speculated about the photoperiodic pigment from the time the Beltsville group had conceived of it to the time they were able to inspect it firsthand. From physiological action spectra, he knew it would be a blue-eyed chromoprotein—a protein joined to a chromophore such as an open-chain tetrapyrrole. Proposing that it could exist in two different forms, he had calculated how much light each form absorbed, how efficiently the two interconverted, and the order of their reaction rates. He was convinced that another reactant took part in the phototransformation, that Pfr was an enzyme acting at a metabolic crossroads, and that its reversion to Pr in darkness allowed photoperiodic plants to measure the passage of time. These proposals could now be tested, using Siegelman's preparations.

NO ADDITIONAL REACTANT

Confirmation of the proteinaceous nature of phytochrome and its photo-reversibility immediately after the first extraction supported the two most basic deductions (see Chapter 10). But when extracts retained photore-versibility after dialysis (Butler et al., 1959), which removes small molecules, it became apparent that the photoreaction involved only the pigment itself. Moreover, strong oxidizing or reducing agents failed to affect photo-transformation. Thus the group abandoned its hypothesis that a second reactant, such as a redox reagent, was required (Borthwick and Hendricks, 1961; Hendricks, 1961). Bonner, whose earlier physiological experiments had argued against this idea, reached the same conclusion after making similar tests on pea extracts (Bonner, 1960). The most likely transformation mechanism was thus isomerization between the pigment's two forms:

$$Pr \underset{\text{far-red max}}{\overset{\text{red max}}{\rightleftharpoons}} Pfr$$

SPECTROSCOPIC PROPERTIES

The extracted pigment's R/FR reversibility identified it as the photoreversible photoreceptor, as did its difference spectrum (Fig. 11-1), which
closely matched that for living maize seedlings (see Fig. 9-4). When Butler
placed the extract in a commercial spectrophotometer that simultaneously
passed R through one sample and FR through another (Butler, 1961), he
observed that the photoconversion was independent of temperature, at
least down to $-20°C$, confirming the temperature independence observed
in vivo (Borthwick et al., 1954). Photoreversibility was gradually lost as the
temperature fell further and was not observable at all at $-196°C$, suggesting that the molecule had to change shape during its phototransformation.

The action spectra had suggested that the photoperiodic pigment absorbed not only R but also B. Butler was initially unable to extend the
difference spectrum into this region of the spectrum because of B-
absorbing impurities in the preparations. In 1961, however, Bonner observed absorption differences between the two forms in the near UV and
B, as well as in the expected R and FR, regions of the spectrum (Bonner,
1961). He also showed that B, if given at adequate intensities, could
convert Pr in solution to Pfr (Bonner, 1963).

A difference spectrum obtained by the Beltsville group the next year
(Hendricks et al., 1962) clearly showed a peak at 360 nm and dip at 420 nm

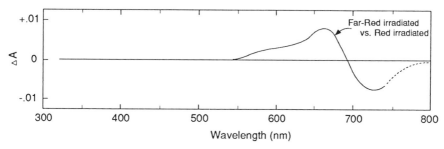

FIGURE 11-1   Difference spectra of R-irradiated versus FR-irradiated soluble extract of
dark-grown maize seedlings. (From Butler, 1961.)

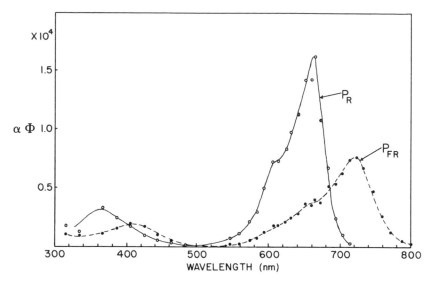

FIGURE 11-2 Action spectra for photoreactions of partially purified oat phytochrome. (From Butler et al., 1964a.)

in addition to the larger differences in the longer wavelengths. A later action spectrum on oat phytochrome (Fig. 11-2) supported Bonner's conclusion.

Butler studied the photoreaction itself with the dual-wavelength monochromator and, later, the difference meter, using beams at 650 nm and 730 nm to measure relative amounts of phytochrome in samples. When $\triangle$A between these two bands was measured after R and then after FR, the difference between the two readings gave the value $\triangle$ ($\triangle$A). The magnitude of this value depended on the average concentration of phytochrome in a sample, the sample's thickness, the absorption coefficient of the pigment, and the degree to which the sample scattered light. If the latter three values were constant, $\triangle$ ($\triangle$A) values represented the comparative concentrations of phytochrome (Pr plus Pfr) in samples.

To assay Pr and Pfr separately, Butler set the two beams to 650 and 800 nm for Pr and to 730 and 800 nm for Pfr. The 800-nm beam was absorbed by neither the two phytochrome forms nor chlorophyll.

In 1960, Butler studied the kinetics of the photoreaction. After irradiating a preparation of maize phytochrome with FR, he turned up the photometer lamp to full brightness to trigger the photoreaction. With the aid of the same light source, he then monitored the disappearance of Pr. He followed conversion of Pfr to Pr in a similar way by irradiating initially with

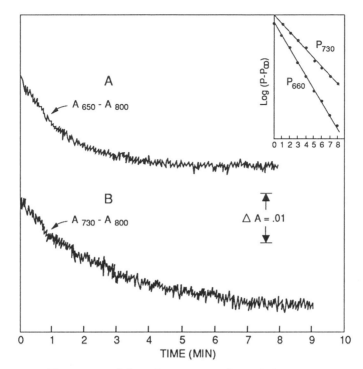

FIGURE 11-3   Time course of phytochrome conversion: (A) after initial FR irradiation; (B) after initial R irradiation. (From Butler, 1961.)

R. The recording pen traced two curves (Fig. 11-3), but when logarithms of $\triangle A$ were plotted against time, straight lines were obtained (Fig. 11-2, insert). This showed that the reversible photoreaction had first-order kinetics in both directions, as Hendricks had deduced. Bonner reached the same conclusion in an unpublished paper (Bonner, 1963).

The slopes of Butler's curves were the rate constants for the forward and backward reactions, and the biophysicist derived an equation that related each rate constant to light energy, mole fraction of the phytochrome form generated at steady state, and the product of the absorption coefficient (α)[1] and quantum yield (Φ). This allowed him to derive absorption coefficients as Hendricks had done earlier, but with data derived from the pigment itself rather than from the measurable results of its action.

1. The literature of this era is confusing. Whereas current terminology assigns α to absorption coefficient ($cm^2\ mol^{-1}$) and ϵ to molar absorption coefficient ($dm^3\ mol^{-1}\ cm^{-1}$), Butler defined ϵ as absorption coefficient and α as molar absorption coefficient. Moreover, the symbol ω also represents absorption coefficient ($cm^2\ mol^{-1}$) in some papers (e.g., Hendricks, 1956—see References, Chapter 7). $\omega = \alpha\ log_e\ 10 = 2.3\alpha$

Butler made the necessary measurements by irradiating a very thin layer of maize phytochrome solution on the spectrograph with measured amounts of R and FR. He then determined photometrically how much pigment had been converted. Expressed as a molar absorption coefficient, ϵ, the minimal value was 22,000 dm^3 mol^{-1} cm^{-1} for both Pr and Pfr (Butler, 1961). This estimate, the first for phytochrome in solution, is of historical importance only because it was based on two assumptions that subsequently proved to be invalid: that R converts 100% Pr to Pfr and that Pr and Pfr have roughly equal absorption coefficients.

Butler and Hendricks eventually derived action spectra for the photo-reaction by obtaining values for $\alpha\Phi$ at many different wavelengths, first for maize phytochrome in the R and FR (Hendricks et al., 1962) and then throughout the visible spectrum using the purer oat preparation (Fig. 11-2). The curves for the conversion of Pr to Pfr closely resembled the action spectra for floral induction, germination, stem elongation, and other R responses, showing a strong peak in the R, a dip in the green, and a weaker peak in the B (Butler et al., 1964a). Thus the spectrograph was driving the very photoreaction that controlled these vital stages of plant development *in vivo*.

The absorption spectra of the extracted oat phytochrome (Fig. 11-4) had maxima of 665 nm for Pr and 725 nm for Pfr—5 nm toward the shorter wavelengths than the peaks in the difference spectra. The isosbestic point was 695 nm (Butler et al., 1965), and there was a peak near 670 nm after R

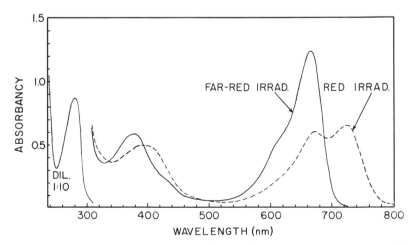

FIGURE 11-4   Absorption spectra of partially purified phytochrome from oat seedlings. (From Butler et al., 1965.)

irradiation, which the Beltsville researchers had not anticipated (see text below). Because both Pr and Pfr absorbed appreciably between 300 and 800 nm, both forms would be present in WL such as sunlight, their relative quantities depending on the quality of light and on the absorbancies of Pr and Pfr in light of that quality.

DETECTION IN OTHER SPECIES

The Beltsville workers also scanned a variety of plants with their rapid photometric assay. After detecting phytochrome in maize, they found the pigment in several types of etiolated grass seedlings and in cauliflower florets, avocado fruit, zucchini fruit, and mustard seedlings (Borthwick and Hendricks, 1961). It was also detectable in artichoke flowers, cabbage and pumpkin cotyledons (Hendricks, 1961), wheat germ, and dark-grown seedlings of oat, barley, rice, wheat, soybean, peanut, lima bean, squash, sunflower, red cabbage, morning glory, watermelon, cucumber, cotton, cowpea, and lupin (Butler et al., 1965).

The pigment was undetectable in green leaves because of massive interference from chlorophyll. Neither could it be found in leaf extracts of cocklebur, soybean, chrysanthemum, and *Perilla,* although the photoperiodic responses of these species betrayed its presence. It was detectable in leaf extracts of spinach (Borthwick and Hendricks, 1960) and several other green plants, including Maryland Mammoth tobacco (Lane et al., 1962, 1963). Siegelman and Harry Lane, a Beltsville sorghum researcher on in-house sabbatical, tried to purify the pigment from green leaves, making sufficient headway to conclude that the "green" pigment was similar to the "etiolated" pigment. But because the concentration of phytochrome was always much lower in green than etiolated materials, dark-grown seedlings remained the preferred source.

DESTRUCTION

This discrepancy suggested that much of the phytochrome in dark-grown seedlings is inactivated or destroyed when seedlings see light, as had happened en route to Montreal. In 1960, Hendricks and Butler demonstrated this in the lab by briefly irradiating dark-grown maize seedlings with R and returning them to darkness. Taking samples at various times thereafter, they photometrically measured both total phytochrome and Pfr (Fig. 11-5). During the next 4 h, Pfr disappeared to a barely detectable level and total phytochrome declined to about one-third of the original level (Hendricks et al., 1962). The loss[2] occurred only in the presence of oxygen

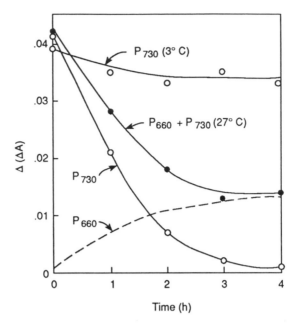

FIGURE 11-5   Changes in the relative amounts of phytochrome in dark-grown maize seedlings immediately after a single R irradiation. (From Hendricks et al., 1962. Reprinted with permission from the *Journal of Physical Chemistry* vol. 66, pp. 2550–2555. Copyright 1962 American Chemical Society.)

(Butler et al., 1963), and it was slowed by respiratory inhibitors (Butler and Lane, 1965) and chilling (Fig. 11-5, top curve).

After 4 h at room temperature, all the phytochrome was present as Pr (Fig. 11-5), whose level did not decline subsequently in the dark. So the researchers concluded that photoreversible phytochrome disappeared only as Pfr. When Butler and colleagues continuously converted any Pr in dark-grown maize seedlings to Pfr through constant R irradiation, the amount of reversible phytochrome declined to a very low level—and at the same rate as Pfr. Using a variety of R and FR filters to maintain different proportions of the pigment as Pfr, they were able to measure disappearance rates even when Pfr accounted for only 1% of the total pigment. The rate was the same for all light sources maintaining at least 10% of phytochrome molecules as Pfr but slower when a smaller fraction of the pigment was in the FR-absorbing form (Butler et al., 1963).

2. The process was variously named *destruction* and *decay,* although both names were initially unsatisfactory. *Destruction,* the term used today, implied breakdown of the protein, for which there was no evidence at that time, whereas *decay* was sometimes applied to reversion as well as destruction.

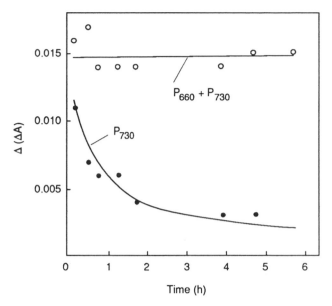

FIGURE 11-6   Phytochrome levels immediately after a cauliflower head was exposed to R and then placed in the dark. (From Butler et al., 1963. Reproduced with permission from the American Society of Plant Physiologists.)

LIGHT-GROWN PLANTS

Because phytochrome was present and active in green plants, the Beltsville workers knew it did not disappear entirely when plants were exposed to light. They concluded instead that it fell to undetectable levels. Looking for chlorophyll-free material that would allow them to measure much lower concentrations of the pigment, they bought a cauliflower head from the local supermarket. Because this had been sitting on a shelf, any light-induced loss of phytochrome had already occurred. After exposing the head to R for 2 min, they placed it in the dark and assayed samples over the next few hours. The phytochrome that was present did not decline in total; there was only a decrease in Pfr (Fig. 11-6). Thus phytochrome in this light-grown tissue was stable. It underwent only reversion, which, unlike phytochrome disappearance, occurred even in the absence of oxygen. The stable pigment appeared to have the same spectroscopic properties as phytochrome in dark-grown plants (Butler et al., 1963).

"Apparently, a small fraction of the phytochrome in dark-grown seedlings is different, or potentially different, from the rest," Butler concluded. "This small fraction might be more stable by virtue of different binding sites, which might, or might not be related to its presumed enzymatic activity. The question arises as to whether all of the phytochrome in dark-grown seedlings is physiologically functional or whether the only active phytochrome is the small amount that remains in continuous illumination'' (Butler and Lane, 1965).

INCOMPLETE CONVERSION OF PR

When the researchers obtained the data in Fig. 11-5, they were seeking evidence for their hypothesis that Pfr reversion is the photoperiodic timer (see Chapter 6). Thus until 1963 (Butler, 1964), they believed that the decline in Pfr observed after dark-grown maize seedlings were exposed to R and then returned to darkness (Fig. 11-5) was due to dark reversion of Pfr to Pr as well as to Pfr's destruction. According to this interpretation, the Pr remaining after 4 h was the end result of this reversion.

At Yale, however, Bonner realized that R fails to convert phytochrome entirely to Pfr. Commenting on the absorption spectrum he obtained for phytochrome extracted from dark-grown pea seedlings, he wrote that "the most obvious unexpected characteristic of the pigment preparations is the high minimum 665 nm absorption at red saturation. There is an obvious Pr peak remaining with the same 665 nm peak. Comparison of the slopes on the sides of this peak with the comparable slopes when Pr absorption is maximum indicates a ratio of about one to five, quite reproducible from one preparation to another. A reasonable estimate from this is that about one-fifth of Pr remains at red saturation" (Bonner, 1963).

This statement never appeared in print because the paper was returned for rewriting before publication. Bonner, "a very uncertain individual, who never felt confident about what he was doing," was discouraged, and he failed to resubmit the manuscript. In 1963, however, he informed Butler of his findings when the biophysicist visited him at Harvard, to where he had moved in 1961. When Butler heard the news, "the sweat poured off his forehead."

In Wageningen, meanwhile, Carel J. P. Spruit (see Fig. 20-4) had designed and built a dual-wavelength spectrophotometer for photosynthesis research. In 1960, doctoral candidate Pieter J. A. L. de Lint (see Fig. 20-4), who had been studying the spectral sensitivity of stem elongation and flowering in *Hyoscyamus niger* (de Lint, 1960), begged him to assay phytochrome in dark-grown maize seedlings. "As soon as a seedling reaches the surface of the soil and sees light, the mesocotyl stops elongating and the nodes make roots," Spruit explains. "So de Lint suggested there might be phytochrome in the mesocotyl. We found quite a lot. But if we kept the mesocotyl in the dark after exposing it to light, a large fraction of the phytochrome apparently disappeared."

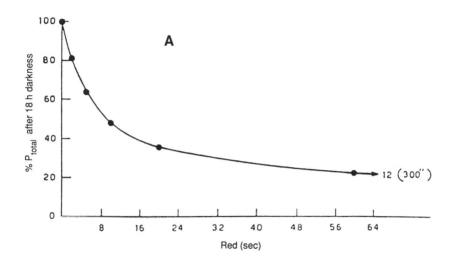

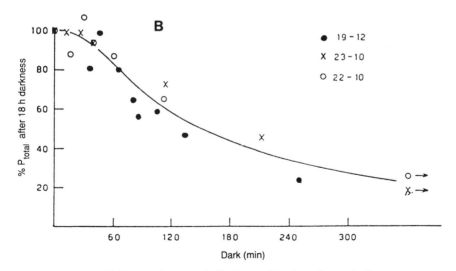

FIGURE 11-7   (A) Spectrophotometrically detectable phytochrome in *Zea mays* meso-cotyls, measured after 18 h darkness at 20°C followed irradiation with low dosages of R. (B) Time course of total spectrophotometrically detectable phytochrome after saturating R irradiation of mesocotyls. (From de Lint and Spruit, 1963.)

Pfr disappearance began 30–45 min after a light exposure, although it was prevented by a temperature drop from 20° to 2°C. When it ceased, one-fifth of the original phytochrome remained (Fig. 11-7A and B), although repeated R irradiation further reduced the level.

Gracing the literature with the correct explanation for the continued presence of Pr, de Lint and Spruit concluded that "P_{730} in the mesocotyls of dark-grown corn seedlings is [unstable] and converted into a photochemically inactive form by a temperature sensitive dark process. In view of the fact that purified solutions of both forms of phytochrome, isolated from corn shoots, proved to be about equally stable in the dark, one is led to the conclusion that the intact tissue contains an enzyme, specific for the destruction of P_{730}. . . . We have found no evidence for a dark reconversion of P_{730} into P_{660} as is suggested by Hendricks. It is true that, after a saturating irradiation of mesocotyls with 'red' light, the total concentration of phytochrome does not reach zero even after a long dark period and that the 20% phytochrome remaining is in the 660 [nm] form. In our opinion, however, this must be due to the impossibility to completely interconvert the two forms of phytochrome" (de Lint and Spruit, 1963).

Later that year, Butler, Siegelman, and Carlos O. Miller, a sabbatical visitor from Indiana University at Bloomington, obtained their own evidence (Butler et al., 1964b). The oat phytochrome was now sufficiently pure that the absorption curve after R irradiation showed a peak in the R as well as in the FR (Fig. 11-8A). When this solution was treated with *para*-mercuribenzoate, a chemical that reacts with the sulfhydryl (-SH) groups of proteins, the peak in the FR disappeared while the R peak remained (Fig. 11-8B). Thus *para*-mercuribenzoate destroyed the absorbance of Pfr while having little effect on that of Pr. When this altered preparation was irradiated with R, the absorbance in the R decreased as more Pr was converted. Complete absorption spectra derived later that year clearly showed a peak in the R as well as in the FR after oat phytochrome was exposed to R (see Fig. 11-4).

The researchers thus had to revise the interpretation of their 1960 data. They realized that loss of Pfr in R-illuminated, dark-grown maize seedlings (see Fig. 11-5) "was due almost entirely to the loss of photoreversible phytochrome" (Butler, 1964). Thus reversion did not occur in maize, although it did in pea (Bonner, 1962) and cauliflower. The estimates of quantum yields and absorption coefficients also had to be revised. Determining that R light converted 81% of phytochrome to Pfr, leaving 19% in the Pr form, Butler and Hendricks obtained molar absorption coefficients of at least 16,000 dm^3 mol^{-1} cm^{-1} for Pr and 10,000 dm^3 mol^{-1} cm^{-1} for Pfr. And they deduced that R was 1.5 times more effective than FR in phototransforming the pigment (Butler et al., 1964a).

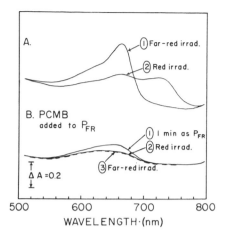

FIGURE 11-8   Absorption curves obtained on a single-beam spectrophotometer for (A) FR-irradiated and R-irradiated solution of oat phytochrome, and (B) the same solution treated with *para*-mercuribenzoate. (From Butler et al., 1964b. Reprinted with permission from *Biochemistry* vol. 3, pp. 851–857. Copyright 1964 American Chemical Society.)

CHEMICAL STUDIES

Although maximal absorbance change in the FR region of the spectrum equaled that in the R region in seedlings and some extracts, this ratio ($\Delta A_{fr}/\Delta A_r$) declined when extracts sat at room temperature. It was also always less then 1.0 in barley extracts. A decrease in solubility of the phytochrome accompanied the FR decline, which therefore appeared to result from denaturation of the Pfr form of the protein.

Butler, Siegelman, amd Miller were able to mimic this change with a strong solution of urea, which decreased Pfr absorbance within 1 min but merely shifted the absorption band of Pr from 665 to 660 nm. Thus the two forms of phytochrome appeared to have different protein conformations. *p*-Mercuribenzoate mimicked the effects of urea, suggesting that a sulfhydryl group exposed only in Pfr contributed to chromophore integrity.

A conformational change would presumably confer physiological activity on the phytochrome molecule. The Beltsville group believed that phytochrome "is undoubtedly an enzyme" (Butler and Downs, 1960), and many enzymes were known to be activated by small molecules that induce conformational changes. Thus the detected shape change perhaps represented the activation of an enzyme (Butler et al., 1964b), although by light rather than a metabolite.

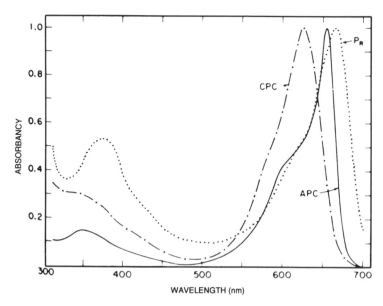

FIGURE 11-9   Absorption spectra of allophycocyanin (APC), C-phycocyanin (CPC), and oat Pr. (From Siegelman et al., 1966. Reproduced with permission from the American Society of Plant Physiologists.)

THE CHROMOPHORE

Hendricks had predicted that phytochrome was a biliprotein with a phycocyanin-like chromophore. Comparison of the absorption spectra of oat Pr, allophycocyanin, and C-phycocyanin (Fig. 11-9) supported this hypothesis.

Siegelman removed the chromophore by denaturing phytochrome with trichloracetic acid and then boiling in methanol, a combination of techniques used with algal biliproteins (Fujita and Hattori, 1962; Ó hEocha, 1963). The isolated chromophore was blue but incapable of phototransformation. Only minute quantities—10 μg/kg seedlings—were obtained, making chemical analysis impossible (Siegelman and Hendricks, 1965).

The phytochrome chromophore and that of allophycocyanin migrated similar distances during thin-layer chromatography. Moreover, when the chromatograms were placed under UV, the biladiene—a red spot—fluoresced, whereas the blue spots formed by the phytochrome and allophycocyanin chromophores did not; neither did the known bilatrienes,[3] which were also blue. This difference persisted when the chroma-

3. A bilatriene is a linear tetrapyrrole with three methine bridges between the pyrrole rings (see Fig. 28-4).

togram was sprayed with zinc acetate. And when an alkaline solution of iodine was sprayed on top of that, the biladiene spot turned orange and fluoresced green, whereas the other pigments fluoresced bright red (Siegelman et al., 1966).

These observations strongly suggested that the phytochrome chromophore was a bile pigment and that it was a bilatriene rather than a biladiene. Thus it seemed to be very similar to the allophycocyanin chromophore except for its photoreversibility in the intact molecule, although bilatrienes appeared capable of isomerization (Hendricks and Siegelman, 1967). Also the molar absorption coefficient of phytochrome was characteristic of a strongly absorbing pigment with only one chromophore (Siegelman and Firer, 1964), whereas allophycocyanin was thought to have several (Ó Carra, 1962).

Later studies would reveal that the phytochrome chromophore is very similar, although not identical, to two of the phycocyanin chromophores (see Chapter 27). Thus Hendricks had deduced the nature of the chromophore, its attachment to a protein, the photoreversibility of the resulting chromoprotein, its absorption characteristics, the approximate values of its absorption coefficients, and the orders of the reactions that converted it between its two forms. "Surely, if Alfred Nobel had included the plant sciences among his prizes," Butler decided, "Sterling Hendricks would have been a recipient" (Butler, 1982).

REFERENCES

Bonner, B. A. (1960). Partial purification of the photomorphogenic pigment from pea seedlings. *Plant Physiol.* 35(suppl. xxxii).

Bonner, B. A. (1961). Properties of phytochrome from peas. *Plant Physiol.* 36(suppl. xliii).

Bonner, B. A. (1962). *In vitro* dark conversion and other properties of phytochrome. *Plant Physiol.* 37(suppl. xxvii).

Bonner, B. A. (1963). Some properties of phytochrome from etiolated peas. Unpublished manuscript.

Borthwick, H. A., and Hendricks, S. B. (1960). Photoperiodism in plants. *Science* 132,1223–1228.

Borthwick, H. A., and Hendricks, S. B. (1961). Effects of radiation on growth and development. *In* "Handbuch der Pflanzenphysiologie" (W. Ruhland, ed.), vol. XVI, pp. 299–330. Springer, Berlin-Göttingen, Heidelberg.

Borthwick, H. A., Hendricks, S. B., Toole, E. H. and Toole, V. K. (1954). Action of light on lettuce-seed germination. *Bot. Gaz.* 115,205–225.

Butler, W. L. (1961). Some photochemical properties of phytochrome. *Prog. Photobiol. Proc. Int. Congr. Photobiol. 3rd,* pp. 569–571.

Butler, W. L. (1964). Dark transformations of phytochrome *in vivo. Q. Rev. Biol.* 39,6–10.

Butler, W. L. (1982). Sterling Hendricks: A molecular plant biologist. *Plant Physiol.* 70,

S1–S3. Quotation reprinted with permission from the American Society of Plant Physiologists.

Butler, W. L., and Downs, R. J. (1960). Light and plant development. *Sci. Am.* 203,56–63.

Butler, W. L., and Lane, H. C. (1965). Dark transformations of phytochrome *in vivo*. II. *Plant Physiol.* 40,13–17. Quotation reprinted with permission from the American Society of Plant Physiologists.

Butler, W. L. Norris, K. H., Siegelman, H. W., and Hendricks, S. B. (1959). Detection, assay, and preliminary purification of the pigment controlling photoresponsive development of plants. *Proc. Natl. Acad. Sci. USA* 45,1703–1708.

Butler, W. L., Lane, H. C., and Siegelman, H. W. (1963). Nonchemical transformations of phytochrome *in vivo*. *Plant Physiol.* 38,514–519.

Butler, W. L., Hendricks, S. B., and Siegelman, H. W. (1964a). Action spectra of phytochrome *in vitro*. *Photochem. Photobiol.* 3,521–528.

Butler, W. L., Siegelman, H. W., and Miller, C. O. (1964b). Denaturation of phytochrome. *Biochemistry* 3,851–857.

Butler, W. L., Hendricks, S. B., and Siegelman, H. W. (1965). Purification and properties of phytochrome. *In* "Chemistry and Biochemistry of Plant Pigments" (T. W. Goodwin, ed.), pp. 197–210. Academic Press, New York, London.

de Lint, P. J. A. L. (1960). An attempt to analysis of the effect of light on stem elongation and flowering in *Hyoscyamus niger*. *Meded. Landbouwhogesch. Wageningen* 60(14),1–59.

de Lint, P. J. A. L., and Spruit, C. J. P. (1963). Phytochrome destruction following illumination of mesocotyls of *Zea mays* L. *Meded. Landbouwhogesch. Wageningen* 63(14),1–7.

Fujita, Y., and Hattori, A. (1962). Preliminary note on a new phycobilin isolated from blue-green algae. *J. Biochem. (Tokyo)* 51,89–91.

Hendricks, S. B. (1961). Rates of change of phytochrome as an essential factor determining photoperiodism in plants. *Cold Spring Harbor Symp. Quant. Biol.* (1960) 25,245–248.

Hendricks, S. B., and Siegelman, H. W. (1967). Phytochrome and photoperiodism in plants. *Compr. Biochem.* 27,211–235.

Hendricks, S. B., Butler, W. L., and Siegelman, H. W. (1962). A reversible photoreaction regulating plant growth. *J. Phys. Chem.* 66,2550–2555.

Lane, H. C., Siegelman, H. W., Butler, W. L., and Firer, E. M. (1962). Extraction and assay of phytochrome from green plants. *Plant Physiol.* 37(suppl. xxvii).

Lane, H. C., Siegelman, H. W., Butler, W. L., and Firer, E. M. (1963). Detection of phytochrome in green plants. *Plant Physiol.* 38,414–416.

Ó Carra, P. (1962). Doctoral thesis. The National University of Ireland.

Ó hEocha, C. (1963). Spectral properties of the phycobilins. I. Phycocyanobilin. *Biochemistry* 2,375–382.

Siegelman, H. W., and Firer, E. M. (1964). Purification of phytochrome from oat seedlings. *Biochemistry* 3,418–423.

Siegelman, H. W., Hendricks, S. B. (1965). Purification and properties of phytochrome: a chromoprotein regulating plant growth. *Fed. Proc.* 24,863–867.

Siegelman, H. W., Turner, B. C., and Hendricks, S. B. (1966). The chromophore of phytochrome. *Plant Physiol.* 41,1289–1292.

CHAPTER 12

In Vivo Spectrophotometric Studies

The spectrophotometric studies of maize extracts confirmed the existence of a photoreversible pigment with many of the properties predicted from physiological studies. But when the photometric assay was applied to living material, attempts to correlate physiological responses with phytochrome content were bewilderingly unsuccessful. The resulting confusion prompted William S. Hillman to ask: "Is phytochrome the photoreceptor for red, far-red reversible processes?—It is as well to face this question squarely, since the less successful attempts at physiological correlation have suggested to some—though not to those most familiar with the field—the possibility of a negative answer" (Hillman, 1967).

The paradoxes, as the noncorrelations were known, surfaced after the Ratiospect went on the market in 1963. "Everyone had just started to measure *in vivo* phytochrome," recalls Masaki Furuya, who left Japan to obtain a Ph.D. in 1962 with Galston[1] before working with Hillman at Brookhaven. "We thought we could show some nice correlations," he recalls, "but none of us could." At a meeting in Colorado in August 1964, Furuya (Furuya and Hillman, 1964a) was "worried because our data were entirely negative. But to my surprise, other people reported similar findings."

Furuya had detected phytochrome in intact roots of dark-grown pea seedlings (Furuya and Hillman, 1964b), even though low-irradiance R had no effect on auxin-dependent root initiation in root segments (Furuya and Torrey, 1964). Perversely, the pigment was undetectable in isolated roots older than 7 days (Furuya and Hillman, 1964b), even though auxin-dependent initiation was in this case strongly R/FR-reversible (Furuya and Torrey, 1964).

---

1. During his 35 years at Yale, Galston supervised 24 graduate students and 67 postdocorates. By introducing Furuya to phytochrome and etiolated peas, Galston had a major impact on the phytochrome field.

THE PISUM PARADOX

Hillman (Fig. 12-1), who obtained his Ph.D. at Yale with David Bonner and became Galston's first postdoctorate in 1955, described another pea paradox (Hillman, 1965), which with a later *Zea* paradox became the two "most subversive" (Hillman, 1967) in the literature. Using filters to obtain mixtures of R and FR, he exposed stem segments from dark-grown pea seedlings to brief light treatments that would produce different photo-

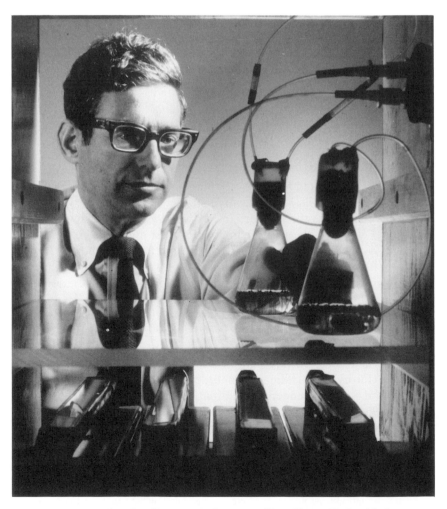

FIGURE 12-1   William S. Hillman 1977. Courtesy of Brookhaven National Laboratory.

stationary states of phytochrome. Thus 12% of the phytochrome in one batch of segments was initially present as Pfr, whereas other batches had 20, 44, 64, 88, or 100%.[2] Measuring the elongation that occurred over 20 h in darkness, Hillman observed that all the light-treated segments elongated less than the dark controls and that inhibition was greatest when half or more of the phytochrome was Pfr (Fig. 12-2A). Until maximal inhibition, the response correlated well with the phytochrome photoequilibrium.

Never one to leave well alone, Hillman gave pea seedlings 15 min R 8–9 h before cutting segments. This protocol, known to cause Pfr destruction, was expected to leave the tissue with a small amount of Pr, which the mixed light sources would fractionally convert to Pfr. Thus the stem pieces would differ from those in the first experiment only in total phytochrome content. Hillman wanted to know whether a greater percentage of the pigment now had to be Pfr to obtain maximal inhibition.

Each photostationary state was less inhibitory after the total concentration of the pigment was lowered (Fig. 12-2B). But those segments exposed to light sources establishing 20% or less Pfr actually elongated more than controls cut from R-pretreated plants and then kept in darkness.

The logical explanation was that segments from R-pretreated plants contained 20% of their phytochrome as Pfr at cutting. Thus some of the light sources would reduce this percentage and permit more growth than in the control segments. But Hillman detected only Pr in R-pretreated tissue. Thus "these data present a substantial contradiction between phytochrome state as 'assayed' physiologically and as assayed by *in vivo* spectrophotometry" (Hillman, 1965).

Apparently "the phytochrome active in regulating elongation is only a very small fraction of the total," Hillman suggested, and "in this active fraction, the rate of change of Pfr in darkness is much slower than in the remainder" (Hillman, 1965). Hillman dubbed the possibly inactive fraction "bulk" phytochrome, raising the fear that data gathered from *in vitro* studies might not apply to the physiologically active pigment.

THE ZEA PARADOX

The *Zea* paradox was uncovered at Stanford University when Briggs and a Korean graduate student, Hyangju P. Chon, were studying the phototropic sensitivity of maize coleoptiles. Pretreatment with R usually altered this complex response, the effect was FR-reversible, and the action spectrum resembled those of other phytochrome-mediated processes (Chon and Briggs, 1966). "But when we looked at how much red light it took to shift

---

2. The percentages were not corrected for incomplete conversion of Pr to Pfr.

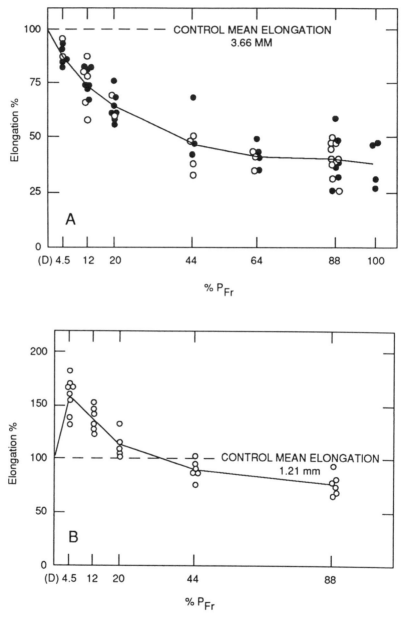

FIGURE 12-2 (A) Effects of maintaining the indicated photostationary state (as percent Pfr) for 5 (○) or 15 min (●) on subsequent growth in darkness of stem segments of dark-grown pea seedlings. (B) Similar to (A) except that stem segments were cut from seedlings given 15 min of R fluorescent light 8–9 h before cutting. (From Hillman, 1965.)

phototropic sensitivity," Briggs recalls, "we had a devil of a time because it looked to us as if the amount of light needed was far lower than that needed to transform a measurable amount of phytochrome."

Chon exposed 80-h-old *Zea* seedlings to R, placed them in darkness, and induced a phototropic response with B 1 h after the beginning of the R treatment. Without the pretreatment, the seedlings developed a phototropic curvature of about 3°. A saturating R treatment increased the curvature to more than 20°, being most effective at the coleoptile tip, where phytochrome was most abundant. When the R treatment preceded the B treatment by 2 or 3 days, however, phototropic sensitivity was just as pronounced, even though half the phytochrome had disappeared before the B exposure. "It seemed reasonable to assume," Briggs wrote, "that only a portion of the coleoptilar phytochrome might be involved in the observed physiological response" (Briggs and Chon, 1966).

From the amount of R that saturated the physiological response, Briggs and Chon calculated the size of this portion. This was possible because Lee H. Pratt had obtained a very precise action spectrum (Fig. 12-3) for phytochrome phototransformation *in vivo*, also in dark-grown maize seedlings (Pratt and Briggs, 1966). The results were unnerving, for the curves

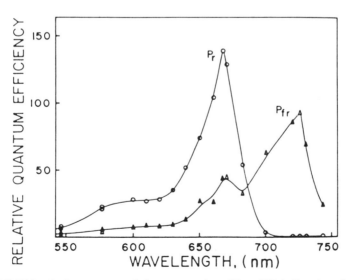

FIGURE 12-3 Action spectrum of photoconversion of Pr and Pfr in R region of spectrum. Pr to Pfr after sample was initially irradiated with FR (○); Pfr to Pr after sample was initially irradiated with R (△). Units of relative quantum efficiency are (percent transformation/ Einstein) (10^{-8}). (From Pratt and Briggs, 1966. Reproduced with permission from the American Society of Plant Physiologists.)

for physiological response and phytochrome transformation did not even overlap on the light dosage axis (Fig. 12-4). The physiological response (left curve) became saturated at light dosages two orders of magnitude too small to induce measurable production of Pfr (right curve). Moreover, the threshold values of the two curves were at least three orders of magnitude apart, whereas four orders of magnitude separated their saturation values (80% for phytochrome phototransformation, owing to the impossibility of fully converting Pr to Pfr).

After discussions with Hillman, Briggs and Chon decided that "the results . . . are open to two interpretations. First, induction of the physiological response might merely require transformation of a minute fraction of a percent of the total phytochrome. . . . Second, the phytochrome associated with the physiological response might be in some way sequestered from the bulk of phytochrome measurable, and physically and kinetically independent from it" (Briggs and Chon, 1966).

They favored the second explanation for three reasons. First, FR reversed the effect of the R treatment although it left at least 1% of the phytochrome as Pfr. Second, the differing slopes of the two curves (Fig. 12-4) suggested that the physiologically active phytochrome might have different kinetics from the phytochrome that had been extracted. Third, there was no relationship between the amount of Pfr formed and the size and duration of the physiological change. The researchers found it

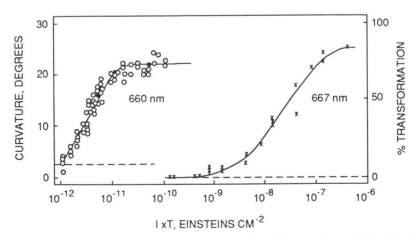

FIGURE 12-4 Dosage–response curves for alteration of phototropic sensitivity (circles) and phytochrome transformation (crosses). Dotted lines, dark controls. (From Briggs and Chon, 1966. Reproduced with permission from the American Society of Plant Physiologists.)

"likely that this active phytochrome is sequestered in some manner that facilitates its transformation in either direction" (Briggs and Chon, 1966).

Thus the phytochrome that sensitized the seedlings to B would be vastly more sensitive to R (or more able to transduce the light signal) than bulk phytochrome, perhaps as a result of its location or orientation. If it accounted for only a small part of the cell's total phytochrome, its transformation to sufficient Pfr to saturate the response would not be seen in a spectrophotometric assay.

Summarizing the paradoxes in a brilliant review, Hillman stated that "the discussion above has noted only the possibility that 'bulk' and 'active' phytochrome might differ in location or in relationship to substrate, but one can envisage many other situations. . . . A general knowledge of other pigment-protein complexes suggests the likelihood that several phytochromes may exist, differing only slightly in structure but significantly in biological activity. Thus allowing, or rather, anticipating, under the term phytochrome the same limited heterogeneity accepted in terms such as chlorophyll or cytochrome, it becomes completely reasonable not only to conclude that phytochrome is indeed the photoreceptor for red/far-red photo-reversible processes in plants, but also to assume that the heterogeneity in question will satisfactorily account for presently inexplicable results" (Hillman, 1967).

Two decades would pass before the emergence of concrete evidence for different phytochrome pools. "No one had a way to test Hillman's ideas at the biochemical level," Pratt says. "Since we couldn't resolve the situation, we ignored it for a while."

A few satisfying correlations between response and levels of spectrophotometrically detectable Pfr came to light, meanwhile. Smithsonian researchers observed a nonparadoxical relationship for hook opening in the Black Valentine bean system (Klein et al., 1967) and mesocotyl inhibition in etiolated oat seedlings (Loercher, 1966). There was also a close correlation between initially detectable phytochrome and growth inhibition of rice coleoptiles (Pjon and Furuya, 1968).

The rice study was a personal triumph for Furuya, who moved back to Japan in 1965. At Nagoya University, he became an associate professor, with many misgivings. "It is almost impossible to imagine the situation in Japan at that time," he explains. "It was like China now. Most of my American friends thought I would not be able to publish anymore—there was nothing in the laboratory when I arrived. But before I left the States, Warren Butler gave me a blueprint of the Ratiospect, and I asked a Japanese company, Hitachi, Ltd., to build one." Hitachi later built a commercial model, using two monochromators instead of filters.

REVERSION AND DESTRUCTION

With the Ratiospect at Brookhaven, Hillman surveyed a range of fruits and vegetables from the local market, detecting phytochrome in about one-third. Brussels sprouts, globe artichoke hearts, and parsnip roots had the highest levels, even though the latter two tissues displayed no known responses to light. Total phytochrome in artichoke heart proved to be light-stable, but the tissue exhibited "Pfr disappearance and Pr appearance," as Hillman (1964) described reversion.

The phytochrome in Brussels sprouts disappeared after R treatment, with no evidence of reversion. This was odd because the sprouts had already been exposed to light in the field and market. But Hillman realized that the sample came from the center of the sprout, which, swaddled in layers of leafy tissue, was essentially dark-grown tissue (Hillman, 1964).

Hillman and Furuya detected reversion in dark-grown pea stems and root tips (Furuya and Hillman, 1964b), even after correction for incomplete photoconversion of Pr (Hillman, 1967). Dark reversion was subsequently observed in many dicot tissues (e.g., Hopkins and Hillman, 1965) but not in coleoptiles of cereal grasses (e.g., Pjon and Furuya, 1968). Thus dark reversion was apparently confined to dicotyledons (Hillman, 1967), whereas phytochrome destruction occurred in both monocots and dicots(Hopkins and Hillman, 1965).

Little was known about reversion and destruction at this point except that the latter required oxygen and was sensitive to respiratory inhibitors, such as sodium azide (see Chapter 11). Furuya and William G. Hopkins at Brookhaven inhibited the process in segments of dark-grown oat coleoptiles with metal-complexing agents, such as EDTA (ethylenediamine tetraacetic acid) and 2-mercaptoethanol, which were used to stabilize phytochrome during purification. They confirmed that azide prevented destruction but noted that it was also a metal-complexing agent. EDTA inhibited destruction while having little effect on respiration. Thus it seemed likely that "Pfr decay is a metal-dependent process, possibly oxidative, and not directly linked with respiration" (Furuya et al., 1965). Furuya later showed that metal ions, such as zinc and iron, partially reversed this inhibition by metal-complexing agents (Manabe and Furuya, 1971).

In the monocot *Zea*, azide protected both total phytochrome and Pfr from destruction, whereas it prevented only total pigment loss in the dicot *Pisum*. This suggested that azide affected destruction but not reversion (Furuya et al., 1965) and could perhaps distinguish between the two (see Chapter 20).

Testing the hypothesis that phytochrome destruction and respiration were unrelated, Matthews O. Bradley at Brookhaven exposed oat coleoptile segments to 2,4-dinitrophenol at concentrations that uncoupled respiration from ATP production. Because phytochrome loss was not inhibited, he concluded that "respiratory chain energy is not intimately related to phytochrome decay. This finding supports the view that decay requires oxygen and metal activity, not as a function of the respiratory chain, but rather as a function of other, as yet unknown, reactions" (Bradley and Hillman, 1966).

SYNTHESIS

De novo synthesis of phytochrome was first detected by Butler and Lane in Beltsville. In maize seedlings, synthesis occurred primarily in regions of

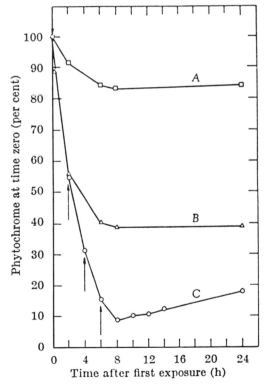

FIGURE 12-5   Phytochrome levels in pea hook segments. Values are percentages of those in plants grown in dark at zero hours. Arrows indicate 15-min R treatments. (*A*) dark control; (*B*) one R exposure; (*C*) four R exposures. (From Clarkson and Hillman, 1967a. Reprinted by permission from *Nature* vol. 213, pp. 468–470. Copyright © 1967 Macmillan Magazines Ltd.)

dividing cells (primary leaves) rather than in regions growing by cell expansion (mesocotyl). And the pigment appeared as Pr, not Pfr. R irradiation decreased not only the total amount of phytochrome but also the rate at which the pigment continued to be made after the light treatment (Butler and Lane, 1965).

Early in 1967, Hillman reported that apparent synthesis could be divorced from growth and that it was triggered when the pigment's concentration fell below a critical level. When David T. Clarkson briefly treated dark-grown pea hook segments with R and returned them to darkness for 6–8 h, phytochrome levels were half those in nonirradiated hook segments (Fig. 12-5A and B). But after a series of 15-min exposures, total amount of phytochrome fell but then increased (Fig. 12-5C), even when segment elongation was inhibited with mannitol (Clarkson and Hillman, 1967a). The process seemed to involve protein synthesis, because it was inhibited by cycloheximide and actinomycin D (Clarkson and Hillman, 1967b).

This increase occurred only after phytochrome levels decreased to one-fifth the initial concentration. Moreover, total synthesis depended on how far below this critical concentration the level had plunged, so that equal

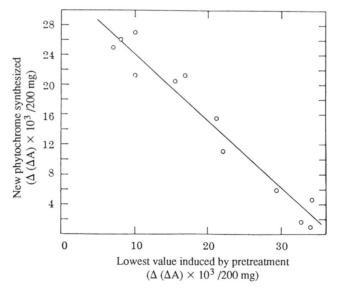

FIGURE 12-6 Amount of new phytochrome synthesized in apical hook segments in the 24 h after last illumination as a function of lowest level of phytochrome induced by illumination pretreatment. (From Clarkson and Hillman, 1967a. Reprinted by permission from *Nature* vol. 213, pp. 468–470. Copyright © 1967 Macmillan Magazines Ltd.)

levels of phytochrome were maintained despite different illumination schedules (Fig. 12-6). This suggested that light-treated plants kept the pigment at a fixed concentration, striking a careful balance between phytochrome synthesis and destruction (Clarkson and Hillman, 1967a). The researchers were unsure, however, whether the pigment was synthesized from scratch or whether "destruction" removed the chromophore, leaving a pool of protein to which it could later be reattached. "If *de novo* synthesis does, indeed, occur in repeatedly illuminated, but not in dark grown tissue," they suggested, "it is necessary to propose that at some point before our observations begin synthesis has been switched off, and that the receipt of a red light signal followed by the destruction of a large proportion of the Pfr, is active in some way in switching synthesis back on. If *de novo* synthesis were switched off in the dark grown tissue when a critical concentration of phytochrome or some inhibitor was reached, phytochrome synthesis, like the synthesis of porphyrins, might be subject to negative feedback or end product inhibition" (Clarkson and Hillman, 1967b).

REFERENCES

Bradley, M. O., and Hillman, W. S. (1966). Insensitivity of phytochrome decay *in vivo* to respiratory uncoupling by 2,4-dinitrophenol. *Nature* 210,838.

Briggs, W. R., and Chon, H. P. (1966). The physiological versus the spectrophotometric status of phytochrome in corn coleoptiles. *Plant Physiol.* 41,1159–1166. Quotation reproduced with permission from the American Society of Plant Physiologists.

Butler, W. L., and Lane, H. C. (1965). Dark transformations of phytochrome *in vivo*. *II. Plant Physiol.* 40,13–17.

Chon, H. P., and Briggs, W. R. (1966). Effect of red light on the phototropic sensitivity of corn coleoptiles. *Plant Physiol.* 41,1715–1724.

Clarkson, D. T., and Hillman, W. S. (1967a). Apparent phytochrome synthesis in *Pisum* tissue. *Nature* 213,468–470.

Clarkson, D. T., and Hillman, W. S. (1967b). Modification of apparent phytochrome synthesis in *Pisum* by inhibitors and growth regulators. *Plant Physiol.* 42,933–940. Quotation reprinted with permission from the American Society of Plant Physiologists.

Furuya, M., and Hillman, W. S. (1964a). Observations on spectrophotometrically assayable phytochrome in etiolated peas. *Plant Physiol.* 39(suppl.1).

Furuya, M., and Hillman, W. S. (1964b). Observations on spectrophotometrically assayable phytochrome *in vivo* in etiolated *Pisum* seedlings. *Planta* 63,31–42.

Furuya, M., and Torrey, J. G. (1964). The reversible inhibition by red and far-red light of auxin-induced lateral root initiation in isolated pea roots. *Plant Physiol.* 39,987–991.

Furuya, M., Hopkins, W. G., and Hillman, W. S. (1965). Effects of metal-complexing and sulfhydryl compounds on nonphotochemical phytochrome changes *in vivo*. *Arch. Biochem. Biophys.* 112,180–186.

Hillman, W. S. (1964). Phytochrome levels detectable by *in vivo* spectrophotometry in plant parts grown or stored in the light. *Am. J. Bot.* 51,1102–1107.

Hillman, W. S. (1965). Phytochrome conversion by brief illumination and the subsequent elongation of etiolated *Pisum* stem segments. *Physiol. Plant.* 18,346–358.

Hillman, W. S. (1967). The physiology of phytochrome. *Annu. Rev. Plant. Physiol.* 18,301–324.

Hopkins, W. G., and Hillman, W. S. (1965). Phytochrome changes in tissues of dark-grown seedlings representing various photoperiodic classes. *Am. J. Bot.* 52,427–432.

Klein, W. H., Edwards, J. L., and Shropshire, W., Jr. (1967). Spectrophotometric measurements of phytochrome *in vivo* and their correlation with photomorphogenic responses of *Phaseolus*. *Plant Physiol.* 42,264–270.

Loercher, L. (1966). Phytochrome changes correlated to mesocotyl inhibition in etiolated *Avena* seedlings. *Plant Physiol.* 41,932–936.

Manabe, K., and Furuya, M. (1971). Effects of metallic ions on nonphotochemical decay of Pfr in *Avena* coleoptiles. *Bot. Mag.* 84,417–423.

Pjon, C. J., and Furuya, M. (1968). Phytochrome action in *Oryza sativa* L. II. The spectrophotometric versus the physiological status of phytochrome in coleoptiles. *Planta* 81,303–313.

Pratt, L. H., and Briggs, W. R. (1966). Photochemical and nonphotochemical reactions of phytochrome *in vivo*. *Plant Physiol.* 41,467–474.

CHAPTER 13

Seed Germination and Flowering

The Beltsville workers greatly expanded their knowledge of phytochrome's action in seed germination and flowering during the 1960s. Eben Toole, who retired from his position as principal physiologist with the Crops Research Division in 1959, continued to collaborate. He died at age 78 on August 2, 1967.

SEED GERMINATION

The researchers had shown that lettuce seed germinated quickly after a single, brief light exposure and that phytochrome was clearly the photoreceptor. But they wanted to know how widely the pigment was involved, because the dormancy of many species was much less easily broken. In hundreds of experiments, they demonstrated the operation of the photoreversible reaction in the seeds of several pine species (Toole et al., 1961, 1962), and in those of many angiosperms, both monocots (Hendricks et al., 1968; Toole and Borthwick, 1968a,b; Toole and Borthwick, 1971; Danielson and Toole, 1976; Toole, 1976; Toole and Koch, 1977) and dicots (Toole et al., 1953, 1955a, 1957; Toole, 1961; Borthwick et al., 1964).

LIGHT AND TEMPERATURE

Lepidium seed, for which the group obtained action spectra (see Chapter 7), illustrated that germination in many seeds is controlled not only by phytochrome but also by a temperature block. Not a single *Lepidium* seed germinated without exposure to light, and at 15°C, only one-third of the seeds germinated after a R exposure. The remaining ones would sprout if they were (1) imbibed at 15°C and then held at 25°C after a light treatment or (2) irradiated at the end of the second of 8–16 daily alternations between the two temperatures or (3) imbibed at 20°C on blotters wetted with potassium nitrate. Holding water-imbibed *Lepidium* seed at 35°C for 2 h just before or just after a light treatment in an otherwise 20°C regime also made the seeds highly sensitive to R and FR, although the effect was most dramatic if the high-temperature treatment preceded the irradiation. "High temperature before irradiation significantly increases sensitivity to red energy, but it does not affect the reversible photoreaction," Vivian Toole explains. "High temperature after irradiation removes a block that otherwise prevents the germination of a large percentage of the seeds, even after a saturating exposure to red light."

Exploring interactions between light, temperature, and nitrogen supply in seeds of several temperate-zone species (Toole et al., 1955b), the Tooles determined that, although some seeds required a high temperature, others failed to germinate, either in constant darkness or after a R exposure, if they were maintained at or above 30°C. The most favorable regime was usually a daily alternation or a single shift between a lower and higher temperature, even if the latter was inhibitory by itself. In nature, such fluctuating temperatures mark the advent of spring, the season that favors seedling survival and heralds the longest growing season.

Although temperature changes altered the balance of metabolism, nitrate probably acted at a specific control point, for the ion did not affect respiration. Thus "light is obligatory for the germination of each seed of *Lepidium virginicum*, but some of the seeds have a further requirement. A temperature change and the presence of nitrate are two alternative means for initiating further necessary adjustment among component reactions of germination" (Toole, et al., 1955b)

The photoreversible reaction was thus just one of several control mechanisms that operated simultaneously (see Toole, 1973). Moreover, its participation in the germination process was not necessary for all seeds or under all conditions. Tobacco seed, which required light at constant temperature, germinated well in total darkness if the temperature alternated daily between 20°C and 30°C. White Boston lettuce seed was light-requiring if the temperature remained constant at 15°C or 20°C but showed a high-percentage germination at 10°C (Toole et al., 1957).

PROLONGED IRRADIATION

Whereas many seeds germinate after a brief glimpse of light, others require prolonged irradiation, a behavior that some researchers once classified as photoperiodic (e.g., Isikawa, 1954; Nagao et al., 1959). Wanting to see if phytochrome was involved in this response, the Tooles and Borthwick collected *Paulownia tomentosa* (the empress tree) seeds in Beltsville and the Blue Ridge Mountains. Although some lots had to be irradiated for as long as 4 days, phytochrome was obviously the photoreceptor, because if 5-min alternations of R and FR followed a long light treatment, R induced and FR prevented germination.

Terminating prolonged periods of fluorescent illumination with darkness or FR, the researchers determined how long Pfr continued to act after the light was turned off (Fig. 13-1). Because periods of dark action ranged from 42 h (75% germination) to 49 h (20% germination), Pfr acted in total from 48 h to more than 108 h, depending on the individual seeds in a sample. Although some atypical seeds germinated after a single brief irradiation, others were best promoted when one saturating pulse followed another after at least 24 h. Thus in the early stages of germination, a single production of Pfr satisfied the requirement for active phytochrome for 24 h. If additional Pfr was required, it could be supplied by an additional pulse even after 72 h. So Pfr simply needed to be present during the appropriate stages of germination and "the phytochrome relationships of *Paulownia* seeds . . . do not need to be fitted to a flowering model" (Borthwick et al., 1964).

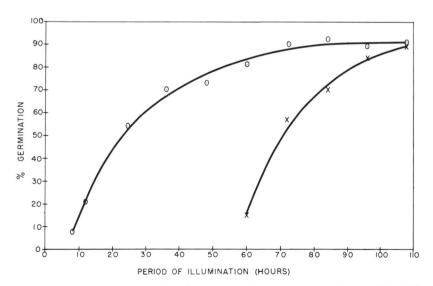

FIGURE 13-1   Germination of *Paulownia* seeds in response to various periods of fluorescent illumination with (x) or without (o) subsequent FR irradiations. (From Borthwick et al., 1964.)

INHIBITION BY PROLONGED ILLUMINATION

Some seeds fail to germinate if they are exposed to light for long periods, although they will germinate in darkness or after one or more brief R exposures. *Poa pratensis,* Kentucky bluegrass, germinated well in darkness after daily 15–25°C alternations or at a constant 20°C if exposed briefly each day to high-irradiance fluorescent light. The response obeyed the reciprocity law, and phytochrome was the photoreceptor because FR reversed the effects of the fluorescent treatments. The promotive effects of intermittent light and alternating temperature were additive in one seed lot.

When seeds were exposed to continuous, lower-irradiance fluorescent light at 20°C during or between the brief light treatments—or during alternating temperature treatments—germination was inhibited. Thus prolonged light exposure overrode the effects of promotive treatments. The inhibition response was irradiance-dependent and maximal between 700 and 750 nm (Hendricks et al., 1968; Toole and Borthwick, 1971).

PRECHILLING

Nurserymen often prepare pine seeds for germination by moistening and chilling them for several weeks. Studying *Pinus virginiana* (Toole et al., 1961), *Pinus taeda,* and *Pinus strobus* (Toole et al., 1962), the Beltsville workers determined that very few pine seeds germinated unless they had been held at 5°C, but that R treatment drastically reduced the required prechilling period. Conversely, lengthening the cold, moist period increased sensitivity to R, decreased sensitivity to FR, and caused more rapid escape from control by FR. For *P. taeda* and *P. strobus,* brief R exposures twice daily or continuous light for several days was the most effective light treatment. Because of their thick seed coats, *P. virginiana* seeds required 10 times more energy than *Lepidium* seeds for 50% conversion of phytochrome to Pfr and about five times the energy for 50% conversion in the opposite direction.

DARK-GERMINATING SEEDS

Great Lakes lettuce and tomato seed germinated perfectly well in total darkness but became reversibly responsive to light after exposure for a few days to FR at 30°C or 35°C (Hendricks et al., 1959; Toole, 1961). Thus phytochrome was present in these seeds.

Alberto L. Mancinelli, who left the University of Rome for a sabbatical in Beltsville in 1962, explored the possibility that the pigment is active during dark germination. He studied three lettuce varieties that germinated in darkness between 15°C and 20°C but became light-dependent at 25°C. At 20°C, the seeds germinated equally well in continuous R, fluorescent WL, and darkness. But incandescent light reduced the percentage below the dark controls, and continuous FR was almost completely inhibitory. A brief R exposure, applied at the end of prolonged FR exposures, reversed the FR inhibition unless FR was applied for 3 days or more.

Germination was also inhibited by 4 days of cyclic FR (1 min FR/9 min dark) but not cyclic FR-R (1 min FR/1 min R/8 min dark). The R/FR reversibility observed in cyclic treatments suggested that the active form of phytochrome operated during normal germination in darkness in lettuce (Mancinelli and Borthwick, 1964) and tomato (Mancinelli et al., 1966).

A single, brief FR treatment did not inhibit germination in lettuce. Thus "the necessity of continuous or intermittent far-red irradiation over a prolonged period indicates that Pfr is constantly reappearing during such a period. Eventually, a time is reached when far red is no longer required to prevent germination, indicating that further production of Pfr has ceased. The Pfr that reappears might be from either rehydration of a reserve pool or *de novo* synthesis" (Mancinelli and Borthwick, 1964).

Similar results were obtained with tomato and cucumber seeds (Mancinelli et al., 1966, 1967; Yaniv et al., 1967; Mancinelli and Tolkowsky, 1968) as Mancinelli continued this work at Columbia University in New York. The presence of Pfr in dark-germinating seeds and its reappearance in darkness after FR irradiation would be confirmed spectrophotometrically and dubbed "inverse dark reversion" (see Chapter 20).

ARABIDOPSIS ACTION SPECTRA

In July 1959, biophysicist Walter Shropshire, Jr. returned from Caltech to the Smithsonian, where he had completed doctoral studies with Withrow. He and Klein, who was directing the lab by this time, agreed "that there had never been a detailed action spectrum of seed germination other than the Flint and McAlister one." Shropshire recalls, "So we worked on the spectrum for germination of *Arabidopsis thaliana,* a mustard, which you can grow in test tubes." The seeds were absolutely light-requiring.

Shropshire derived curves for induction and reversal of germination (Shropshire et al., 1961). The reversal spectrum had double peaks in the FR—at 720 and 740 nm (Fig. 13-2) (Shropshire et al., 1961), in contrast to the absorption spectra derived by the Beltsville group for isolated phytochrome, which had a single peak in this region (see Fig. 11-4). The Smithsonian workers had also noted two peaks in the action spectrum for reversal of bean hook opening (see Fig. 6-9). Because double peaks have

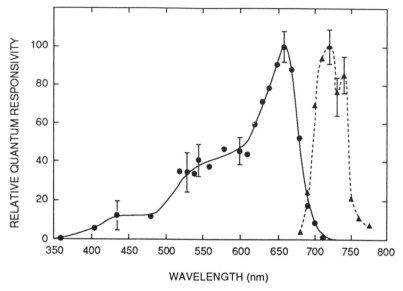

FIGURE 13-2   Action spectrum for photoinduction (solid curve) and photoinactivation (dashed curve) of germination of *Arabidopsis thaliana*. (From Shropshire et al., 1961.)

not been noted for other responses nor in absorption spectra of subsequent phytochrome preparations, this observation remains to be explained. "My suspicion," says Shropshire, "is that this could be an anomalous result due to some peculiarity in the interference filters that were used in this spectral region."

SEED COAT ENZYMES

Experiments at Harvard University revealed one possible link in the chain between the Pfr action and germination. Thimann and Hiroshi Ikuma determined that Grand Rapids lettuce seed, which normally does not germinate well in darkness, would sprout if cut in half or slit longitudinally. But it would not germinate in the dark if only the tip was removed. Moreover, promotion by R and subsequent inhibition by FR was just as marked in de-tipped as in normal seeds. These findings suggested that mechanical factors, rather than chemical inhibitors, restricted radicle elongation in darkness or after FR radiation and that R (Pfr) somehow overcame this restriction.

Pfr might act by promoting the secretion of an enzyme able to break down cell wall materials in the tissue surrounding the embryo, Thimann and Ikuma reasoned. Injections of such enzymes into lettuce seed increased percentage germination, whereas injections of inactivated enzymes did not (Ikuma and Thimann, 1963). "I think the coat-imposed dormancy of lettuce is one of the clearer phytochrome cases," Thimann says. "It's quite obvious that it has to release the enzyme that softens the seed coat."

FLORAL INDUCTION IN CHRYSANTHEMUM

From 1956 onward, Borthwick studied the photoperiodic responses of chrysanthemum in collaboration with Henry Marc Cathey (Fig. 13-3) a research horticulturist in Ornamentals Investigations. "The chrysanthemum was a very difficult system," explains Cathey at the U.S. National Arboretum, where he has been director since 1981. "It took 2 weeks of exposures and then you had to wait another week for dissection."

Cathey and Borthwick applied light breaks "in total darkness," Cathey explains, "so each box was marked with nails, which you would feel. Nails on the left meant a red treatment, while those on the right meant far-red."[1] A 1-min R light break drastically reduced floral initiation in Indianapolis Yellow and Shasta cultivars, and 27 min R completely prevented floral induction. FR repromoted the plants, although not to the stage of the controls, and its efficiency decreased with the number of reversals, failing completely if 90 min elapsed between R and FR pulses. Short periods of FR alone had little effect, but 81 min FR inhibited flowering in plants not treated with R and failed to repromote those that had been treated. This R action of FR had first been noted in *Xanthium* (see Fig. 7-3). Thus "the adage, 'If a little is good, more is better,' is apparently not applicable to the use of far-red radiant energy" (Cathey and Borthwick, 1957).

Hoping to explain why chrysanthemum needs such a long light break, Cathey and Borthwick applied 12-min incandescent light treatments during 15-h dark periods. When they gave the light break all at once, flower bud development was only slightly less than in control plants. But when they split the break into halves, the response decreased as the interval between the two 6-min treatments increased, although a 60-min interval did not completely prevent flowering.

Wondering if the light break could be further subdivided, Cathey and Borthwick exposed plants to a series of light–dark cycles that ranged in length from 1 to 120 min but contained equal periods of irradiation. This

1. "Borthwick and Parker devised this nail-braille system during the first action spectrum studies," Downs explains. "If a lot of treatments were involved, two nail sizes were used. Almost all the treatments were run in the dark, and you could identify the Photoperiod Project members by their dimpled shins."

FIGURE 13-3   Henry Marc Cathey.

Herculean project revealed that intervals between successive light pulses could not greatly exceed 30 min when incandescent lamps were used. Although ruby-red, incandescent filament, and fluorescent lamps were equally satisfactory in 15-min cycles, ruby-red light was ineffective in cycles longer than 30 min, incandescent light became ineffective when cycles were lengthened to 60 min, but fluorescent light prevented flowering even in longer cycles (80–120 min) (Borthwick and Cathey, 1962). Thus "in chrysanthemum the time required for the flower-inhibiting reactions catalyzed by Pfr is greater than the dark reversion time of Pfr to an inactive form. . . . Effective cycle length depends on the ratio of red to far-red energies in the light. As the relative amount of red increases, the amount of Pfr formed and the time required for its reversion in darkness to a level ineffective for flower inhibition also increase" (Borthwick and Cathey, 1962).

Studies with different light sources offered a possible explanation for why Pfr disappears so quickly in this plant. A 1-min irradiation with at least 600 footcandles of fluorescent light (largely R) in the middle of the dark period completely prevented flowering of Honeysweet chrysanthemum, whereas 32 min of sunlight (R plus FR) was required for complete inhibition. And 16,000 footcandles of incandescent light (R plus FR, Sungunquartz iodide) was only partially inhibitory, although it would have been adequate over a longer period. "It seems clear," Cathey and Borthwick concluded, "that the equilibrium level of Pfr left in the plants when they are taken from sunlight or incandescent light to darkness is inadequate to permit completion of the flower-inhibiting function before the Pfr reverts to Pr" (Cathey and Borthwick, 1964).

Both cocklebur and chrysanthemum leaves transmitted about 50% of incident FR, but chrysanthemum let through very little R because chloroplast-packed cells covered its entire area. If the chloroplasts screened phytochrome, Pfr levels would be very low in incandescent light or sunlight. When irradiation ceased, the level of active phytochrome would therefore quickly plunge below the threshold required for floral inhibition (Cathey and Borthwick, 1964).

Experiments by Michael J. Kasperbauer, a research associate in the Plant Physiology Lab from 1961 to 1963, revealed that cyclic lighting could also regulate the flowering time of LDP. Using sweet clover (*Melilotus alba*) as an example, he induced flowering by irradiating with incandescent light for only 10% of 16-h dark periods in cycle lengths up to 1 h. Some effective cycles were 1.5 min of light every 15 min or 6 min every hour (Kasperbauer et al., 1963a). Longer cycles were effective when cool-white fluorescent lamps placed a higher percentage of phytochrome in the Pfr form during the light portion of each cycle (Kasperbauer and Borthwick, 1964).

Application of the studies at Beltsville (Cathey et al., 1961; Cathey and Borthwick, 1961, 1962; Kasperbauer et al., 1963a) and other institutions had a dramatic impact on the floral industry. At the University of Connecticut at Storrs, Sidncy Waxman, who first reported the effects of cyclic lighting (flashlighting), showed that it was both very effective (Waxman, 1961a) and could cut utility costs by more than 85% (Waxman, 1961b, 1963). This cost reduction greatly benefited less affluent growers. "Cyclic lighting allowed flower production to escape to the tropics," Cathey explains, "because growers could use a little bit of photoperiodic lighting to delay the flowering of SDP such as chrysanthemum or accelerate the flowering of LDP such as carnation with very low levels of energy. Thus chrysanthemums, carnations, and many other photoperiodic plants are now grown in Colombia, Ecuador, Peru, and Costa Rica, where natural daylengths are nearly constant throughout the year and different temperatures are available at different elevations. So the concept of phytochrome brought about total changes in the way flowers are produced. . . . By manipulating lighting along with temperature and the application of chemicals, you have a whole model with which you can time and tailor plants."

FLORAL INDUCTION IN CARYOPTERIS

Piringer left Beltsville in 1962 to become assistant director of the U.S. National Arboretum. His last research project focused on blue mist spirea (*Caryopteris* x *clandonensis* A. Simmonds), a woody, ornamental shrub from Hendricks's garden that bloomed in mid August, almost to the day.

Plants developed visible flower buds on 8-, 12-, or 16-h photoperiods, even when a 3-h light break interrupted 8-h nights. But buds matured into flowers, a process known as *anthesis,* only when exposed to days shorter than 16 h.

Plants exposed to 10-h days flowered 22–25 days after the beginning of SD treatment. But if as many as 30 days of continuous light were inserted between days 7 and 8, only about 26 SD were required in total. Thus the plants seemed to "remember" inductive treatments even though the buds that opened were not even present during the first 7 days.[2]

Short days failed to induce anthesis if dark periods were interrupted with a 1-min R light break, and this effect was FR-reversible. Brief irradiation with FR alone delayed the response, whereas 60 min FR inhibited anthesis nearly as effectively as 1 min R. "Far-red radiation maintains a small but effective fraction of phytochrome in the active form," the researchers explained. "In darkness this small fraction of active phytochrome quickly reverts to the inactive form, but if it is maintained by irradiating the plants with far red for an hour or more, it continues to act and effectively inhibits anthesis" (Piringer et al., 1963).

RED ACTION OF FAR-RED

The absorption spectrum of isolated Pr supported this conclusion because its tail extended into the FR region (see Fig. 11-4). Thus Butler, Lane, and Siegelman proposed that such a small amount of Pfr acting over a long period might be just as effective as a larger amount that acted briefly before disappearing (Butler et al., 1963).

Other members of the group tested this proposal with tiny seedlings of *Chenopodium rubrum,* a SD Canadian weed. Studying different latitudinal ecotypes at the Central Experimental Farm in Ottawa, Bruce G. Cumming had found that plants were extremely sensitive to photoperiod and other environmental factors, even in the seedling stage (Cumming, 1959a,b). Thus flower primordia would form between the cotyledons just a few days

---

2. This phenomenon, known as fractional induction (Hillman, 1962), had previously been observed in several species. Walter W. Schwabe at Imperial College, London University, noted that, in *Kalanchoë blossfeldiana,* interruption of inductive SD with a LD suppressed the action of 1.5–2.0 subsequent SD. Obtaining similar results with soybean, *Perilla,* and *Chenopodium*—although not with *Xanthium* and *Pharbitis,* which respond to a single long night—Schwabe proposed that both a floral stimulus and a floral inhibitor regulate flowering (Schwabe, 1956, 1959). LD produced an additive inhibitory effect only if they were inserted singly into a SD regime and not if several were given consecutively, as in Piringer's study. Thus the effects of SD were additive, whereas those of LD were not (see Vince-Prue, 1975).

after germination, even—in some ecotypes—if just a single dark period interrupted continuous light. So large populations could be grown from seeds to flower in small petri dishes, and individual experiments could be completed within 10 to 14 days (Cumming, 1967).

Cumming sent seeds of an early-responding strain to Borthwick, and Kasperbauer (Fig. 13-4) later developed a miniplant system for spectrographic studies. "We wanted a large number of plants in the relatively small spectrum," Kasperbauer explains. Using a template, Kasperbauer planted rows of seeds in rectangular plastic trays. A week later, the seedlings were treated on the spectrograph in the middle of the 16-h nights of five consecutive diurnal cycles. During the many experiments, Hendricks determined energy levels at each row position, Kasperbauer positioned the plants in the spectrum, and Borthwick manually opened and closed the spectrograph door at signals from an old Gra-Lab timer.

The resulting action spectra (Fig. 13-5) were much more precise than those obtained with larger plants, such as cocklebur, because each row of tiny *Chenopodium* seedlings fitted into a very narrow wave band. They

FIGURE 13-4   Michael J. Kasperbauer (*left,* holding tray of *Chenopodium* seedlings) and Harry A. Borthwick (*right*), Beltsville, 1962.

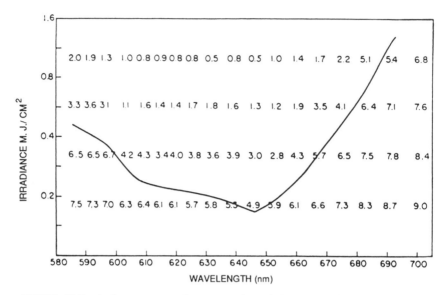

FIGURE 13-5 Action spectrum for suppression of floral induction in *Chenopodium rubrum* seedlings. Solid curve shows irradiance required to reduce flowering stage from 9.0 to 5.0 (From Kasperbauer et al., 1963b. Copyright by the University of Chicago.)

revealed that floral inhibition by a light break was greatest at 640–650 nm, rather than at 660 nm, the *in vitro* absorption maximum of phytochrome from etiolated seedlings. The group attributed this difference to competitive absorption by chlorophyll at 660 nm. Short exposures to wavelengths between 720 and 740 nm most effectively reversed the inhibitory effects of R.

When the researchers applied FR light breaks in the middle of 16-h dark periods, the response varied drastically with duration of irradiation. Thus brief treatments with FR alone or after a pulse of R allowed floral induction. But exposure for more than 60 min at about 730 nm completely suppressed the flowering response, although calculations suggested that the FR source converted less than 2% of the phytochrome to Pfr.

Sixty-minute FR light breaks were as inhibitory as 5 min R/55 min FR, 5 min R/51 min darkness/4 min FR, or 56 min R/4 min FR. This indicated that the same degree of floral inhibition resulted from the placement of the phytochrome photoequilibrium near 2% Pfr by FR or at more than 80% by R before reduction at the end of the irradiation to below the 2% level. Thus only a threshold level of Pfr was required over a prolonged period for the light-break response. Although this level was also generated by a short FR exposure, it was not maintained long enough to inhibit floral induction

(Kasperbauer et al., 1963b). Prolonged FR is not effective in all plants, however. Floral induction in *Pharbitis* is not prevented by such a low Pfr level.

The researchers used the *Chenopodium* response to middle-of-the-night FR light breaks to estimate phytochrome reversion rates *in vivo*. After establishing a low level of Pfr with a 5-min FR exposure, they compared the effects of prolonged FR and various light–dark cycles on floral induction. When plants were FR irradiated for 6 s every min, suppression of floral induction was slightly less than after exposure to an equal period of continuous FR. But it was greater than after irradiation for 1 min every 10 min (Table 13-1). This suggested rapid Pfr dark reversion to an ineffective level. Analysis of the data obtained with R light breaks (Kasperbauer et al., 1963b) yielded a similar value—about 1% per minute at the temperatures used in these experiments (Kasperbauer et al., 1964). Thus, although phytochrome remained in its Pfr form for long periods when suppressing stem extension and promoting seed germination, it reverted fairly rapidly during the light-break inhibition of floral induction. This was consistent with Borthwick and Cathey's explanation of why cyclic lighting lost its effectiveness in chrysanthemum as cycles lengthened.

TABLE 13-1

Average Inflorescence Stages of *Chenopodium rubrum* After Several Patterns of FR Irradiation Following an Initial 5-Min FR Exposure[a][b]

| Time of treatment after initial exposure (min) | Inflorescence stages[c] induced by various exposures[d] | | | | |
| | Continuous exposure | Cyclic—far-red 10%, darkness 90% | | | |
		1 min	10 min	20 min	30 min
0	9.0				
10	8.3	8.7	8.9		
20	7.4	8.5	8.8	8.9	
30	6.5	8.1	8.7		8.7
40	5.7	7.4	7.9	8.6	
50	2.7	7.1	7.5		
60	1.2	6.6	7.4	7.8	8.0
70	0.1	6.1	6.9		
80	0.0	4.3	6.7	7.6	
90	0.0	2.4	5.8		7.6

[a] From Kasperbauer et al., 1964. Copyright by the University of Chicago.

[b] Irradiance $710 - 800$ nm $= 0.8$ mW cm^{-2}.

[c] Inflorescence stages are described in Kasperbauer et al., 1963b.

[d] Stage of development of controls: darkness, 9.0; 5 min R, 0.0; FR first 9 min, 8.5; FR at 81–90 min, 8.4; 5 min R followed by 3 min FR, 8.7.

PFR-REQUIRING REACTION IN PHARBITIS

A quite separate effect of FR was uncovered in Japan and Beltsville. In the 1950s, Japanese workers began to study *Pharbitis nil* (Japanese morning glory), a SDP that prepares to flower as soon as its cotyledons emerge from the seed case if it experiences a single inductive dark period (Imamura, 1953). Unlike *Xanthium, Pharbitis* was greatly inhibited by a brief FR light break, and R reversed this inhibition (Nakayama, 1958; Takimoto and Ikeda, 1959a, 1960a,b). Thus *Pharbitis* appeared to require the presence of Pfr during the night, whereas *Xanthium* seemed to flower in its absence.

The discrepancy, as Shidai Nakayama suspected at Miyazaki University (Nakayama, 1958), was partly due to experimental design, for the Japanese workers gave the light break at the beginning of the dark period, whereas the Beltsville researchers routinely gave it in the middle. When Nakayama later spent a sabbatical at Beltsville, he explained that there were no growth rooms or time switches in Japan; researchers therefore preferred to irradiate plants at the end of the day rather than stay up until midnight. "Thus the Japanese experiments . . . revealed . . . that the response to Pfr action depends on conditions in the plant at the time it acts and responses can be diametrically opposite in the same plant at different times" (Borthwick, 1972).

In Beltsville, Nakayama subjected *Pharbitis* seedlings grown in continuous light to 3 SD and 3 long nights, irradiating batches on the spectrograph at different times during the dark periods. The resulting action spectra indicated that FR was the most inhibitory part of the spectrum at the beginning of the dark period, when R (maximum 660 nm) repromoted floral induction. But when seedlings were irradiated in the middle of the night, R was most inhibitory.[3] This effect was not reversed by FR, but the energy required for half-maximal effect was the same as at the beginning of the dark period, implicating phytochrome photoconversion. Thus floral induction in *Pharbitis* appeared to require the presence of Pfr at the beginning of the dark period and its absence later.

Nakayama, Borthwick, and Hendricks discovered that R became inhibitory about 4 h into 16-h dark periods, reaching maximal effectiveness at about 8 h. Realizing that the change back to reversible FR inhibition must occur sometime during the photoperiod, they studied a morning glory that could be induced by just one photoperiod. Removing seedlings from continuous light, they subjected them to 8 h darkness, a short photoperiod, and then a 16-h dark period. If the photoperiod was at least 2 h long, the

---

3. Takimoto and Ikeda, meanwhile, had reported that FR applied 8 h after the beginning of the dark period inhibits flowering maximally (Takimoto and Ikeda, 1960a). This effect was due to the impurity of their FR source, which was contaminated with R (see Takimoto and Hamner, 1965a).

FIGURE 13-6   Henri Fredericq, 1989. Photo by Linda Sage.

seedlings flowered and were fully FR/R-reversible at the beginning of the dark period (Nakayama et al., 1960; Borthwick et al., 1961). Suppression of flowering with end-of-day FR was also observed in *Lemna perpusilla* (Purves, 1961).

At the University of Ghent, plant physiologist Henri Fredericq (Fig. 13-6) discovered in 1961 that FR also inhibited *Kalanchoë blossfeldiana* after a very short (5-min) photoperiod (Fredericq, 1963). During a sabbatical at Beltsville from 1962 to 1963 (Fredericq, 1965), he studied *Pharbitis* also, obtaining this inhibition only if the photoperiod was as short as 4 h or irradiance was low during normal SD. Under these conditions, "Pfr has not completed its flower-promoting function at the close of photoperiod and must continue to act during the first part of the dark period," Fredericq decided. "This explanation is supported by the fact that the far-red inhibition is fully reversed by red at the beginning of the dark period, which reestablishes Pfr" (Fredericq, 1964).[4]

In the course of this study, Fredericq determined why the reverse response—the middle-of-the-night light-break reaction—was apparently irreversible in this species, even after normal SD. Whereas a 2-min R pulse did irreversibly inhibit flowering in the middle of a 16-h night, a 30-s light break was less effective if 30 s FR followed immediately. The apparent lack of reversibility thus appeared to "be due to the very rapid action of Pfr

4. This hypothesis, although generally accepted, does not explain why terminal FR after continuous light and before a single dark period inhibits floral induction in *Pharbitis* if the dark period exceeds 14–15 h and promotes floral induction if the dark period is shorter than 12 h (Reid et al. 1967, Evans and King, 1969).

instead of some peculiar qualitative difference between *Pharbitis* and other short-day species such as *Xanthium*'' (Fredericq, 1964).

But the quantitative difference was striking. Xanthium lost photoreversibility when 30–45 min elapsed between R and FR treatments (Downs, 1956). In *Pharbitis* seedlings, escape time was only 1–4 min (Fredericq, 1964). In *Kalanchoë*, there was considerable loss of photoreversibility after 6 min (Fredericq, 1965). "In *Chenopodium album*," says Downs, "Mancinelli and I found the escape time to be zero—we couldn't get the far-red on fast enough to get reversal even when the far-red was turned on before the red was turned off. But when the plants were chilled to 8° or 10°C during the interruption, the red still inhibited and we obtained far-red reversibility.''

XANTHIUM

The *Pharbitis* work inspired Borthwick and Downs to embark on a similar study with *Xanthium*. When they subjected plants to 1½- or 2-h photoperiods of fluorescent light or sunlight (followed by 22½ or 22 h of darkness), a brief FR exposure after the photoperiod drastically inhibited floral induction. As in *Pharbitis* seedlings, R almost completely reversed this effect. Moreover, R repromoted flowering as long as 5 h after the FR treatment. R reversal was less marked after 6–8 h and impossible after 8 h. Thus Pfr could be removed at the beginning of the night but had to be present at hour 8.

FR itself was inhibitory anytime during the first 8 h of darkness. R light breaks, in contrast, were relatively ineffective during the first 8 h but completely inhibitory 2 h later. Thus Pfr appeared to promote flowering between hours 8 and 10 and then suddenly to become inhibitory. And although FR treatments at the beginning of the dark period were reversible, both R and FR were inhibitory at 8 h.

"Under these experimental conditions (a) Pfr was present at the close of the photoperiod, and (b) Pfr, if not removed, acted during the dark period to promote flowering," the researchers concluded. Because a R light break later in the dark period was inhibitory, "(a) Pfr was not present late in the dark period, and (b) it acted to inhibit flowering if it was introduced into the plant by red light at that time. Successful flowering of a cocklebur plant thus depends on the presence and flower-promoting action of Pfr sometime during the first part of the dark period and on its absence and, therefore, lack of action later in the same dark period" (Borthwick and Downs, 1964).

CHENOPODIUM

Cumming had also suspected that Pfr was required for floral induction. His *Chenopodium* ecotypes showed optimal photoperiod responses—earliest floral initiation in different daylengths—according to their latitude of origin. Moreover, floral initiation was delayed by daylengths both shorter and longer than optimal. But he was able to compensate by changing the fraction of phytochrome that was in the Pfr form as the plant went into darkness. Thus daylengths that would otherwise be too short became more optimal if plants were irradiated at the end of the photoperiod by light with a high R:FR ratio. Conversely, overly long daylengths became more favorable if the end-of-day light pulse had a low R:FR ratio (Cumming, 1963).

Light quality during the photoperiod also influenced optimal daylength. When plants had a low proportion of their phytochrome in the Pfr form during the day, optimal induction required a longer day and shorter night than when the daytime Pfr/P_{total} was high (Cumming, 1963). In fact, two ecotypes flowered earlier if continuous, low R:FR light was not interrupted by nightly 8-h dark periods. One of these was induced to flower when low R:FR light replaced a single dark period interrupting continuous WL—even though the low R:FR light supplied sufficient energy for photosynthesis (Cumming, 1969). Cumming concluded that a specific quality and quantity of light, as well as darkness, could be inductive in at least one SDP, and he questioned "whether there is any essential dark-requiring reaction in the induction not only of long-day but also of short-day plants." He decided that "photosynthate and phytochrome-Pfr are implicated as having a positive (promotive) effect during darkness" and that the magnitude of the effect depends on the duration and concentrations of these two interacting factors (Cumming, 1969).

To pursue aspects of this work further, Cumming had visited Beltsville. Working there from 1963 to 1964, he performed 40 experiments, involving 500,000 *Chenopodium* plants—of which he dissected 100,000! The study revealed that the Pfr-requiring reaction and the light-break response may occur simultaneously during extended darkness. Scanning a 72-h dark period with a 4-min R light break, he discovered that the light-break response entailed a rhythmic promotion and inhibition of flowering by Pfr (Fig. 13-7). The Pfr-requiring reaction, revealed by a 10-s saturating FR pulse, was not rhythmic. The pulse was strongly inhibitory for the first 40 h and detrimental for up to 66 h. Thus the light break was not reversible at its most inhibitory times (4 and 34 h) because FR itself inhibited flowering (Cumming et al., 1965).

To determine how the Pfr requirement changed during a 72-h dark period, Cumming applied single 1-min light breaks with various R/FR

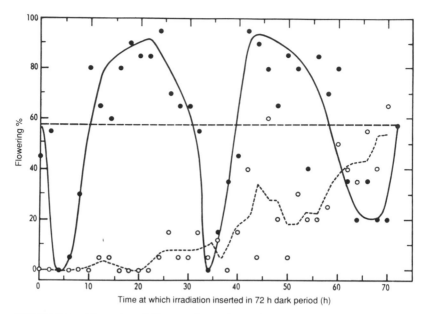

FIGURE 13-7   Flowering of *Chenopodium rubrum* after a 72-h dark period interrupted by
4 min R (solid line) or intense FR (dotted line) at one of the indicated 2-h intervals. Horizontal
line is 72-h dark control (57% flowering). (From Cumming et al., 1965.)

mixtures and noted which resulted in a null response (in which test plants
flowered to the same extent as plants not illuminated during the dark
period). The data (Fig. 13-8) suggested that Pfr levels optimal for 50%
flowering were high for the first few hours of the dark period, declined
sharply between 5 and 11 h and even more sharply between 35 and
41 h—two periods during which R light breaks were most inhibitory.
"Thus the rhythmic response of flowering according to the time that
darkness is interrupted by light," Cumming concluded, "is paralleled by a
rhythmic change in the optimum Pfr level that would be required for the
maximum flowering" (Cumming et al., 1965).

    Null response measurements in several other laboratories suggested
that high Pfr levels are required during the first few hours of darkness
(Evans and King, 1969; King and Cumming, 1972; Kato, 1979) but that low
Pfr levels are needed at certain subsequent points. Null measurements
were difficult to interpret, however, and changing sensitivity to Pfr was
often confused with Pfr levels themselves, leading to the unfounded conclusion that Pfr persists for the first few hours of darkness and then
suddenly reverts to Pr.

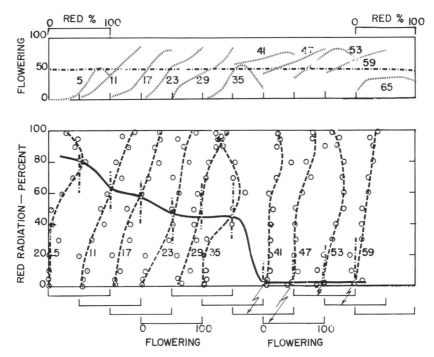

FIGURE 13-8 Flowering of *Chenopodium rubrum* in response to single 1-min exposures of narrow band R and FR at hours 5 to 65 of a 72-h dark period (broken lines). Solid horizontal (*upper figure*) and vertical (*lower figure*) lines represent the flowering of dark controls (50%). Solid curve in lower figure is drawn through the percentage R at each time that resulted in the same flowering as dark controls. Abscissa for each of the top set of curves represent R as a percentage of total R + FR mixture. Data for hour 65 omitted from lower figure because no radiation mixture resulted in flowering equivalent to dark control. (From Cumming et al., 1965.)

At Beltsville, Cumming was unable to obtain a null response at hours 5, 35, and 65. Although this result seemed inexplicable at the time (Cumming et al., 1965), it would occur if both Pfr formation and Pfr removal inhibits flowering at these times.

Salisbury had suggested that "optimal flowering in *Pharbitis* may require a mixture or balance of the pigment forms near the middle of the dark period. . . . Thus it seems possible that shifting the pigment system in either direction in the middle of the dark period might result in something incompatible with flowering" (Salisbury, 1961).

Deducing from photoreversibility experiments with *Pharbitis* that Pfr was present during the first half of a 48-h dark period for floral induction, that its concentration remained constant, but that saturation of the light-break effect required enough R to convert a much greater amount of phytochrome to Pfr, Hamner and Atsushi Takimoto decided that "red radiant energy absorbed by some pigment other than phytochrome is also effective for flower inhibition, and this effect may not be reversed by subsequent irradiation with far-red" (Takimoto and Hamner, 1965b).

A quarter century later, these two suggestions would merge into a new hypothesis: that there are two separate pigment pools, one containing the Pr involved in the light-break reaction and the other containing the flower-promoting Pfr (see Chapter 33).

Takimoto and Hamner were puzzled because, although both R and FR were almost equally inhibitory at hour 8 of a 48-h dark period, FR followed by R acted like R, whereas R followed by FR was more inhibitory than the other treatments (Takimoto and Hamner, 1965a). A possible explanation is that FR followed by R gives a R light break but only momentarily removes the Pfr needed for the Pfr-requiring reaction, whereas R followed by FR gives a light break *and* permanently removes the necessary Pfr.

LONG-DAY PLANTS

It had become apparent that flowering in LDP was also more complex than the first light-break experiments had suggested. In 1949, Borthwick received a letter from C. W. Doxtator, a sugar beet grower in Colorado who for years had produced seed during winter with the aid of supplementary incandescent greenhouse lighting. When Doxtator upgraded his facilities by building a new house with fluorescent lighting, his beets stubbornly refused to make seed stalks.

Trying to solve this puzzle, Borthwick and Parker compared the effects of supplemental fluorescent and incandescent radiation. After $2\frac{1}{2}$ months, most of the beet plants in the incandescent group had developed seed stalks, whereas hardly any of those under fluorescent lighting were seeding (Borthwick and Parker, 1952). This was unsettling because the fluorescent lamps delivered ample R, which supposedly best induced flowering.

Other workers had observed that day extensions with incandescent light or R/FR mixtures were more effective than those with fluorescent or R (Wassink et al., 1951; Takimoto, 1957; Downs et al., 1958, 1959; Lona, 1959; Meyer, 1959; Piringer and Cathey, 1960; Friend et al., 1961). In Wageningen, Stolwijk and Jan A. D. Zeevaart determined that *Hyoscyamus niger* elongated stems[5] in supplementary high-irradiance B, violet, or

---

5. Most LD form leaf rosettes under noninductive conditions. When they are exposed to LD, stem elongation and flower bud initiation occur simultaneously. Thus stem elongation is often used as an index of floral induction.

R only if infrared was added (Stolwijk and Zeevaart, 1955). Thus the quality of a daylength extension clearly had to be different from that of a flower-promoting light break. Even a 2-h FR extension at the end of a short photoperiod induced stem elongation, suggesting that high Pfr levels inhibited stem elongation under these conditions (de Lint, 1958).

At Reading University in England, Daphne Vince was also studying LDP. At the university's horticultural facility, Shinfield Grange, Vince observed that supplementary R was unable to promote flowering in carnation and lettuce unless FR was added (Vince et al., 1964). She also used a strain of darnel grass (*Lolium temulentum*) that required only a few LD for floral induction. Allowing greenhouse plants to see sunlight from 8:00 A.M. to 4:00 P.M. each day, she then "drew black covers around and over the top of the plants. With lights programmed by a time clock, we had about six fairly crude compartments (Fig. 13-9) that would give light at different times after the photoperiod."

Because equal energies of R and FR provided an effective photoperiod extension, Vince compared the effects of 8 h sunlight and then (1) 8 h R, (2) 8 h R + FR, (3) 7 h R followed by 1 h FR, (4) 7 h FR followed by 1 h R, and (5) 8 h darkness. Each of these treatments was followed by 8 h darkness (Vince, 1965). Only (2) and (4) strongly promoted flowering.

FIGURE 13-9   Daphne Vince in one of the greenhouse compartments at Shinfield Grange.

Moreover, an "extension" of 7 h darkness/1 h R had some promotive effect. This suggested that the Pfr level needed to be low during the early part of the night for floral promotion. "This low concentration of [Pfr] would be achieved during the dark period prior to the red night break as a result of the thermal reaction, [Pfr→[Pr]," Vince explained. "Keeping the [Pfr] concentration low by irradiating with far-red during the whole of the 7 hours prior to the night break with red light is equally effective in promoting flowering. Far-red during the whole of the 8-hour period is, however, without effect on flowering so that a high concentration of [Pfr] is necessary at some stage" (Vince, 1965).

The application of single 1-h R light breaks between 9:00 P.M. and 2:00 A.M. revealed that R promoted flowering most effectively between midnight and 2:00 A.M. (Fig. 13-10), suggesting that "following a short day in sunlight, [Pfr] becomes increasingly inhibitory to flowering up to about 8 P.M., and then is increasingly promoting" (Vince, 1965). Thus LDP appeared to change their sensitivity to both R and FR.

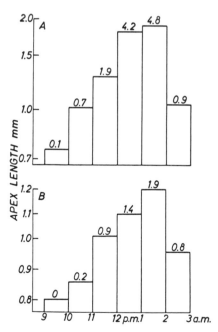

FIGURE 13-10   Effect of timing of a R light break. (A) *Lolium* plants were exposed to FR from 4:00 P.M. until the beginning of the light break; (B) darkness preceded the light break. Numbers indicate mean floral stage reached in each treatment. (From Vince, 1965.)

When Vince submitted this paper in September 1964, the Beltsville group was studying the LD responses of sugar beet, annual henbane, dill, barley, petunia, and *Lolium*. Using 15-min light breaks in the middle of 16-h dark periods, Lane, Cathey, and an Australian visitor named Lloyd T. Evans were unable to induce flowering, even in henbane (*Hyoscyamus niger*) and barley (*Hordeum vulgare*), in which the Beltsville workers had first demonstrated the effects of a light break on LDP (see Chapter 4). Thus a brief light break was effective only when plants were grown in near-inductive light–dark cycles (13/11 for *Hyoscyamus*) rather than on 8-h days and 16-h nights. Under the latter conditions, 4-h light breaks in the middle of the dark period were required, although 8-h extensions to the photoperiod were more effective. This suggested that "optimal induction in long-day plants requires the continued action of light over extended periods" (Lane et al., 1965).

Light sources with intermediate R/FR ratios provided the most effective photoperiod extension, but effectiveness also depended on extension length. Thus "the optimum Pfr level may vary in the course of a photoperiod extension" (Lane et al., 1965). The responsiveness of barley and the Ceres strain of *Lolium temulentum* on one-cycle induction did not change with time—only R/FR mixtures or sequential application of R and FR in 10–30-min cycles caused rapid flowering (Evans et al., 1965; Lane et al., 1965). But Evans later showed, through spectrographic and null response experiments, that Ceres *Lolium* requires phytochrome to be mostly in the Pr form during the early hours of darkness and then in the Pfr form later, although the requirement was less marked than with the *Lolium* strain studied by Vince (Evans, 1976). This sequence was the reverse of that required by SDP.

ACTION OF FAR-RED

Vince proposed that FR day extensions promote flowering by lowering Pfr levels. Thus 15 min FR at the beginning of a day extension followed by 105 min darkness was just as effective as 2 h FR, and the effect was R-reversible (Vince, 1966). Moreover, the insertion of a dark period alone, without any FR, into a R extension promoted flowering in *Lolium,* and spike length was proportional to the duration of the dark period. If a FR pulse preceded the interruption, it replaced 45 min darkness, suggesting that Pfr decayed or reverted to a noninhibitory level in that time. This level was established by a 25% R/75% FR mixture. R reversed the effect of the initial FR pulse, and FR reversed the inhibitory effect of a R pulse in the middle of a 2-h dark interruption (Holland and Vince, 1971).

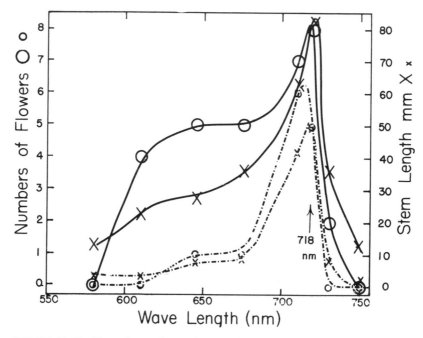

FIGURE 13-11 Flowering and stem lengthening of *Hyoscyamus niger* in response to indicated wavelengths at a single fluence during middle 8 h of daily 16-h dark periods. Solid lines and broken lines represent separate lots of plants in an experiment. (From Schneider et al., 1967.)

By 1966, the Beltsville workers were "beginning to think in terms of other photoreactions in connection with long-day plants" (Borthwick, 1966). Two years earlier, floral promotion in wheat by FR had been attributed to the HER (Friend, 1964). Two light reactions had also been suggested for floral induction in *Lemna* (Esashi and Oda, 1966).

The Beltsville group obtained the first action spectrum for LDP induction under prolonged irradiation by irradiating *Hyoscyamus niger* on the spectrograph for 8 h in the middle of 16-h dark periods. The curve contrasted sharply with those obtained using a brief light break and a shorter photoperiod (See Chapter 4), because the most effective wavelengths lay between 710 and 720 nm (Fig. 13-11). The response also failed to saturate at the highest fluence tested, another characteristic of an HER (Schneider et al., 1967).

At the Phytotron of the Centre National de la Recherche Scientifique, Francois Blondon and Roger Jacques induced flowering in *Lolium temulentum* with 15-h photoperiod extensions at 700, 710, or 725 nm. Because added R displaced the action peak further into the FR, the researchers decided that "Il semble donc logique d'associer l'assez faible pourcent-age en phytochrome actif et l'induction florale des plantes de jours longs"[6] (Blondon and Jacques, 1970). Thus, despite the bizarre action spectrum, phytochrome appeared to be involved in the flowering re-sponse, although it seemed not to act simply as Pfr.

Phytochrome thus controlled LDP flowering "in a manner not under-stood at present" (Evans et al., 1965). "Our experiments," Evans says, "often required Hendricks and I to run the spectrograph *throughout* the night, so we had plenty of time to discuss everything between changing the carbon arcs at regular intervals. We debated many possible explanations of the light requirement by LDP and explored several of them in our papers. But we were not convinced by any of them. I revisited this prob-lem subsequently. Sterling also kept thinking about it in later years, and it is some comfort to know that its solution escaped even his sharp mind."

REFERENCES

Blondon, F., and Jacques, R. (1970). Action de la lumière sur l'initiation florale du *Lolium temulentum* L.: Spectre d'action et rôle du phytochrome. *C. R. Acad. Sci. Ser. D* 270,947–950.

Borthwick, H. A. (1966). Letter to Bill Hillman, July 25. Unpublished.

Borthwick, H. A. (1972). The biological significance of phytochrome. *In* "Phytochrome" (K. Mitrakos and W. Shropshire, Jr., eds.), pp. 27–44. Academic Press, London, New York.

Borthwick, H. A., and Cathey, H. M. (1962). Role of phytochrome in control of flowering of chrysanthemum. *Bot. Gaz.* 123,155–162. Copyright by the University of Chicago.

Borthwick, H. A., and Downs, R. J. (1964). Roles of active phytochrome in control of flowering of *Xanthium pensylvanicum*. *Bot. Gaz.* 125,227–231. Copyright by the Univer-sity of Chicago.

Borthwick, H. A., and Parker, M. W. (1952). Light in relation to flowering and vegetative development. *Proc. Int. Hort. Congr. 13th*, pp. 801–810.

Borthwick, H. A., Nakayama, S., and Hendricks, S. B. (1961). Failure of reversibility of the photoreaction controlling plant growth. *Prog. Photobiol. Proc. Int. Congr. Photobiol. 3rd*, pp. 394–398.

Borthwick, H. A., Toole, E. H., and Toole, V. K. (1964). Phytochrome control of *Paulownia* seed germination. *Isr. J. Bot.* 13,122–133.

Butler, W. L., Lane, H. C., and Siegelman, H. W. (1963). Nonchemical transformations of phytochrome *in vivo*. *Plant Physiol.* 38,514–519.

Cathey, H. M., and Borthwick, H. A. (1957). Photoreversibility of floral initiation in chry-santhemum. *Bot. Gaz.* 119,71–76.

Cathey, H. M., and Borthwick, H. A. (1961). Cyclic lighting for controlling flowering of chrysanthemums. *Proc. Am. Soc. Hort. Sci.* 78,545–552.

Cathey, H. M., and Borthwick, H. A. (1962). Cyclic lighting found economical and effective. *Florists' Rev.* 131.

6. It seems logical to associate a fairly low percentage of Pfr with floral induction in LD.

Cathey, H. M., and Borthwick, H. A. (1964). Significance of dark reversion of phytochrome in flowering of *Chrysanthemum morifolium*. *Bot. Gaz.* 125,232–236.

Cathey, H. M., Bailey, W. A., and Borthwick, H. A. (1961). Cyclic lighting—To reduce cost of timing chrysanthemum flowering. *Florists' Rev.* 129,21.

Cumming, B. G. (1959a). Extreme sensitivity of germination and photoperiodic reaction in the genus *Chenopodium* (Tourn.) L. *Nature* 184,1044–1045.

Cumming, B. G. (1959b). Light and temperature sensitivity in germination and very early floral initiation of some *Chenopodium* species. *Proc. Int. Bot. Congr. 9th* 2,83.

Cumming, B. G. (1963). Evidence of a requirement for phytochrome-Pfr in the floral initiation of *Chenopodium rubrum*. *Can J. Bot.* 41,901–926.

Cumming, B. G. (1967). Early-flowering plants. *In* "Methods in Developmental Biology" (F. H. Wilt and N. K. Wessells, eds.), pp. 277–299. Thomas Y. Crowell, New York.

Cumming, B. G. (1969). Photoperiodism and rhythmic flower induction: Complete substitution of inductive darkness by light. *Can J. Bot.* 47,1241–1250.

Cumming, B. G., Hendricks, S. B., and Borthwick, H. A. (1965). Rhythmic flowering responses and phytochrome changes in a selection of *Chenopodium rubrum*. *Can. J. Bot.* 43,825–853.

Danielson, H. R., and Toole, V. K. (1976). Action of temperature and light on the control of seed germination in Alta tall fescue (*Festuca arundinacea* Schreb.) *Crop Sci.* 16,317–320.

de Lint, P. J. A. L. (1958). Stem formation in *Hyoscyamus niger* under short days including supplementary irradiation with near infra-red. *Meded. Landbouwhogesch. Wageningen* 58(10),1–6.

Downs, R. J. (1956). Photoreversibility of flower initiation. *Plant Physiol.* 31,279–284.

Downs, R. J., Borthwick, H. A., and Piringer, A. A. (1958). Comparison of incandescent and fluorescent lamps for lengthening photoperiods. *Proc. Am. Soc. Hort. Sci.* 71,568–578.

Downs, R. J., Piringer, A. A., and Wiebe, G. A. (1959). Effects of photoperiod and kind of supplemental light on growth and reproduction of several varieties of wheat and barley. *Bot. Gaz.* 120,170–177.

Esashi, Y., and Oda, Y. (1966). Two light reactions in the photoperiodic control of flowering of *Lemna perpusilla* and *Lemna gibba*. *Plant Cell Physiol.* 7,59–74.

Evans, L. T. (1976). Inflorescence initiation in *Lolium temulentum* L. XIV. The role of phytochrome in long day induction. *Austr. J. Biol. Sci.* 3,207–217.

Evans, L. T., and King, R. W. (1969). Role of phytochrome in photoperiodic induction of *Pharbitis nil*. *Z. Pflanzenphysiol.* 60,277–288.

Evans, L. T., Borthwick, H. A., and Hendricks, S. B. (1965). Inflorescence initiation in *Lolium temulentum* L. VII. The spectral dependence of induction. *Austr. J. Biol. Sci.* 18,745–762.

Fredericq, H. (1963). Flower formation in *Kalanchoë blossfeldiana* by very short photoperiods under light of different quality. *Nature* 198,101–102.

Fredericq, H. (1964). Conditions determining effects of far-red and red irradiations on flowering response of *Pharbitis nil*. *Plant Physiol.* 39,812–816. Quotation reprinted by permission from the American Society of Plant Physiologists.

Fredericq, H. (1965). Action of red and far-red light at the end of the short day, and in the middle of the night, on flower induction in *Kalanchoë blossfeldiana*. *Biol. Jaarb.* 33,66–91.

Friend, D. J. C. (1964). The promotion of floral initiation of wheat by far-red radiation. *Physiol. Plant* 17,909–920.

Friend, D. J. C., Helson, V. A., and Fisher, J. E. (1961). The influence of the ratio of incandescent to fluorescent light on the flowering response of Marquis wheat grown under controlled conditions. *Can. J. Plant Sci.* 41,418–427.

Hendricks, S. B., Toole, E. H., Toole, V. K., and Borthwick, H. A. (1959). Photocontrol of plant development by the simultaneous excitations of two interconvertible pigments. III. Control of seed germination and axis elongation. *Bot. Gaz.* 121,1–8.

Hendricks, S. B., Toole, V. K., and Borthwick, H. A. (1968). Opposing actions of light in seed germination of *Poa pratensis* and *Amaranthus arenicola*. *Plant Physiol.* 43,2023–2028.

Hillman, W. S. (1962). "The Physiology of Flowering." Holt, Rinehart and Winston, New York.

Holland, R. W. K., and Vince, D. (1971). Floral initiation in *Lolium temulentum* L.: the role of phytochrome in the responses to red and far-red light. *Planta* 98,232–243.

Ikuma, H., and Thimann, K. V. (1963). The role of seed-coats in germination of photosensitive seeds. *Plant Cell Physiol.* 4,169–185.

Imamura, S. (1953). Photoperiodic initiation of flower primordia in Japanese morning glory, *Pharbitis nil* Chois. *Proc. Jpn. Acad. Ser. B* 29,368–373.

Isikawa, S. (1954). Light sensitivity against germination. I. "Photoperiodism" of seeds. *Bot. Mag.* 67,51–56.

Kasperbauer, M. J., and Borthwick, H. A. (1964). Photoreversibility of stem elongation in *Melilotus alba* Desr. *Crop Sci.* 4,42–44.

Kasperbauer, M. J., Borthwick, H. A., and Cathey, H. M. (1963a). Cyclic lighting for promotion of flowering of sweetclover, *Melilotus alba* Desr. *Crop Sci.* 3,230–232.

Kasperbauer, M. J., Borthwick, H. A., and Hendricks, S. B. (1963b). Inhibition of flowering of *Chenopodium rubrum* by prolonged far-red radiation. *Bot. Gaz.* 124,444–451.

Kasperbauer, M. J., Borthwick, H. A., and Hendricks, S. B. (1964). Reversion of phytochrome 730 (Pfr) to P660 (Pr) assayed by flowering in *Chenopodium rubrum. Bot. Gaz.* 125,75–80.

Kato, A. (1979). Effect of interruption of the nyctoperiod with an R/FR mixture of various ratios of red and far-red light in *Lemna gibba* G3. *Plant Cell Physiol.* 20,1273–1283.

King, R. W., and Cumming, B. G. (1972). The role of phytochrome in photoperiodic time measurement and its relation to rhythmic timekeeping in the control of flowering in *Chenopodium rubrum. Planta* 108,39–57.

Lane, H. C., Cathey, H. M., and Evans, L. T. (1965). The dependence of flowering in several long-day plants on the spectral composition of light extending the photoperiod. *Am. J. Bot.* 52,1006–1014.

Lona, F. (1959). Some aspects of photothermal and chemical control of growth and flowering. In "Photoperiodism" (R. B. Withrow, ed.), pp. 351–358. American Association for the Advancement of Science, Washington, D.C.

Mancinelli, A. L., and Borthwick, H. A. (1964). Photocontrol of germination and phytochrome reaction in dark-germinating seeds of *Lactuca sativa* L. *Ann. Bot. (Rome)* 28,9–24.

Mancinelli, A. L., and Tolkowsky, A. (1968). Phytochrome and seed germination. V. Changes of phytochrome content during the germination of cucumber seeds. *Plant Physiol.* 43,489–494.

Mancinelli, A. L., Borthwick, H. A., and Hendricks, S. B. (1966). Phytochrome action in tomato-seed germination. *Bot. Gaz.* 127,1–5.

Mancinelli, A. L., Yaniv, Z., and Smith, P. (1967). Phytochrome and seed germination. I. Temperature dependence and relative Pfr levels in the germination of dark-germinating tomato seeds. *Plant Physiol.* 42,333–337.

Meyer, G. (1959). The spectral dependence of flowering and elongation. *Acta Bot. Neerl.* 8,189–246.

Nagao, M., Esashi, Y., Tanaka, T., Kumagai, T., and Fukumoto, S. (1959). Effects of photoperiod and gibberellin on the germination of seeds of *Begonia Evansiana* ANDR. *Plant Cell Physiol* 1,39–47.

Nakayama, S. (1958). Photoreversible control of flowering at the start of inductive dark period in *Pharbitis nil. Ecol. Rev.* 14,325–326.

Nakayama, S., Borthwick, H. A., and Hendricks, S. B. (1960). Failure of photoreversible control of flowering in *Pharbitis nil. Bot. Gaz.* 121,237–243.

Piringer, A. A., and Cathey, H. M. (1960). Effects of photoperiod, kind of supplemental light, and temperature on growth and flowering of petunia plants. *Proc. Am. Soc. Hort. Sci.* 76,649–660.

Piringer, A. A., Downs, R. J., and Borthwick, H. A. (1963). Photocontrol of growth and flowering of *Caryopteris. Am. J. Bot.* 50,86–90.

Purves, W. K. (1961). Dark reactions in the flowering of *Lemna perpusilla* 6746. *Planta* 56,684–690.

Reid, H. B., Moore, P. H., and Hamner, K. C. (1967). Control of flowering of *Xanthium pensylvanicum* by red and far-red light. *Plant Physiol.* 42,532–540.

Salisbury, F. B. (1961). Photoperiodism and the flowering process. *Annu. Rev. Plant Physiol.* 12,293–326.

Schneider, M. J., Borthwick, H. A., and Hendricks, S. B. (1967). Effects of radiation on flowering of *Hyoscyamus niger. Am. J. Bot.* 54,1241–1249.

Schwabe, W. W. (1956). Evidence for a flowering inhibitor produced in long days in *Kalanchoë blossfeldiana. Ann. Bot. (Lond.)* 20,1–14.

Schwabe, W. W. (1959). Studies of long-day inhibition in short-day plants. *J. Exp. Bot.* 10,317–329.

Shropshire, W., Jr., Klein, W. H., and Elstad, V. B. (1961). Action spectra of photomorphogenic induction and photoinactivation of germination in *Arabidopsis thaliana. Plant Cell Physiol.* 2,63–69.

Stolwijk, J. A. J., and Zeevaart, J. A. D. (1955). Wave length dependence of different light reactions governing flowering in *Hyoscyamus niger. Proc. K. Ned. Akad. Wet. Ser. C.* 58,386–396.

Takimoto, A. (1957). Photoperiodic induction in *Silene armeria* as influenced by various light sources. *Bot. Mag.* 70,312–321.

Takimoto, A., and Hamner, K. C. (1965a). Effect of far-red light and its interaction with red light in the photoperiodic response of *Pharbitis nil. Plant Physiol.* 40,859–864.

Takimoto, A., and Hamner, K. C. (1965b). Kinetic studies on pigment systems concerned with the photoperiodic response in *Pharbitis nil. Plant Physiol.* 40,865–872.

Takimoto, A., and Ikeda, K. (1959a). Studies on the light controlling flower initiation of *Pharbitis nil.* I. Intensity and quality of the light preceding the inductive dark period. *Bot. Mag.* 72,137–145.

Takimoto, A., and Ikeda, K. (1959b). Studies on the light controlling photoperiodic induction of *Pharbitis nil.* II. Effect of far-red light preceding the inductive dark period. *Bot. Mag.* 72,181–189.

Takimoto, A., and Ikeda, K. (1959c). Studies on the light controlling flower initiation of *Pharbitis nil.* III. Light-sensitivity of the first process of inductive dark period. *Bot. Mag.* 72,388–396.

Takimoto, A., and Ikeda, K. (1959d). Studies on the light controlling flower initiation in *Pharbitis nil.* IV. Further studies on the light preceding the inductive dark period. *Bot. Mag.* 73,37–43.

Takimoto, A., and Ikeda, K. (1960a). Studies on the light controlling flower initiation of *Pharbitis nil.* VII. Light break. *Bot. Mag.* 73,341–348.

Takimoto, A., and Ikeda, K. (1960b). Studies on the light controlling flower initiation of

Pharbitis nil. VIII. Light-sensitivity of the inductive dark process. *Bot. Mag.* 73,468–473.

Toole, E. H. (1961). The effect of light and other variables on the control of seed germination. *Proc. Int. Seed Test. Assoc.* 26,659–673.

Toole, E. H., Borthwick, H. A., Hendricks, S. B., and Toole, V. K. (1953). Physiological studies of the effects of light and temperature on seed germination. *Proc. Int. Seed Test. Assoc.* 18,267–276.

Toole, E. H., Toole, V. K., Borthwick, H. A., and Hendricks, S. B. (1955a). Photocontrol of *Lepidium* seed germination. *Plant Physiol.* 30,15–21.

Toole, E. H., Toole, V. K., Borthwick, H. A., and Hendricks, S. B. (1955b). Interaction of temperature and light in germination of seeds. *Plant Physiol.* 30,473–478.

Toole, E. H., Toole, V. K., Hendricks, S. B., and Borthwick, H. A. (1957). Effect of temperature on germination of light-sensitive seeds. *Proc. Int. Seed Test. Assoc.* 22,196–204.

Toole, V. K. (1973). Effects of light, temperature and their interactions on the germination of seeds. *Seed Sci. Technol.* 1,339–396.

Toole, V. K. (1976). Light and temperature control of germination in *Agropyron smithii* seeds. *Plant Cell Physiol.* 17,1263–1272.

Toole, V. K., and Borthwick, H. A. (1968a). The photoreaction controlling seed germination in *Eragrostis curvula*. *Plant Cell Physiol.* 9,125–136.

Toole, V. K., and Borthwick, H. A. (1968b). Light responses of *Eragrostis curvula* seed. *Proc. Int. Seed Test. Assoc.* 33,515–530.

Toole, V. K., and Borthwick, H. A. (1971). Effect of light, temperature, and their interactions on germination of seeds of Kentucky bluegrass (*Poa pratensis* L.). *J. Am. Soc. Hort. Sci.* 96,301–304.

Toole, V. K., and Koch, E. J. (1977). Light and temperature controls of dormancy and germination in bentgrass seeds. *Crop Sci.* 17,806–811.

Toole, V. K., Toole, E. H., Hendricks, S. B., Borthwick, H. A., and Snow, A. G., Jr. (1961). Reponses of seeds of *Pinus virginiana* to light. *Plant Physiol.* 36,285–290.

Toole, V. K., Toole, E. H., Borthwick, H. A., and Snow, A. G., Jr. (1962). Responses of seeds of *Pinus taeda* and *P. strobus* to light. *Plant Physiol.* 37,228–233.

Vince, D. (1965). The promoting effect of far-red light on flowering in the long day plant *Lolium temulentum*. *Physiol. Plant.* 18,474–482.

Vince, D. (1966). An interpretation of the promoting effect of far-red light on the flowering of long-day plants. *Photochem. Photobiol.* 5,449–450.

Vince-Prue, D. (1975). "Photoperiodism in Plants." McGraw-Hill, London.

Vince, D., Blake, J., and Spencer, R. (1964). Some effects of wave-length of the supplementary light on the photoperiodic behavior of the long-day plants, carnation and lettuce. *Physiol. Plant.* 17,119–125.

Wassink, E. C., Stolwijk, J. A. J., and Beemster, A. B. R. (1951). Dependence of formative and photoperiodic reactions in *Brassica Rapa* var., *Cosmos* and *Lactuca* on wavelength and time of irradiation. *Proc. K. Ned. Akad. Wet. Ser. C* 54,421–432.

Waxman, S. (1961a). Connecticut tests reveal effectiveness of mum flashlighting. *Florists' Rev.* 127,142–144.

Waxman, S. (1961b). Flashing lights cut lighting costs by 86 per cent. *The Grower* 56,252–253.

Waxman, S. (1963). Flashlighting chrysanthemums. Progress Report 54, Univ. Conn. Agric. Exp. Stn.

Yaniv, Z., Mancinelli, A. L., and Smith, P. (1967). Phytochrome and seed germination. III. Action of prolonged far red irradiation on the germination of tomato and cucumber seeds. *Plant Physiol.* 42,1479–1482.

CHAPTER 14

Photoperiodic Timing

" 'Photo' is clean-cut. . . . 'Period' befuddled us," Hendricks admitted. "We were overly impressed by the operation of the photoreversible system. To view the whole is possibly the first principle of physiology, the neglect of which is to one's peril" (Hendricks, 1970).

Hendricks and Borthwick had proposed in the 1950s that dark reversion of phytochrome was the timing mechanism in photoperiodism (see Chapter 6). According to their elegant scheme, Pfr behaved like sand in an hourglass, reverting to Pr after the end of a photoperiod. During a long night, reversion pushed the Pfr level low enough to permit floral induction in SDP but prevent induction in LDP. During short nights, the level would remain in the range that induced LDP but would always be too high for SDP induction.

The hourglass hypothesis was a major theme in the early 1960s (see Borthwick and Hendricks, 1960; Hendricks, 1961, 1964; Borthwick, 1964). But there were several objections. First, dark reversion had never been observed in monocots (see Chapter 12). Second, spectrophotometric studies—albeit with etiolated seedlings—suggested that reversion would occupy only a small fraction of the night. Third, hastening Pfr to Pr conversion in *Pharbitis* with end-of-day FR appeared to have little effect on critical night length or time of maximal sensitivity to a light break (LB_{max}) (Takimoto and Hamner, 1965). Flowering response did increase with the length of the dark period, and LB_{max} was less temperature-sensitive than critical night length, suggesting that, in some plants an "hourglass" might be a subsidiary component of the timing mechanism (Takimoto and Hamner, 1964).

During a 1962–1963 sabbatical at the University of Tübingen, Salisbury did "what seemed to me," he says, "to be an absolutely conclusive experiment demonstrating that hourglass timing simply would not account for time measurement in photoperiodism." Placing *Xanthium* plants at

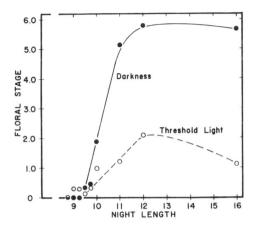

FIGURE 14-1   Flowering response of *Xanthium* to night length as influenced by threshold light. (From Salisbury, 1963. Copyright 1963, Springer-Verlag.)

different distances from a small incandescent lamp during the whole of a 16-h "dark" period, he determined a "threshold light" level that maintained enough phytochrome in its active form to reduce floral induction drastically—from stage 6.0 to 1.5. When Salisbury then exposed plants to threshold light during dark periods that differed in duration, the critical night length was the same for both the test plants and the controls that received genuine dark periods (Fig. 14-1). So although threshold light prevented Pfr from converting completely to Pr, it had no effect whatsoever on dark timing. "This is incompatible with a mechanism of time measurement involving conversion of one form of phytochrome to the other in the dark," Salisbury pointed out (Salisbury, 1963). He believed (Salisbury, 1961) that photoperiodic time measurement involved an internal oscillator, as Bünning had suggested (see Chapter 2).

CIRCADIAN CLOCK

Bünning's proposal was generally scorned by plant physiologists until the late 1950s because several studies appeared to argue against it (e.g., Hussey, 1954; Wareing, 1954; Kribben, 1955). But at the Max-Planck Institute for Biology in Tübingen (the renamed Kaiser Wilhelm Institute), Lang and H. Claes had subjected the LDP *Hyoscyamus niger* to 7 h light/41 h dark cycles. Two-hour light breaks applied to successive groups of plants induced flowering most effectively near the beginning and end of the dark period, but not in the middle, as would happen in a 24-h cycle (Claes and

Lang, 1947). Dennis J. Carr at the University of Manchester obtained similar results with the SDP *Xanthium saccharatum* and *Perilla ocymoides,* except that the light breaks were inhibitory (Carr, 1952a).

Two possible explanations were that either a plant's response to light changed rhythmically or a light break interacted with a main photoperiod when the two were sufficiently close together (Claes and Lang, 1947). In the summer of 1952, Carr visited Tübingen to distinguish between these possibilities. According to Bünning's hypothesis, there would be three opportunities in a 65-h dark period for a light break to inhibit maximally the flowering of an SDP. But if Claes and Lang's explanation were correct, there would still be only two.

With *Kalanchoë blossfeldiana* (Fig. 14-2), Carr obtained three peaks of inhibition (Carr, 1952b). Thus he declared Bünning's theory "finally and decisively proved" (Carr, 1952b). But many botanists disagreed. One of the problems was that "the theory of Bünning was something of an exercise in mysticism. Biological phenomena were to be explained in physicochemical terms. Thus unknown precursors, intermediates, and final flowering substances were postulated. Bünning, on the other hand, spoke only of phases that loved either the light or the darkness" (Salisbury, 1961).

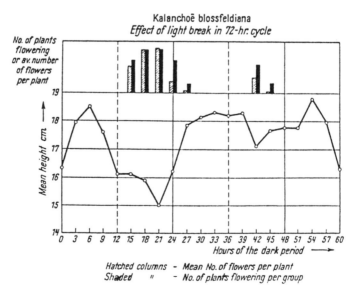

FIGURE 14-2   Effects of 2-h light breaks given to *Kalanchoë blossfeldiana* during the 65-h dark period of a 72-h cycle. (From Carr, 1952b.)

At the Gatlinburg symposium, Bünning offered a modified version of his hypothesis. "The regulation of the endogenous cycles by the light–dark periods, as is well known, makes the short-day plants scotophil in the second half of the day, the long-day plants photophil," he said. "But now we understand that this is not due to a 12-hr phase difference of the basic cycle in the two types. All our experiments with different species of both types show that in long-day and short-day plants the same biochemical features prevail in parts of the cycle which are comparable with respect to their time position within the light–dark periods" (Bünning, 1959). He proposed that LDP flower when daylight extends into the scotophilic half of each cycle and that SDP flower when it is restricted to the photophilic half (Fig. 14-3). This left the problem of why the two groups reacted differently to illumination of the "scotophilic" phase.

In 1959, Bünning attended the International Botanical Congress in Montreal. "Several leaders in plant physiology got up and said, in essence, that while the theory might be true, it wasn't very important" (Hamner and Takimoto, 1964).

At the 1960 Cold Spring Harbor Symposium, Bünning (Bünning, 1960a,b) heard a lecture by Colin Pittendrigh from Princeton University.

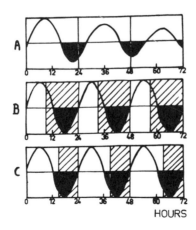

FIGURE 14-3 Bünning's revised hypothesis of photoperiodic timing. The clock causes alternation of half cycles with qualitatively different light sensitivity (white versus black). (A) Free-running clock in continuous light or continuous darkness. Cycles not exactly 24 h; (B) short-day conditions; (C) long-day conditions, which allow light to fall into the "black" half cycle. (From Bünning, 1960b. Copyright 1961 The Biological Laboratory, Long Island Biological Association, Inc.)

After extensive studies of a directly observable rhythm (pupal emergence) in *Drosophila*, Pittendrigh had concluded that "it is a fact—not a matter of discussion—that the circadian system does measure photoperiod" (Pittendrigh, 1960). He later decided that *scotophilic phase* was an imprecise term because only a fraction of that half cycle was sensitive to the inductive (or inhibitory) effects of light (Pittendrigh and Minis, 1964). Pittendrigh termed this fraction the *photoperiodically inducible phase* (Pittendrigh, 1966).

Pittendrigh emphasized that light has two quite separate effects: to induce or inhibit a response at this specific point in the cycle, and to entrain the rhythm (Pittendrigh and Minis, 1964). Regarding entrainment, he complained that "it is not clear that the problem has been fully recognized as especially pertinent." Bünning's model (Fig. 15-3) assumed "that the rhythm which measures the photoperiod is locked to dawn (the dark–light transition) and is itself photoperiod-independent. In fact, it is not: the phase of circadian systems is strongly dependent on the photoperiod of the driving light cycle" (Pittendrigh and Minis, 1964).

EVIDENCE FOR CIRCADIAN TIMING

Evidence for the operation of a circadian clock in plants came from

1. application of light breaks at different times during a long dark period;
2. alteration of the length of the "daily" cycle—the Nanda-Hamner protocol (Hamner, 1961);
3. use of skeleton photoperiods; and
4. studies of rhythms that could be observed in real-time.

Applying the first approach, Hamner's group at the University of California, Los Angeles, exposed soybean to seven 8 h light/64 h dark cycles, interrupting each extended period of darkness once with 4 h high-irradiance WL. The ingenious use of "photocyclers"—electronically controlled filing cabinets—automated the irradiations (Sirohi and Hamner, 1960). Thus drawers containing plants rolled open when a light break was required. This approach revealed (Fig. 14-4) that "the amount of flowering in Biloxi soybean plants is determined by when the plant receives light in relation to an endogenous circadian rhythm. . . . If phytochrome is the critical component, the results clearly indicate that the effects of Pfr may be inhibitory, innocuous, or stimulatory to flowering depending upon its time of occurrence in the cycle" (Coulter and Hamner, 1964). Cumming's work with *Chenopodium* proved this point more rigorously, because he

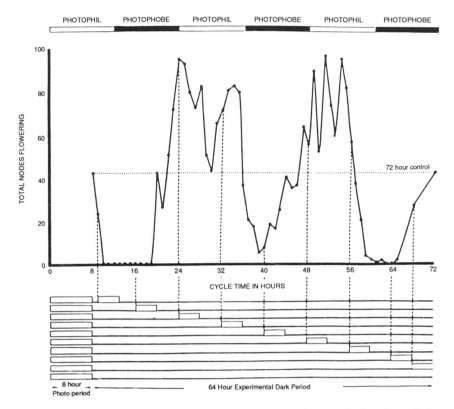

FIGURE 14-4   Four-hour high-irradiance WL breaks applied to Biloxi soybean during dark period of a 72-h cycle. Each point is plotted at beginning of a light break. (From Coulter and Hamner, 1964. Reproduced with permission from the American Society of Plant Physiologists.)

obtained a rhythmic response to light breaks using R, rather than WL (see Fig. 13-7).[1]

The UCLA researchers also exposed Biloxi soybean to cycles of different lengths in which the photoperiod always lasted for 8 h. Whereas 35- and 60-h cycles inhibited flowering, 24-, 48-, and 72-h cycles were optimal (Hamner and Takimoto, 1964). They obtained similar results with the LDP *Hyoscyamus niger,* but the rhythm was 12 h out of phase with that in soybean (Hsu and Hamner, 1967). Thus light–dark cycles apparently had to synchronize with an internal rhythm in both photoperiodic classes of plants.

1. This figure clearly shows that, although a light break inhibits flowering in SDP, it may also promote it if suitably timed.

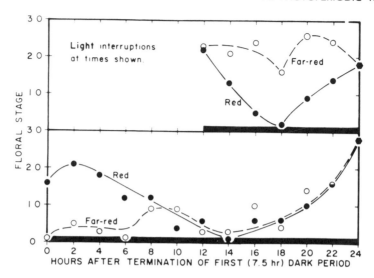

FIGURE 14-5   Floral stage as a function of times of 5-min R or FR light interruptions after a 7.5-h subcritical dark period terminated by 12 h (*top*) or 2 min (*bottom*) light. (From Salisbury, 1965. Copyright 1965, Springer-Verlag.)

Xanthium did not respond rhythmically to light breaks but exhibited a single inhibitory point 8 h after transfer to darkness (Moore et al., 1967). But the dark period did not appear to be timed simply by Pfr and Pr reversion. At Colorado State University, Salisbury subjected cocklebur to a suboptimal dark period followed by 2 min saturating WL (group a) or 12 h WL (group b) and then extended darkness interrupted by a R or FR light break. *Xanthium* is usually most sensitive to a light break 8 h after the beginning of an inductive dark period. But in this case it reached maximal sensitivity after 14 h in group a (Fig. 14-5, lower curve) but after only 6 h in group b (Fig. 14-5, upper curve). Thus "counting from the *beginning of the second dark period,* timing (the time required to reach maximum sensitivity to a red light interruption) is delayed in the long dark period compared to the shorter one, but counting from the *end of the first dark period* (real time), the reverse is true!" Salisbury concluded that "time measurement in the flowering process may well involve an oscillating system, during one phase of which light promotes, during the other phase of which light is inhibitory" (Salisbury, 1965).

When Salisbury gave cocklebur a noninductive dark period of at least 6 h followed by an intervening light period and then a test dark period, the beginning of the light period acted as dawn, creating maximum sensitivity to light 14 h later, at 22 h real-time, regardless of whether the light period lasted for seconds or up to 5 h (Fig. 14-6). But if the light period exceeded

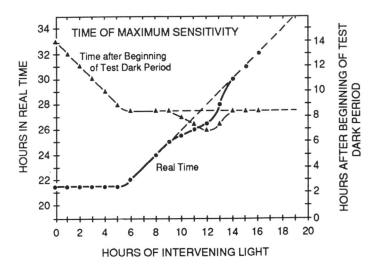

FIGURE 14-6   Time of maximum sensitivity to light during a test dark period in real-time (heavy line/left ordinate) or after beginning of the test period (lighter line/right ordinate) as a function of length of intervening light period. (From Papenfuss and Salisbury, 1967. Reproduced with permission from the American Society of Plant Physiologists.)

about 5 h, the clock was rephased at "dusk," producing maximum sensitivity about 8 h after the return to darkness. "We might imagine that the clock has been 'suspended' after 5 hours until it is 'started' by a dusk signal," Salisbury suggested (Papenfuss and Salisbury, 1967). This "concept of rephasing makes it possible to explain most light interruption experiments on the basis of effects on the clock alone." He pointed out that if a clock allowed the flowering stimulus to become available only at 14 h after dawn or 8.5 h after dusk, a light break might rephase the rhythm so that the stimulus would not be produced before the rhythm was rephased again at the beginning of the next photoperiod (Papenfuss and Salisbury, 1967). This suggestion was controversial, but the notion of entrainment at dawn, suspension during the light period, and then release at dusk made Salisbury "the first person who really got to grips with how the rhythm measures time" (D. Vince-Prue, personal communication).

SKELETON PHOTOPERIODS

Studying *Lemna perpusilla 6746,* Hillman initially discounted Bünning's hypothesis, producing instead a complex scheme involving both phytochrome reversion and destruction to explain photoperiodic responses to abnormal light–dark cycles (Hillman, 1963).

Lemna perpusilla (now *Lemna paucicostata*) is a SD duckweed that fails to flower when photoperiods exceed 11 h. Cultured in sucrose-containing medium, it can survive on very brief periods of light each day. Following Pittendrigh's *Drosophila* protocol, Hillman exposed *Lemna* to "skeleton" photoperiods—stretches of darkness sandwiched between two brief light exposures. Experimenting with modified 11-h light/13-h dark cycles [11(13)], in which he applied 15 min R at the beginning and end of each "light" period [continuous WL and then $\frac{1}{4}(10\frac{1}{2})\frac{1}{4}$ (13) $\frac{1}{4}(10\frac{1}{2})$ . . .], Hillman wondered if the sequence of the light and dark periods affected flowering. Surprisingly, the plants distinguished between skeleton 11(13) and skeleton (13)11 schedules because, when the "light" period came first, only 12–17% of the fronds formed flower primordia, whereas about 48% were induced when the 13-h dark period was first [continuous WL and then (13) $\frac{1}{4}(10\frac{1}{2})\frac{1}{4}$ (13) . . .]. Thus the time of the first light exposure after transfer to darkness appeared to determine flowering levels.

When Hillman inserted dark periods of differing lengths before the skeleton 11(13) schedule, he observed an obvious 24-h periodicity (Fig. 14-7, open circles), because 15 h initial darkness promoted flowering much more effectively than 0 and 24 h. When varying periods of darkness

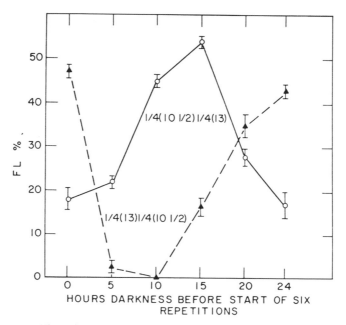

FIGURE 14-7   Flowering of *Lemna perpusilla* given various durations of darkness before the start of six repetitions of indicated skeleton schedules. (From Hillman, 1964.)

preceded the skeleton (13)11 schedule (Fig. 15-7, closed triangles), 10 h initial darkness was least effective, whereas maximal promotion occurred at 0 and 24 h. With longer periods of initial darkness, the rhythm persisted for at least 48 h (Hillman, 1964).

Pittendrigh regarded this as probably "the most impressive single set of data supporting Bünning's proposition" (Pittendrigh, 1966) and pointed out that it perfectly fitted his own interpretation of rhythm entrainment. He had deduced that, in a 24-h cycle, a light-off signal entrained the *Drosophila* rhythm to the beginning of the subjective night (Bünning's scotophilic phase) or circadian time (CT) 12. Because the photoinducible phase occurred at a set time in the period, it therefore would be illuminated only at certain times of year. If the transition from continuous light to darkness also set the rhythm at CT 12 in *Lemna*, the effects of skeleton photoperiods would depend on the interval between the light–dark transition and the first light pulse, as Hillman had observed, because this would determine whether the pulse at the end of the photoperiod fell in the photoinducible phase (Pittendrigh, 1966).

LIGHT-ON AND LIGHT-OFF SIGNALS

In the late 1960s, Hillman detected a circadian rhythm in carbon dioxide (CO_2) output from *L. perpusilla 6746*. On certain nitrogen sources the rhythm's response to skeleton periods mimicked that of the flowering rhythm, so he concluded that the two were controlled by the same circadian timer. Therefore he used the CO_2 rhythm, detectable in real-time, as the visible hands of the circadian clock.[2]

The rhythm began when plants were transferred from continuous light to darkness. But when 24-h light–dark cycles followed, the rate of CO_2 output peaked at a set time after the start of each cycle, regardless of photoperiod length. And in contrast to the results obtained with skeleton photoperiods enclosing two nearly equal dark periods, schedules with highly unequal portions [$\frac{1}{4}(5\frac{1}{2})\frac{1}{4}$ (18)] entrained the CO_2 rhythm to a 24-h periodicity that peaked 14–16 h after the first light pulse, regardless of whether the "photoperiod" preceded or followed the dark period. Thus a light-off signal initiated an endogenous rhythm after growth in continuous light, but a subsequent light-on signal entrained the rhythm, even if the signal persisted only for 15 min (Hillman, 1970).

At the University of Western Ontario, Cumming and Rod W. King detected a rhythmic sensitivity to the time at which a single dark period

2. Another overt rhythm, leaf movement in *Xanthium*, definitely did not correlate with the photoperiodic clock (Salisbury and Denney, 1971).

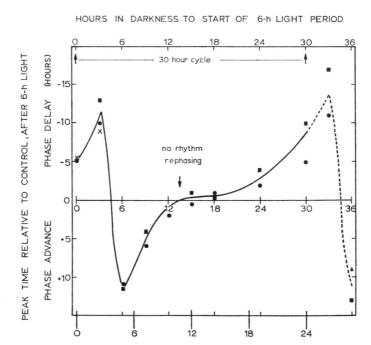

FIGURE 14-8   Rephasing of rhythmic flowering response of *C. rubrum* by a 6-h photoperiod given at different times after entry into darkness. (From King and Cumming, 1972b. Copyright 1972, Springer-Verlag.)

ended when *Chenopodium rubrum* was otherwise kept in continuous WL. The period length was about 30 h, and the rhythm peaked at 13, 43, and 73 h of darkness. When a R photoperiod interrupted the dark period, both light-on and light-off signals phased the rhythm if the photoperiod lasted for 6 h. Thus, the rhythm was delayed if the photoperiod began early in the *subjective* night and advanced if it began late in the subjective night (Fig. 14-8). It was suspended during 12-h and 18-h photoperiods, starting up again after a light-off signal and peaking 13 h later. Because it was not released immediately after a light–dark transition (King and Cumming, 1972) and because Pfr was thought to disappear only after several hours of darkness in this species (Evans and King, 1969; King, 1971), the researchers suggested that an hourglass mechanism operated sometime during the first part of the dark period (King and Cumming, 1972), determining the onset of dark timing—although not acting as the timer itself.

ROLES OF PHYTOCHROME

Hillman decided that phytochrome could rephase the endogenous rhythm in *Lemna* as well as interact with its photoinducible phase. Thus red (R/R/R) or far-red (F/F/F) light every 8 h failed to entrain the CO_2 output rhythm, but entrainment was observed when the two types of light were combined into a schedule that would impart 24-h periodicity if they had different effects. Moreover, F,R/R/R behaved like R/R/R and R,F/F/F like F/F/F, whereas R,F/R/R and F,R/F/R gave good entrainment. Hillman concluded that "the entraining factor here is the regular change in imposed Pfr level, since reversibility occurs in the classic phytochrome manner. . . . The zeitgeber[3] is the sharpest transition from low Pfr (darkness) to high Pfr among the events recurring with 24-h periodicity. This is the only hypothesis that explains why schedules such as R/−/− and F/−/−, R/R/− and F/F/− establish similar phase relationships with respect to the first signal, and also accounts for the fact that in the schedule F/F/R the phase relationship to the red signal is almost like that in −/−/R, yet very different from that in F/F/− with respect to the far-red signals" (Hillman, 1971).

Because entrainment under FR generated earlier maxima and minima than entrainment under R, Pfr levels had to be lowered before timing could begin. "In short, although the concept of phytochrome reversion—or any simple hour-glass system—as the sole timer in higher plant photoperiodism has fared badly," Hillman said, "its function as a subsidiary timer . . . is fully consistent with the data discussed here" (Hillman, 1971).

At the University of Hawaii, Douglas J. C. Friend observed that a photoperiod consisting of just two R pulses sometimes allowed a subsequent 24-h dark period to induce flowering in dark-grown *Pharbitis* seedlings. Because a FR pulse partially reversed the effect of the first R pulse and completely reversed that of the second, phytochrome appeared to be the photoreceptor. And because the effectiveness of the pulses varied rhythmically with the length of the dark interval between them, the first pulse—acting through Pfr—appeared to set the phase of an endogenous rhythm (Friend, 1975).

EXTERNAL VERSUS INTERNAL COINCIDENCE

The concept that light inhibits or promotes flowering by illuminating a certain phase of an internal rhythm formed the basis of *external coincidence* models of photoperiodism (Pittendrigh, 1972), of which Bünning's

---

3. The zeitgeber is the time-giver or synchronizing signal.

was the first. Pittendrigh also suggested that *internal coincidence* models be investigated (Pittendrigh, 1960, 1972; Danilevskii, 1965), because it was apparent that several semiautonomous rhythms coexist in living organisms. If one of these was entrained by dawn and one by dusk, two components that had to interact to produce a physiological response—such as two reactants or an enzyme and its substrate—might become available at different times in the daily cycle, and no response would occur. "Only under some photoperiods would critical phase-points of two (internal) oscillations coincide, thus closing a switch and effecting photoperiodic induction," Pittendrigh explained (Pittendrigh, 1972).

Hamner interpreted data from *Pharbitis* with the internal coincidence hypothesis (e.g., Hamner and Hoshizaki, 1974) because the peaks and valleys of the flowering rhythm related to the beginnings and ends of photoperiods when successive light and dark periods varied in length. "It is not obvious, however," Hillman pointed out, "that such results require a two-oscillator hypothesis rather than merely being consistent with it" (Hillman, 1976a).

Hillman himself, despite his important contributions, was unable to develop a satisfactory model. Plagued by increasingly poor health, he continued to study rhythms of CO_2 output (Hillman, 1976b) and O_2 uptake (Hillman, 1977). But he died in 1981, at age 51.

REFERENCES

Borthwick, H. A. (1964). Phytochrome action and its time displays. *Am. Nat.* 98,347–355.
Borthwick, H. A., and Hendricks, S. B. (1960). Photoperiodism in plants. *Science* 132,1223–1228.
Bünning, E. (1959). Additional remarks on the role of the endogenous diurnal periodicity in photoperiodism. *In* "Photoperiodism and Related Phenomena in Plants and Animals" (R. B. Withrow, ed.), pp. 531–535. Copyright 1961 by the American Association for the Advancement of Science, Washington, D.C. Reprinted with permission of the publisher.
Bünning, E. (1960a). Opening address: Biological clocks. *Cold Spring Harbor Symp. Quant. Biol.* 25, 1–9.
Bünning, E. (1960b). Circadian rhythms and time measurement in photoperiodism. *Cold Spring Harbor Symp. Quant. Biol.* 25,249–256.
Carr, D. J. (1952a). The photoperiodic behavior of short-day plants. *Physiol. Plant.* 5,70–84.
Carr, D. J. (1952b). A critical experiment on Bünning's theory of photoperiodism. *Z. Naturforsch.* 7b,570–571.
Claes, H., and Lang, A. (1947). Die Blütenbildung von *Hyoscyamus niger* in 48-stündigen Licht-Dunkel-Zyklen und in Zyklen mit aufgeteilten Licht-phasen. *Z. Naturforsch.* 2b,56–63.
Coulter, M. W., and Hamner, K. C. (1964). Photoperiodic flowering response of Biloxi soybean in 72-hour cycles. *Plant Physiol.* 39,848–856. Quotation reprinted with permission from the American Society of Plant Physiologists.

Danilevskii, A. S. (1965). "Photoperiodism and Seasonal Development of Insects." Oliver and Boyd, Edinburgh.

Evans, L. T., and King, R. W. (1969). Role of phytochrome in photoperiodic induction of *Pharbitis nil*. *Z. Pflanzenphysiol*. 60,277–288.

Friend, D. J. C. (1975). Light requirements for photoperiodic sensitivity in cotyledons of dark-grown *Pharbitis nil*. *Physiol. Plant*. 35,286–296.

Hamner, K. C. (1961). Photoperiodism and circadian rhythms. *Cold Spring Harbor Symp. Quant. Biol*. 25,269–277.

Hamner, K. C., and Hoshizaki, T. (1974). Photoperiodism and circadian rhythms: An hypothesis. *BioScience* 24,407–414.

Hamner, K. C., and Takimoto, A. (1964). Circadian rhythms and plant photoperiodism. *Am. Nat*. 98,295–322.

Hendricks, S. B. (1961). Rates of change of phytochrome as an essential factor determining photoperiodism in plants. *Cold Spring Harbor Symp. Quant. Biol*. 25,245–248.

Hendricks, S. B. (1964). Photochemical aspects of plant photoperiodicity. *In* "Photophysiology" (A. C. Geise, ed.), vol. 1, pp. 305–331. Academic Press, New York, London.

Hendricks, S. B. (1970). The passing scene. Quotation reproduced, with permission, from the *Annual Review of Plant Physiology*, vol. 21, pp. 1–10. © 1970 by Annual Reviews Inc.

Hillman, W. S. (1963). Photoperiodism: An effect of darkness during the light period on critical night length. *Science* 140, 1397–1398.

Hillman, W. S. (1964). Endogenous circadian rhythms and the response of *Lemna perpusilla* to skeleton photoperiods. *Am. Nat*. 98,323–328.

Hillman, W. S. (1970). Carbon dioxide output as an index of circadian timing in *Lemna* photoperiodism. *Plant Physiol*. 45,273–279.

Hillman, W. S. (1971). Entrainment of *Lemna* CO_2 output through phytochrome. *Plant Physiol*. 48,770–774. Reprinted with permission from the American Society of Plant Physiologists.

Hillman, W. S. (1976a). Biological rhythms and physiological timing. *Annu. Rev. Plant Physiol*. 27,159–179.

Hillman, W. S. (1976b). Light/timer interactions in photoperiodism and carbon dioxide output patterns: Towards a real-time analysis of photoperiodism. *In* "Light and Plant Development" (H. Smith, ed.), pp. 383–391. Butterworths, London.

Hillman, W. S. (1977). Control of plant respiration through non-photosynthetic light action. *Nature* 266,833–835.

Hsu, J. C. S., and Hamner, K. C. (1967). Studies on the involvement of an endogenous rhythm in the photoperiodic response of *Hyoscyamus niger*. *Plant Physiol*. 42,725–730.

Hussey, G. (1954). Experiments with two long-day plants designed to test Bünning's theory of photoperiodism. *Physiol. Plant*. 7,253–260.

King, R. W. (1971). Time measurement in the photoperiodic control of flowering. Ph.D. Thesis. Univ. Western Ontario, Canada.

King, R. W., and Cumming, B. G. (1972). Rhythms as photoperiodic timers in the control of flowering in *Chenopodium rubrum*. L. *Planta* 103,281–301.

Kribben, F. J. (1955). Zu den Theorien des Photoperiodismus. *Beitr. Biol. Pflanz*. 31,297–311.

Moore, P. H., Reid, H. B., and Hamner, K. C. (1967). Flowering responses of *Xanthium pensylvanicum* to long dark periods. *Plant Physiol*. 42,503–509.

Papenfuss, H. D., and Salisbury, F. B. (1967). Aspects of clock resetting in flowering of *Xanthium*. *Plant Physiol*. 42,1562–1568.

Pittendrigh, C. S. (1960). Circadian rhythms and the circadian organization of living systems. *Cold Spring Harbor Symp. Quant. Biol*. 25,159–184.

Pittendrigh, C. S. (1966). The circadian oscillation in *Drosophila pseudoobscura* pupae: a model for the photoperiodic clock. *Z. Pflanzenphysiol.* 54,275–307.

Pittendrigh, C. S. (1972). Circadian surfaces and the diversity of possible roles of circadian organization in photoperiodic induction. *Proc. Natl. Acad. Sci. USA* 69,2734–2737.

Pittendrigh, C. S., and Minis, D. H. (1964). The entrainment of circadian oscillations by light and their role as photoperiodic clocks. *Am. Nat.* 38,261–294.

Salisbury, F. B. (1961). Photoperiodism and the flowering process. *Annu. Rev. Plant Physiol.* 12,293–326.

Salisbury, F. B. (1963). Biological timing and hormone synthesis in flowering of *Xanthium*. *Planta* 59,518–534.

Salisbury, F. B. (1965). Time measurement and the light period in flowering. *Planta* 66,1–26. Copyright 1965, Springer-Verlag.

Salisbury, F. B., and Denney, A. (1971). Separate clocks for leaf movements and photoperiodic flowering in *Xanthium strumarium* L. (cockelbur). *In* "Biochronometry" (M. Menaker, ed.), pp. 292–311. National Academy of Science, Washington, D.C.

Sirohi, G. S., and Hamner, K. C. (1960). Automatic device for controlling the length of light and dark periods in cycles of any desired duration. *Plant Physiol.* 35,276–278.

Takimoto, A., and Hamner, K. C. (1964). Effect of temperature and preconditioning on photoperiodic response of *Pharbitis nil. Plant Physiol.* 39,1024–1030.

Takimoto, A., and Hamner, K. C. (1965). Effect of far-red light and its interaction with red light in the photoperiodic response of *Pharbitis nil. Plant Physiol.* 40,859–864.

Wareing, P. F. (1954). Experiments on the 'light-break' effect in short-day plants. *Physiol. Plant.* 7,157–172.

CHAPTER 15

Mougeotia

In 1956, just after Mohr completed his doctorate at the University of Tübingen, dozent Wolfgang Haupt began to study chloroplast movement in *Mougeotia*. Each cylindrical cell of this filamentous alga has one enormous, flat chloroplast, which turns on its long axis to face dim light, assuming an optimal position for photosynthesis. In brighter light, the face swings away, avoiding photochemical damage.[1]

Looking for the photoreceptor for these responses, Haupt expected to implicate chlorophyll. "But as a joke, I tried to get reversion of the low-intensity chloroplast response by far red," he decided, after discussing fern spore germination with Mohr. Much to his surprise, induction of the movement proved to be R/FR reversible.

Haupt (Fig. 15-1), completed doctoral work with Bünning in 1952, on floral induction in peas. While he was working toward his habilitation, he wrote a chapter on phototaxis (movement toward or away from light) for a plant physiology handbook (Haupt, 1956b) and realized he should also discuss chloroplast phototaxis (Haupt, 1959a).

A major question (Senn, 1908) was whether chloroplasts turn themselves or are moved by structures in the cytoplasm. Haupt realized that location of the photoreceptor for the response might resolve this issue.

CHLOROPLAST MOVEMENTS

Haupt illuminated *Mougeotia* filaments on a microscope slide from one side with weak WL, making the chloroplasts turn to face the light and

---

1. These movements are captured in the movie: Schönbohm, E., and Haupt, W. (1975). Lichtorientierte Chloroplastenbewegung bei *Mougeotia* spec. Institut für den wissenschaftlichen Film, Nonnenstieg 72, D-3400, Göttingen.

FIGURE 15-1 Wolfgang Haupt, 1989. Photo by
Linda Sage.

present their profiles to the microscope eyepiece. He then applied a pulse
of monochromatic light from below, inducing the chloroplasts to reveal
their faces.

Just one pulse was sufficient, for once induced, the chloroplasts turned
in the dark. After 10 min, they had clearly changed position, and after half
an hour, they stopped moving. Using a green safelight to record their
positions, Haupt summed the movements in a given number of cells. Using
a range of monochromatic pulses, he obtained action spectra for induction
and reversal of the low-energy response. As with floral induction and seed
germination, wavelengths between 600 and 700 nm were most effective,
whereas reversal was maximal at 717 nm. Thus phytochrome appeared to
be the photoreceptor.[2]

The response was repeatedly reversible, provided R and FR followed
each other immediately (Fig. 15-2). But there was a 50% loss of photoreversibility when 5 min elapsed between the two treatments. Thus induction of chloroplast movement in *Mougeotia* was realized much faster
than any phytochrome-mediated response previously studied (Haupt,
1959c).[3]

Unlike the low-energy response, the light-avoidance response appeared
to be induced by B, Haupt's student Ekkehard Schönbohm determined
(Schönbohm, 1963). Continuing to investigate this system at the University of Münster, he found that simultaneous irradiation with R and B

2. In this respect, *Mougeotia* differs from many other algae whose chloroplast movements
are induced by B.
3. Movement was later detected within 60 s of the onset of R (Haupt and Übel, 1975).

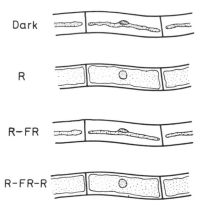

Dark

R

R—FR

R—FR—R

FIGURE 15-2   Response of *Mougeotia* chloroplasts to low-irradiance R and FR. After R, chloroplasts turn to face the direction from which light beam came. If FR follows R, chloroplasts continue to present their profiles. (From Haupt, 1970a.)

caused the chloroplast to turn its face away from the R beam, regardless of the direction of the B (Schönbohm, 1967). Thus the high-energy response was really a response to R. B switched the response from a low-energy response to a high-energy one, changing the direction of chloroplast movement.

MOLECULAR ORIENTATION

Bünning's student Helmut Etzold had begun to study the growth responses of fungi to polarized light in the mid 1950s. "Etzold asked me one day," Haupt recalls, " 'Have you tried polarized light on *Mougeotia?*' I said, 'I don't know why I should, but just for fun, why don't I try it?' "

Etzold had observed that certain fungal spores germinate parallel to the plane of vibration of the electrical vector of polarized B and that mature hyphae and fruiting bodies bend perpendicularly to that plane. He deduced that the unknown pigment mediating these responses must have a fixed orientation in the cell, presumably in or close to the wall (Etzold, 1961). He also discovered (Bünning and Etzold, 1958) that filaments emerging from moss and fern spores exhibit a polarotropic response to R.[4]

When Haupt exposed *Mougeotia* filaments to a pulse of polarized R from above, chloroplasts in cells with long axes perpendicular to the vibration plane markedly changed position during 30 min in darkness, whereas cells with long axes parallel to that plane behaved as if they had

4. Polarotropic response is a directional growth response to polarized light.

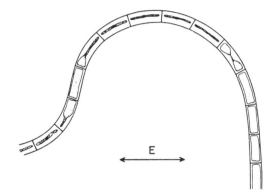

FIGURE 15-3 *Mougeotia* filament after irradiation with polarized R. Arrow indicates electrical vector of the light. (From Haupt, 1970a.)

seen 100 times less light (Fig. 15-3). This suggested that the photoreversible pigment molecules were precisely oriented in *Mougeotia* (Haupt, 1960). "As a first approximation," Haupt later explained, "we could suggest that the axis of main absorption or the dipole of each molecule is oriented parallel to the surface and perpendicularly to the cell axis" (Haupt, 1970a).

Additional experiments, in which an induction was followed by a light pulse from a different direction, supported this deduction. For this work, Haupt enclosed single *Mougeotia* filaments in capillary tubes mounted between tiny paraffin blocks.

PARTIAL IRRADIATION

Haupt's student Gudrun Bock was able to irradiate one end of a cell with a R beam, causing only that part of the chloroplast to turn to its face, an effect that had been discovered more than 50 years previously (Senn, 1908). Because *Mougeotia* cells are 100–200 μm long and 20–25 μm wide, and the cholorplast is about 6 μm thick, it was possible —though technically difficult—to focus the beam within the chloroplast's profile or to irradiate a small portion of cytoplasm without illuminating the chloroplast at all. Irradiating the cytoplasm with an unpolarized microbeam 6 μm in diameter was nearly as effective as irradiating the chloroplast. But although a 3-μm polarized microbeam directed at the chloroplast induced a response when its vibration plane was across or along the cell axis, only cross-vibrating polarized light was effective when the beam was directed

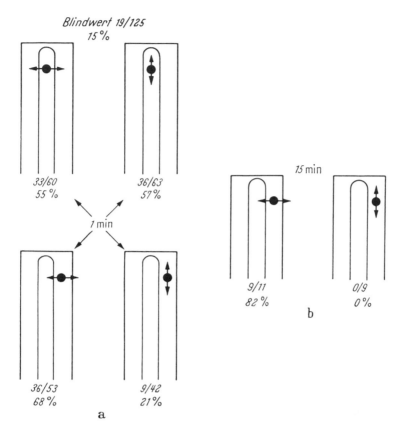

FIGURE 15-4 Partial irradiation with a microbeam 3 μm in diameter. Blindwert = control. Fractions indicate number of cells that responded out of total number irradiated. (a) Irradiation for 1 min; (b) irradiation for 15 min—not necessarily comparable with (a). (From Bock and Haupt, 1961. Copyright 1961, Springer-Verlag.)

at the cytoplasm (Fig. 15-4).[5] The researchers concluded that "eine Absorption von Rotlicht durch das Phytochromsystem in Cytoplasma stattfindet"[6] (Bock and Haupt, 1961).

Phytochrome in the cytoplasm would have to be fixed to a stable structure to maintain a precise orientation. "The most reasonable assumption . . . ." said Haupt, "is a localization at or near the outer cytoplasmic membrane (plasmalemma)" (Haupt, 1973).

5. A wider longitudinally vibrating beam decreased the response, showing that this type of light was absorbed.
6. The absorption of red light takes place through the phytochrome system in the cytoplasm.

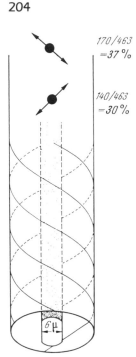

170/463
=37 %

140/463
=30 %

FIGURE 15-5 Schematic presentation of helical orientation of phytochrome molecules in peripheral cytoplasm of a
Mougeotia cell. Molecules are oriented so that they absorb R
when it vibrates at a given point in direction of spiral line.
Above (arrows) are two planes of vibration that were compared experimentally. Figures at the side are the observed
reaction values. (From Haupt and Bock, 1962. Copyright
1962, Springer-Verlag.)

HELICAL ARRANGEMENT

During his daily bicycle ride between his home and the botanical institute
of the University of Tübingen, Haupt had a flash of insight. "When we
found that phytochrome was dichroic,"[7] he explains, "we had to assume
that Pr was oriented longitudinally as well as perpendicularly in order to
explain several of our experiments." Thus polarized R vibrating parallel to
the cell axis antagonized induction of chloroplast movement by cross-
vibrating light (Haupt, 1960), and microbeams of parallel-vibrating light
were just as effective as those of cross-vibrating light (Bock and Haupt,
1961).

"This could really mean," Haupt says, "that (1) we had two families of
photoreceptors, each with a different orientation, (2) the photoreceptors
were randomly oriented so that the total components in each direction
were about equal, (3) that all the photoreceptor molecules were oriented in
the same direction, namely obliquely, which also would give a component
in both directions. During my bicycle ride I thought of a way to test which
possibility was correct."

7. Dichroism is the absorption of light vibrating in one plane more effectively than light
vibrating in the plane at right angles.

TABLE 15-1

Irradiation of Individual *Mougeotia* Cells with a Polarized Microbeam Whose Vibration Plane Was at a 45° Angle to the Longitudinal Axis of the Cell. Time of Irradiation Was 1 min. Irradiance Was Constant. Light Field Was 3 μm in Diameter[a]

	Number of cells responding	Number of cells irradiated	Percent		Difference
	Left 45°	Right 45°	Left 45°	Right 45°	
Plane of vibration		a) Lightfield, ⌀ 3 μ			
	13/26	11/27	50	41	9
	13/25	13/33	52	39	13
	11/25	7.5/26	44	29	15
	10.5/29	9/30	36	30	6
	12.5/31	9.5/31	40	31	9
	10.5/40	9/37	26	24	2
	12/42	9/37	29	24	5
	13.5/40	14.5/40	34	36	−2
	16/60	14/58	27	24	3
	18.5/46	14/44	40	32	8
	23/59	16.5/60	39	28	11
	17/40	13.5/40	43	34	9
Total	170.5/463	140.5/463→	37	30	
		b) Lightfield, ⌀ 6 μ			
	17.5/30	10.5/30	58	35	23
	18.5/41	13.5/41	45	33	12
	18.5/43	19/44	43	43	0
	9/27	8/28	33	29	4
	17/42	17/42	41	41	0
	25/50	18.5/49	50	38	12
	20.5/47	15/50	44	30	14
Total	126/280	101.5/284→	45	36	

[a] From Haupt and Bock, 1962. Copyright 1962 Springer-Verlag.

He realized that the ratios of the responses to R microbeams with different planes of vibration would distinguish possibility (3) from (1) and (2). If the main absorption axis of the pigment molecules was diagonal (3), light vibrating in that plane (upper double arrow, Fig. 15-5) would be absorbed more efficiently by the phytochrome on the upper side of the cell than on the lower; the reverse would be true for light polarized in the other plane (bottom double arrow, Fig. 15-5). Because only 30% of the light that hit the top of the cell reached the underside, light polarized in the plane of the upper molecules would therefore theoretically be three times as effective as light polarized in the plane of the molecules on the underside. If the molecules were oriented in two different directions (1) or oriented randomly (2), the two forms of light would evoke equal responses.

"This was a very time-consuming experiment," Haupt recalls, "because we had to do each cell individually." Comparing chloroplast movements in response to the two planes of light, he did not obtain a 3:1 effectiveness ratio. But he did consistently observe a greater response when the vibration plane was aligned on one diagonal than when it was aligned on the other (Table 15-1). This difference disappeared when cells were irradiated in the less effective position for twice as long as in the most effective position, and it increased under the reverse condition. Moreover, setting the vibration planes at 22.5° and 67.5° to the long axis of the cell rather than at +45° and −45° abolished the difference. "Damit kann qualitativ an der Tatsache, daβ eine Differenz in definierter Richtung besteht, kaum mehr gezweifelt werden,"[8] Haupt decided (Haupt and Bock, 1962). He proposed that phytochrome molecules in *Mougeotia* are oriented helically around the cell. Such an arrangement appeared feasible because *Spirogyra*, a relative of *Mougeotia*, has a helical chloroplast (and presumably a corresponding helical arrangement of the cytoplasm), which winds in the same direction as the putative phytochrome helix in *Mougeotia*. Evidence for a helical cytoplasmic structure in *Mougeotia* surfaced later (Haupt and Wirth, 1967).

NEWLY FORMED PFR

Starting one experiment, Haupt mistakenly aligned *Mougeotia* filaments with their faces rather than their profiles to the microscope eyepiece. But when he irradiated with low-irradiance polarized R with the usually ineffective vibration parallel to the cell axis, the chloroplasts turned unexpectedly, now swinging their faces, rather than their edges, away from the light. "This so-called reverse low-intensity movement," Haupt says, "brought me to the interpretation that newly formed Pfr is more effective than old Pfr."

Doctoral candidate Walter Fischer showed that phytochrome induced the reverse movement, although less effectively than the normal low-energy movement. He was able to cancel the effect by inserting a FR treatment between the WL pretreatment and the inductive R pulse. Thus there seemed to be a requirement for both the Pfr formed during the pretreatment and the Pfr formed by the pulse. Haupt and Fischer proposed that the chloroplast moved from a region of newly formed Pfr to a region of "aged" Pfr, turning to profile from face position and to face from profile— and even from one oblique position to the other (Fischer, 1963). Unlike the orientation movement, this was not a response to light direction but simply

---

8. In qualitative terms, it can no longer be doubted that there is a difference in defined direction.

a turning of 90° regardless of whether light was striking the chloroplast's edge or face or somewhere in between.

The unicellular alga *Mesotaenium,* a relative of *Mougeotia,* also appeared to distinguish between newly formed and already formed Pfr. Its chloroplast responded to low-irradiance R, but irradiations of about half an hour were required. Single flashes every 3–5 min were also effective, suggesting that the phytochrome that induces chloroplast movement must be renewed at regular intervals (Haupt and Thiele, 1961). Pfr in *Mesotaenium* appeared to be stable in the dark but to quickly lose physiological activity. Potency was restored when the Pfr was converted to Pr and then back to Pfr (Haupt and Reif, 1979), suggesting that newly formed Pfr is the active phytochrome in this alga.

ORIENTATION OF PFR

Haupt left Tübingen in 1962 to become a professor at the University of Erlangen, and his research was severely disrupted. But he toyed with the idea that the orientation of the phytochrome molecule might change when Pr became Pfr. "When I was writing the 1960 *Planta* paper," he says, "I made a statement to that effect in the first draft. But then an American biophysicist visiting my lab said, 'That is impossible. If you make this suggestion, people will say in the future, Don't trust Haupt.' So I left it out. And I had no firm evidence. It was just that some experiments couldn't be explained if Pr and Pfr had the same orientation."

Etzold had moved to the University of Freiburg, where he continued to study polarotropism in *Dryopteris*. Measuring the reorientation of filaments after a 50° change in the vibration plane of monochromatic polarized light, he had obtained an action spectrum that implicated both phytochrome and a B receptor. Both pigments were dichroic and therefore likely to be located close to or in the wall (Etzold, 1965).

By dropping starch grains on growing filament tips, Etzold showed that growth direction changed in B (phototropism) because the side that absorbed the most light bulged rather than because the shaded side increased its growth rate. Applying this principle to phytochrome-mediated polarotropism, he deduced that the tip of an apical cell would grow most rapidly where Pfr concentration was highest.

Studying the photoreversibility of the reorientation growth response, Etzold observed that although unpolarized FR partly reversed the effect of polarized R, polarized FR enhanced it. Pretreatment with unpolarized R, moreover, reduced the sensitivity to polarized R but enhanced the effects of polarized FR compared with darkness. But if unpolarized FR followed the R pretreatment, the decrease was prevented. So he concluded that,

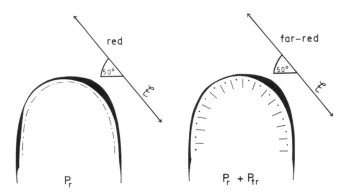

FIGURE 15-6 Etzold's model of orientation of main absorption axes of Pr and Pfr in apical cells of fern chloronema. Thickness of outer line represents relative amounts of Pfr that would be present after exposure to polarized light. *Light:* dark-adapted tip about to be exposed to polarized R; *right:* R-adapted tip about to be exposed to polarized FR. Dashes indicate main absorption axes of Pfr molecules. Dots represent cross section through main absorption axes of Pr molecules that were not converted to Pfr during R. (From Etzold, 1965. Copyright 1965, Springer-Verlag.)

with polarized light also, bending occurred at the site of maximal light absorption. But this would be possible only if the axis of maximum absorption of Pfr molecules was perpendicular to that of Pr molecules (Etzold, 1965).

Etzold proposed that Pr molecules were arranged around the periphery of an apical cell of a dark-adapted sporeling, each having its main absorption axis parallel to the cell surface (Fig. 15-6, left). Thus when the tip was irradiated from above with polarized R oriented at 50° to the plane toward which the tip was growing, the Pr molecules in the right-hand part of the tip would absorb the most light. This region would consequently acquire the most Pfr and grow more than the other regions, reorienting the tip. If, however, the tip was adapted to unpolarized R (Fig. 15-6, right), the phytochrome would be largely (>80%) in the Pfr form. If Pfr molecules were oriented perpendicularly to Pr molecules, those in the right-hand portion of the tip would be *less* able to absorb 50° polarized FR than those in other regions, and the Pfr concentration would remain higher in this region, which therefore would also grow toward the plane of vibration. Thus the two treatments would have similar and not opposite effects.

Haupt was fascinated by Etzold's proposal, but he was unable to apply it to *Mougeotia*. Then in 1967, Haupt went to Freiburg to perform an evaluation for the Deutsche Forschungsgemeinschaft, the government agency that funds German scientific research. "We worked hard for a few days and then I left on an overnight train," Haupt recalls. "My visit had been very stimulating. Thus on the train, in the middle of the night, I suddenly awoke

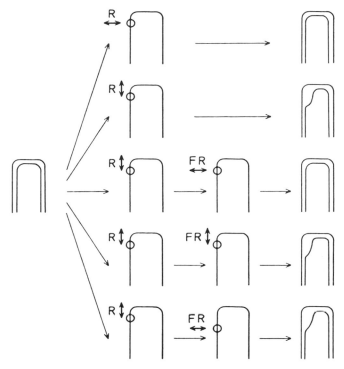

FIGURE 15-7   Partial irradiation of *Mougeotia* cell with polarized microbeams. *Double arrows,* electrical vector of polarized R and FR. (From Haupt et al., 1969. Copyright 1969, Springer-Verlag.)

and realized that I had mistakenly interpreted my data when thinking about Etzold's model. Before, I had always considered that I had Pr when I started an experiment. But I had not—because before inducing chloroplast movement, I always oriented the chloroplasts toward white light, which of course converts half the Pr to Pfr.''

Haupt had accumulated several puzzling observations. One was that, although 680–690 nm light strongly induced chloroplast movement whether it was polarized or not, light of 670 nm had to be polarized for maximal saturation value. Also, low energies of FR induced responses that eventually equaled R-induced responses. Moreover, the action spectrum for induction of chloroplast movement by polarized light was shifted to the longer wavelengths and dropped off much less steeply than normal, indicating that FR was an effective region. The assumption that Pr molecules were oriented parallel to the surface of the cell (as he had previously proposed) but that Pfr was oriented perpendicularly to the surface explained these anomalies (Haupt, 1968).

Haupt and students Gertraud Mörtel and Irmgard Winkelnkemper soon obtained evidence that phytochrome changes its orientation during phototransformation (Haupt et al., 1969). First they irradiated *Mougeotia* with FR to put phytochrome in its Pr form throughout the cell. Then they briefly irradiated the side of the cell with an R microbeam, inducing just the corner of the chloroplast to turn. When the microbeam was polarized, turning occurred only when the plane of vibration was parallel (Fig. 15-7 part I) rather than perpendicular (part II) to the cell axis. Thus Pr molecules were oriented parallel to the cell surface.

When the same site was subsequently pulsed with a polarized FR microbeam, the R effect was cancelled—if the FR vibration plane was perpendicular (Fig. 15-7 part III)—but not parallel (part IV)—to the cell axis, and not if the beam was moved to a different site (part V). This clearly showed that Pfr molecules were perpendicular to the cell surface and therefore that Pr and Pfr had different orientations (Fig. 15-8). Further experiments supported this conclusion (Haupt, 1970b).

MOUGEOTIA IN NATURE

Haupt argued that a cell with helically wound phytochrome molecules would also show a directional response to nonpolarized light—a mixture of perpendicularly and longitudinally vibrating rays. Because the transition moments of helically arranged Pr molecules have longitudinal and perpendicular components, cross-vibrating polarized light (vibration plane perpendicular to the cell axis) would be absorbed by Pr on the top and bottom of a cell but not on the flanks (Fig. 15-9A), resulting in a Pfr gradient and therefore induction of chloroplast movement. Polarized light

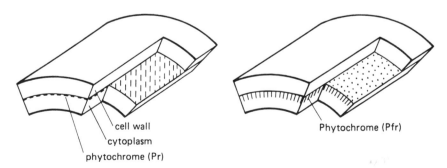

FIGURE 15-8   Part of cylindrical *Mougeotia* cell with dichroic absorption of phytochrome (dashes) parallel (Pr) or normal (Pfr) to the surface. (From Haupt, 1970a.)

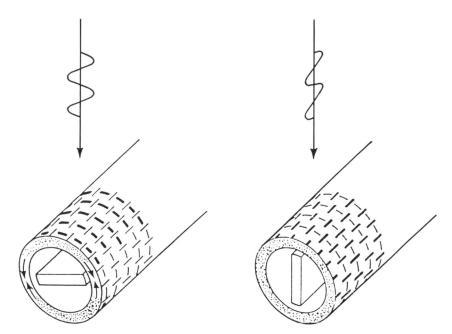

FIGURE 15-9   Absorption of polarized light by oriented photorecptors in *Mougeotia*. Heavy and thin lines show photoreceptors that can or cannot absorb light because of their orientation. Resulting absorption is indicated by dots near periphery; absorption gradient is shown by arrows. Electrical vector of the light is indicated by waves. (From Haupt and Schönbohm, 1970, after Haupt, 1960. Copyright 1960, Springer-Verlag.)

vibrating parallel to the cell axis would be absorbed by Pr all around the cell (B), so no Pfr gradient would be formed. Thus if a cell were exposed unilaterally or bilaterally to unpolarized light, a Pfr gradient would be established, because the two absorption patterns would be integrated.

This gradient would be expected to disappear under saturating light conditions that eventually convert Pr to Pfr. But chloroplast orientation occurs in nature under continuous light, so the gradient must be maintained. Considering only the electrical vector of the light that vibrates perpendicularly to the cell axis, Haupt proposed that Pr molecules at the front and rear of the cell would be in a favorable position to absorb the light, whereas Pfr molecules would not. Thus the photoequilibrium would shift in favor of Pfr. At the flanks, Pfr molecules would be optimally aligned but Pr molecules would not, so the photoequilibrium would shift toward Pr (Fig. 15-10). A persistent Pfr gradient would therefore be established (Haupt, 1972, 1981).

Although *Mougeotia* is an obscure alga, Haupt's work had a major

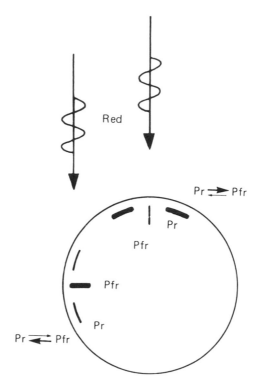

FIGURE 15-10   Dependence of phytochrome photoequilibrium on dichroic orientation of Pr and Pfr in a *Mougeotia* cell irradiated with polarized light vibrating perpendicularly to the cell axis. Molecules with orientation favorable for absorption are boldfaced. (From Haupt, 1972.)

impact on the phytochrome field because it supported a hypothesis, proposed by the Beltsville group in 1967, that the photoreversible pigment is membrane-associated (See Chapter 16). In recognition of his influence and ingenuity, the Finsen Foundation[9] awarded Haupt a medal in 1984 for "fundamental contributions to photobiology. More specifically, Professor Haupt has worked primarily with the light reactions of the green alga *Mougeotia* and, using mostly very simple equipment, has derived information about the regulator pigment phytochrome, which is of great general significance for plant photobiology" (Urbach, 1984).

9. Operated by the International Photobiology Association, the foundation honors Niels Finsen, a Danish photodermatologist who won the 1903 Nobel Prize in Medicine and Physiology.

REFERENCES

Bock, G., and Haupt, W. (1961). Die Chloroplastendrehung bei *Mougeotia* III. Die Frage der Lokalisierung des Hellrot-Dunkelrot-Pigmentsystems in der Zelle. *Planta* 57,518–530.

Bünning, E., and Etzold, H. (1958). Über die Wirkung von polarisiertem Licht auf keimende Sporen von Pilzen, Moosen und Farnen. *Ber. Dtsch Bot. Ges.* 71,304–306.

Etzold, H. (1961). Die Wirkungen des linear polarisierten Lichtes auf Pilze und ihre Benziehungen zu den tropistischen Wirkungen des einseitigen Lichtes. *Exp. Cell Res.* 25,229–245.

Etzold, H. (1965). Der Polarotropismus und Phototropismus der Chloronemen von *Dryopteris filix mas* L. Schott. *Planta* 64,254–280.

Fischer, W. (1963). Untersuchungen über die Inversion der Schwachlichtbewegung des *Mougeotia*-Chloroplasten. *Z. Bot.* 51,348–387.

Haupt. W. (1959a). Chloroplastenbewegung. *In* "Handbuch der Pflanzenphysiologie" (W. Ruhland, ed.), vol. 17/1, pp. 278–317. Springer, Berlin-Göttingen-Heidelberg.

Haupt. W. (1959b). Die Phototaxis der Algen. *In* "Handbuch der Pflanzenphysiologie" (W. Ruhland, ed.), vol. 17/1, pp. 318–370. Springer, Berlin-Göttingen-Heidelberg.

Haupt, W. (1959c). Die Chloroplastendrehung bei *Mougeotia* I. Über den quantitativen und qualitativen Lichtbedarf der Schwachlichtbewegung. *Planta* 53,484–501.

Haupt, W. (1960). Die Chloroplastendrehung bei *Mougeotia* II. Die Induktion der Schwachlichtbewegung durch linear polarisiertes Licht. *Planta* 55,465–479.

Haupt, W. (1968). Die Orientierung der Phytochrom-Moleküle in der *Mougeotia*-Zelle: Eine neues Modell zur Deutung der experimentellen Befunde. *Z. Pflanzenphysiol.* 58,331–346.

Haupt, W. (1970a). Localization of phytochrome in the cell. *Physiol. Veg.* 8,551–563.

Haupt, W. (1970b). Über den Dichroismus von Phytochrom$_{660}$ und Phytochrom$_{730}$ bei *Mougeotia*. *Z. Pflanzenphysiol.* 62,287–298.

Haupt, W. (1972). Perception of light direction in oriented displacements of cell organelles. *Acta Protozool.* 11,179–188.

Haupt, W. (1973). Role of light in chloroplast movement. *BioScience* 23,289–296.

Haupt, W. (1981). The perception of light direction and orientation responses in chloroplasts. *Symp. Soc. Exp. Biol.* 36,423–442.

Haupt, W., and Bock, G. (1962). Die Chloroplastendrehung bei *Mougeotia*. IV. Die Orientierung der Phytochrom-Moleküle im Cytoplasma. *Planta* 59,38–48.

Haupt, W., and Reif, G. (1979). "Ageing" of phytochrome in *Mesotaenium*. *Z. Pflanzenphysiol.* 92,153–161.

Haupt, W., and Schönbohm, E. (1970). Light-oriented chloroplast movements. *In* "Photobiology of Microorganisms" (P. Halldal, ed.), pp. 283–307. Wiley, London.

Haupt, W., and Thiele, R. (1961). Chloroplastenbewegung bei *Mesotaenium*. *Planta* 56,388–401.

Haupt, W., and Übel, H. (1975). Zum Mechanismus der Phytochromwirkung bei der Chloroplastenbewegung von *Mougeotia*. *Z. Pflanzenphysiol.* 75,165–171.

Haupt, W., and Wirth, H. (1967). Nachweis einer Schraubenstruktur in der *Mougeotia*-Zelle. *Plant Cell Physiol.* 8,541–543.

Haupt, W., Mörtel, G., and Winkelnkemper, I. (1969). Demonstration of different dichroic orientation of phytochrome Pr and Pfr. *Planta* 88,183–186.

Schönbohm, E. (1963). Untersuchungen über die Starklichtbewegung des *Mougeotia*-Chloroplasten. *Z. Bot.* 51,233–276.

Schönbohm, E. (1967). Die Bedeutung des Gradienten von Phytochrom 730 und die tonische Wirkung von Blaulicht bei der negativen Phototaxis des *Mougeotia*-Chloroplasten. *Z. Pflanzenphysiol.* 56,282–291.

Senn, G. (1908). ''Die Gestalts- und Lageveränderungen der Pflanzen-Chromatophoren.'' Wilhelm Engelmann, Leipzig.

Urbach, F. (1984). Introduction of Finsen medalists. *Photobiol. 1984. Proc. Int. Congr. Photobiol. 9th,* pp. 4–7.

CHAPTER 16

Mode of Action

Hendricks and Borthwick were convinced that phytochrome was an enzyme that controlled the flow of carbon through an intersection of major metabolic pathways, such as the crossroads at which acetyl coenzyme A arrives from carbohydrate or fat breakdown and enters the tricarboxylic acid cycle or the pathway that synthesizes fats (Fig. 16-1). "Such phenomena as the photo-induced reddening of the apple, red cabbage, and many seedlings, which require far-red absorbing pigment for formation of color," Hendricks asserted, "indicate that the control point is an acyl formation or a function of coenzyme A" (Hendricks, 1963).

Despite hopes for an imminent revelation of the pigment's action, studies of the extracted pigment gave no clue. Thus by the mid 1960s, there was still no evidence that phytochrome was an enzyme. Hillman expressed "disappointment that the discovery and extraction of phytochrome, and particularly the availability of methods to assay it in intact tissue, have so far proved so uninformative with respect to light responses" (Hillman, 1967).

GENE ACTIVATION

At the Institut Pasteur in Paris, meanwhile, François Jacob and Jacques Monod had developed a model for bacterial gene regulation, which earned them a Nobel Prize in 1965. They proposed that small effector molecules interact with specific proteins, enabling the latter to combine with regulatory regions of genes (Jacob and Monod, 1961).

The notion that a signal from one part of a cell controlled gene transcription in another seemed applicable to higher organisms; thus in 1963, Peter Karlson of the University of Munich proposed that the hormone ecdysone prompts insects to molt by activating genes (Karlson, 1963). Mohr (Fig. 16-2), whom Karlson taught in Tübingen, thought that "it was absolutely

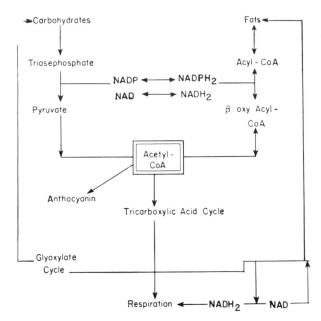

FIGURE 16-1 Intersection of metabolic pathways involving acetyl coenzyme A. Hendricks proposed that phytochrome acted within this network. (From Hendricks, 1963. Copyright 1963 by the American Association for the Advancement of Science.)

FIGURE 16-2 Hans Mohr, 1966.

clear that phytochrome could only operate to produce anthocyanin by activating genes.''

Graduate student Bertold Hock had observed that R increased oxygen uptake equally in mustard cotyledons and hypocotyls. But cotyledon growth was promoted by light, whereas hypocotyl growth was light-inhibited. Because phytochrome itself seemed unlikely to have opposite effects, Mohr attributed the specificity of the responses to differential gene activation (Hock and Mohr, 1964).

At a symposium in Rouen in November 1965, Mohr presented a list (Table 16-1) of phytochrome-controlled responses his group had uncovered in mustard seedlings. Making sense of this bewildering array, he proposed that ''the primary physiological functon of this particular chromoprotein [Pfr] is the same in all cells in which it has been formed. . . . Apparently it depends on the 'specific status of differentiation' of the cells and tissues as to which response can take place'' (Mohr, 1966a).

The Freiburg workers had observed that photoresponses may be positive, negative, or complex (Mohr, 1966b), and Mohr's initial model (Fig. 16-3) accounted only for those that were triggered or enhanced by light. It assumed that each cell of a dark-grown seedling contains active, inactive, and potentially active genes. A light exposure that formed Pfr would

TABLE 16-1

Photoresponses of the Mustard Seedling, *Sinapis alba,* Investigated in Mohr's Laboratory by the Mid 1960s[a]

Inhibition of hypocotyl lengthening
Enlargement of cotyledons
Hair formation along the hypocotyl
Opening of the plumular hook
Formation of leaf primordia
Differentiation of the primary leaves
Increase of negative geotropic reactivity
Formation of tracheary elements
Differentiation of stomata
Changes of the rate of cell respiration
Increase of protein synthesis
Increase of RNA synthesis
Synthesis of anthocyanin
Increase of synthesis of ascorbic acid
Changes in the rate of fat degradation
Changes in the rate of degradation of reserve protein
Increase in the rate of protochlorophyll formation

[a] From Mohr, 1966a.

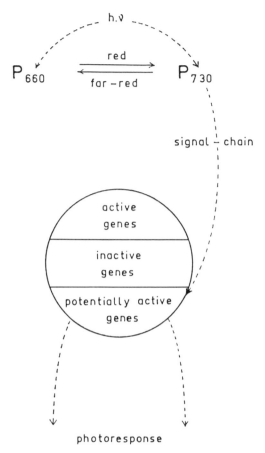

FIGURE 16-3   Mohr's hypothesis to explain action of Pfr in photomorphogenesis. (From Mohr, 1966a.)

switch on the latter. "Which genes are potentially active had been determined earlier," Mohr explained, "by the regulating factors of primary differentiation; the signal is only the 'trigger' " (Mohr, 1966a).

Pfr formation would thus alter RNA and protein levels, because genes are transcribed into messenger-RNA molecules, which carry information for polypeptide synthesis. The Freiburg group reported preliminary evidence for such changes in mustard seedlings (Weidner et al., 1965) because actinomycin D, an antibiotic that specifically inhibits gene transcription, blocked phytochrome-induced physiological responses that normally occur in continuous FR. These responses included enlargement of mustard cotyledons (Mohr et al., 1965), anthocyanin synthesis (Lange and Mohr, 1965), opening of the hypocotyl hook, and cotyledon unfolding (Schopfer, 1967). Puromycin, an antibiotic that inhibits protein synthesis, also inhibited light-dependent anthocyanin synthesis (Lange and Mohr, 1965).

In Belfast, Carr and D. M. Reid had discovered that actinomycin D inhibits the phytochrome-mediated unrolling of the barley seedling's first leaf and the opening of pea and bean hooks. Whereas Mohr envisioned an indirect interaction between phytochrome and genes, they were tempted to suggest "that the phytochrome may be very closely associated with the DNA as a suppressor; also, that red light may cause disruption of the DNA–phytochrome complex, whereas far-red immediately following red causes re-association" (Carr and Reid, 1966).

MEMBRANE PERMEABILITY

Hendricks and Borthwick were unable to accept Mohr's hypothesis because certain examples of phytochrome action seemed too rapid to involve gene transcription, which was thought to take several hours. These included chloroplast movement in *Mougeotia* (see Chapter 15) and floral induction in *Pharbitis nil* (see Chapter 13).

J. C. Fondeville of the Faculté des Sciences in Poitiers uncovered another very rapid phytochrome response during a visit to Beltsville. He studied phytochrome action in *Mimosa pudica,* whose leaflets fold when touched. They also fold at night and reopen in the morning, exhibiting the nyctinastic response that Bünning had studied in beans (see Chapter 2).

The response was observable 5 min after the onset of darkness and was complete in 30 min. The leaflets remained open for hours if irradiated with 2 min FR at the start of the dark period, but they closed within 30 min if a R pulse followed the FR. Spectrographic studies revealed that light near 660 nm best promoted the closing response, whereas the opening response was maximal between 720 and 750 nm, implicating phytochrome as the photoreceptor.

"The rapidity (minutes) of the *M. pudica* and *Mougeotia* responses . . . eliminates gene activation as the primary action of phytochrome," the researchers pronounced, pointing out that gene activation would require time for both de-repression of the gene and protein synthesis. "Controls of major metabolic pathways, on the other hand, can have time constants of the order of seconds or a few minutes and changes in cell permeabilities can be even faster" (Fondeville et al., 1966).

The *M. pudica* sleep movements seemed to result from sudden changes in cell permeability and hence turgor in the cells of the pulvinus, an organ at the base of the leaflet (Weintraub, 1950). So in the spring of 1967, the Beltsville group proposed that "phytochrome action is on membrane permeability rather than on enzyme action as a metabolic control, or on gene activation, which would be much slower" (Fondeville et al., 1967).

Forty-four years earlier, Garner and Allard had thought it "possible that highly refined regulation of the degree of hydration of the protoplasm furnishes a basis for the various responses of the plant to changes in the light period. . . . This implies, perhaps, control of permeability" (Garner and Allard, 1923).

Haupt was impressed with the membrane hypothesis, which fit his *Mougeotia* data so well. "I perhaps went too far," he says, "when I proposed—not in my papers but in lectures—that phytochrome is generally a membrane effector. Hans Mohr did not like this, and he was right. He convinced me after several years that there may be different responsivities in different systems and that one should not expect to find a master reaction. But I nevertheless thought that it was not impossible that the first step of gene activation might occur at a membrane, such as the nuclear membrane, although there was no evidence for such an assumption."

In October 1967, Hendricks and Borthwick submitted a paper to the National Academy of Sciences proposing that "an early consequence of phytochrome action is a change in permeability of the cells involved and possibly of intracellular components." Although Siegelman had extracted a soluble pigment, it was possible that some phytochrome remained attached to membrane, accounting for active phytochrome. Moreover, the hypothesis did not "require the presence of phytochrome in a membrane as contrasted to action of Pfr on a membrane" (Hendricks and Borthwick, 1967).

THE TANADA EFFECT

A colleague in the Soil and Water Conservation Division was at that time investigating the mechanism of radiation damage to plants. "I was using excised root tips," Takuma Tanada explains, "and I noticed that they stuck to the bottoms of Pyrex beakers when I rinsed them. This aroused my curiosity, and I soon found that certain chemicals prevented the sticking. This got me interested in whether red or far-red light could affect the adhesion. After many hours of work, I found a mixture of chemicals which would allow red/far-red reversibility of root-tip attachment."

The mixture contained ATP, indoleacetic acid, ascorbic acid, and ions of manganese, magnesium, and potassium. Moreover, it had to be swirled with just the right force and with no jerky movements. If all went well, the excised barley root tips stuck to glass within 30 s of R and detached just as quickly after FR. Because adhesion could be detected within 15 s, Tanada (1968a) proposed that it "is quite likely an initial expression of phytochrome action."

He pointed out that "because phosphate causes the glass surface to be negatively charged, adherence of the tips to the surface suggests that they are positively charged. Detachment takes place when the tips become either neutral or negatively charged. . . . The system reported in this paper also indicates that phytochrome is located near or in the cell membrane. Its reaction to external compounds and its quick response upon their addition point to such a location" (Tanada, 1968a). Mung bean root tips also exhibited R/FR-reversible attachment to glass—if calcium ions were added to the bathing solution (Tanada, 1968b).

BIOELECTRIC POTENTIAL CHANGES

In 1967, Mordecai (Mark) J. Jaffe and Galston at Yale and Hillman and Willard L. Koukkari at Brookhaven demonstrated phytochrome control of nyctinasty in *Albizzia julibrissin* (the silk tree), whose leaflets are nyctinastic but insensitive to touch (Hillman and Koukkari, 1967; Jaffe and Galston, 1967). Jaffe and Galston detected a greater electrolyte efflux from cut pinnules (leaflet stems) after R than FR treatment (Jaffe and Galston, 1967), suggesting that phytochrome altered membrane permeablility, although not necessarily directly.

Moving to Ohio University, Jaffe compared adhesion kinetics with those of bioelectric potential changes on the surface of a root tip immersed in "Tanada solution" and exposed to R or FR. The almost identical curves supported Tanada's hypothesis that adhesion resulted from the generation of a positive charge on the apical end of the tip. Jaffe concluded that "the development of a bioelectric potential . . . is one of the first events, if not the first, following photoreception and photoconversion of . . . phytochrome" (Jaffe, 1968), though he neglected to show that FR alone was ineffective (see Newman and Briggs, 1972).

Later studies suggested that hydrogen ions are extruded as the root tip becomes positive (Yunghans and Jaffe, 1970) and that acetylcholine, known to control ion movements across nerve cell membranes, accumulates both inside and outside the tissue. But a proposal that Pfr specifically stimulates acetylcholine production, which, in turn, increases membrane permeability, thus generating a positive bioelectric potential (Jaffe, 1970), was not supported by experimental evidence (Gressel et al., 1971; Kasemir and Mohr, 1972; Mohr, 1972; Satter et al., 1972).

Mohr argued that, although Pfr production could obviously change bioelectric potentials, that ability was not relevant to plant development. "Rather, this type of phytochrome-mediated response must be classified under the term 'fully-reversible modulations of highly-specialized cells,' " he said. "Unfortunately, these phenomena have been greatly overestimated as far as their significance for the action mechanism of Pfr in *morphogenesis* is concerned" (Mohr, 1972). Mohr opposed the general belief that the primary action of Pfr was the same in all responses, arguing that the pigment might activate genes in some cells but act via a different mechanism in other cells or cell compartments (Mohr, 1972).

NAD$^+$ KINASE

Ironically, the enzyme hypothesis was resurrected just after the Beltsville group laid it to rest. In 1968, Takafumi Tezuka and Yukio Yamamoto at Nagoya University proposed that phytochrome either possessed NAD$^+$ kinase activity or controlled the activity of such an enzyme, which phosphorylates the coenzyme nicotinamide adenine dinucleotide (NAD$^+$) to make nicotinamide dinucleotide phosphate (NADP$^+$). Thus a 5-min R light break after 8 h darkness tripled the NADP$^+$ level in *Pharbitis* cotyledons and increased NAD$^+$ kinase activity in cotyledon extracts by 50%. The first response was partly reversed by FR. Moreover, blue fractions from brushite and Sephadex columns had kinase activity that increased after exposure to R (Tezuka and Yamamoto, 1969; Yamamoto and Tezuka, 1972). The idea was appealing because NAD$^+$ is a cofactor for reactions that dismantle carbon compounds, whereas NADPH is involved in synthetic processes, such as photosynthesis. By converting NAD$^+$ to NADP+, phytochrome could cause a major shift in metabolism—a fitting action for a developmental regulator.

Unfortunately, the work proved unrepeatable (Hopkins and Briggs, 1973). But the kinase hypothesis surfaced again in the late 1980s (see Chapter 29), serving as a reminder that the enzyme hypothesis remains tenable, despite emphasis on phytochrome's possible action on membranes or genes.

REFERENCES

Carr, D. J., and Reid, D. M. (1966). Actinomycin-D inhibition of phytochrome-mediated responses. *Planta* 69,70–78.
Fondeville, J. C., Borthwick, H. A., and Hendricks, S. B. (1966). Leaflet movement of *Mimosa pudica* L. indicative of phytochrome action. *Planta* 69,357–364.

Fondeville, J. C., Schneider, M. J., Borthwick, H. A., and Hendricks, S. B. (1967). Photo-control of *Mimosa pudica* L. leaf movement. *Planta* 75,228–238.

Garner, W. W., and Allard, H. A. (1923). Further studies in photoperiodism, the response of the plant to relative length of day and night. *J. Agric. Res.* 23,871–920.

Gressel, J., Strausbauch, I., and Galun, E. (1971). Photomimetic effect of acetylcholine on morphogenesis in *Trichoderma*. *Nature* 232,648–649.

Hendricks, S. B. (1963). Metabolic control of timing. *Science* 141,21–27.

Hendricks, S. B. and Borthwick, H. A. (1967). The function of phytochrome in regulation of plant growth. *Proc. Natl. Acad. Sci. USA* 58,2125–2130.

Hillman, W. S. (1967). The physiology of phytochrome. *Annu. Rev. Plant. Physiol.* 18,301–324.

Hillman, W. S., and Koukkari, W. L. (1967). Phytochrome effects in the nyctinastic leaf movements of *Albizzia julibrissin* and some other legumes. *Plant Physiol.* 42,1413–1418.

Hock, B., and Mohr, H. (1964). Die Regulation der O_2-Aufnahme von Senfkeimlingen (*Sinapis alba* L.) durch Licht. *Planta* 61,209–228.

Hopkings, D. W., and Briggs, W. R. (1973). Phytochrome and NAD kinase: A reexamination. *Plant Physiol.* 51(suppl. 52).

Jacob, F., and Monod, J. (1961). Genetic regulatory mechanisms in the synthesis of proteins. *J. Mol. Biol.* 3,318–356.

Jaffe, M. J. (1968). Phytochrome-mediated bioelectric potentials in mung bean seedlings. *Science* 162,1016–1017.

Jaffe, M. J. (1970). Evidence for the regulation of phytochrome-mediated processes in bean roots by the neurohumor, acetylcholine. *Plant Physiol.* 46,768–777.

Jaffe, M. J., and Galston, A. W. (1967). Phytochrome control of rapid nyctinastic movements and membrane permeability in *Albizzia julibrissin*. *Planta* 77,135–141.

Karlson, P. (1963). New concepts on the mode of action of hormones. *Perspect. Biol. Med.* 6,203–214.

Kasemir, H., and Mohr, H. (1972). Involvement of acetylcholine in phytochrome-mediated processes. *Plant Physiol.* 49,453–454.

Lange, H., and Mohr, H. (1965). Die Hemmung der Phytochrom-induzierten Anthocyan-synthese durch Actinomycin D und Puromycin. *Planta* 67,107–121.

Mohr, H. (1966a). Differential gene activation as a mode of action of phytochrome 730. *Photochem. Photobiol.* 5,469–483.

Mohr, H. (1966b). Untersuchungen zur phytochrominduzierten Photomorphogenese des Senfkeimlings (*Sinapis alba* L.). *Z. Pflanzenphysiol.* 54,63–83.

Mohr, H. (1972). "Lectures on Photomorphogenesis." Springer, New York, Heidelberg, Berlin.

Mohr, H., Schlickewei, I., and Lange, H. (1965). Die Hemmung des Phytochrom-induzierten Kotyledonenwachstums durch Actinomycin D. *Z. Naturforsch.* 20b,819–821.

Newman, I. A., and Briggs, W. R. (1972). Phytochrome-mediated electric potential changes in oat seedlings. *Plant Physiol.* 50,687–693.

Satter, R. L., Applewhite, P. B., and Galston, A. W. (1972). Phytochrome controlled nyctinasty in *Albizzia julibrissin*. V. Evidence against acetylcholine participation. *Plant Physiol.* 50,523–525.

Schopfer, P. (1967). Die Hemmung der phytochrominduzierten Photomorphogenese ("positive" Photomorphosen) des Senfkeimlings (*Sinapis alba* L.) durch Actinomycin D und Puromycin. *Planta* 72,297–305.

Tanada, T. (1968a). A rapid photoreversible response of barley root tips in the presence of 3-indoleacetic acid. *Proc. Natl. Acad. Sci. USA* 59,376–380.

Tanada, T. (1968b). Substances essential for a red, far-red light reversible attachment of mung bean root tips to glass. *Plant Physiol.* 43,2070–2071.

Tezuka, T., and Yamamoto, Y. (1969). NAD kinase and phytochrome. *Bot. Mag.* 82,130–133.

Weidner, M., Jakobs, M., and Mohr, H. (1965). Über den Einfluβ des Phytochroms auf den Gehalt an Ribonucleinsäure und Protein in Senfkeimlingen (*Sinapis alba* L.). *Z. Naturforsch.* 20b,689–693.

Weintraub, M. (1950). Leaf movement in *Mimosa pudica* L. *New Phytol.* 50,357–382.

Yamamoto, Y., and Tezuka, T. (1972). Regulation of NAD kinase by phytochrome and control of metabolism by variation of NADP level. *In* "Phytochrome" (K. Mitrakos and W. Shropshire, Jr., eds.), pp. 407–429. Academic Press, London, New York.

Yunghans, H. O., and Jaffe, M. J. (1970). Phytochrome-controlled adhesion of mung bean root tips to glass: A detailed characterization of the phenomenon. *Physiol. Plant* 23,1004–1016.

CHAPTER **17** _____

Hartmann Model

"I have delayed replying to your letter . . . because of the unsettled immediate future of my laboratory," Borthwick explained to a prospective visitor in September 1966. " I must retire because of age in January, 1968, and upon my retirement the laboratory, I find, will be discontinued" (Borthwick, 1966). Unwilling to be reassigned, Downs and Siegelman accepted positions in other institutions.

Siegelman joined the staff of Brookhaven National Laboratory in 1965, where, after visiting Haxo and Colm Ó hEocha, he developed methods for purifying large amounts of algal biliproteins. After samples were subjected to nuclear magnetic resonance spectroscopy at Argonne National Laboratory (Chapman et al., 1967; Crespi et al., 1967), Siegelman predicted the structure of the phytochrome chromophore (Siegelman et al., 1968). The proposed formula (Fig. 17-1) was correct except in the D-ring of the tetrapyrrole (see Fig. 27-4).

Siegelman left the phytochrome field at that point but continued to work on algal biliproteins. "Sterling Hendricks and I decided" he says, "that, at that point in time, the tools were not available to make reasonable progress."

Downs left Beltsville in 1965 to direct the phytotron at North Carolina State University in Raleigh. He put his knowledge of phytochrome to practical use (Downs, 1975, 1983) "to enable us to grow plants under electric lights that resemble those in the field." Butler had left Norris's lab in 1964, spending a sabbatical year with Britton Chance at the Johnson Foundation before moving to the University of California, San Diego.

HIGH-ENERGY REACTION

During the mid 1960s, interest in the high-energy reaction revived, both at Beltsville and in Europe. Although Hendricks had identified phytochrome

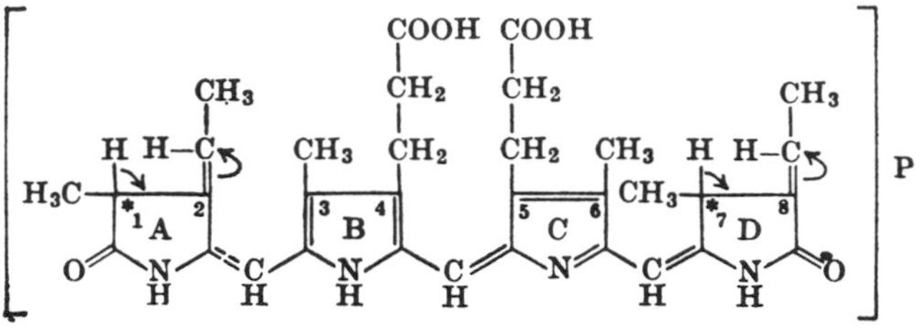

FIGURE 17-1   Siegelman's predicted structure of phytochrome chromophore. He proposed that photoconversion of Pr to Pfr occurred by conversion of ethylidene groups in rings A and D to ethyl groups. This would create a double bond in each ring. (From Siegelman et al., 1968.)

as the photoreceptor in 1959, he changed his mind after an unexpected finding. Downs was studying anthocyanin synthesis in milo sorghum (*Sorghum vulgare*) seedlings (Downs, 1961), which produce the red pigment only when illuminated. A high-energy reaction appeared to trigger synthesis (Downs and Siegelman, 1963), but its action spectrum (Fig. 17-2) "turned out to be more than we anticipated," says Downs, "because it showed that anthocyanin is only produced in blue light. You get none in the longer wavelengths."[1]

The group discounted its previous proposal and invoked the involvement of a photosynthetic pigment in the HER (Hendricks and Borthwick, 1965). Downs had reached a similar conclusion for anthocyanin synthesis in apple skin on the basis of the action spectrum (maximum near 650 nm and subsidiary peak between 420 and 470 nm) and response to DCMU, an inhibitor of photosynthetic electron transport (Downs, 1964; Downs et al., 1965).

The hypothesis appeared not to be applicable to inhibition of hypocotyl lengthening in lettuce and petunia seedlings. Deriving action spectra, the researchers considered "the possibility that the high-energy reaction is mediated by two photoprocesses, one activated by blue alone or by blue and far-red light, and the other by red and far-red light. . . . As to action in the red, far-red region, it is possible that this is mediated by phytochrome" (Evans et al., 1965).

In Reading, Vince had long suspected the involvement of a separate B receptor (Sale and Vince, 1959; Heath and Vince, 1962; Vince and Grill,

1. Subsequent studies at Freiburg revealed that anthocyanin synthesis in milo sorghum requires the sequential action of a blue/UV-A photoreceptor (an unidentified pigment named *cryptochrome*) and phytochrome and that phytochrome also regulates the expression of the B effect (Drumm and Mohr, 1978). A third photoreceptor, which absorbs in the UV-B region (280–320 nm) interacts synergistically with cryptochrome (Drumm-Herrel and Mohr, 1981).

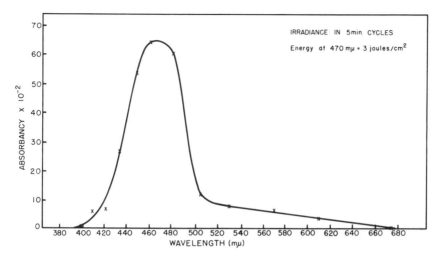

IRRADIANCE IN 5min. CYCLES

Energy at 470 mμ = 3 joules/cm^2

FIGURE 17-2   Action spectrum for first photoreaction that induces anthocyanin formation in Wheatland milo seedlings. (From Downs and Siegelman, 1963. Reproduced with permission from the American Society of Plant Physiologists.)

1966). During an extensive study of high-energy responses, she reported differential sensitivity of Great Lakes lettuce seedlings to the two ends of the spectrum. Whereas 24 h B inhibited hypocotyl growth in dark-grown seedlings regardless of age, 24 h FR became progressively less inhibitory, having little effect on seedlings more than 60 h old. And when dark-grown seedlings were illuminated for 72 h, their growth rate became and remained sharply reduced in B but declined in FR only for 24 h. Once seedlings had become insensitive to FR, moreover, they could still be inhibited by B. Thus "two separate reactions are involved, and we conclude that a distinct blue light absorbing pigment is present, which mediates a reaction inhibitory to elongation" (Turner and Vince, 1969).

KARL M. HARTMANN

The first convincing evidence for phytochrome's participation in high-energy responses came from Freiburg, where opposition to Hendricks' original proposal had been strongest. At the Rouen symposium where Mohr outlined his gene activation theory, Karl M. Hartmann (Fig. 17-3) proposed "a general hypothesis to interpret 'high energy phenomena' of photomorphogenesis on the basis of phytochrome" (Hartmann, 1966).

After receiving his doctorate in Tübingen in 1962, Hartmann followed Mohr to Freiburg. "The 1959 proposal by Hendricks and Borthwick—that the noticeably variable action spectra of the [HER] might result from the

FIGURE 17-3   Karl M. Hartmann, 1967.

simultaneous excitation of the two interconvertible forms of phytochrome—attracted my attention," he explains. "The main hindrance
against this hypothesis was that both synergistic and antagonistic interactions between phytochrome and the [HER] had been documented con-
vincingly by Hans Mohr (Mohr, 1964)."

The way to test the Beltsville hypothesis "was simply that if you could
achieve simultaneous excitation of the interconvertible pigment phytochrome," Hartmann adds, "you should be able to destroy and also
imitate the spectral band around 720 nm of the HER. . . . Already in my
thesis (on the regulation of gametogenesis in *Chlamydomonas*) I had been
working with simultaneous exposure with two spectral bands. This principle was derived from the basic work of Otto Warburg (one of my grandfa-
ther's colleagues at the Kaiser Wilhelm Institute in Berlin), on the catalytic
effects of blue light combined with red light on photosynthesis—the enhancement effect of Emerson. And so I had the idea that if there was an
interaction of the two pigment forms, I would get some information by
irradiation with not one but two spectral bands and changes in the quantum
flux ratio."

Germination of Grand Rapids lettuce seed was known to be promoted by
the low-energy reaction but still inhibited by the HER during its later
stages (Mohr and Appuhn, 1963). Although this appeared to argue against
a common photoreceptor, Hartmann realized that phytochrome might
mediate both responses if

   1.   the later stages of germination are promoted by a low level of Pfr
that persists in darkness, which seemed likely because the response has a
negative temperature dependence,

FIGURE 17-4   Part of Freiburg Xenosol-irradiation unit to make high-power monochromatic and dichromatic light. Note Hartmann's sign on the wall.

2. high-energy exposure reduces Pfr concentration to a suboptimal level for germination—Hartmann had observed common characteristics for Pfr destruction and the HER, including oxygen dependence and negative temperature dependence (Hartmann, 1966).

If these two assumptions were valid, the HER would operate most efficiently at a specific, intermediate level of Pfr, which could be determined experimentally.

Hartmann began his experiments on July 27, 1964, while Mohr was attending the Fourth International Photobiology Congress in Oxford. He placed dishes of seeds in a black room in the basement; monochromatic beams from light sources in an adjacent room were conveyed through a wall. The sources were low-voltage incandescent and xenon-arc projectors plus double interference filters, which produced extremely pure bands of monochromatic light (Fig. 17-4).

Four hours after R induction, when escape from FR reversibility was accomplished and about 97% germination was expected, Hartmann exposed Grand Rapids lettuce seed to 1 h high-energy light mixtures that established various phytochrome photoequilibria by superimposing

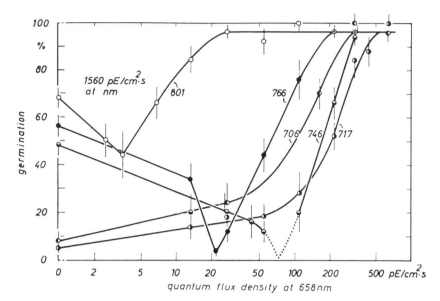

FIGURE 17-5   Irradiation of Grand Rapids lettuce seed with different flux densities of R and a constant high-irradiance background of different FR bands. Imbibed seeds were induced to germinate with 1 min R. After a further 3 h, they were given a 1-h exposure to the R + FR mixtures. (From Hartmann, 1966.)

different low intensities of R (658 nm) on high-intensity background FR. He used various FR wavelengths but fixed the photon flux to about 1,560 pmol cm^{-2} s^{-1}. He thus obtained fractions of Pfr ranging from near zero (766 or 801 nm alone, because these wavelengths are absorbed almost exclusively by Pfr) to the maximum value.

Between 57 and 68% of the seeds still germinated (Fig. 17-5, vertical axis) when the photoequilibrium was kept near zero for 1 h because escape from FR reversibility was realized, almost no Pfr destruction was expected to occur, and according to the model, the HER was feeble because the pigment was only weakly cycling between its two forms. As the initial proportion of Pfr rose, germination percentages fell, reaching a minimum when the photon flux of R was 22 pmol cm^{-2} s^{-1} (about 1.4% of the deep-red at 766 nm) and the fraction of Pfr was about 9% (Table 17-1). As Pfr levels increased further, the inhibitory high-energy effect decreased, and germination reached 90% when more than 30% of the phytochrome was in the Pfr form.

Background radiation of 706 nm, which by itself establishes a Pfr/P$_{total}$ of about 12%, was inhibitory (Fig. 17-5, vertical axis). Again, the 90%

TABLE 17-1

Photoequilibria [Pfr]/[P$_{total}$] Calculated for Minimal and 90% Germination in Fig. 17-5[a]

1,560 pE/cm^2·sec at λ_1 (nm)	N (pE/cm^2·sec) at $\lambda_2 = 658$ nm	[Pfr]/[P] for minimal germination (%)
801	3.5	9.4
766	22	8.8
746	70	10.1
717	<100	<12.8
706	< 0	<12.2
		for 90% germination
801	17	31.3
766	150	36.5
746	310	31.3
717	420	32.2
706	250	30.3
445	< 12	<47.0

[a] From Hartmann, 1966.

germination level was reached when this wavelength combined with R gave a Pfr/P$_{total}$ of about 30%. A background of 801 nm, which is poorly absorbed by phytochrome, was less inhibitory even when combined with adequate R, because the energy absorbed by phytochrome was insufficient for a pronounced HER. But fewest seeds again germinated when the Pfr/P$_{total}$ was near 9%—a value similar to that obtained with the 766-nm background. Thus a typical Pfr level—about 9%—was optimal for HER operation in this system. This supported Hartmann's initial assumptions, implicated phytochrome as a photoreceptor, and tied the HER to phytochrome photodestruction.

In parallel, Hartmann performed experiments to explain why the low-energy reaction and HER could be synergistic, as in hypocotyl lengthening in lettuce seedlings, which showed a typical HER spectrum with strong peaks in the B and FR (Mohr and Wehrung, 1960). The assumptions were, again, that Pfr is prone to destruction but, second, that already low levels of Pfr inhibit hypocotyl lengthening. To determine optimal Pfr levels for the HER, Hartmann irradiated hypocotyls simultaneously with R and FR. With a FR of 717 nm, which by itself forms about 3% Pfr and gives a strong HER, inhibition of lengthening faded out when the Pfr/P$_{total}$ photoequilibrium reached about 28%. With deep-red radiation of 768 nm, the HER was most marked at the irradiance ratio (R/deep-red about 1/200) that gave a very similar Pfr/P$_{total}$ of about 3% (Fig. 17-6). No HER was evident with this pure deep-red band alone, however (vertical axis).

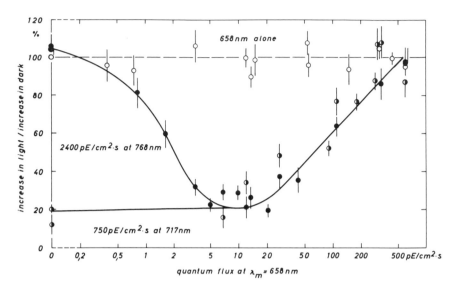

FIGURE 17-6   Hypocotyl increase of *Lactuca sativa* under simultaneous irradiation with different quantum fluxes in R at a constant high quantum flux background of different FR bands. (From Hartmann, 1967b.)

The demonstration that two wavelengths that were separately unable to support an HER could do so together if they produced a suitable phytochrome equilibrium was strong evidence indeed for phytochrome involvement.

In December, Hartmann revealed his data at an open lecture in Freiburg. Then he refined his model, assigning rate constants to the known reactions of phytochrome (Fig. 17-7). The mathematical model revealed that Pfr would disappear more quickly under high-energy conditions as Pfr/P_{total} increased.

Mathematical elaboration supported other features of Hartmann's model, predicting that the action ratio of two different wavelengths would be a function of irradiation time. This had already been observed in several systems, including the lettuce hypocotyl. Moreover the action spectra calculated for different irradiation times conformed remarkably well to experimental data.

Hartmann considered the possibility that an intermediate in phytochrome photoconversion might function as an HER effector molecule. Computing the spectral cycling rate of phytochrome, which should reflect the spectral accumulation of intermediates, he found no conformity with high-energy spectra.

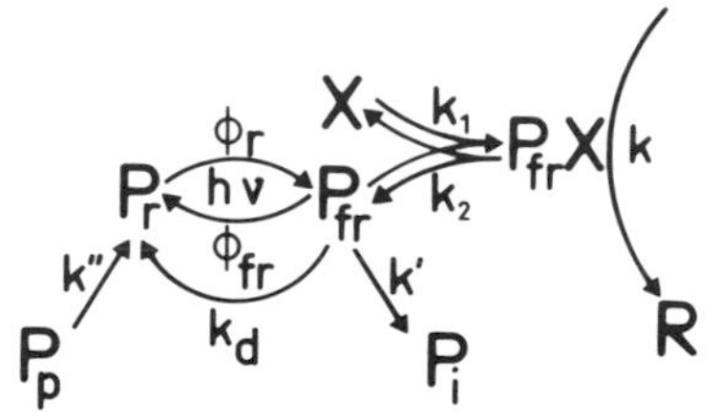

FIGURE 17-7   Hartmann's summary of phytochrome reactions. (From Hartmann, 1966.)

He demonstrated later that the FR action at about 716 nm could be nullified by a mixture of sufficiently high irradiances of deep-red at about 759 nm (Fig. 17-8) (Hartmann, 1967b). In this case, the photoconversion or cycling rate of Pfr and Pr—and consequently the stationary concentration of all intermediates—is higher than for FR alone, whereas the stationary concentration of Pfr approaches the theoretical minimum for deep-red— below 1% Pfr/P_{total}. The data led Hartmann to conclude that Pfr in the electronic ground state is the specific effector of the HER in the FR region and to exclude intermediates of the photoconversion itself (Hartmann, 1967b; Hartmann and Cohnen Unser, 1973).

The model had to account for the irradiance dependency of the HER, for the fraction of phytochrome present as Pfr is mainly dependent on wavelength not quantum flux. In further bichromatic experiments, Hartmann exposed lettuce seedlings to R of increasing irradiance while increasing the irradiance of the added deep-red partner accordingly. This way the photoequilibrium for maximum inhibition of hypocotyl lengthening remained the same, but the magnitude of the response varied. When response was plotted against R photon flux, the slope paralleled that for 716 nm light alone (Fig. 17-9). Thus the total energy absorbed by Pr (from R) and Pfr (at each deep-red wavelength) determined both the fixed photoequilibrium and the extent of the HER (Hartmann, 1967b).

Calculations revealed that the first step of the biological response would be just as dependent on irradiance as was phytochrome destruction, provided there were more Pfr molecules than molecules of X, the hypothetical substance with which Pfr interacted. Thus the postulate, derived from Hillman's *Pisum* data (Hillman, 1965), that the phytochrome pool of etiolated seedlings was larger than the available pool of binding sites was essential to the model (Hartmann, 1966).

Because Pfr formation saturates at energies below those required for the HER, Hartmann proposed that the HER is driven by the effector complex Pfr$X$—but only as long as free X is available as a reaction partner for newly

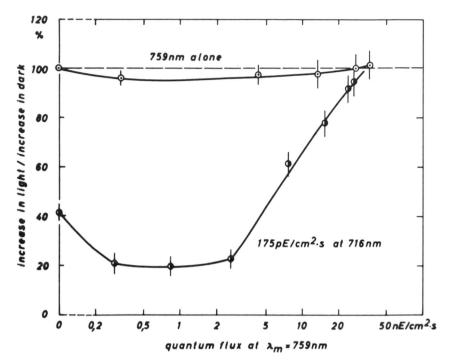

FIGURE 17-8 Hypocotyl lengthening of *Lactuca sativa*. Irradiation with difference quantum fluxes of deep-red (759 nm) at a constant quantum flux background of FR (716 nm). (From Hartmann, 1967b.)

formed Pfr just resulting from photoconversion; he named this short-lived species *overcritical activated Pfr*. Pfr would thus induce a biological response in two separate ways. At the start of photoconversion in low irradiances, it would combine with X directly to form Pfr$X$; resulting responses would be most marked in R, which produces a high level of Pfr. During long-term exposures at high irradiances, phytochrome would be repeatedly converted to overcritically activated Pfr, which would form an efficient effector complex only if sufficient free X were available as a binding partner. HER action spectra would therefore have peaks at low fractions of Pfr in the FR and B rather than in the R, because these wavelengths would favor the formation of overcritically activated Pfr. In the R, slower cycling and less free X would result in competitive self-inhibition.

Hartmann's model was roundly criticized. "It should be clear by now," said Hillman, "that its very basis—the idea that all, or at least a constant proportion of the phytochrome present in a tissue is active and that thus

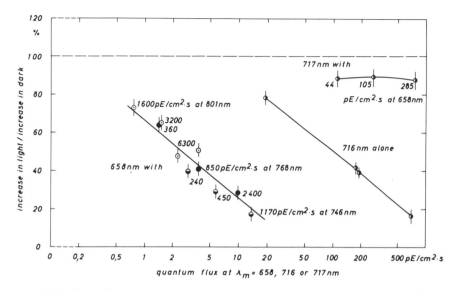

FIGURE 17-9   Hypocotyl lengthening in lettuce seedlings as a function of irradiance during (*left slope*) bichromatic irradiation with different quantum fluxes of R combined with deep-red wavelengths of constant flux and (*right slope*) monochromatic irradiation at 716 nm. (From Hartmann, 1967b).

destruction as observed in etiolated tissues is of direct physiological signif-icance—has little, if any, experimental footing" (Hillman, 1967).

Few people, moreover, grasped the concept of overcritically activated Pfr, which was mistakenly assumed to be an electronically excited species. "The concept comes from Arrhenius's theory," explains Hartmann, "that only a small fraction of molecules in any population has enough energy to react. I had the idea that newly formed Pfr just coming out from photocon-version is more likely to be overcritically activated than a molecule that has already been Pfr for some time. It might be a little bit warmer and therefore more able to react if it contacted a partner. But this has been repeatedly misunderstood."

During the Fifth International Congress on Photobiology in Hanover in 1968, Hartmann answered his critics with new spectrophotometric data on phytochrome, obtained with Spruit's dual-wavelength spectrophotometer in Wageningen (Hartmann, 1968; Hartmann and Spruit, 1968).

If the hypothesis was not accepted in its entirety, it was very influential. Whereas Hendricks' original proposal could be dismissed as wild specula-tion, Hartmann had compelling evidence for the involvement of phy-tochrome in the HER.

Hartmann made use of the superb light sources at Freiburg to derive an extremely detailed action spectrum for the response of lettuce seedling hypocotyls to prolonged irradiation (from 54 to 72 h after sowing) (Fig. 17-10). The curve extended all the way from 320 to 1,100 nm. Its first version had a sharp peak centered at 716 nm—later refined to 720 nm—which "agrees with the conclusion that the far-red peak of action in the case of long-time irradiation is exclusively due to phytochrome," Hartmann reported. "The fine structure of the action spectrum in the blue and ultraviolet range (especially the peak around 363 nm) points to a flavoprotein as the photoreceptor" (Hartmann, 1967a). But despite this similarity, "phytochrome is not excluded to act as photoreceptor in this range" (Hartmann, 1967b).

Simultaneous irradiation with B and deep-red (Hartmann and Menzel, 1969) led to the proposal that "there was an interaction between the two so that one was sure that the short wavelength part of the HER cannot be interpreted solely on the basis of a separate flavoprotein but that there must also be some contribution from phytochrome." Later, while Hartmann was a visiting associate professor at Purdue University, he argued that phytochrome could be solely responsible for some plant re-

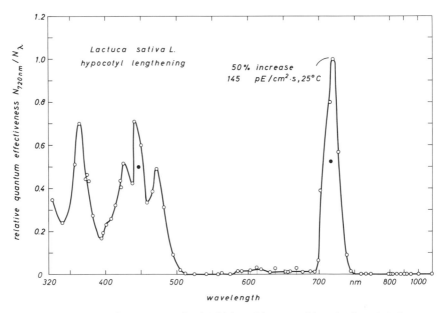

FIGURE 17-10   Action spectrum for inhibition of hypocotyl lengthening of dark-grown *Lactuca sativa* irradiated from 54 to 72 h after sowing. (From Hartmann, 1967a).

sponses in the B and UV and "that Pfr is an effector molecule for the blue-UV-mediated photoresponses too" (Hartmann and Cohnen Unser, 1973).

END OF AN ERA

The Beltsville researchers were not at first convinced by Hartmann's evidence that phytochrome mediates the FR peak of the HER. Deriving action spectra for suppression of germination of *Poa pratensis* and *Amaranthus arenicola* seeds by prolonged irradiation, they obtained a peak at 720 nm. But when they applied R + FR mixtures, the response appeared not to relate to the proportion of phytochrome present as Pfr (Hendricks et al., 1968).

Hartmann attributes this to the impurity of the radiation sources. "The problem was that if you don't work with sufficiently clean deep-red," he explains, "you cannot do enhancement experiments by admixing a small amount of red because the response is already enhanced. Conversely, you cannot nullify the peak of action at 720 nm by simultaneously irradiating at longer wavelengths with deep-red."

During a visit to the United States in 1968, Hartmann explained to Hendricks "that very pure monochromatic beams can only be obtained by additional filtering through 3 or even 6 nm of black SCHOTT RG9 glass." Presumably Hendricks took this advice, because the group later observed maximal suppression of germination in *Amaranthus* at a specific proportion of Pfr, namely, 16% (Fig. 17-11). This convinced Hendricks and Borthwick that Hartmann was correct, at least in identifying the photoreceptor. "Attempts by us and others to involve light absorption in the HER by pigments other than phytochrome have generally led back to phytochrome," they conceded (Borthwick et al., 1969).

They offered an embroidered version of Hartmann's model, arguing that the original failed to account for HER responses under conditions where the proportion of Pfr was high, for the irradiance dependence of the HER, and for the photochemical properties of Pfr$X$.[2] They proposed that Pfr$X$ had a higher absorbancy between 710 and 720 nm than Pfr and a much lower one between 600 to 680 nm.

The model (Fig. 17-12) contained two additional components: a photoreaction (5) that produced a protonated form of Pfr,[3] and a photoreaction

2. Paul Rollin, a plant physiologist in Rouen, first drew attention to the photochemical properties of Pfr$X$ (Rollin and Maignan, 1966).

3. The existence of such a protonated form had just been proposed by researchers in Delaware (see Chapter 21).

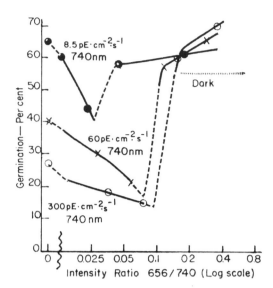

FIGURE 17-11   Germination of *Amaranthus arenicola* seed under continuous irradiation at 740 nm at three irradiances with accompanying 656-nm radiation of various irradiances. (From Borthwick et al., 1969.)

(4) that dissociated Pfr$X$ and was assumed to be irradiance-dependent and to occur much faster than the association of Pfr and X in the dark. The physiological response obtained would therefore depend on the balance between Pfr$X$ formation (reaction 3_d) and photodissociation (reaction 4). If the latter predominated, the plant would exhibit an HER.

The Beltsville workers plucked examples from their vast bank of data on flowering, seed germination, and leaflet movement. They had determined, for example, that leaflets of *Mimosa pudica* maintained under conditions that establish high Pfr/P_{total} ratios closed within 20 min of darkening, a repeatedly reversible response (reaction 1) (Fondeville et al., 1966). Maintenance of the open position in the light showed a typical HER action spectrum with a maximum near 720 nm (Fondeville et al., 1967). Thus reaction 1 was opposed by a second photoreaction, such as reaction 4, they argued. Inhibition of germination by prolonged irradiation suggested a similar interplay between two light reactions. In short, the HER operated most effectively near 720 nm because the Pfr$X$ needed for the low-energy reaction photodissociated maximally at that wavelength (Borthwick et al., 1969).

"This proposal," Hartmann says, "cannot explain why critical mixtures of red and deep-red can mimic the HER. It also fails to predict that

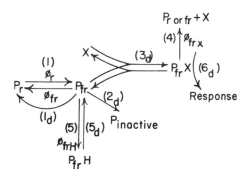

FIGURE 17-12   Beltsville 1968 HER model. Photoreactions 1, 4, and 5 are accompanied by dark reactions 1_d, 2_d, 3_d, and 6_d. Reaction 5 is probably dark-reversible but not photo-reversible. (From Borthwick et al., 1969.)

the most efficient band around 720 nm can be nullified by high-intensity admixture of deep-red around 720 nm.''

Although the model was not widely accepted, it drew attention to the photochemical properties of Pfr$X$. And the paper became a memento to Borthwick and Hendricks' long collaboration, for it was their last joint publication. Borthwick retired in 1968 on his 70th birthday. He was honored[4] by a Symposium on Photoperiodism at Beltsville.

The closing of the Plant Physiology Lab disbanded an exemplary group of scientists who had pooled their knowledge of plant physiology, chemistry, physics, mathematics, electronics, and horticulture to unfold "a whole new area of knowledge about plants" (Hendricks, 1976). During 25 years, they had deduced the existence of a photoreversible photoreceptor, detected the pigment in living tissues, uncovered reversion and destruction, named, extracted, and partly purified phytochrome, established its basic properties, and explored many of its actions in whole plants. Their hypotheses were not foolproof, for they clung stubbornly to their notion of hourglass photoperiodic timing and refused to believe that phytochrome might regulate genes. But with their daring ideas and scrupulous experimental technique, they turned the pigment of the imagination into a molecule that could be studied in the laboratory—and one whose mode of action in the plant would surely one day be revealed.

4. Borthwick received many honors during the latter part of his career. He was elected to the National Academy of Sciences in 1961, and in 1962 he and Hendricks shared both the Hoblitzelle Award in Agricultural Sciences and the Stephen Hales Award from the American Society of Plant Physiologists. He was also President of that society from 1955 to 1956. Other honors included the Joachem-Hafiz Award from Switzerland and the USDA Distinguished Service Award.

REFERENCES

Borthwick, H. A. (1966). Letter to S. C. Bhargava in New Delhi, September 28. Unpublished.

Borthwick, H. A., Hendricks, S. B., Schneider, M. J., Taylorson, R. B., and Toole, V. K. (1969). The high-energy light action controlling plant responses and development. *Proc. Natl. Acad. Sci. USA* 64,479–486.

Chapman, D. J., Cole, W. J., and Siegelman, H. W. (1967). Chromophores of allophycocyanin and R-phycocyanin. *Biochem. J.* 105,903–905.

Crespi, H. L., Boucher, L. J., Norman, G. D., Katz, J. J., and Dougherty, R. C. (1967). Structure of phycocyanobilin. *J. Am. Chem. Soc.* 89,3642–3643.

Downs, R. J. (1961). Photocontrol of anthocyanin synthesis in dark-grown milo seedlings. *Plant Physiol.* 36(suppl. xlii).

Downs, R. J. (1964). Photocontrol of germination of seeds of the Bromeliaceae. *Phyton* 21,1–6.

Downs, R. J. (1975). "Controlled Environments for Plant Research." Columbia University Press, New York, London.

Downs, R. J. (1983). Climate simulations. *In* "Beltsville Symposia in Agricultural Research 6: Strategies of Plant Reproduction" (W.J. Meudt, ed.), pp. 351–368. Allanheld, Osmun & Co., London, Toronto, Sydney.

Downs, R. J., and Siegelman, H. W. (1963). Photocontrol of anthocyanin synthesis in milo seedlings. *Plant Physiol.* 38,25–30.

Downs, R. J., Siegelman, H. W., Butler, W. L., and Hendricks, S. B. (1965). Photoreceptive pigments for anthocyanin synthesis in apple skin. *Nature* 205,909–910.

Drumm, H., and Mohr, H. (1978). The mode of interaction between blue (UV) light photoreceptor and phytochrome in anthocyanin formation of the *Sorghum* seedling. *Photochem. Photobiol.* 27,241–248.

Drumm-Herrel, H., and Mohr, H. (1981). A novel effect of UV-B in a higher plant (*Sorghum vulgare*). *Photochem. Photobiol.* 33,391–398.

Evans, L. T., Hendricks, S. B., and Borthwick, H. A. (1965). The role of light in suppressing hypocotyl elongation in lettuce and *Petunia*. *Planta* 64,201–218. Copyright 1965, Springer-Verlag.

Fondeville, J. C., Borthwick, H. A., and Hendricks, S. B. (1966). Leaflet movement of *Mimosa pudica* L. indicative of phytochrome action. *Planta* 69,357–364.

Fondeville, J. C., Schneider, M. J., Borthwick, H. A., and Hendricks S. B. (1967). Photocontrol of *Mimosa pudica* L. leaf movement. *Planta* 75,228–238.

Hartmann, K. M. (1966). A general hypothesis to interpret 'high energy phenomena' of photomorphogenesis on the basis of phytochrome. *Photochem. Photobiol.* 5,349–366.

Hartmann, K. M. (1967a). Ein Wirkungsspektrum der Photomorphogenese unter Hochenergiebedingungen und seine Interpretation auf der Basis des Phytochroms (Hypokotylwachstumshemmung bei *Lactuca sativa* L.). *Z. Naturforsch.* 22b,1172–1175.

Hartmann, K. M. (1967b). Photoreceptor problems in photomorphogenic responses under high-energy conditions (UV-blue-far-red). Book of abstracts. European Photobiology Symposium, Hvar, Yugoslavia, pp. 29–32.

Hartmann, K. M. (1968). Identification of the photoreceptor of the 'high energy photomorphogenic reaction' for the far-red region. *Int. Congr. Photobiol. 5th.* Abstracts, p. 2.

Hartmann, K. M., and Cohnen Unser, I. (1973). Carotenoids and flavins versus phytochrome as the controlling pigment for blue-uv-mediated photoresponses. *Z. Pflanzenphysiol.* 69,109–124.

Hartmann, K. M., and Menzel, H. (1969). Phytochrome, a photoreceptor for blue-uv-mediated responses? *Int. Bot. Congr. 11th Abstracts,* p. 85.

Hartmann, K. M., and Spruit, C. J. P. (1968). Difference spectra and photostationary states for phytochrome *in vivo. Plant Physiol.* 43(suppl. 15).

Heath, O. V. S., and Vince, D. (1962). Some non-photosynthetic effects of light on higher plants with special reference to wave-length. *Symp. Soc. Exp. Biol.* 16,110–137.

Hendricks, S. B. (1976). Harry Alfred Borthwick, 1898–1974. *Biogr. Mem. Natl. Acad. Sci. USA* 48,105–122.

Hendricks, S. B., and Borthwick, H. A. (1965). The physiological functions of phytochrome. *In* "Chemistry and Biochemistry of Plant Pigments" (T. W. Goodwin, ed.), pp. 405–436. Academic Press, New York, London.

Hendricks, S. B., Toole, V. K., and Borthwick, H. A. (1968). Opposing actions of light in seed germination of *Poa pratensis* and *Amaranthus arenicola. Plant Physiol.* 43,2023–2028.

Hillman, W. S. (1965). Phytochrome conversion by brief illumination and the subsequent elongation of etiolated *Pisum* stem segments. *Physiol. Plant.* 18,346–358.

Hillman, W. S. (1967). The physiology of phytochrome. *Annu. Rev. Plant Physiol.* 18,301–324.

Mohr, H. (1964). The control of plant growth and development by light. *Biol. Rev. Cambridge Philos. Soc.* 39,87–112.

Mohr, H., and Appuhn, U. (1963). Die Keimung von *Lactuca*-Achänen unter dem Einfluß des Phytochromsystems und der Hochenergie Reaktion der Photomorphogenese. *Planta* 60,274–288.

Mohr, H., and Wehrung, M. (1960). Die Steuerung des Hypokotylwachstums bei den Keimlingen von *Lactuca sativa* L. durch sichtbare Strahlung. *Planta* 55,438–450.

Rollin, P., and Maignan, G. (1966). La nécessité du phytochrome Prl (=P_{730}) pour la germination des akènes de *Lactuca sativa* L. variété "Reine de Mai." *C. R. Acad. Sci.* 263(3),756–757.

Sale, P. J. M., and Vince, D. (1959). Effects of wavelength and time of irradiation on internode length in *Pisum sativum* and *Tropaeolum majus. Nature* 183,1174–1175.

Siegelman, H. W., Chapman, D. J., and Cole, W. J. (1968). The bile pigments of plants. *In* "Porphyrins and Related Compounds" (T. W. Goodwin, ed.), pp. 107–120. Academic Press, London, New York.

Turner, M. R., and Vince, D. (1969). Photosensory mechanisms in the lettuce seedling hypocotyl. *Planta* 84,368–382.

Vince, D., and Grill, R. (1966). The photoreceptors involved in anthocyanin synthesis. *Photochem. Photobiol.* 5,407–411.

PART II

CHAPTER 18

Large Phytochrome

Phytochrometrists gathered from August 30 to September 17, 1971 near the ancient city of Eretria, on the Greek island of Euboea. "By the time the meeting was finished, all the big groups of the future were established and their areas of research informally defined," Harry Smith recalls.

Borthwick (Fig. 18-1), whose health was failing by this time, discussed "The History of Phytochrome" (Borthwick, 1972a) and "The Biological Significance of Phytochrome" (Borthwick, 1972b). These were his last scientific publications, for he died on May 21, 1974 at age 76.

The Eretria symposium found the field at a low ebb. The major research center had closed, the paradoxes were unexplained, and phytochrome's action was still unknown.

SIZE HETEROGENITY

The molecule's size was also in dispute. Siegelman and Firer had reported a molecular weight of 90,000–150,000 for phytochrome in 1964 (see Chapter 10). But when two chemists at E. I. du Pont de Nemours and Company in Wilmington, Delaware, purified the pigment to apparent homogeneity, their estimate was much lower.

Franklin E. Mumford and Edward L. Jenner followed Siegelman and Firer's procedure of working in a cold room under fluorescent light. They then subjected the preparation to continuous flow electrophoresis (to remove impurities that differed from phytochrome in overall electrical charge) and ran it twice through a Bio-Gel column (to remove impurities that differed in molecular size). Because phytochrome had a strong tendency to adsorb to any support material used for electrophoresis, they pumped partially purified phytochrome into an electrophoresis chamber consisting of a film of liquid between two 2 × 2-ft plates. As the film

FIGURE 18-1   Harry A. Borthwick (*left*) at Eretria. Courtesy of Walter Shropshire, Jr.

moved from top to bottom, the materials in the solution separated according to charge and flowed into a series of ports at the bottom (Fig. 18-2).

The Bio-Gel preparation was at least 90% pure, and it was 740 times purer than the crude extract, although only 3% of the original was retained. It had 10 to 11 sulfhydryl groups per molecule and molar absorption coefficients of 76,000 dm^3 mol^{-1} cm^{-1} at 644 nm for Pr and 46,000 dm^3 mol^{-1} cm^{-1} at 724 nm for Pfr. The molecular weight, estimated on Bio-Gel and Sephadex columns, was only about 60,000. But a small proportion emerged from the preparative Bio-Gel column in higher molecular weight fractions (Mumford and Jenner, 1966).

IN VITRO DESTRUCTION

After returning to Stanford from Beltsville in 1963, Briggs also began to purify oat phytochrome. Although his preparations were not as pure as those of Mumford and Jenner, they contained 30–50% of the original pigment, which was maintained as Pr during ammonium sulfate precipitation, a step that tended to destroy Pfr.

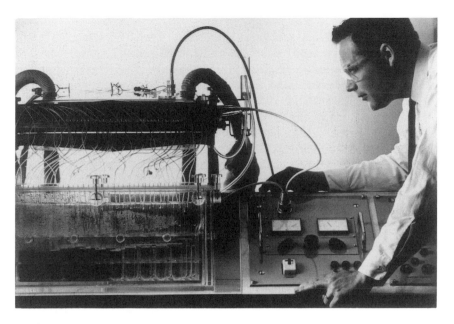

FIGURE 18-2   Franklin E. Mumford observing purification of oat phytochrome in an
Elphor model FF apparatus (Brinkmann Instruments, Westbury, NY).

Briggs also noticed two different sizes of phytochrome, which he named
small and *large* (Fig. 18-3). Their roughly estimated molecular weights
were 80,000 and 180,000, and the large species disappeared on standing
(Briggs et al., 1968).

Freshly prepared oat phytochrome absorbed maximally at 667 nm, as
did oat Pr *in vivo*. But when purified samples of either size stood overnight,
the R absorption maximum fell to 660 nm if the pigment was stored as Pfr,
even in a dark cold room (Fig. 18-4). "Thus rapid procedure, and maintenance of Pr during as much of the isolation procedure as possible are
required," Briggs warned, "if what one desires is material spectrally
similar to the native material" (Briggs et al., 1968).

Even with these precautions, the spectral characteristics of purified oat
and pea phytochromes always differed from those of phytochrome *in
vivo*, having difference spectrum maxima at 665 and 725 nm and a minimum at 413 nm rather than at 665, 735, and 430 nm (Everett and Briggs,
1970).

The paper by Briggs, Zollinger, and Pratt, along with the Beltsville
group's demonstration that Pfr was more sensitive to attack by proteases
and sulfhydryl-reacting chemicals than Pr (see Chapter 11) led to the

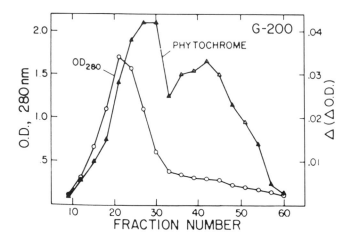

FIGURE 18-3   Elution pattern of Briggs's phytochrome (fresh sample) on a Sephadex G-200 column. (From Briggs et al., 1968. Reproduced with permission from the American Society of Plant Physiologists.)

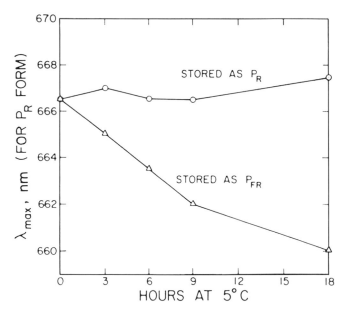

FIGURE 18-4   Change in R absorption maximum for extracted phytochrome stored either as Pr (top curve) or Pfr (lower curve). (From Briggs et al., 1968. Reproduced with permission from the American Society of Plant Physiologists.)

general use of dim green light during phytochrome purification, because this region of the spectrum appeared not to convert Pr to Pfr.

At Brookhaven, Furuya discovered an additional reason for extracting pea phytochrome under safelight, because the absorbance of crude extracts decreased instantly on light exposure, sabotaged by a mysterious small molecule dubbed *Pfr killer* (Furuya and Hillman, 1966). This reaction occurred much faster and under different chemical conditions than the gradual destruction of phytochrome *in vivo*. Much later, Furuya identified the offending substance as soyasaponin I, a triterpenoid saponin[1] (Yokota et al., 1982).

MESOTAENIUM PHYTOCHROME

Bonner, who moved from Harvard to the University of California, Davis, in 1963, gave up purifying pea phytochrome (see Chapter 11). He and Anthony O. Taylor chose to work instead with the alga *Mesotaenium caldariorum,* which seemed likely to have a shorter signal transduction pathway. They had to extract phytochrome from chlorophyll-filled cells because *Mesotaenium* refused to grow in the dark on organic substrates. The preparation was the first extracted phytochrome from lower plants and the cleanest preparation from green material (Taylor and Bonner, 1967).

Unexpectedly, the optical properties of the algal phytochrome differed from that extracted from dark-grown higher plants, because Pr absorbed maximally at only 649 nm and Pfr at 710 nm. But a Harvard undergraduate had observed a peak at 650 nm rather than 660 nm on a detailed action spectrum for chloroplast movement in *Mesotaenium* (Machemer, 1964). Liverwort (Taylor and Bonner, 1967), and moss (Giles and von Maltzahn, 1968) preparations also had unusual absorption characteristics.

SMALL OAT PHYTOCHROME

Three graduate students purified oat phytochrome to apparent homogeneity, meanwhile (Hopkins and Butler, 1970; Hopkins 1971; Rice, 1971; Rice et al., 1973; Roux, 1971). The preparations had molecular weights of 60,000 to 70,000 and absorbed about equally at 280 nm (where proteins absorb maximally) and 660 nm (the Pr chromophore peak). In contrast, Siegelman and Firer's preparation had absorbed about 10 times more

---

1. Widespread in plants, saponins are glycolysated steroids, such as the pharmaceutical compound digitonin.

strongly at 280 nm than at 660 nm because it was contaminated with other proteins.

Hanno H. Kroes of the Unilever Research Laboratory at Duiven performed spectrophotometric studies on extracted phytochrome under Wassink's supervision at Wageningen. His product (Kroes et al., 1969) had a molecular weight of about 55,500. There was also a heavier fraction, "which must be an association product with a molecular weight above 200,000" (Kroes, 1970).

Stanley J. Roux detected 70,000- and 110,000-dalton[2] species (Roux, 1971). But Twyford Laboratory researchers obtained species ranging in molecular weight from 18,000 to 359,000 (Walker and Bailey, 1970a,b).

Apart from the Twyford data, the results appeared to suggest that purified phytochrome had a molecular mass of about 60,000 daltons but aggregated into a larger species.

CHARACTERIZATION

Circular dichroism (CD), which probes the secondary structures of proteins,[3] provided evidence for conformational differences between Pr and Pfr. Early data (Anderson et al., 1970; Hopkins and Butler, 1970; Kroes, 1970) were discrepant but suggested that at least small changes in protein conformation occurred during phototransformation.

Roux reported that, although certain chemicals attack Pfr in preference to Pr (see Chapter 10), pea Pr was much more sensitive to aldehydes than Pfr, possibly because conversion to Pfr concealed and therefore protected the side chains of certain lysine residues (Roux and Hillman, 1969).

Butler and Hopkins obtained rabbit antibodies[4] to the phytochrome molecule, reasoning that, if some sites were more exposed in Pr or Pfr, the preparation would react differentially. The difference was unconvincing (Hopkins and Butler, 1970).

2. One dalton is one-twelfth the molecular mass of the most abundant form of carbon, ^{12}C.

3. Circular dichroism (CD) measurements are made in the same manner as regular absorption measurements, but the measuring beam passes through a linear polarizer before hitting the sample. The polarizer resolves the beam into left-handed polarized light and right-handed polarized light, which strike the sample alternately. Because detection is also modulated, the instrument produces separate absorption spectra for the two beams, and the sample is said to show CD at wavelengths where there is a difference.

Coiled regions of polypeptide chains (α-helices) absorb right-handed polarized light more efficiently than left-handed polarized light. Thus if one form of a protein has a higher helical content than another, the difference may be detectable on CD spectra.

4. Antibodies are proteins that appear in the bloodstream after the introduction of a foreign protein (an antigen). A mixture of the products of different clones of antibody-secreting cells is said to be a polyclonal preparation.

RYE PHYTOCHROME

Although most workers studied oat phytochrome in the 1960s, the Smithsonian group extracted the pigment from annual rye (*Secale cereale* cv *Balbo*), hoping to prepare massive amounts of chromophore. Germinated on batting-lined trays covered with cheesecloth, kilogram lots yielded 5–10 kg of yellow-white seedlings. "The whole process for isolating phytochrome took a further 3 days," Shropshire (Fig. 18-5) says, "and once it began, we had to work 24 h around the clock, standing in the cold room just above freezing." David L. Correll (Fig. 18-6), who had joined the Division of Radiation and Organisms in 1963, provided the biochemical expertise.

By April 1966, the group achieved a 200-fold purification, but the primary product appeared to be a large protein that denatured into subunits (Correll et al., 1966a). After a further brief report later that year (Correll et al., 1966b) and violent arguments between Correll and Butler over the validity of these results, there was a long silence, which caused workers in other labs to accuse the group of secrecy. Three papers then appeared in 1968.

Correll and colleagues worked under a green safelight, adding FR to maintain phytochrome as Pr. They purified the pigment 1,200-fold, with a 25% yield (Correll et al., 1968a). The main product sedimented in the ultracentrifuge at 9 Svedberg (S) units and had a molecular weight of about 180,000. When denatured chemically, the 9-S protein readily dissociated into smaller subunits, each sedimenting at 2 S and having a molecular weight of about 42,000. A 14-S species accounted for up to 25% of the total protein and disappeared with aging.

FIGURE 18-5 Walter Shropshire, Jr., 1988. Photo by Linda Sage.

FIGURE 18-6   David L. Correll with phytochrome
bibliography.

Electron micrographs revealed several species, including small (4-nm diameter) particles thought to be the 2-S monomers, plus tetramers and hexamers (Fig. 18-7). The researchers equated the tetramers with the 9-S species and the hexamers with the 14-S species (Correll et al., 1968b). Thus rye phytochrome appeared to be most stable at 180,000 but could be dissociated into 42,000 subunits. Oat phytochrome was most stable at 60,000 but appeared to aggregate into a larger species.

The molar absorption coefficients of the rye subunit were also much lower than those for oat reported by Mumford and Jenner, and amino acid analyses were also at variance (Mumford and Jenner, 1966; Correll et al., 1968b), especially since cysteine seemed to be absent from rye phytochrome. Later electron micrographs of oat phytochrome from du Pont showed parallel aggregates of 6-nm particles—microcrystals of small phytochrome (Stasny and Mumford, 1971).

In view of these discrepancies, Correll wanted to compare oat and rye phytochromes directly. Crude extracts of both contained two photoactive molecules that separated on Sephadex G-200 and DEAE-cellulose. On dialysis, "a shift toward the small aggregate occurred. This is believed to be due to a shift to a lower pH within the dialysis tubing (Gibbs-Donnan effect)" (Correll and Edwards, 1970). The major protein, which comprised the bulk of the phytochrome, was about twice the size of the minor component. Thus perhaps both oat and rye phytochromes might be large molecules that separated into subunits.

In 1970, the Division of Radiation and Organisms was closed for about a year while a new Radiation Biology Laboratory was constructed in Rockville, Maryland. Unable to continue lab work, Correll, Edwards, and Shropshire compiled a phytochrome bibliography (see Fig. 18-6), which appeared in 1977 (Correll et al., 1977).

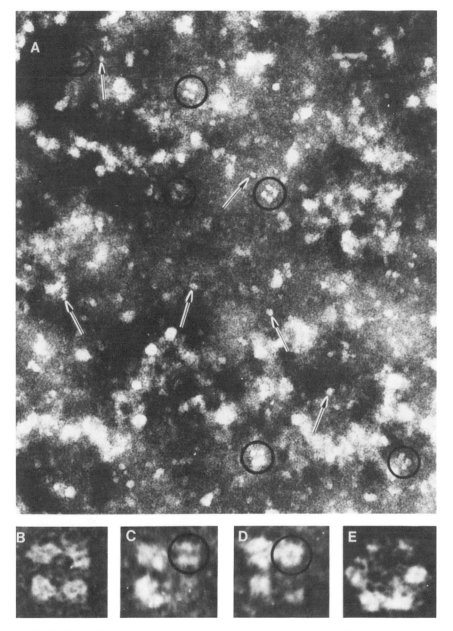

FIGURE 18-7   Electron micrograph of rye phytochrome prepared at Smithsonian Institution.   (A) × 490,000.   Monomers are indicated with arrows; tetramers are circled.
(B–D) Tetramers × 1,960,000. (E) Hexamer × 1,470,000. (From Correll et al., 1968b.)

FIGURE 18-8   Winslow R. Briggs.

PROTEOLYTIC DEGRADATION

Briggs (Fig. 18-8) had completed a Ph. D. at Harvard in 1956. Returning to his alma mater in 1967, he was disquieted by the molecular weight discrepancies. Then graduate student John R. Pringle discovered that commercial yeast malate dehydrogenase contained a proteolytic enzyme that, even in minute amounts, degraded the protein. Because the resulting fragment retained malate dehydrogenase activity, it was readily confused with the native enzyme. Phenylmethanesulfonyl fluoride (PMSF), known to inhibit some proteases that require sulfhydryl groups for activity, protected the dehydrogenase from degradation (Pringle, 1970).

Briggs wondered if small phytochrome was a degradation product rather a subunit of large phytochrome. When oat preparations were exposed to a wide range of pH values and salt concentrations, the molecules never clumped into larger molecules.

Modifying Rice's oat protocol, Gary Gardner isolated rye phytochrome and immediately subjected a sample to sucrose gradient centrifugation. Two sizes of molecules sedimented, at approximately 8 S and 4 S, whereas a 4-day-old sample contained only the 4-S species (Fig. 18-9). Rapidly prepared oat phytochrome also came in two molecular sizes, and pH, salt concentration, or phytochrome concentration failed to affect the relative proportions of each. On standing at 4°C, the large molecule converted quantitatively into the small one, and PMSF inhibited this conversion (Gardner et al., 1971).

In another corner of the lab, Carl S. Pike was isolating a protease from dark-grown oat seedlings, hoping it might react differentially with Pr and Pfr. The enzyme proved to be a neutral protease that remained active at

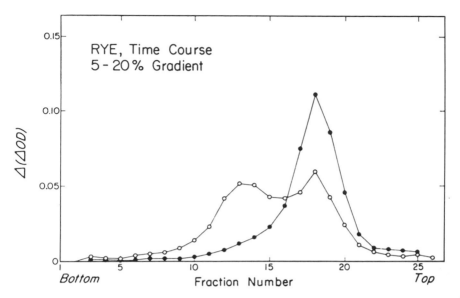

FIGURE 18-9   Ultracentrifugation of partially purified rye phytochrome. Open circles, fresh sample; closed circles, sample incubated at 4°C for 100 h. (From Gardner et al., 1971. Reproduced by permission from the American Society of Plant Physiologists.)

pH 8.0, was activated by mercaptoethanol, and inhibited by PMSF (Pike and Briggs, 1972). It would thus be active in the buffers used for phytochrome purification.

Added to purified rye phytochrome that lacked endogenous proteases, the enzyme accelerated small phytochrome appearance, as did a series of commercially available endoproteases.[5] Thus small phytochrome did indeed appear to be a degradation product of the larger molecule.

Subjecting an undegraded phytochrome sample to SDS-polyacrylamide gel electrophoresis,[6] Gardner obtained a major band with an estimated molecular mass of 120,000 daltons; this was almost absent from trypsin-treated samples, which generated bands at the 62,000 and 42,000 positions.

Misleadingly, this size reduction occurred "without any appreciable effect on the chromophore absorbancy. . . . The photoreversible product of limited proteolysis is extremely stable" (Gardner et al., 1971).

5. Endoproteases attack the internal peptide bonds of polypeptide chains.

6. Sodium dodecyl sulfate–polyacrylamide gel electrophoresis (SDS-PAGE) was a recently introduced method of estimating molecular weights of protein subunits. Polypeptides electrophoresed on acrylamide gel separate according to both charge and molecular weight, but the anionic detergent SDS converts them to anions, causing them to migrate strictly according to molecular weight.

In Eretria, Briggs outlined the method for isolating large, pure rye phytochrome (Briggs et al., 1972). An important review summarized the molecular weight discrepancies and attempts at phytochrome characterization (Briggs and Rice, 1972).

"Our feeling at that time," says Briggs, "was that now we could really get on with characterizing the native molecule rather than studying a proteolytic fragment. But we later had to eat our (spoken) words about the 'nativeness' of our protein."

REFERENCES

Anderson, G. R., Jenner, E. L., and Mumford, F. E. (1970). Optical rotatory dispersion and circular dichroism spectra of phytochrome. *Biochim. Biophys. Acta* 221,69–73.

Borthwick, H. A. (1972a). The history of phytochrome. *In* "Phytochrome" (K. Mitrakos and W. Shropshire, Jr., eds.), pp. 3–23. Academic Press, London, New York.

Borthwick, H. A. (1972b). The physiological significance of phytochrome. *In* "Phytochrome" (K. Mitrakos and W. Shropshire, Jr., eds.), pp. 27–44. Academic Press, London, New York.

Briggs, W. R., and Rice, H. V. (1972). Phytochrome: Chemical and physical properties and mechanism of action. *Annu. Rev. Plant Physiol.* 23,293–334.

Briggs, W. R., Tocher, R. D., and Wilson, J. F. (1957). Phototropic auxin redistribution in corn coleoptiles. *Science* 126,210–212.

Briggs, W. R., Zollinger, W. D., and Platz, B. B. (1968). Some properties of phytochrome isolated from dark-grown oat seedlings (*Avena sativa* L. *Plant Physiol.* 43,1239–1243.

Briggs, W. R., Gardner, G., and Hopkins, D. W. (1972). Some technical problems in the purification of phytochrome. *In* "Phytochrome" (K. Mitrakos and W. Shropshire, Jr., eds.), pp. 142–158. Academic Press, London, New York.

Correll, D. L., and Edwards, J. L. (1970). The aggregation states of phytochrome from etiolated rye and oat seedlings. *Plant Physiol.* 45,81–85.

Correll, D. L., Steers, E., Jr., Edwards, J. L., Suriano, J. R., and Shropshire, W., Jr. (1966a). Phytochrome: Isolation and partial characterization. *Fed. Proc.* 25,736.

Correll, D. L., Steers, E., Jr., Edwards, J. L., Suriano, J. R., and Shropshire, W. Jr. (1966b). Phytochrome: Isolation from rye and partial characterization. *Plant Physiol.* 41(suppl. xvi–xvii).

Correll, D. L., Edwards, J. L., Klein, W. H., and Shropshire, W., Jr. (1968a). Phytochrome in etiolated annual rye. III. Isolation of photoreversible phytochrome. *Biochim. Biophys. Acta* 168,36–45.

Correll, D. L., Steers, E., Jr., Towe, K. M., and Shropshire, W., Jr. (1968b). Phytochrome in etiolated annual rye. IV. Physical and chemical characterization of phytochrome. *Biochim. Biophys. Acta* 168,46–57.

Correll, D. L., Edwards, J. L., and Shropshire, W. Jr. (1977). "Phytochrome. A Bibliography with Author, Biological Materials, Taxonomic, and Subject Indexes of Publications Prior to 1975." Smithsonian Institution, Washington, D.C.

Everett, M. S., and Briggs, W. R. (1970). Some spectral properties of pea phytochrome *in vivo* and *in vitro*. *Plant Physiol.* 45,679–683.

Furuya, M., and Hillman, W. S. (1966). Rapid destruction of the Pfr form of phytochrome by a substance in extracts of *Pisum* tissue. *Plant Physiol.* 41,1242–1244.

Gardner, G., Pike, C. S., Rice, H. V., and Briggs, W. R. (1971). "Disaggregation" of phytochrome *in vitro*—A consequence of proteolysis. *Plant Physiol.* 48,686–693.

Giles, K. L., and von Maltzahn, K. E. (1968). Spectrophotometric identification of phytochrome in two species of *Mnium*. *Can. J. Bot.* 46,305–306.

Hopkins, D. W. (1971). Protein conformational changes of phytochrome. Ph.D. thesis. University of California, San Diego.

Hopkins, D. W., and Butler, W. L. (1970). Immunochemical and spectroscopic evidence for protein conformational changes in phytochrome transformations. *Plant Physiol.* 45,567–570.

Kroes, H. H. (1970). A study of phytochrome, its isolation, structure and photochemical transformations. *Meded. Landbouwhogesch. Wageningen* 70(18), 1–112.

Machemer, C. (1964). Senior Honors Thesis, Harvard University.

Mumford, F. E., and Jenner, E. L. (1966). Purification and characterization of phytochrome from oat seedlings. *Biochemistry* 5,3657–3662.

Pike, C. S., and Briggs, W. R. (1972). Partial purification and characterization of a phytochrome-degrading neutral protease from etiolated oat shoots. *Plant Physiol.* 49,521–530.

Pringle, J. R. (1970). The molecular weight of the *undegraded* polypeptide chain of yeast hexokinase. *Biochem. Biophys. Res. Commun.* 39,46–52.

Rice, H. V. (1971). Purification and partial characterization of oat and rye phytochrome. Ph. D. thesis. Harvard University.

Rice, H. V., Briggs, W. R., and Jackson-White, C. J. (1973). Purification of oat and rye phytochrome. *Plant Physiol. 51,917–926.*

Roux, S. J. (1971). Chemical approaches to the structural properties of phytochrome. Ph.D. thesis. Yale University.

Roux, S. J., and Hillman, W. S. (1969). The effect of glutaraldehyde and two monoaldehydes on phytochrome. *Arch. Biochem. Biophys.* 131,423–429.

Stasny, J. T., and Mumford, F. E. (1971). Electron microscopic observations of purified oat phytochrome. *Proc. Electron. Microsc. Soc. Am.* 29,362–363.

Taylor, A. O., and Bonner, B. A. (1967). Isolation of phytochrome from the alga *Mesotaenium* and liverwort *Sphaerocarpos*. *Plant Physiol.* 42,762–766.

Walker, T. S., and Bailey, J. L. (1970a). Studies on phytochrome: Two photoreversible chromoproteins from etiolated oat seedlings. *Biochem. J.* 120,607–612.

Walker, T. S., and Bailey, J. L. (1970b). Studies on phytochrome: Some properties of electrophoretically pure phytochrome. *Biochem. J.* 120,613–622.

Yokota, T., Baba, J., Konomi, K., Shimazaki, Y., Takahashi, N., and Furuya, M. (1982). Identification of a triterpenoid saponin in etiolated pea shoots as phytochrome killer. *Plant Cell Physiol.* 23,265–271.

CHAPTER 19

Gene Regulation Hypothesis

In 1966, Mohr gave a seminar at Beltsville. "I tried to convince the people there that they should investigate photomorphogenesis in terms of gene physiology," he recalls. "It was a really horrible experience."

Though well received in Germany, the gene regulation hypothesis faced an icy reception in the United States. "In principle I was not very concerned because we wanted to prove the hypothesis stepwise anyhow," Mohr says.

Mohr studied anthocyanin synthesis in mustard seedling cotyledons, where cell number and DNA content remained constant from 24 to 120 h after sowing. The Freiburg workers put phytochrome into its high-energy mode during most experiments to avoid the decrease in Pfr concentration that complicates light-pulse experiments (Schopfer, 1972). Combining filters with newly developed incandescent tubes, Mohr developed a light source that operated the phytochrome system without promoting chlorophyll formation and photosynthesis (Mohr, 1966). This quickly set the phytochrome photoequilibrium to that obtained at 718 nm—2.5% Pfr (Oelze-Karow et al., 1970).

PAL ACTIVITY

Because it was impossible to measure gene transcription directly at that time, Mohr chose to monitor activity levels of a key enzyme in the anthocyanin-forming pathway, L-phenylalanine ammonia-lyase (PAL), which was known to be light-regulated (Zucker, 1965). PAL catalyzes the conversion of phenylalanine to trans-cinnamic acid, an anthocyanin precursor.

Graduate student Francis Durst detected a greater increase in PAL activity when dark-grown seedlings developed under the FR source than when they continued to grow in darkness (Fig. 19-1) (Durst and Mohr,

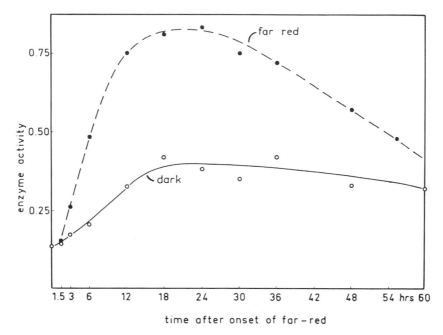

FIGURE 19-1   Increase in PAL activity after onset of FR irradiation. FR source was turned on 36 h after sowing. (From Durst and Mohr, 1966. Copyright 1966, Springer-Verlag.)

1966). Actinomycin D, which blocks gene transcription, partly prevented this increase.

PAL activity did not increase until $1\frac{1}{2}$ h after Pfr appeared in the cell (Fig. 19-1). But when dark-grown seedlings were irradiated for 12 h and then returned to darkness, the increase began as soon as the light was again turned on, even if it had ceased entirely (Fig. 19-2). Because the protein synthesis inhibitors puromycin and cycloheximide prevented this secondary response, Mohr concluded that "reappearance of P_{730} leads to *de novo* synthesis of enzyme protein" (Rissland and Mohr, 1967). The kinetics suggested that "Pfr-mediated induction . . . is a rapid process which can be a matter of a few minutes under appropriate conditions" (Schopfer, 1972). This contradicted the dogma that enzyme induction was a time-consuming process in higher plants.

Anthocyanin accumulation, with an initial lag phase of 3 h, also lacked a secondary lag phase (Lange et al., 1967), as did every other phytochrome response studied in Freiburg. Later studies suggested that mRNA translation rather than mRNA production formed the basis of the rapid response.

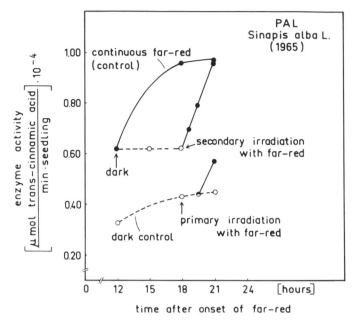

FIGURE 19-2   PAL activity in mustard cotyledons. Thirty-six hours after sowing (time zero), dark-grown seedlings were either (*lower part*) kept in the dark for a further 18 h and then irradiated with FR (primary irradiation) or (*upper part*) irradiated with FR for 12 h, darkened for 6 h, and then reirradiated (secondary irradiation). Note the presence of a 1.5-h lag after the primary irradiation and the absence of a lag after the secondary irradiation. (From Rissland and Mohr, 1967. Copyright 1967, Springer-Verlag.)

LIPOXYGENASE ACTIVITY

Mohr's original model accounted only for positive photoresponses (see Chapter 16) (Mohr, 1966). After Peter Schopfer determined that Pfr and actinomycin or puromycin independently inhibited hypocotyl lengthening under steady-state conditions (Schopfer, 1967), Mohr expanded the model to include repressible as well as inducible genes (Fig. 19-3).

Mohr was able to investigate negative regulation after lipoxygenase (lipoxidase) was found to be under phytochrome control. This enzyme catalyzes the oxidation of certain polyunsaturated fatty acids during storage fat degradation. Lipoxygenase (LOG) activity in the cotyledons of dark-grown squash seedlings peaked on day 5 after sowing and then declined, and the R/FR-reversible reaction hastened its disappearance (Surrey and Barr, 1966; Surrey, 1967).

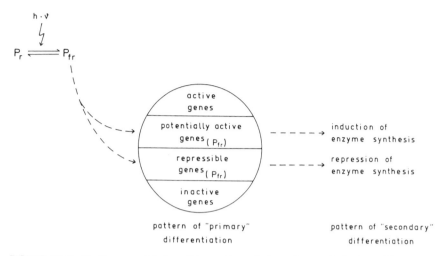

FIGURE 19-3   Freiburg model describing action of phytochrome in both positive and negative photoresponses. (From Schopfer, 1967. Copyright 1967, Springer-Verlag.)

Heidemarie Karow (later H. Oelze-Karow) (Fig. 19-4) assayed LOG activity between 36 and 48 h after sowing, as it increased in darkness (Fig. 19-5, upper curve). FR halted this increase for at least 12 h, after only a short lag (Fig. 19-5, lower line). Because actinomycin D was also inhibitory, she concluded that phytochrome acted at the level of synthesis (Oelze-Karow et al., 1970). Isocitrate lyase, an enzyme that shunts carbon atoms from fatty seed reserves into synthetic pathways, was unaffected (Karow and Mohr, 1966).

FIGURE 19-4   Heidemarie Oelze-Karow, 1973.

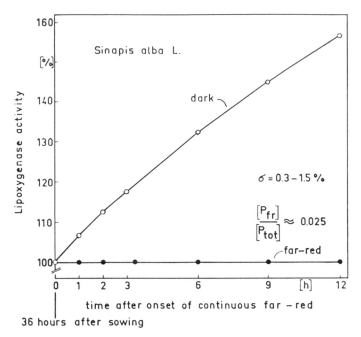

FIGURE 19-5   Increase in lipoxygenase activity in dark-grown mustard seedlings is arrested by continuous FR that maintains 2.5% of phytochrome as Pfr. (From Oelze-Karow et al., 1970.)

Phytochrome operated in its low-energy mode in this instance, so Oelze-Karow could correlate response with Pfr levels. When she set the phytochrome photoequilibrium at 2.5% Pfr (with either 5 min FR or 5 min R/5 min FR or 6 h FR), the activity increase ceased after 45 min (Fig. 19-6). But there was a $4\frac{1}{2}$-h delay when 80% of the phytochrome was initially converted to Pfr (5 min R). Because Pfr's half-life in mustard cotyledons was known to be 45 min (Marmé, 1969), 1.25% would remain after 45 min in the first case, whereas 80% Pfr would halve six times—again to 1.25%—in $4\frac{1}{2}$ h.

Other initial percentages of Pfr generated predicted lag phases. And when the Pfr level was lowered to less than 1.25% of initial total phytochrome, LOG activity immediately increased, stabilizing after Pfr rose once again[1] above the threshold level. Thus "repression and depression of lipoxygenase synthesis occur without a detectable lag after the Pfr threshold is passed in one or the other direction" (Oelze-Karow et al., 1970).

1. Through synthesis of Pr and conversion by the continuous FR source to Pfr.

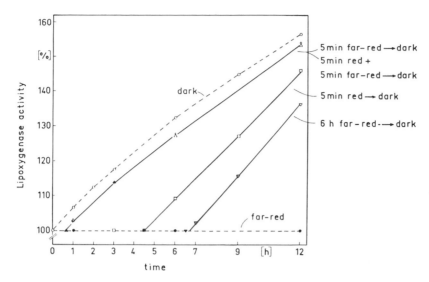

FIGURE 19-6   The course of lipoxygenase formation after a single brief irradiation at time zero with R, FR, or R followed by FR. Identical lag period before resumption of enzyme synthesis in cases 5 min FR/dark and 6 h FR/dark suggests that even in the case of prolonged irradiation with FR only phytochrome is involved. (From Oelze-Karow et al., 1970.)

Hypocotyl extension in mustard seedlings also proved to be a threshold response (Schopfer and Oelze-Karow, 1971).

PATTERN SPECIFICATION

In 1968, while the LOG studies were in progress, the group moved into new facilities. "The Deutsche Forschungsgemeinschaft," Mohr explains, "supported the build up of a center of excellence in Freiburg where people from different fields—including animal development and microbiology—formed a unit."

Mohr aimed to relate phytochrome's actions to genetically patterned development. He stressed that "which genes are active, inactive, potentially active, or repressible in a particular cell at a particular moment is determined by the previous history of the cell. We call this previous history 'primary differentiation.' . . . Pfr acts on the level of 'secondary differentiation.' The pattern of Pfr-dependent secondary differentiation is predetermined by primary differentiation; Pfr is only the trigger" (Mohr, 1969). Thus mustard seedling epidermal cells responded to Pfr by developing hairs, whereas subepidermal cells formed anthocyanin.

In addition to differences in spatial development, photoresponsiveness changed with time. Anthocyanin synthesis was impossible until 24 h after sowing and ceased about 36 h later, no matter what light treatments were given or how much pigment had accumulated (Wagner and Mohr, 1966). And Pfr was unable to decrease LOG activity before 33.25 h after sowing. Apparent synthesis of the enzyme then came under phytochrome control, but, at 48 h, enzyme activity under FR suddenly equaled that in darkness (Oelze-Karow and Mohr, 1970).

INDUCTION OR ACTIVATION?

A major objection to the Freiburg group's proposals was that changes in enzyme activity did not necessarily reflect changes in gene activity—proteins themselves were known to be regulated by smaller molecules. Also, inhibitors of transcription or translation presumably might alter a particular enzyme's activity by preventing the synthesis of a different protein.

The most vocal proponent of an alternative view—that phytochrome might regulate enzyme activity at several levels, including the activation of preexisting enzyme molecules—was Harry Smith, who had obtained a botany degree at the University of Manchester in 1959 and a Ph.D. in 1962 from the University of Aberystwyth. At Queen Mary College, London

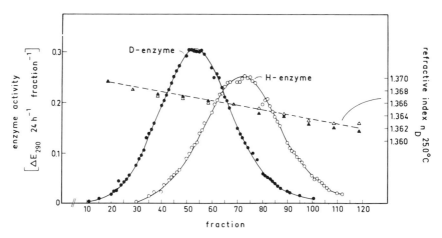

FIGURE 19-7   Density gradient centrifugation of PAL extracted from the cotyledons of mustard seedlings grown on 80% deuterium oxide (D-enzyme) or distilled water (H-enzyme). (From Schopfer and Hock, 1971. Copyright 1971, Springer-Verlag.)

University, Smith and Terry Attridge determined that increase in PAL activity showed R/FR reversibility, a low-energy requirement, and a logarithmic dose response in terminal buds of etiolated pea seedlings. "The results reported here appear to be the first demonstration of an unequivocal phytochrome-mediated increase in this enzyme," the British researchers claimed, ". . . but it constitutes no proof that direct control of gene activation is the primary role of phytochrome as has been recently suggested" (Attridge and Smith, 1967).

Seeking direct proof, Mohr and Hermann Dittes were unable to detect changes in RNA content. But at least the experiments suggested that "Pfr mediates morphogenesis not by dramatic regulatory changes but by induction and repression of a relatively small number of enzymes" (Dittes and Mohr, 1970).

The group also grew mustard in the dark on water or heavy water and then turned on the FR source for several hours. PAL from the cotyledons of labeled seedlings was heavier than that from the controls (Fig. 19-7). "It is confirmed that . . . a true *de novo* synthesis of phenylalanine ammonialyase is induced by active phytochrome (Pfr)," the researchers concluded (Schopfer and Hock, 1971). Heavy PAL also formed if deuterium oxide was present during irradiation of gherkin seedlings (Smith, 1972).

Experiments with *Xanthium* (Zucker, 1969) and gherkin (Engelsma, 1967, 1969, 1970) pointed to the existence of a PAL-inactivating factor, however. Thus Smith could "only conclude that no conclusion is yet possible regarding the mechanism through which the light mediated increases in PAL are brought about" (Smith, 1972).

CONTROL OF TRANSLATION

In 1971, Mohr gave 24 lectures at the University of Massachusetts, Amherst (Mohr, 1972). By this time, he had concluded that phytochrome controlled enzyme synthesis both at the level of transcription (transfer of information from DNA to RNA) and translation (transfer of information from RNA to protein). Thus in mustard seedlings more than 30 h old, anthocyanin synthesis began after a 3-h lag at 25°C regardless of whether it was induced by the low-energy reaction of phytochrome (Fig. 19-8A) or the HER (Fig. 19-8B).

Herbert Lange and Shropshire, who visited Freiburg from 1968 to 1969, found that complete reversal was possible for only a few minutes after the light was turned on but that escape was not rapid. This indicated "a continuous requirement for Pfr during the whole period of anthocyanin accumulation. In addition, Pfr clearly mediates anthocyanin synthesis during the lag-phase in spite of the fact that the actual synthesis of the

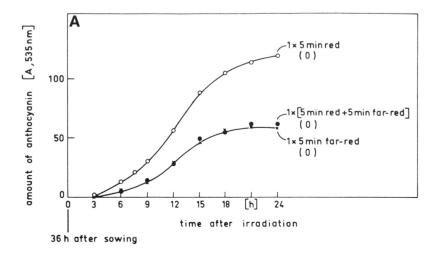

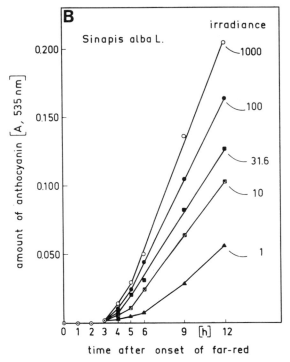

FIGURE 19-8 (A) Time course of anthocyanin synthesis in mustard seedling after brief irradiations with R or FR (5 min each). Irradiations were performed at time zero, which was 36 h after sowing. (B) Time course of anthocyanin synthesis in mustard seedling under continuous FR (as function of irradiance, where 1,000 represents that of the Freiburg standard far-red source). Irradiations were performed at time zero, which was 36 h after sowing. (From Lange et al., 1971. Reproduced with permission from the American Society of Plant Physiologists.)

pigment can proceed only after the lag phase is overcome" (Lange et al., 1971).

The absence of a lag after secondary irradiation (Mohr and Bienger, 1967) suggested that, during the initial lag phase, Pfr first induced the capacity for anthocyanin synthesis. This appeared to involve synthesis of stable mRNA because actinomycin D was inhibitory during this phase only. Subsequently, Pfr mediated anthocyanin synthesis, which was always sensitive to inhibitors of protein synthesis, suggesting that Pfr now affected translation.

Smith continued to propose that phytochrome-mediated increases in enzyme activity could involve the activation of preexisting molecules, at least in some systems (Attridge and Smith, 1973; Iredale and Smith, 1973). The argument peaked in the mid 1970s, when his group and Mohr's reached opposite conclusions from density-labeling experiments on the same material (Attridge et al, 1974; Peter and Mohr, 1974; Acton and Schopfer, 1975). But by this time, Smith had moved to the University of Nottingham School of Agriculture in Sutton Bonington, Leicestershire, where his research began to move in a new direction.

Much later, through immunotitration, "we showed without any ambiguity," says Mohr, "that phytochrome-mediated induction of PAL and chalcone synthase in mustard cotyledons is to be attributed to *de novo* synthesis of the enzyme protein (Brödenfeldt and Mohr, 1986)."

"But I don't think the question we were seeking to answer has been resolved," says Smith. "We did in fact characterize a protein that inactivates PAL in gherkin reversibly (Billett et al., 1978), so there is actually evidence that B acts on a preexisting enzyme."

SIGNAL TRANSMISSION

Oelze-Karow, meanwhile, knew that the standard FR source set the photostationary state at 2.5% Pfr in mustard *cotyledons* (Oelze-Karow and Mohr, 1970). This agreed well with spectrophotometric data obtained from mustard seedling *hooks* by Hartmann (Hartmann and Spruit, 1968; Hanke et al., 1969). But when Freiburg doctoral candidate Eberhard Schäfer assayed phytochrome spectrophotometrically in mustard *cotyledons,* he found that the standard FR source established between 14 and 7.5% Pfr, depending on the degree of deetiolation (Schäfer et al., 1972).

Using the LOG system, Oelze-Karow checked photostationary states (Oelze-Karow and Mohr, 1973). The results again agreed with the spectrophotometric data for mustard hook, which Schäfer confirmed (Schäfer et al., 1973). Yet LOG was synthesized largely in cotyledons. It thus appeared that "the phytochrome responsible for control of apparent LOG synthesis in the cotyledons is located in the hook region, rather than in the cotyledons themselves" (Oelze-Karow and Mohr, 1973).

Testing this hypothesis, Oelze-Karow monitored LOG activity in excised parts of seedlings. LOG "synthesis" occurred to the same extent and with the same kinetics under continuous FR as in the dark in isolated cotyledons (cut-off at point 1 or 2 in Fig. 19-9). But when more than half the hook was still attached (excision point 3 or 4), the increase came under phytochrome control (Oelze-Karow and Mohr, 1974). "The suppression of LOG synthesis in the cotyledons is controlled by phytochrome located in the hypocotylar hook—an inter-organ signal transfer is involved" was the "bold but inevitable conclusion" (Mohr and Oelze-Karow, 1975).

Oelze-Karow then irradiated seedlings for 5 min and excised the cotyledons 1½ h later. Seven minutes after excision, the isolated organs began to make LOG once more, indicating that "the signal transmitted from the hook to the cotyledons is not conserved to any significant extent in the cotyledons" (Oelze-Karow and Mohr, 1974).

This spatial separation of Pfr and LOG argued against a signal transduction chain leading directly from the photoreceptor to genetic material. Mohr interpreted "the strong cooperativity which is indicated by the steepness of the threshold in terms of cooperativity in biological membranes. Pfr would be analogous to a ligand and the primary reactant of Pfr (usually designated X) would be analogous to a preexisting membrane capable of performing fully reversible conformational transitions with a high degree of cooperativity" (Oelze-Karow and Mohr, 1973).

Because the kinetics of anthocyanin synthesis provided no evidence for threshold control (and thus cooperativity), Mohr urged adherence "to the

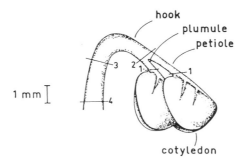

FIGURE 19-9 Upper parts of mustard seedling 36 h after sowing. *Numbers* indicate excision points. (From Oelze-Karow and Mohr, 1974.)

hypothesis that the reaction partner (or site) of Pfr (X) may be different in different cells or in different compartments of a particular cell, and that for this reason we can expect a multiplicity of 'primary reactions of Pfr' '' (Drumm and Mohr, 1974).

This "multiple primary action" hypothesis contradicted the simpler and more appealing notion that phytochrome has only one primary action. Peter H. Quail argued that it was unlikely that Pfr "interacts with more than one molecular species of reaction partner, physically located at sites in the cell as distinctly different as the cellular membrane and the genome, and that the direct consequence of this interaction is different at each site." He argued that "any graded signal is convertible to a cooperative one at any step in a signal cascade," and therefore that "the available data are entirely consistent with the operation of a single molecular action of Pfr" (Quail, 1983).

REFERENCES

Acton, G. J., and Schopfer, P. (1975). Control over activation or synthesis of phenylalanine ammonia-lyase by phytochrome in mustard (*Sinapis alba* L.)? A contribution to eliminate some misconceptions. *Biochim. Biophys. Acta* 404,231–242.

Attridge, T. H., and Smith, H. (1967). A phytochrome-mediated increase in the level of phenylalanine ammonia-lyase activity in the terminal buds of *Pisum sativum. Biochim. Biophys. Acta.* 148,805–807.

Attridge, T. H., and Smith, H. (1973). Evidence for a pool of inactive phenylalanine ammonia-lyase in *Cucumis sativus* seedlings. *Phytochemistry* 12,1569–1574.

Attridge, T. H., Johnson, C. B., and Smith, H. (1974). Density-labelling evidence for the phytochrome-mediated activation of phenylalanine ammonia-lyase in mustard cotyledons. *Biochim. Biophys. Acta* 343,440–451.

Billett, E. E., Wallace, W., and Smith, H. (1978). A specific and reversible macromolecular inhibitor of phenylalanine ammonia-lyase and cinnamic acid-4-hydroxylase in gherkins. *Biochim. Biophys. Acta* 524,219–230.

Brödenfeldt, R., and Mohr, H. (1986). Use of immunotitration to demonstrate phytochrome-mediated synthesis *de novo* of chalcone synthase and phenylalanine ammonia lyase in mustard seedling cotyledons. *Z. Naturforsch.* 41c,61–68.

Dittes, H., and Mohr, H. (1970). MAK-Chromatographie der RNS im Zusammenhang mit der phytochromgesteuerten Morphogenese. *Z. Naturforsch.* 25b,708–710.

Drumm, H., and Mohr, H. (1974). The dose response curve in phytochrome-mediated anthocyanin synthesis in the mustard seedling. *Photochem. Photobiol.* 20,151–157.

Durst, F., and Mohr, H. (1966). Phytochrome-mediated induction of enzyme synthesis in mustard seedlings (*Sinapis alba* L.). *Naturwissenschaften* 53,531–532.

Engelsma, G. (1967). Effect of cycloheximide on the inactivation of phenylalanine deaminase in gherkin seedlings. *Naturwissenschaften* 54,319–320.

Engelsma, G. (1969). Low-temperature dependent development of phenylalanine ammonia-lyase in gherkin hypocotyls. *Naturwissenschaften* 56,563.

Engelsma, G. (1970). Low-temperature effects on phenylalanine ammonia-lyase in gherkin hypocotyls. *Planta* 91,246–254.

Hanke, J., Hartmann, K. M., and Mohr, H. (1969). Die Wirkung von "Störlicht" auf die Blütenbildung von *Sinapis alba* L. *Planta* 86,235–249.

Hartmann, K. M., and Spruit, C. J. P. (1968). Difference spectra and photostationary states for phytochrome *in vivo*. *Plant Physiol.* 43(suppl. 15).

Iredale, S. E., and Smith, H. (1973). Density-labelling studies on the photocontrol of phenyl-alanine ammonia lyase levels in *Cucumis sativus*. *Phytochemistry* 12,2145–2154.

Karow, H., and Mohr, H. (1966). Changes of activity of isocitritase (EC 4.1.3.1.) during photomorphogenesis in mustard seedlings. *Planta* 72,170–186.

Lange, H., Bienger, I., and Mohr, H. (1967). Eine neue Beweisführung für die Hypothese einer differentiellen Genaktivierung durch Phytochrom 730. *Planta* 76,359–366.

Lange, H., Shropshire, W., Jr., and Mohr, H. (1971). An analysis of phytochrome-mediated anthocyanin synthesis. *Plant Physiol.* 47,649–655.

Marcus, A. (1971). Enzyme induction in plants. *Annu. Rev. Plant Physiol.* 22,313–336.

Marmé, D. (1969). Photometrische Messungen am Phytochromsystem von Senfkeimlingen (*Sinapis alba* L.). *Planta* 88,43–57.

Mohr, H. (1966). Untersuchungen zur phytochrominduzierten Photomorphogenese des Senf-keimlings (*Sinapis alba* L.). *Z. Pflanzenphysiol.* 54,63–83.

Mohr, H. (1969). Photomorphogenesis. *In* "An Introduction to Photobiology" (C. P. Swanson, ed.), pp. 99–141. Reprinted by permission of Prentice-Hall, Englewood Cliffs, New Jersey.

Mohr, H. (1972). "Lectures on Photomorphogenesis." Springer, New York, Heidelberg, Berlin.

Mohr, H. (1978). Pattern specification and realization in photomorphogenesis. *Bot. Mag. Spec. Issue* 1,199–217.

Mohr, H., and Bienger, I. (1967). Experimente zur Wirkung von Actinomycin D auf die durch Phytochrom bewirkte Anthocyansynthese. *Planta* 75,180–184.

Mohr, H., and Oelze-Karow, H. (1975). Phytochrome action as a threshold phenomenon. *In* "Light and Plant Development" (H. Smith, ed.), pp. 257–284. Butterworths, London, Boston.

Oelze-Karow, H., and Mohr, H. (1970). Experiments regarding the problem of differentiation in multicellular systems. *Z. Naturforsch.* 25b,1282–1286.

Oelze-Karow, H., and Mohr, H. (1973). Quantitative correlation between spectro-photometric phytochrome assay and physiological response. *Photochem. Photobiol.* 18,319–330.

Oelze-Karow, H., and Mohr, H. (1974). Interorgan correlation in a phytochrome-mediated response in the mustard seedling. *Photochem. Photobiol.* 20,127–131.

Oelze-Karow, H., Schopfer, P., and Mohr, H. (1970). Phytochrome-mediated repression of enzyme synthesis (lipoxygenase): A threshold phenomenon. *Proc. Natl. Acad. Sci. USA* 65,51–57.

Peter, K., and Mohr, H. (1974). Control of phenylalanine ammonia-lyase and ascorbate oxidase in the mustard seedling by light and Hoagland's nutrient solution. *Z. Natur-forsch.* 29c,222–228.

Quail, P. H. (1983). Rapid action of phytochrome in photomophogenesis. *In* "Encyclopedia of Plant Physiology" (W. Shropshire, Jr. and H. Mohr, eds.), vol. 16A, pp. 178–212. Springer, Berlin, Heidelberg. Copyright 1983, Springer-Verlag.

Rissland, I., and Mohr, H. (1967). Phytochrom-induzierte Enzymbildung (Phenylalanindes-aminase), ein schnell ablaufender Prozeβ. *Planta* 77,239–249.

Schäfer, E., Marchal, B., and Marmé, D. (1972). *In vivo* measurements of the phytochrome photostationary state in far red light. *Photochem. Photobiol.* 15,457–464.

Schäfer, E., Schmidt, W., and Mohr, H. (1973). Comparative measurements of phytochrome in cotyledons and hypocotyl hook of mustard (*Sinapis alba* L.). *Photochem. Photobiol.* 18,331–334.

Schopfer, P. (1967). Der Einfluss von Actinomycin D und Puromycin auf die phytochrom-induzierte Wachstumshemmung des Hypokotyls beim Senfkeimling (*Sinapis alba* L.). *Planta* 72,306–320.

Schopfer, P. (1972). Phytochrome and the control of enzyme activity. *In* "Phytochrome" (K. Mitrakos and W. Shropshire, Jr., eds.), pp. 485–513. Academic Press, London, New York.

Schopfer, P., and Hock, B. (1971). Nachweis der Phytochrom-induzierten *de novo* Synthese von Phenylalaninammoniumlyase (PAL, E.C.4.3.1.5.) in Keimlingen von *Sinapis alba* L. durch Dichtemarkierung mit Deuterium. *Planta* 96,248–253.

Schopfer, P., and Oelze-Karow, H. (1971). Nachweis einer Schwellenwertsregulation durch Phytochrom bei der Photomodulation des Hypokotylstreckungswachstums von Senf-keimlingen (*Sinapis alba* L.). *Planta* 100,167–180.

Smith, H. (1972). The photocontrol of flavonoid biosynthesis. *In* "Phytochrome" (K. Mitrakos and W. Shropshire, Jr., eds.), pp. 433–481. Academic Press, London, New York.

Surrey, K. (1967). Action and interaction of red and far-red radiation on lipoxidase metabolism of squash seedlings. *Plant Physiol.* 42,421–424.

Surrey, K., and Barr, E. M. (1966). Light-dependent modifications in the metabolic responses of squash seedlings. *Plant Physiol.* 41,780–786.

Wagner, E., and Mohr, H. (1966). Primäre und sekundäre Differenzierung im Zusammenhang mit der Photomorphogenese von Keimpflanzen (*Sinapis alba* L.). *Planta* 71,204–221.

Zucker, M. (1965). Induction of phenylalanine deaminase by light and its relation to chlorogenic acid synthesis in potato tuber tissue. *Plant Physiol.* 40,779–784.

Zucker, M. (1969). Induction of phenylalanine ammonia-lyase in *Xanthium* leaf disks. Photosynthetic requirement and effect of daylength. *Plant Physiol.* 44,912–922.

Zucker, M. (1971). Induction of phenylalanine ammonia-lyase in *Xanthium* leaf discs. Increased inactivation in darkness. *Plant Physiol.* 47,442–444.

CHAPTER **20** _____

Dark Transformations

At Queen Mary College, London University, Richard E. Kendrick and Barry Frankland (Fig. 20-1) found that dark-germinating *Amaranthus caudatus* (love-lies-bleeding) seeds were inhibited by prolonged irradiation, especially with FR, and repromoted by a subsequent short exposure to R but not R/FR (Kendrick and Frankland, 1969a). Since measurements in seeds with the Ratiospect were "difficult, if not impossible, we had to look at phytochrome in seedlings and try to extrapolate back," Kendrick explains. In seedlings, he observed destruction but not reversion (Kendrick and Frankland, 1968, 1969b), even though *Amaranthus* is a dicot.

Hartmann's model required destruction to be a first-order reaction, with a rate constant proportional to the fraction of phytochrome as Pfr (Hartmann, 1966). "Kendrick and Frankland didn't believe me," Hartmann says. But surprisingly—because destruction in monocots saturates at low Pfr levels (Butler et al., 1963; de Lint and Spruit, 1963; Pratt and Briggs, 1966; Kendrick, 1972)—they obtained the first support for Hartmann's proposal. Under various light sources, the phytochrome in dark-grown *Amaranthus* seedlings disappeared at different rates, but destruction was always exponential with time (Fig. 20-2a). A plot of rate constant for destruction of total phytochrome against photoequilibrium (Fig. 20-2b) revealed that "the rate of phytochrome decay in *Amaranthus* seedlings under continuous illumination is directly related to the proportion in the Pfr form" (Kendrick and Frankland, 1968). This was true even if the phytochrome photoequilibrium was changed while destruction was in progress (Kendrick and Frankland, 1969b).

CYCLED PR

A single, short FR exposure does not cause a large loss of phytochrome. But when Mancinelli tried to prevent R-induced destruction with a subsequent dose of FR, a substantial amount of phytochrome disappeared in darkness over the next few hours, even though only a minute proportion of the pigment was Pfr (Table 20-1). This suggested that the Pr formed after a

FIGURE 20-1   Richard E. Kendrick (*right*), Barry Frankland (*center*), and Paul Rollin (*left*), Eretria, 1971.

R-FR sequence might be subject to destruction (Dooskin and Mancinelli, 1968). A lability of Pr had already been suggested (Chorney and Gordon, 1966).

For several years the light-activated loss of Pr was considered "an anomalous result yet to be confirmed" (Frankland, 1972). But John M. Mackenzie, Jr. also found that light-activated loss of Pr occurs after Pr has been cycled through Pfr (Mackenzie et al., 1978). Destruction of cycled Pr was then observed in both etiolated and light-grown seedlings (Jabben, 1980; Jabben et al., 1980; Schäfer, 1981).

PHYTOCHROME IN SEEDS

In the mid 1960s, Spruit built a spectrophotometer (Fig. 20-3) with greatly increased signal-to-noise ratio (Spruit, 1970).[1] This machine enabled Jean

1. Spruit used a bicycle dynamo to generate the demodulating signal from the amplifier. The dynamo mounted directly on the shaft of the motor, rotating a glass prism that acted as a chopper. Thus it was impossible for chopping and the demodulating oscillations to get out of phase.

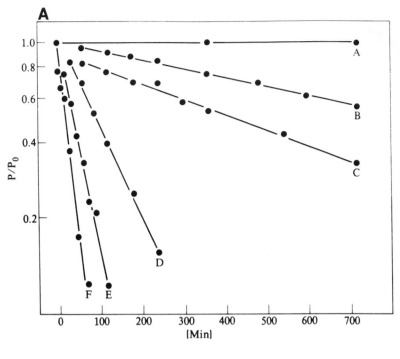

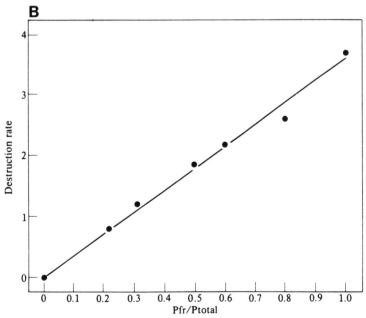

TABLE 20-1

Phytochrome Destruction in Darkness after Short Exposure to Red, Far Red or Both[a]

Treatment	Phytochrome remaining (% of dark control)	% Pfr at end of light treatment[b]
3 min R→3 min FR	90	0–2
3 min R→3 min FR→60 min Dark	83	0–2
3 min R→3 min FR→120 min D	78	0–2
3 min R→3 min FR→180 min D	73	0–2
3 min R→3 min FR→240 min D	71	0–2
6 min FR	96	0–2
6 min FR→240 min D	95	0–2
3 min R	95	75
3 min R→240 min D	37	75

[a] From data of Dooskin and Mancinelli, 1968 (Fig. 1; Tables 2, 3). Courtesy of A. L. Mancinelli.

[b] Measured immediately after 3 min R → 3 min FR, or 6 min FR, or 3 min R.

Boisard to detect phytochrome in dark-germinating *Nemophila insignis* and "May Queen" lettuce seeds 2 to 3 h after the beginning of imbibition. Surprisingly, the pigment was mostly Pfr,[2] which was known to be unstable in etiolated seedlings. As the radicle emerged, concentration increased, but the additional phytochrome was Pr (Spruit et al., 1968).

When Boisard irradiated imbibed seeds with FR, most of the pigment

2. The question of the origin of Pfr in seeds was addressed by Shropshire and James M. McCullough. They discovered that *Arabidopsis thaliana* seed, which can have an absolute requirement for light, would also germinate in the dark if parent plants were grown under continuous R-rich WL. This increased dark germination was prevented by exposure to FR and reinitiated by exposure to R, showing that it was phytochrome-mediated. A smaller percentage of seeds from plants grown under lower R:FR ratios germinated in the dark, moreover. The researchers concluded "that changes in red/far-red ratios in light present during seed production under natural conditions operating through the phytochrome system can significantly influence subsequent germination patterns" (McCullough and Shropshire, 1970; see also Hayes and Klein, 1974).

FIGURE 20-2 (A) Decay of total phytochrome at 25°C in seedlings with different phytochrome photostationary states. P/P_o, proportion of initial total phytochrome. A, darkness; B, continuous low-irradiance FR (734 nm); C, continuous high-irradiance FR; D, continuous B (22% Pfr); E, mixture of continuous R and FR (49% Pfr); F, darkness after a 10-min R treatment (80% Pfr). (B) Relationship between decay constants for total phytochrome and proportion of phytochrome maintained in Pfr form. Decay constants derived from **A** with addition of three points: 31% and 60% Pfr using mixed R/FR, and 80% Pfr using R. (From Kendrick and Frankland, 1968. Copyright 1968, Springer-Verlag.)

FIGURE 20-3   Spectrophotometer built by Spruit in mid 1960s. The study with de Lint (see Chapter 11) was performed on a prototype that had a lower signal–noise ratio.

became Pfr again after about 10 min. This unexpected behavior was tentatively dubbed *inverse dark reversion* (Boisard, 1968; Spruit et al., 1968), because the phytochrome appeared to be reverting in the opposite direction to that in seedlings, despite the thermodynamic improbability of such behavior[3] (Boisard et al., 1968).

Spruit (Fig. 20-4) next detected phytochrome in *dry* cucumber seeds. The pigment was three-fourths Pfr, but the R peak of its difference spectrum was shifted to 675 nm, suggesting perhaps "that in dry cucumber seeds there is a type of phytochrome different from that found in seeds imbibed for 24 h or more" (Spruit and Mancinelli, 1969).

Total phytochrome remained constant after a saturating FR pulse, but the proportion of Pfr increased for several days. Because the rate of inverse dark reversion was similar to that in imbibed seeds, even though there was more phytochrome in total, it appeared "likely that it is only the dry-seed phytochrome that undergoes this inverse dark reversion." (Spruit and Mancinelli, 1969).

3. Pr is more stable than Pfr, because Pfr reverts spontaneously to Pr in darkness.

FIGURE 20-4 European Symposium on Photomorphogenesis organized by Spruit in Wageningen, February 1968. *Back of table, left to right:* K. Hartmann, G. Engelsma, C. J. P. Spruit, P. Rollin, and H. Fredericq. *End of table:* P. J. A. L. de Lint. *Front of table, right to left:* G. Meyer and E. C. Wassink.

HYDRATION AND SYNTHESIS

Elaine M. Tobin, a graduate student in Brigg's lab, detected phytochrome in pine embryos, which were large enough to be assayed with a Ratiospect. Embryos from dry seeds had hardly any spectrophotometrically detectable phytochrome, bur photoreversibility increased as soon as water was added, even at 0°C (Tobin and Briggs, 1969). The pigment was light-stable, undergoing reversion but not destruction during 45 min R (Tobin and Briggs, 1969).

In *Amaranthus* seeds, photoreversibility increased soon after imbibition and again a few hours later (Fig. 20-5) (Kendrick et al., 1969). In a single isolated pumpkin embryo, the rehydration phase took 3 h. After a further 16- to 17-h period of almost constant phytochrome content, spectrophotometrically detectable phytochrome once more increased, although the radicle had not emerged. Further increases occurred after visible germination (Malcoste et al., 1970).

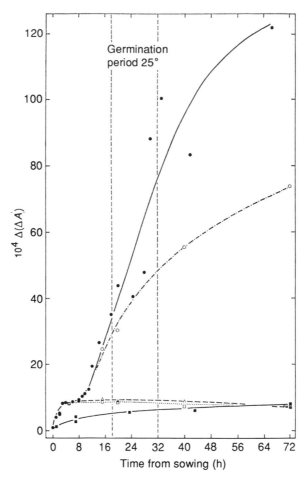

FIGURE 20-5   Time course of phytochrome appearance in darkness and under continuous R, FR, and B. Units expressed as Δ (ΔA) between 735 and 806 nm after actinic irradiation. (●) Dark, 25°C; (■) dark, 0°C; (△) R, 22.5°C; (○) FR, 22.5°C; (□) B, 22.5°C. (From Kendrick et al., 1969. Copyright 1969, Springer-Verlag.)

Phytochrome appearance had also been noted in etiolated barley shoots (Briggs and Siegelman, 1965; Correll and Shropshire, 1968). Studies at Wageningen, Rouen, and Stanford favored synthesis at this second stage (Rollin, 1968; Zouaghi et al., 1972) or even also during imbibition (Mc-Arthur and Briggs, 1970). But observing that radioactive leucine failed to find its way into phytochrome in growing seedlings, Correll decided that "nearly all of the phytochrome protein found in and isolated from the rye shoot was preformed in the dry seed" (Correll et al., 1968a).

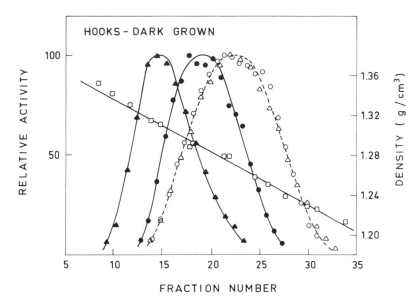

FIGURE 20-6   Equilibrium distribution in cesium chloride gradients of phytochrome from hooks of *Cucurbita pepo* seedlings grown for 120 h in dark on either H_2O (●) or 70% D_2O (▲). Profiles of a marker protein (lactate dehydrogenase) from labeled (△) and unlabeled (○) gradients have been superimposed and drawn as one. (□) indicates gradient density. (From Quail et al., 1973a. Reproduced with permission from the American Society of Plant Physiologists.)

An Australian research associate at Freiburg, Peter H. Quail, questioned this conclusion. Phytochrome from seedlings of yellow zucchini squash (*Cucurbita pepo*) grown in the dark on deuterium oxide was heavier than that from seedlings grown on normal water (Fig. 20-6), suggesting *de novo* synthesis (Quail et al., 1973a).

At Vanderbilt University, Pratt and co-workers followed the appearance of phytochrome apoprotein in oat and rye embryos between 4 and 24 h after the beginning of imbibition at 15°C. The use of polyclonal antibodies showed quite clearly that the pigment was synthesized *de novo* (Coleman and Pratt, 1974).[4]

The Freiburg workers monitored phytochrome synthesis after light-induced destruction. At 96 and 169 h after sowing, 2 h R/15 min FR depleted phytochrome levels in *Cucurbita pepo*. But levels increased again in darkness at rates similar to those observed in the nonirradiated seedlings (Fig. 20-7). Moreover, irradiated seedlings incorporated deuterium into phytochrome, again suggesting *de novo* synthesis. Thus phy-

---

4. These antibodies did not cross-react with the phytochrome in seeds.

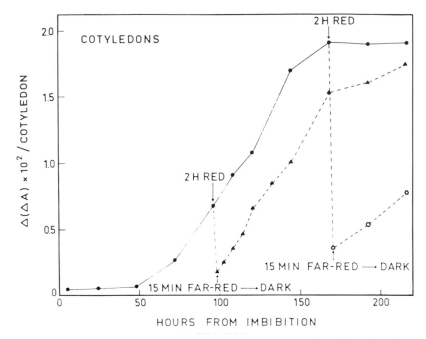

FIGURE 20-7   Time course of level of extractable phytochrome/cotyledon of dark-grown
Cucurbita pepo seedlings (●) and of seedlings irradiated at 96 h (▲) or 168 h (○) with 2 h R +
15 min FR and returned to the dark. (From Quail et al., 1973b. Reproduced with permission
from the American Society of Plant Physiologists.)

tochrome disappearance and recovery appeared "to involve, respectively,
the degradation and synthesis of phytochrome protein" (Quail et al,
1973b).

After 168 h, nonirradiated seedlings continued to make phytochrome,
but the total level remained constant (Fig. 21-7). Thus degradation of Pr
(that had not been cycled through Pfr) was also taking place, although at
rates that were perhaps 100 times slower than those for Pfr destruction.

Clarkson and Hillman had suggested that rates of phytochrome synthesis in pea might change according to how much Pfr was present in the cell
(see Chapter 12). But in zucchini (Fig. 20-7) and mustard (Schäfer et al.,
1972), synthesis was unaffected by Pfr levels (Quail et al., 1973b).

CENTROSPERMAE

The absence of dark reversion in *Amaranthus*, a member of the Centrospermae, challenged Hillman's conclusion that reversion occurs in dicots
but not monocots (see Chapter 13). Workers in Freiburg, moreover, re-

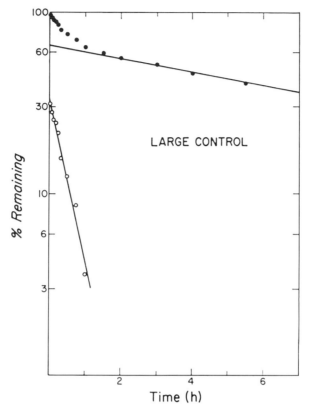

FIGURE 20-8 Resolution of *in vitro* reversion curve for large rye phytochrome into two lines. (●) Data points; (○) replotted points for fast component. (From Pike and Briggs, 1972. Reproduced by permission from the American Society of Plant Physiologists.)

ported its absence in mustard (Marmé, 1969a; Oelze-Karow and Mohr, 1970).

At Brookhaven, Kendrick failed to detect dark reversion in two other members of the Centrospermae (Kendrick and Hillman, 1971), which also show several other monocot features. But spectrophotometric assays on mustard in the presence of sodium azide revealed that reversion did, in fact, occur in this species (Kendrick and Hillman, 1970).

REVERSION *IN VITRO*

Undenatured small oat phytochrome reverted with a half-time of 9 h even though it came from a nonreverting monocot plant (Mumford, 1966). The first-order reaction accelerated strongly below pH 6, in proportion to hydrogen ion concentration, suggesting temporary protonation (Anderson et al., 1969). Reduced ferridoxin increased the rate 800-fold (Mumford and Jenner, 1971), but reducing agents were unable to mimic FR-induced physiological responses in intact higher plants.

Taylor detected dark reversion in partially purified preparations from etiolated oat and pea and light-grown parsnip. A relatively rapid phase and a slow phase suggested ''at least two naturally occurring forms of phytochrome were also observed'' (Taylor, 1968).

Two-phase kinetics were also observed in rye phytochrome (Correll et al., 1968b). Highly purified large rye phytochrome reverted more rapidly than small, and the resolution of its curve into two straight lines (Fig. 20-8) was compatible with a two-pool interpretation (Pike and Briggs, 1972).

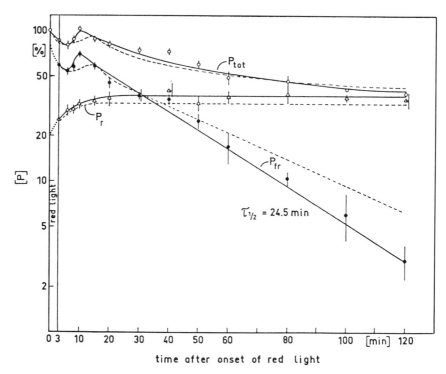

FIGURE 20-9 Phytochrome levels in cotyledons of dark-grown mustard seedlings after 3 min saturating R. (●) Measured values of Pfr; (○) measured values of total phytochrome corrected for incomplete photoconversion by R; (△) calculated values of Pr (P_{total} − Pfr). (—) 72 h from sowing; (---) 48 h from sowing. (From Marmé et al., 1971. Copyright 1971, Springer-Verlag.)

In pea phytochrome, calcium and magnesium ions accelerated both fast and slow phases, an effect that might allow "plant cells to control the rate of dark reversion, thereby controlling processes mediated by Pfr" (Negbi et al., 1975). A naturally occurring inhibitor of reversion in pea crude extracts (Manabe and Furuya, 1971) proved to be a small, ethanol-soluble molecule (Shimazaki and Furuya, 1975).

PHOTOSTATIONARY STATES

The Freiburg workers soon confirmed that mustard phytochrome reverted *in vivo* during the first 15 min after a saturating R pulse (Fig. 20-9). Pfr destruction began after a short delay (Fig. 20-9), and its half-life decreasd as seedlings aged (Marmé et al., 1971; Schäfer et al., 1972). Thus there was a distinction between a phytochrome photoequilibrium established by a light source and a phytochrome photostationary state, which resulted not only from the action of light but also from alterations of Pfr levels by dark reactions.

The group obtaincd first-order kinetics for phytochrome phototransformation (Schmidt et al., 1973), confirming Butler's report for extracted

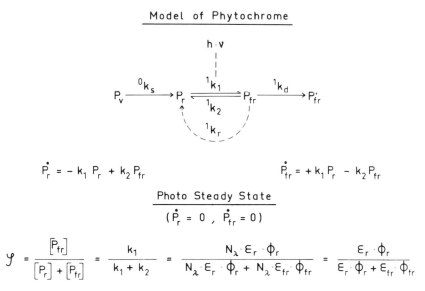

FIGURE 20-10 Simplified model of phytochrome system in etiolated mustard hooks and cotyledons. (From Mohr, 1972. Copyright 1972, Springer-Verlag.)

phytochrome (see Chapter 11) and Pratt's for phytochrome *in vivo* (Pratt and Briggs, 1966). Synthesis appeared to be independent of phytochrome concentrations, fitting the definition of a zero-order process (Schäfer et al., 1972).

Schäfer incorporated these findings into a model (Fig. 20-10) for phytochrome action (Schäfer, 1971; Schäfer et al., 1971, 1972). Mathematical treatment (Drumm and Mohr, 1974; Schäfer and Mohr, 1974) revealed that the model accounted for the pigment's action under inductive conditions and steady-state conditions obtained with low irradiances of continuous FR. Produced as supposedly inactive Pr, the pigment was readily photoconverted to Pfr. The active form did not persist but was removed by destruction and—in dicots—reversion.

"But the model," Schäfer says, "could neither explain high-energy responses (since it predicted that the level of Pfr would be independent of irradiance) nor the fact that dark reversion was faster than destruction but seemed to be restricted to a smaller fraction of phytochrome. The astonishing outcome was that researchers either ignored all models, ignored dark reversion, or used this totally wrong but simple model."

PHOTOPROTECTION

Spruit and Kendrick, meanwhile, discovered that phytochrome may be significantly protected from destruction at high irradiances, which allow a large proportion of the pigment to exist as intermediates.

During a NATO fellowship inWageningen, Kendrick used Spruit's new quasi-continuous spectrophotometer, which delivered alternating millisecond pulses of actinic and measuring beams (Spruit, 1971). Wanting to use high-irradiance WL, he removed the filters. When he turned on the unfiltered light, "we saw phototransformation of phytochrome. But when we put it out, more Pfr appeared."

Irradiating whole, dark-grown *Amaranthus* seedlings, Kendrick recorded absorbance differences between 738 and 806 nm (Pfr) during the actinic sequence dark/R/dark/FR/dark/WL/dark/WL (Fig. 20-11). Predictably, R formed a large amount of Pfr, which persisted in the dark and then disappeared during FR irradiation. But after a WL pulse, Pfr increased in darkness and then returned to its previous level during a second WL treatment. This suggested that WL maintained a pool of one or more intermediates, which converted to Pfr after it was turned off and reformed when it came back on. At 0°C, intermediate accounted for at least 30% of the total phytochrome (Kendrick and Spruit, 1972a). The proportion of phytochrome as intermediate decreased when oxygen was removed (Kendrick and Spruit, 1973a).

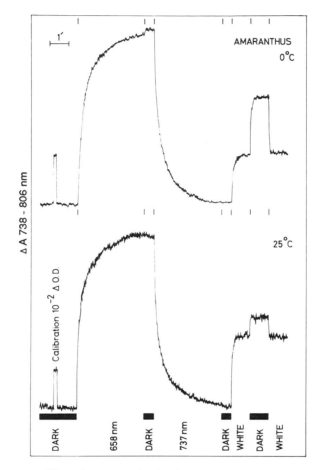

FIGURE 20-11   Differential changes in absorbance between 738 and 806 nm in whole, dark-grown *Amaranthus* seedlings subjected to irradiation sequence shown at bottom of the figure. (From Kendrick and Spruit, 1972a. Reprinted by permission from *Nature New Biology*. vol. 237, pp. 281–282. Copyright © 1972 Macmillan Magazines Ltd.)

In phytochrome-rich pea epicotyls, there was a linear relationship between the proportion of phytochrome maintained as intermediates and irradiance at 698 nm, where both Pr and Pfr are excited. Highest levels were maintained by wavelengths between 690 to 700 nm, but the intermediate itself absorbed only weakly (Kendrick and Spruit, 1973a). Thus it was possibly P_{b1} (later named meta-Rb), a bleached precursor of Pfr (Kendrick and Spruit, 1973b).

Kendrick confirmed that phytochrome destruction in *Amaranthus* deviated from first-order kinetics under high-irradiance WL (Kendrick, 1972).

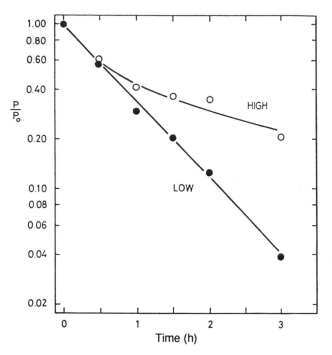

FIGURE 20-12   Total phytochrome in whole *Amaranthus* seedlings as a proportion of that present initially (P/P_o) plotted on a logarithmic scale against time under low- and high-irradiance incandescent light. (From Kendrick and Spruit, 1972b.)

This "accumulation of phytochrome as an intermediate, which itself does not decay, accounts for the gradual decrease in the first-order rate constant of phytochrome (Fig. 20-12) observed under high-irradiance incandescent light in *Amaranthus*" (Kendrick and Spruit, 1972b).

DESTRUCTION MECHANISM

The nature of phytochrome disappearance was still unclear in the early 1970s. Alternatives to protein destruction were "translocations from extended to restricted areas of the cell" (Spruit, 1972), denaturation, and removal of the chromophore without loss of the protein.

Using polyclonal antibodies (see Chapter 21), which collectively recognized many different sites on the phytochrome apoprotein and therefore were capable of detecting proteolytic fragments, Pratt (Fig. 20-13) and co-workers showed that loss of antigenicity paralleled loss of photoreversibility (Fig. 20-14). Immunochemically detectable phytochrome also

FIGURE 20-13   Lee H. Pratt at the University of Geneva in 1981 with the first of his dual-wavelength, microcomputer-operated spectrophotometers. Courtesy of Marie-Michèle Cordonnier.

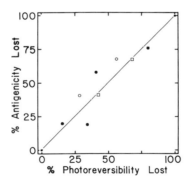

FIGURE 20-14   Relationship between loss of photoreversibility *in vivo* and loss of extractable antigen identifiable as phytochrome. Various symbols indicate data from different experiments. For immunological assay, crude extract was placed in agar wells containing antiserum to small phytochrome. Two days later, staining revealed how far into the agar any phytochrome had diffused. Comparison with diffusion radii of known dilutions of dark control extract revealed how much antigenically determinable phytochrome remained at any time after irradiation. (From Pratt et al., 1974.)

disappeared progressively from electron micrographs during the same period. Thus destruction appeared to involve extensive breakdown of the apoprotein (Pratt et al., 1974). Antigenic activity disappeared at different rates in different tissues (Pratt and Coleman, 1974).

Destruction of Pr that had cycled through Pfr also involved loss of both spectrophotometrically and antigenically detectable phytochrome. It had identical kinetics to Pfr destruction and was similarly sensitive to ethylene, 2-mercaptoethanol, and azide (Stone and Pratt, 1979).

INVERSE DARK REVERSION—AN EXPLANATION

Schäfer's scheme did not include inverse dark reversion, which appeared to be a peculiarity of seeds. Briggs suggested that hydration of a low-

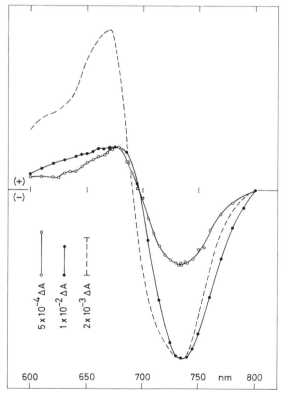

FIGURE 20-15 Difference spectra for phototransformation by R/FR irradiation of phytochrome in fresh (---) and freeze-dried (●—●) pea epicotyl tissue and dry cucumber seeds (○—○). (From Kendrick and Spruit, 1974. Copyright 1974, Springer-Verlag.)

absorbing form of Pfr might account for the spectral shift (Briggs and Rice, 1972). In freeze-dried gelatin films containing phytochrome, an absorption peak at 575 or 590 nm gave way, on partial hydration, to a R absorption peak at 660 nm and a FR peak at 715 nm. The latter was possibly caused by P_{710}, a previously detected intermediate in Pr to Pfr conversion (Tobin et al., 1973).

Kendrick detected this species in frozen tissues using Spruit's quasi-continuous spectrophotometer. At the University of Newcastle, he purchased a Perkin-Elmer 156 dual-wavelength spectrophotometer. Pea epicotyl, freeze-dried with its phytochrome in the Pfr form, generated a difference spectrum (Fig. 20-15) with abnormally low R absorption, a characteristic of the low-temperature intermediate P_{650} (Kendrick and Spruit, 1973b; Spruit and Kendrick, 1973). P_{650} normally converted to Pr (Kendrick and Spruit, 1973b) but, trapped in freeze-dried tissue, it could be reconverted to Pfr (Kendrick, 1974). A difference spectrum, like that obtained from cucumber seed, had a low, broad peak at about 670–680 nm and an isosbestic point at 697 nm (Fig. 21-15).

In Wageningen, Spruit exposed Kendrick's freeze-dried Pfr-containing samples to a saturating FR pulse and then 18 h darkness. A second FR pulse decreased the sample's absorbance at 738 nm, revealing that Pfr had reformed (Fig. 20-16). A subsequent actinic R pulse showed that the

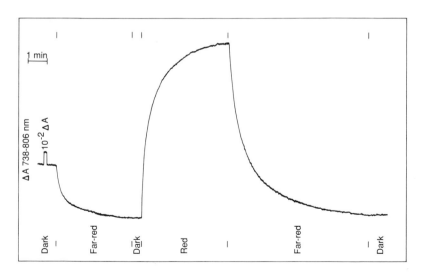

FIGURE 20-16 Differential absorbance changes in a sample of pea epicotyl tissue freeze-dried with its phytochrome largely in Pfr form. The dried sample was previously exposed to saturating FR and then maintained in complete darkness for 18 h. (From Kendrick and Spruit, 1974. Copyright 1974, Springer-Verlag.)

reverted pigment accounted for about a quarter of the phytochrome in the sample. Thus "inverse dark reversion" was not a property of "special seed phytochrome, but of extremely dehydrated normal phytochrome. It represents the reaction of one or more intermediates in the phototransformation of Pfr produced under conditions which prevent complete transformation to Pr" (Kendrick and Spruit, 1974). Rehydration of freeze-dried tissue containing only Pfr restored photoreversibility after a R pulse. But Pfr continued to appear slowly after the irradiation ceased and again after the end of a subsequent FR exposure. Thus a trapped intermediate in the Pr to Pfr pathway converted to Pfr on rehydration. This was presumed to be Pbl (Kendrick and Spruit, 1974).

REFERENCES

Anderson, G. R., Jenner, E. L., and Mumford, F. E. (1969). Temperature and pH studies on phytochrome *in vitro*. *Biochemistry* 8,1182–1187.

Boisard, J. (1968). La photosensibilité des akènes de Laitue (variété "Reine de Mai") et son interprétation par l'étude *in vivo* du photorecepteur. Thesis. University of Rouen.

Boisard, J., Spruit, C. J. P., and Rollin, P. (1968). Phytochrome in seeds: An apparent dark reversion of Pr to Pfr. *Meded. Landbouwhogesch. Wageningen* 68(17),1–5.

Bonner, B. A. (1962). *In vitro* dark conversion and other properties of phytochrome. *Plant Physiol.* 37(suppl. xxvii).

Bonner, B. A. (1963). Some properties of phytochrome from etiolated peas. Unpublished.

Briggs, W. R., and Rice, H. V. (1972). Phytochrome: Chemical and physical properties and mechanism of action. *Annu. Rev. Plant. Physiol.* 23,293–334.

Briggs, W. R., and Siegelman, H. W. (1965). Distribution of phytochrome in etiolated seedlings. *Plant Physiol.* 40,934–941.

Butler, W. L., Lane, H. C., and Siegelman, H. W. (1963). Nonphotochemical transformations of phytochrome *in vivo*. *Plant Physiol.* 38,514–519.

Chorney, W., and Gordon, S. A. (1966). Action spectrum and characteristics of the light activated disappearance of phytochrome in oat seedlings. *Plant Physiol.* 41,891–896.

Coleman, R. A., and Pratt, L. H. (1974). Phytochrome: Immunocytochemical assay of synthesis and destruction. *Planta* 119,221–231.

Correll, D. L., and Shropshire, W., Jr. (1968). Phytochrome is etiolated annual rye. I. Changes during growth in the amount of photoreversible phytochrome in the coleoptile and primary leaf. *Planta* 79,275–283.

Correll, D. L., Edwards, J. L, Klein, W. H., and Shropshire, W., Jr. (1968a). Phytochrome in etiolated annual rye. III. Isolation of photoreversible phytochrome. *Biochim. Biophys. Acta* 168,36–45.

Correll, D. L., Edwards, J. L., and Shropshire, W., Jr. (1968b). Multiple chromophore species in phytochrome. *Photochem. Photobiol.* 8,465–475.

de Lint, P. J. A. L., and Spruit, C. J. P. (1963). Phytochrome destruction following illumination of mesocotyls of *Zea mays* L. *Meded. Landbouwhogesch. Wageningen* 63(14),1–7.

Dooskin, R. H., and Mancinelli, A. L. (1968). Phytochrome decay and coleoptile elongation in *Avena* following various light treatments. *Bull. Torrey Bot. Club* 95,474–487.

Drumm, H., and Mohr, H. (1974). The dose response curve in phytochrome-mediated anthocyanin synthesis in the mustard seedlings. *Photochem. Photobiol.* 20,151–157.

Frankland, B. (1972). Biosynthesis and dark transformations of phytochrome. *In* "Phytochrome" (K. Mitrakos and W. Shropshire, Jr., eds.), pp. 195–225. Academic Press, London, New York.

Hartmann, K. M. (1966). A general hypothesis to interpret 'high energy phenomena' of photomorphogenesis on the basis of phytochrome. *Photochem. Photobiol.* 5,349–366.

Hayes, R. G., and Klein, W. H. (1974). Spectral quality influence of light during development of *Arabidopsis thaliana* plants in regulating seed germination. *Plant Cell Physiol.* 15, 643–653.

Hendricks, S. B., Butler, W. R., and Siegelman, H. W. (1962). A reversible photoreaction regulating plant growth. *J. Phys. Chem.* 66,2550–2555.

Jabben, M. (1980). The phytochrome system in light-grown *Zea mays* L. *Planta* 149,91–96.

Jabben, M., Heim, B., and Schäfer, E. (1980). The phytochrome system in light- and dark-grown dicotyledonous seedlings. *In* "Photoreceptors and Plant Development" (J. de Greef, ed.), pp. 145–158. Antwerpen University Press, Antwerpen.

Kendrick, R. E. (1972). Aspects of phytochrome decay in etiolated seedlings under continuous illumination. *Planta* 102,286–293.

Kendrick, R. E. (1974). Phytochrome intermediates in freeze-dried tissue. *Nature* 250,159–161.

Kendrick, R. E., and Frankland, B. (1968). Kinetics of phytochrome decay in *Amaranthus* seedlings. *Planta* 82,317–320.

Kendrick, R. E., and Frankland, B. (1969a). Photocontrol of germination in *Amaranthus caudatus*. *Planta* 85,326–339.

Kendrick, R. E., and Frankland, B. (1969b). The *in vivo* properties of *Amaranthus* phytochrome. *Planta* 86,21–32.

Kendrick, R. E., and Hillman, W. S. (1970). Dark reversion of phytochrome in *Sinapis alba*. *Plant Physiol.* 46,596–598.

Kendrick, R. E., and Hillman, W. S. (1971). Absence of phytochrome dark reversion in seedlings of the Centrospermae. *Am. J. Bot.* 58,424–428.

Kendrick, R. E., and Spruit, C. J. P. (1972a). Light maintains high levels of phytochrome intermediates. *Nature (Lond.)* New Biol. 237,281–282.

Kendrick, R. E., and Spruit, C. J. P. (1972b). Phytochrome decay in seedlings under continuous incandescent light. *Planta* 107,341–350.

Kendrick, R. E., and Spruit, C. J. P. (1973a). Phytochrome intermediates *in vivo*—I. Effects of temperature, light intensity, wavelength and oxygen on intermediate accumulation. *Photochem. Photobiol.* 18,139–144.

Kendrick, R. E., and Spruit, C. J. P. (1973b). Phytochrome intermediates *in vivo*—III. Kinetic analysis of intermediate reactions at low temperature. *Photochem. Photobiol.* 18,153–159.

Kendrick, R. E., and Spruit, C. J. P. (1974). Inverse dark reversion: An explanation. *Planta* 120,265–272.

Kendrick, R. E., Spruit, C. J. P., and Frankland, B. (1969). Phytochrome in seeds of *Amaranthus caudatus*. *Planta* 88,293–302.

Mackenzie, J. M., Jr., Briggs, W. R., and Pratt, L. H. (1978). Phytochrome photoreversibility: Empirical test of the hypothesis that it varies as a consequence of compartmentalization. *Planta* 141,129–134.

Malcoste, R., Boisard, J., Spruit, C. J. P., and Rollin, P. (1970). Phytochrome in seeds of some Cucurbitaceae: *In vivo* spectrophotometry. *Meded. Landbouwhogesch. Wageningen* 70(16),1–16.

Manabe, K., and Furuya, M. (1971). Factors controlling rates of nonphotochemical transformation of *Pisum* phytochrome *in vitro*. *Plant Cell Physiol.* 12,95–101.

Mancinelli, A. L., and Tolkowsky, A. (1968). Phytochrome and seed germination. V. Changes of phytochrome content during the germination of cucumber seeds. *Plant Physiol.* 43,489–494.

Marmé, D. (1969a). Photometrische Messungen am Phytochromsystem von Senfkeimlingen (*Sinapis alba* L.). *Planta* 88,43–57.

Marmé, D. (1969b). An automatic recording device for measuring phytochrome with a dual wavelength Ratiospect. *Planta* 88,58–60.

Marmé, D., Marchal, B., and Schäfer, E. (1971). A detailed analysis of phytochrome decay and dark reversion in mustard cotyledons. *Planta* 100,331–336.

McArthur, J. A., and Briggs, W. R. (1970). Phytochrome appearance and distribution in the embryonic axis and seedling of Alaska peas. *Planta* 91,146–154.

McCullough, J. M., and Shropshire, W., Jr. (1970). Physiological predetermination of germination responses in *Arabidopsis thaliana* (L.) HEYNH. *Plant Cell Physiol.* 11,139–148.

Mohr, H. (1972). "Lectures on Photomorphogenesis." Springer, New York.

Mumford, F. E. (1966). Studies on the phytochrome dark reaction *in vitro*. *Biochemistry* 5,522–524.

Mumford, F. E., and Jenner, E. L. (1971). Catalysis of the phytochrome dark reaction by reducing agents. *Biochemistry* 10,98–101.

Negbi, M., Hopkins, D. W., and Briggs, W. R. (1975). Acceleration of dark reversion of phytochrome *in vitro* by calcium and magnesium. *Plant Physiol.* 56,157–159.

Oelze-Karow, H., and Mohr, H. (1970). Experiments regarding the problem of differentiation in multicellular systems. *Z. Naturforsch.* 25b,1282–1286.

Pike, C. S., and Briggs, W. R. (1972). The dark reactions of rye phytochrome *in vivo* and *in vitro*. *Plant Physiol.* 49,514–520.

Pratt, L. H., and Briggs, W. R. (1966). Photochemical and nonphotochemical reactions of phytochrome *in vivo*. *Plant Physiol.* 41,467–474.

Pratt, L. H., and Coleman, R. A. (1974). Phytochrome distribution in etiolated grass seedlings as assayed by an indirect antibody-labelling method. *Am. J. Bot.* 61,195–202.

Pratt, L. H., Kidd, G. H., and Coleman, R. A. (1974). An immunochemical characterization of the phytochrome destruction reaction. *Biochim. Biophys. Acta* 365,93–107.

Quail, P. H., Schäfer, E., and Marmé, D. (1973a). *De novo* synthesis of phytochrome in pumpkin hooks. *Plant Physiol.* 52,124–127.

Quail, P. H., Schäfer, E., and Marmé, D. (1973b). Turnover of phytochrome in pumpkin cotyledons. *Plant Physiol.* 52,128–131.

Rollin, P. (1968). La photosensibilité des graines. *Bull. Soc. Fr. Physiol. Veg.* 14,47–63.

Schäfer, E. (1971). Detaillierte photometrische Messungen *in vivo* am Phytochromsystem von *Sinapis alba* L. und *Cucurbita pepo* L. Dissertation, University of Freiburg.

Schäfer, E. (1981). Phytochrome and daylight. *In* "Plants and the Daylight Spectrum" (H. Smith, ed.), pp. 461–480. Academic Press, London.

Schäfer, E., and Mohr, H. (1974). Irradiance dependency of the phytochrome system in cotyledons of mustard (*Sinapis alba* L.). *J. Math. Biol.* 1,9–15.

Schäfer, E., Marchal, B., and Marmé, D. (1971). On the phytochrome phototransformation kinetics of mustard seedlings. *Planta* 101,265–276.

Schäfer, E., Marchal, B., and Marmé, D. (1972). *In vivo* measurements of the phytochrome photostationary state in far red light. *Photochem. Photobiol.* 15,457–464.

Schmidt, W., Marmé, D., Quail, P., and Schäfer, E. (1973). Phytochrome: First-order phototransformation kinetics *in vivo*. *Planta* 111,329–336.

Shimazaki, Y., and Furuya, M. (1975). Isolation of a naturally occurring inhibitor for dark Pfr reversion from etiolated *Pisum* epicotyls. *Plant Cell Physiol.* 16,623–630.

Spruit, C. J. P. (1970). Spectrophotometers for the study of phytochrome *in vivo. Meded. Landbouwhogesch. Wageningen* 70(14),1–18.

Spruit, C. J. P. (1971). Sensitive quasi-continuous measurement of photoinduced transmission changes. *Meded. Landbouwhogesch. Wageningen* 71(21),1–6.

Spruit, C. J. P. (1972). Estimation of phytochrome by spectrophotometry *in vivo:* Instrumentation and interpretation. *In* "Phytochrome" (K. Mitrakos and W. Shropshire, Jr., eds.), pp. 77–104. Academic Press, London, New York.

Spruit, C. J. P., and Kendrick, R. E. (1973). Phytochrome intermediates *in vivo*—II. Characterisation of intermediates by difference spectrophotometry. *Photochem. Photobiol.* 18,145–152.

Spruit, C. J. P., and Mancinelli, A. L. (1969). Phytochrome in cucumber seeds. *Planta* 88,303–310.

Spruit, C. J. P., Boisard, J., and Rollin, P. (1968). Spectrophotometric phytochrome in imbibed seeds of dark germinating plants. *Plant Physiol.* 43(suppl. 15).

Stone, H. J., and Pratt, L. H. (1979). Characterization of the destruction of phytochrome in the red-absorbing form. *Plant. Physiol.* 63,680–682.

Taylor, A. O. (1968). *In vitro* phytochrome dark reversion process. *Plant Physiol.* 43,767–774.

Tobin, E. M., and Briggs, W. R. (1969). Phytochrome in embryos of *Pinus palustris. Plant Physiol.* 44,148–150.

Tobin, E. M., Briggs, W. R., and Brown, P. K. (1973). The role of hydration in the phototransformation of phytochrome. *Photochem. Photobiol.* 18,497–503.

Zouaghi, M., Malcoste, R., and Rollin, P. (1972). Etude du phytochrome detectable, *in vivo* dans les graines de *Cucurbita pepo* L. au cours des differentes phases de la germination. *Planta* 106,30–43.

CHAPTER **21**

Localization

Butler and Bernard Epel arrived at the National Institutes of Health, where a military guard ushered them into a room, retrieved a secret device from a safe, and locked them in. It was August 1966, and the goal was phytochrome localization.

Using an image intensifier that was being developed for gunsights, Butler and his first graduate student looked under a fluorescence microscope for phytochrome in oat coleoptile cells. "Warren played around with different wavelengths of light for about an hour," Epel recalls, "and concluded that all we were seeing was chlorophyll fluorescence."

Galston tried to locate the pigment by scanning for R/FR reversibility, using a universal microspectrophotometer at the Biophysics Department of Kings College, London University. Alternating R and FR exposures triggered reversible absorption shifts along the contours of nuclear membranes, strengthening "the view that the pigment interacts in some way with genetic material, although the localization at or near the nuclear membrane may indicate control of passage of materials between nucleus and cytoplasm" (Galston, 1968).

Skepticism greeted this interpretation, because there appeared to be too little phytochrome to generate such an absorbance change. Spruit estimated that, with such short light paths and tiny measuring beam, the signal-to-noise ratio would be at least 10 million times less than with a bulk sample. Because it was impossible to intensify the beam without photoconverting phytochrome, the prospects for success appeared grim (Spruit, 1972).

IMMUNOLOCALIZATION

In La Jolla, Hopkins and Butler attempted to make phytochrome more detectable by reacting it with antibodies. But a technique developed for hemoglobin proved to be inadequate for the scarcer plant pigment.

FIGURE 21-1 Lee H. Pratt, 1989. Photo by Linda Sage.

At Vanderbilt, Pratt (Fig. 21-1) adopted the immunological techniques he had observed in Butler's lab, intending to identify evolutionarily conserved parts of the phytochrome molecule. But student assistant Richard Coleman encouraged him to learn some immunochemistry from Ludwig Sternberger, a summer visitor to Vanderbilt.

Coleman and Pratt localized phytochrome in the cells of dark-grown oat coleoptiles within a couple of months. The immunochemical staining appeared in parenchyma cells in some regions of the coleoptile and in epidermal cells in others. It was also seen in the node and root cap[1] (Pratt and Coleman, 1971). Distribution patterns varied in oat, rye, barley, rice, and maize (Pratt and Coleman, 1974). But "the central message," says Pratt, "was that the pigment wasn't being made by all cells at the same time."

Adding osmium, Coleman could examine the stained sections under the electron microscope. The detailed images showed only "a possible specific association with membranes such as the plasma membrane, endoplasmic reticulum and nuclear membrane" (Coleman and Pratt, 1974). "I don't think we made much of that," Pratt says, "although other people did. We said it could just have been an artifact due to fixation."

1. Pratt was surprised to find phytochrome in root caps, although Tanada had evidence for such a location (see Chapter 16). It was later found that the roots of certain plants respond normally to gravity only after the phytochrome in their caps has seen light (Tepfer and Bonnett, 1972). The root cap but not the main root of mung bean becomes electrically positive after treatment with R (Racusen and Etherton, 1975).

CELL FRACTIONATION

Other workers tried to localize phytochrome by breaking open cells and separating fractions by centrifugation. Solon A. Gordon claimed to find phytochrome in a mitochondrial fraction (Gordon, 1961), but "the marked influence of pH on the solubility of the chromoprotein near neutrality makes any decision on the location of the pigment tenuous" (Siegelman and Firer, 1964).

At the University of California, Berkeley, Bernard Rubinstein took care to maintain neutral or slightly alkaline conditions. He detected about 1% of extracted phytochrome in 1,500–40,000 g pellets—enough, perhaps, to be physiologically potent (Rubinstein et al., 1969).

PELLETABLE PFR

Quail (Fig. 21-2) had arrived in Freiburg shortly before the Eretria symposium. There he took over the work of a student who was already fractionating cells but "doing some strange things, so all the phytochrome was precipitating out. What I consider to be my major contribution," he says, "was not so much finding suitable buffers as asking the question: 'If you convert phytochrome to the Pfr form, will it associate with membranes differently than the Pr form?' There was this vague idea that phytochrome was membrane-bound, but nobody had addressed the separate concept of Pr being a soluble molecule that, in its active form, Pfr, binds to something on a membrane."

FIGURE 21-2   Peter H. Quail, 1989. Photo by Linda Sage.

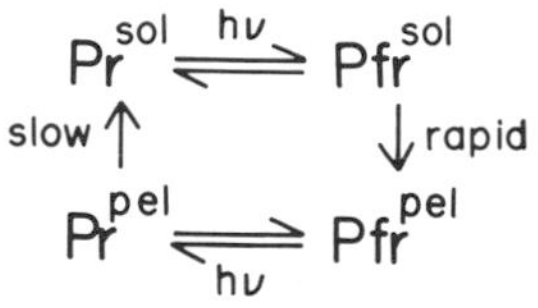

$$Pr^{sol} \underset{}{\overset{h\nu}{\rightleftharpoons}} Pfr^{sol}$$

slow $\uparrow$ $\quad\quad$ $\downarrow$ rapid

$$Pr^{pel} \underset{h\nu}{\rightleftharpoons} Pfr^{pel}$$

FIGURE 21-3 Summary of *in vivo*-induced phytochrome pelletability. (From Pratt, 1978.)

Quail resuspended 20,000 g maize coleoptile pellets, keeping the pH above 6.8. Pellets prepared from unirradiated or FR-treated seedlings contained only a small proportion of total extractable phytochrome. But when seedlings were irradiated with R before homogenization and centrifugation, "the result was spectacular," Quail recalls. "I put the sample in the spectrophotometer, and it seemed to malfunction. Dieter Marmé figured out that it was on a high sensitivity setting and that there was so much phytochrome in the sample that it had gone off the scale."

Surprisingly, phytochrome pelletability increased when Pfr was converted back to Pr before extraction. "So in dark-grown tissue, the phytochrome molecule is primarily soluble and in the Pr form," Quail explains. "When you convert it to the Pfr form, it's momentarily soluble and then becomes pelletable. If you convert it back to Pr before extraction, you don't immediately regain the soluble form but go through a Pr form that remains pelletable" (Fig. 21-3). The half-life of the latter was 50 min.

In the dicot *Cucurbita pepo* (Table 21-1), Pfr was also more pelletable than Pr, but photoconversion back to Pr before homogenization decreased, rather than increased, pelletability, although not to the level of the FR control. In both tissues a second conversion to the Pfr form returned levels to those after a single R treatment (Quail, et al., 1973).

TABLE 21-1

Pelletable Phytochrome (%) in Extracts of *Zea mays* Coleoptiles and *Cucurbita pepo* Hooks in the Dark and After Irradiation with R (660 m) and FR (756 nm)[a,b]

Irradiation program	(%) Pelletable phytochrome	
	Maize coleoptiles	Pumpkin hypocotyl hooks
Dark control	5.6 ± 0.4	7.5 ± 0.4
5 min 756 nm	5.7 ± 0.5	7.4 ± 0.7
3 min 660 nm	15.9 ± 0.5	26.0 ± 1.3
3 min 660 nm + 5 min 756 nm	26.8 ± 0.6	16.9 ± 1.1
3 min 660 nm + 5 min 756 nm + 3 min 660 nm	15.6 ± 0.3	25.2 ± 1.7

[a] pH was 7.2–7.4.
[b] From Quail et al., 1973. Reproduced by permission from *Nature New Biology* vol. 245, pp. 189–191. Copyright © 1973 Macmillan Magazines Ltd.

Quail left Freiburg in 1973 to take a temporary fellowship at the Australian National University in Canberra.

Marmé and Boisard visited Briggs, meanwhile, where they induced pelletability by irradiating *extracts* of dark-grown zucchini squash (Marmé et al., 1973), and intact seedlings (Boisard et al., 1974). The group also observed membranous vesicles in electron micrographs of the Pfr-containing particles (Marmé et al., 1974).

SEQUESTERING

The Harvard lab closed at the end of 1973 when Briggs moved to the Carnegie Institution of Washington in Stanford. While he took an initial sabbatical, Mackenzie spent a year with Pratt. Wondering if pelletability involved intracellular movement of phytochrome, Mackenzie irradiated oat coleoptile tips immediately before fixation and then stained sections immunochemically. Under the light microscope, he saw plenty of pigment in the cells of unirradiated tips. But there appeared to be hardly any after R or R/FR treatments.

Under greater magnification, Mackenzie found the "vanished" pigment, clumped into discrete areas of each cell that were not mitochondria, plastids, or nuclei (Fig. 21-4). The clumps (which escaped being named *phytochromosomes*) were larger in FR-treated tips, and they disappeared over the next 2 h in darkness, while phytochrome slowly appeared again throughout the cytoplasm. FR alone did not affect phytochrome distribution, and the clumps persisted after R alone. This phenomenon, which became known as *sequestering,* also occurred in parenchyma cells from dark-grown rice shoots. Possible explanations were "that phytochrome, as Pfr, might aggregate with itself or with other proteins and thereby form a massive protein aggregate" or "that Pfr may associate with a specific site on a membrane (or region of a membrane) or with an as yet unidentified organelle" (Mackenzie et al., 1975).

ASSOCIATION WITH RIBONUCLEOPROTEIN

Marmé had returned to Freiburg, meanwhile, where he continued to study pelletability. In December 1974, he was joined by Pratt, and the two observed R-enhanced pelletability *in vivo* in Garry oat, winter rye, maize, zucchini, pea, sunflower, radish, soybean, kidney bean, lupin and mustard. *In vitro* binding was confined to zucchini and mustard however. And whereas the latter was completely reversed by high pH and high salt concentrations, *in vivo* binding was not in most species (Pratt and Marmé,

1976). "So we were learning," Pratt explains, "that there were two ways in which phytochrome could associate with pelletable material."

Quail had already reached this conclusion. Subjecting homogenates of R-irradiated zucchini hooks to density gradient centrifugation, he identified the cellular components that sedimented with the pigment. Without

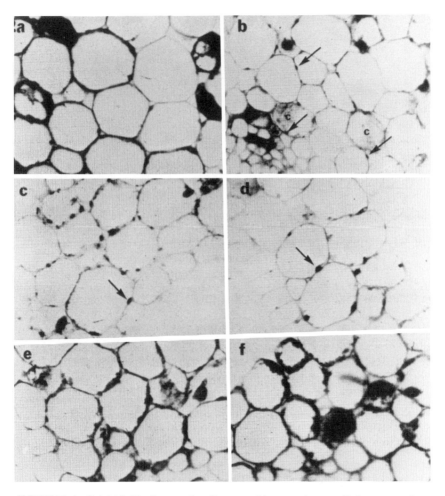

FIGURE 21-4   Bright field micrographs of comparable parenchyma cells from oat coleoptile sections. Arrows indicate discrete areas of dense phytochrome-associated stains. c, unstained cytoplasm. (a) Before light exposure, (b) immediately after exposure to R; (c) immediately after exposure to R followed by FR; (d) same as (c) but after 30 min darkness at 24°C; (e) same as (c) but after 60 min darkness; (f) same as (c) but after 120 min darkness. ×493. (From Mackenzie et al., 1975.)

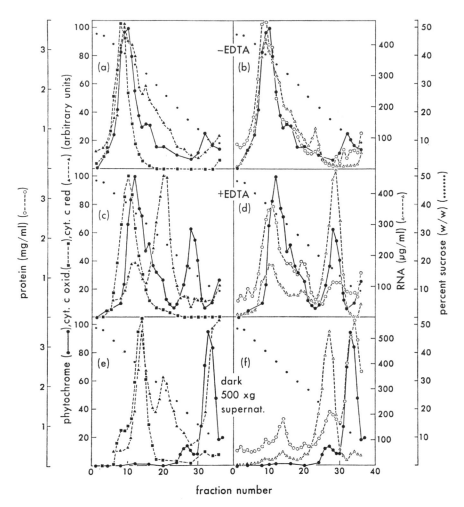

FIGURE 21-5   Effects of EDTA and *in vivo* R irradiation on sucrose gradient profiles of phytochrome and other cellular components from hypocotyl hooks of *Cucurbita* seedlings. Each horizontal pair of panels represents a single gradient, but profiles of membrane markers and phytochrome are compared on the *left,* whereas those of RNA, protein, and phytochrome are compared on the *right.* (For interpretation of symbols, see vertical axes.) (a/b) and (c/d) were resuspended 20,000 g pellets from R-irradiated tissue. (e/f) was a 500 g supernatant from nonirradiated tissue. The medium used throughout lacked magnesium. (From Quail, 1975a. Copyright 1975, Springer-Verlag.)

EDTA and magnesium, pelletable material sedimented into a broad band containing phytochrome, protein, RNA, and enzyme markers for mitochondrial membrane and endoplasmic reticulum (Fig. 21-5a,b). With EDTA but not magnesium, there were two phytochrome-containing bands, and the less dense one contained most of the RNA and associated protein but not the membrane markers (Fig. 21-5c,d). The predominantly nonpelletable phytochrome from unirradiated hooks sedimented separately from RNA (Fig. 21-5,e,f). Thus R irradiation produced a form of phytochrome (Pfr) that bound to RNA during cell homogenization. Because ribosomes sedimented more quickly than the phytochrome-ribonucleoprotein band, the phytochrome appeared to be stuck to small pieces of degraded ribosomal material. It was released by high levels of magnesium, suggesting that it adsorbed electrostatically but nonspecifically to RNA when cells were broken open (Quail, 1975a). This association accounted for *in vitro* binding of zucchini phytochrome to pelletable material.

Low magnesium levels were required for pelletability in *Zea mays,* but the pelleted phytochrome sedimented separately from RNA after resuspension in a magnesium-free medium (Quail, 1975b). Thus *in vivo* pelletability in *Zea* differed from the nonspecific association between Pfr and ribonucleoprotein observed in *Cucurbita* extracts (Quail and Gressel, 1976).

IN VIVO PELLETABILITY

At age 33 Quail left Australia to become a postdoctorate with Briggs at the Carnegie Institution.

Using high-intensity lamps to photoconvert phytochrome in 1 s, Quail monitored the rapid dark reaction that made soluble Pfr molecules pelletable (Fig. 21-6a). Pratt had obtained a half-life of about 40 s for this process at 0.5°C (Pratt and Marmé, 1976).

After a 1-s R flash, Quail exposed oat shoots and maize coleoptiles to 10 s FR. The extent of the rate-limiting dark reaction was exponential to the interval between these two flashes—the time Pfr remained in the tissue. Because its half-time was only 2 s in both oat and maize (Fig. 21-6b), the induction of pelletability became the fastest known Pfr-mediated cellular response "with the potential to be an integral part of the primary molecular mechanism of phytochrome action" (Quail and Briggs, 1978).

This rapidity suggested that induction of pelletability and sequestering might be different aspects of the same intracellular process. Sequestering

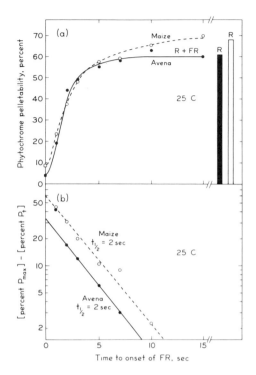

FIGURE 21-6   (a) Time course of increase in phytochrome pelletability in *Avena* shoots
(●———●) and maize coleoptiles (○---○) in air at 25°C as a function of an increasing dark period
interposed between 1 s R and 10 s FR. Because R began at time zero, the 1-s time point on
each curve is for application of FR immediately after the end of R. (b) Semilogarithmic plot of
data from (a). $\%P_{max}$, maximum percent pelletable phytochrome observed for R + FR
treatment; $\%Pr$, percent pelletable phytochrome at time t on the abscissa. (From Quail and
Briggs, 1978. Reproduced with permission from the American Society of Plant Physiologists.)

of Pfr was complete within 1–2 min at 3°C, and relaxation from the sequestered state took 1–2 h at 25°C (Mackenzie, 1976; Kass and Pratt, 1978).
Thus both sequestering and pelletability displayed an initial requirement
for Pr to Pfr conversion, a subsequent and rapid dark phase leading to a
complex that failed to dissociate in FR, and a slow dark phase (after FR
treatment) that regenerated free Pr.

REFERENCES

Boisard, J., Marmé, D., and Briggs, W. R. (1974). *In vivo* properties of membrane-bound
    phytochrome. *Plant Physiol.* 54,272–276.

Coleman, R. A., and Pratt, L. H. (1974). Electron microscopic localization of phytochrome in plants using an indirect antibody-labeling method. *J. Histochem. Cytochem.* 22,1039–1047.

Epel, B. L., Butler, W. L., Pratt, L. H., and Tokuyasu, K. T. (1980). Immunofluorescence localization studies of the Pr and Pfr forms of phytochrome in the coleoptile tips of oats, corn and wheat. *In* "Photoreceptors and Plant Development" (J. de Greef, ed.), pp. 121-133. Antwerpen University Press, Antwerpen.

Galston, A. W. (1968). Microspectrophotometric evidence for phytochrome in plant nuclei. *Proc. Natl. Acad. Sci. USA* 61,454–460.

Gordon, S. A. (1961). The intracellular distribution of phytochrome in corn seedlings. *Prog. Photobiol. Proc. Int. Congr. Photobiol. 3rd,* pp. 441–443.

Kass, L. B., and Pratt, L. H. (1978). Immunocytochemical assay of the time-course of phytochrome sequestering. *Plant Physiol.* 61(suppl. 12).

Mackenzie, J. M., Jr. (1976). Phytochrome distribution and redistribution as revealed by immunochemistry. Ph.D. dissertation. Harvard University.

Mackenzie, J. M., Jr., Coleman, R. A., Briggs, W. R., and Pratt, L. H. (1975). Reversible redistribution of phytochrome within the cell upon conversion to its physiologically active form. *Proc. Natl. Acad. Sci. USA* 72,799–803.

Marmé, D., Boisard, J., and Briggs, W. R. (1973). Binding properties *in vitro* of phytochrome to a membrane fraction. *Proc. Natl. Acad. Sci. USA* 70,3861–3865.

Marmé, D., Mackenzie, J. M., Jr., Boisard, J., and Briggs, W. R. (1974). The isolation and partial characterization of a membrane fraction containing phytochrome. *Plant Physiol.* 54,263–271.

Pratt, L. H., and Coleman, R. A. (1971). Immunocytochemical localization of phytochrome. *Proc. Natl. Acad. Sci. USA* 68,2431–2435.

Pratt, L. H., and Coleman, R. A. (1974). Phytochrome distribution in etiolated grass seedlings as assayed by an indirect antibody-labeling method. *Am. J. Bot.* 61,195–202.

Pratt, L. H., and Marmé, D. (1976). Red light-enhanced phytochrome pelletability. Reexamination and further characterization. *Plant Physiol.* 58,686–692.

Quail, P. H. (1975a). Particle-bound phytochrome: Association with a ribonucleoprotein fraction from *Cucurbita pepo* L. *Planta* 123,223–234.

Quail, P. H. (1975b). Particle-bound phytochrome: The nature of the interaction between pigment and particulate fractions. *Planta* 123,235–246.

Quail, P. H. (1978). Irradiation-enhanced phytochrome pelletability in *Avena: In vivo* development of a potential to pellet and the role of Mg^{2+} in its expression. *Photochem. Photobiol.* 27,147–153.

Quail, P. H., and Gressel, J. (1976). Particle-bound phytochrome: Interaction of the pigment with ribonucleoprotein material from *Cucurbita pepo* L. *In* "Light and Plant Development" (H. Smith, ed.), pp. 111-128. Butterworths, London, Boston.

Quail, P. H., Marmé, D., and Schäfer, E. (1973). Particle-bound phytochrome from maize and pumpkin. *Nature (Lond.) New Biol.* 245,189–191.

Racusen, R., and Etherton, B. (1975). Role of membrane-bound, fixed-charge changes in phytochrome-mediated mung bean root tip adherence phenomenon. *Plant Physiol.* 55,491–495.

Rubinstein, B., Drury, K. S., and Park, R. B. (1969). Evidence for bound phytochrome in oat seedlings. *Plant Physiol.* 44,105–109.

Siegelman, H. W., and Firer, E. M. (1964). Purification of phytochrome from oat seedlings. *Biochemistry* 3,418–423.

Spruit, C. J. P. (1972). Estimation of phytochrome by spectrophotometry *in vivo:* Instrumentation and interpretation. *In* "Phytochrome" (K. Mitrakos and W. Shropshire, Jr., eds.), pp. 77–104. Academic Press, London, New York.

Tepfer, D. A., and Bonnett, H. T. (1972). The role of phytochrome in the geotropic behavior of roots of *Convolvulus arvensis*. *Planta* 106,311–324.

Tokuyasu, K. T., and Singer, S. J. (1976). Improved procedures for immunoferritin labeling of ultrathin sections. *J. Cell Biol.* 71,894–906.

Verbelen, J.-P., Pratt, L. H., Butler, W. L., and Tokuyasu, K. T. (1982). Oat phytochrome distribution and redistribution, an electron microscopic study. *Plant Physiol.* 70,867–871.

CHAPTER 22

Properties of Large Phytochrome

Brigg's group was able to produce large rye phytochrome in 5 days (Rice et al., 1973). The preparation was 3,250 times purer than the crude extract and contained 5% of the original activity.

HARVARD RYE PREPARATIONS

Although this rye pigment had a molecular mass of 120,000 on SDS-PAGE, G-200 Sephadex chromatography (Rice and Briggs, 1973) and sucrose density centrifugation (Garner et al., 1971) gave much higher values, pointing to a 240,000-dalton dimer.

Because the mobility of rye phytochrome during SDS-PAGE differed markedly from the mobilities of phycocyanin and allophycocyanin (Rice and Briggs, 1973), the close resemblance between the phytochrome chromophore and those of the algal pigments did not extend to the apoprotein.

Tobin wondered whether the conformational changes proposed for Pr/Pfr interconversion also applied to large phytochrome. "The thinking at that time," she says, "was that, in the Pr form, the active site was hidden and, when you did the phototransformation, the whole protein conformation would change so that the active site would become exposed." She failed to see any difference in the fluorescence spectrum of Pr and Pfr, however, which meant "that a substantial change is unlikely" (Tobin and Briggs, 1973). She also obtained identical circular dichroism spectra for Pr and Pfr below 240 nm. An ultraviolet difference spectrum revealed "some changes around a tyrosine residue" (Tobin and Briggs, 1973).

Assuming a molecular mass of 120,000 and a photoequilibrium of 81% Pr in saturating R, Tobin calculated molar absorption coefficients of 70,000 dm^3 mol^{-1} cm^{-1} for Pr and 40,000 dm^3 mol^{-1} cm^{-1} for Pfr (Tobin and Briggs, 1973)—similar to those obtained at du Pont for small oat phytochrome but very different from values (16,400 and 8,800) reported by Correll.

Photoconversion of large rye phytochrome displayed first-order kinetics in both directions. Using Tobin's extinction coefficients, Gardner calculated the quantum yield for the conversion of Pr to Pfr to be 0.28 mol Einstein^{-1}, whereas that for the reverse reaction was 0.20 mol Einstein^{-1} (Gardner and Briggs, 1974). Thus the ratio of the quantum yields, $\Phi r/\Phi fr$ was 1.4—close to the 1.5 value reported by Butler, Hendricks, and Siegelman in 1964 for small phytochrome (See Chapter 11).

Gardner issued the first report of a chemical difference between large Pr and Pfr. Radioactive N-ethyl maleimide, which reacts mainly with the sulfhydryl groups of proteins, introduced 70% more label into Pfr than Pr (Gardner et al., 1974).

SMITHSONIAN RYE PREPARATIONS

William Owen Smith, Jr. (Fig. 22-1), who joined the Radiation Biology Laboratory in 1972, finally explained the discrepancies between the Smithsonian and Harvard rye preparations. His preparation generated a major peak on sucrose density gradients at 8.0 S and a shoulder at 6.5 S. Both bands were photoreversible (Fig. 22-2), but a minor peak at 11.5 S was not (Smith, 1975). The 11.5-S species appeared as tetramers under the electron microscope, whereas the 8.0-S and 6.5-S species were amorphous. Thus the tetramer that Correll had equated with native phytochrome was a contaminant that copurified until ultracentrifugation.

FIGURE 22-1 William Owen Smith, Jr., 1988. Photo by Linda Sage.

Because both phytochrome bands migrated to the 120,000-dalton position on SDS-PAGE (whereas the contaminating protein migrated to the
32,000- and 48,000-dalton positions), Smith equated the 6.5-S species with
the 120,000 monomer and the 8.0-S species with the dimer (Smith and
Correll, 1975). Briggs had already proposed that Correll's 9-S species was
the phytochrome dimer (Briggs and Rice, 1972).

VANDERBILT OAT PREPARATIONS

Pratt was purifying large oat phytochrome, meanwhile, hoping to repeat
the basic studies performed at Beltsville. The photochemical properties of
the resulting preparation differed markedly from those of small oat phytochrome. The photoequilibrium in saturating R was only 75%, and quantum yields were 0.17 in each direction (Pratt, 1975).

Assays with polyclonal antibodies revealed immunological similarities
among large oat phytochrome and other monocot phytochromes but little
cross-reactivity between monocot and dicot phytochromes (Pratt, 1973;
Cordonnier and Pratt. 1982).

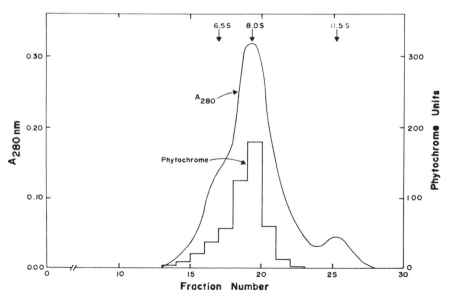

FIGURE 22-2   Sucrose density gradient profile of phytochrome purified by W. O. Smith.
(From Smith and Correll, 1975. Reproduced by permission of the American Society of Plant
Physiologists.)

GROSS STRUCTURE

The immunological assays allowed Pratt to ask whether large phytochrome, with a molecular mass of 120,000, was composed of two covalently linked units of small—60,000-dalton—phytochrome. Antisera against large or small oat phytochrome would contain identical immunoglobulins if large phytochrome duplicated small; otherwise it would contain some unique immunoglobulins. When doctoral student Susan Cundiff electrophoresed a mixture of large and small phytochrome between agar troughs of antilarge and antismall sera, a spur of precipitated phytochrome developed only against the former, "suggesting that there are determinants on large phytochrome which are absent from small phytochrome" (Cundiff and Pratt, 1973).

Extract incubated in darkness for long periods generated a less striking precipitin band, but several additional bands appeared. These presumably contained proteolytic products, but only one was electrophoretically similar to small phytochrome. Obtaining qualitatively similar results with trypsin-degraded phytochrome, the researchers concluded "that small phytochrome does not arise by simple disaggregation or hydrolysis of large phytochrome into identical subunits" (Cundiff and Pratt, 1973).

At Harvard, former Vanderbilt undergraduate George Kidd labeled Pratt's large and small oat phytochromes with radioactive iodide, which primarily marks tyrosine residues. Two-dimensional electrophoresis of proteolytic digests revealed that large phytochrome generated almost twice as many radioactive fragments as small phytochrome. Thus "[large] phytochrome is apparently composed of only one copy of the [small] molecule and an approximately equal amount of unrelated primary structure" (Kidd et al., 1978).

AFFINITY PURIFICATION

Inspired by work with the insulin receptor, Pratt and Robert E. Hunt developed an immunopurification procedure for phytochrome (Hunt and Pratt, 1979). The preparation approached 100% purity and was usually free of proteolytic enzymes (Pratt, 1984).

Oat large phytochrome was a dimer in solution, with molecular weights of 118,000 on SDS-PAGE and 233,000 on sedimentation equilibrium centrifugation. Like the rye pigment, it appeared to be water-soluble with a hydrophobic core. It also had multiple N-terminal residues—in this case alanine and lysine—and formed multiple bands during gel electrophoresis (Hunt and Pratt, 1980). Hunt also detected chemical differences between

FIGURE 22-3   Pill-Soon Song, 1984.

phytochrome's two forms (Hunt and Pratt, 1981). Pratt moved to the University of Georgia, Athens, in 1979.

Smith and Susan M. Daniels also developed an affinity purification that produced purer and more easily obtainable large phytochrome than the conventional method. But they employed a commercially available dye, Cibacron Blue 3GA (Affi-gel Blue), which bound tightly to both Pr and Pfr, although more tightly to the latter. By placing brushite-purified phytochrome on Affi-gel Blue columns, they separated the pigment from proteins that failed to bind to the dye. FMN[1] eluted phytochrome from the column (Hahn et al., 1980; Smith and Daniels, 1981; Song et al., 1981).

SONG'S 1979 MODEL

At Texas Tech University in Lubbock Pill-Soon Song (Fig. 22-3) confirmed, through circular dichroism studies, the absence of major structural differences between large Pr and Pfr (Song et al., 1979).

Looking for subtler changes, Tae-Ryong Hahn and Soon-Seon Kang exposed Affi-gel Blue preparations to potassium permanganate. Pfr bleached more rapidly and to a greater extent than Pr (Hahn et al., 1980), suggesting, Song says, "that one of the things that is happening in the Pr to Pfr phototransformation is the relative movement and therefore accessibility of the chromophore."

The chemical probe 8-anilinonaphthalene-1-sulfonate (ANS) decreased the absorbance of phytochrome more profoundly at 725 nm than at 660 nm, suggesting a greater affinity for Pfr. R irradiation increased fluorescence,

1. FMN is flavin mononucleotide (riboflavin 5'-phosphate).

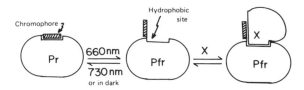

FIGURE 22-4   Song's first model of phytochrome phototransformation. X is the putative Pfr receptor. (From Song, 1983, redrawn from Song et al., 1979. Reprinted by permission from Cambridge University Press and Elsevier Science Publishers.)

leading to the same conclusion (Hahn and Song, 1981). Song explained these results "in terms of occupancy by ANS of the hydrophobic surface on the Pfr form" (Song, 1983; Choi et al., 1990). Because the probe speeded photoconversion of Pr to Pfr while inhibiting the reverse process and *in vitro* dark reversion, a hydrophobic site that was more exposed in Pfr than Pr appeared to lie close to the chromophore (Hahn and Song, 1981). Studies with liposomes (artificial vesicles of lipid) also indicated the presence of a hydrophobic site on Pfr's surface (Kim and Song, 1981).

Song proposed that Pr to Pfr transformation made the chromophore swing away from the surface of the protein, exposing a hydrophobic site to which the phytochrome receptor could bind (Fig. 22-4) (Song et al., 1979). But a difference in hydrophobicity (judged by affinity to long fatty acyl chains) between large pea Pfr and Pr was not retained in small phytochrome (Tokutomi et al., 1981). Thus the hydrophobic site was unlikely to flank the chromophore.

REFERENCES

Briggs, W. R., and Rice, H. V. (1972). Phytochrome: Chemical and physical properties and mechanism of action. *Annu. Rev. Plant Physiol.* 23,293–334.

Choi, J.-K., Kim, I.-S., Kwon, T.-I., Parker, W., and Song, P.-S. (1990). Spectral perturbations and oligomer/monomer formation in 124-kilodalton *Avena* phytochrome. *Biochemistry* 29,6883–6891.

Cordonnier, M.-M., and Pratt, L. H. (1982). Comparative phytochrome immunochemistry as assayed by antisera against both monocotyledons and dicotyledons. *Plant Physiol.* 70,912–916.

Cundiff, S. C., and Pratt, L. H. (1973). Immunological determination of the relationship between large and small sizes of phytochrome. *Plant Physiol.* 51,210–213.

Gardner, G., and Briggs, W. R. (1974). Some properties of phototransformation of rye phytochrome *in vitro*. *Plant Physiol.* 54,263–271.

Gardner, G., Pike, C. S., Rice, H. V., and Briggs, W. R. (1971). "Disaggregation" of phytochrome *in vitro*—A consequence of proteolysis. *Plant Physiol.* 48,686–693.

Gardner, G., Thompson, W. F., and Briggs, W. R. (1974). Differential reactivity of the red- and far-red-absorbing forms of phytochrome to [^{14}C] N-ethyl maleimide. *Planta* 117,367–372.

Hahn, T.-R., and Song, P.-S. (1981). Hydrophobic properties of phytochrome as probed by 8-anilinonaphthalene-1-sulfonate fluorescence. *Biochemistry* 20,2602–2609.

Hahn, T.-R., Kang, S.-S., and Song, P.-S. (1980). Difference in the degree of exposure of chromophores in the Pr and Pfr forms of phytochrome. *Biochem. Biophys. Res. Commun.* 97,1317–1323.

Hunt, R. E., and Pratt, L. H. (1979). Phytochrome immunoaffinity purification. *Plant Physiol.* 64,332–336.

Hunt, R. E., and Pratt, L. H. (1980). Partial characterization of undegraded oat phytochrome. *Biochemistry* 2,390–394.

Hunt, R. E., and Pratt, L. H. (1981). Physicochemical differences between the red- and the far-red-absorbing forms of phytochrome. *Biochemistry* 20,941–945.

Kidd, G. H., Hunt, R. E., Boeshore, M. L., and Pratt, L. H. (1978). Asymmetry in the primary structure of undegraded phytochrome. *Nature (Lond.)* 276,733–735.

Kim, I. S., and Song, P.-S. (1981). Binding of phytochrome to liposomes and protoplasts. *Biochemistry* 20, 5482–5489.

Pratt, L. H. (1973). Comparative immunochemistry of phytochrome. *Plant Physiol.* 51,203–209.

Pratt, L. H. (1975). Photochemistry of high molecular weight phytochrome *in vitro. Photochem. Photobiol.* 22,33–36.

Pratt, L. H. (1984). Phytochrome purification. *In* "Techniques in Photomorphogenesis"(H. Smith, ed.), pp. 175–200. Academic Press, London.

Rice, H. V., and Briggs, W. R. (1973). Partial characterization of oat and rye phytochrome. *Plant Physiol.* 51,927–938.

Rice, H. V., Briggs, W. R., and Jackson-White, C. J. (1973). Purification of oat and rye phytochrome. *Plant Physiol.* 51,917–926.

Smith, W. O., Jr. (1975). Purification and physicochemical studies of phytochrome. Ph.D. thesis. University of Kentucky.

Smith, W. O., Jr. (1981). Probing the molecular structure of phytochrome with immobilized Cibacron blue 3GA and blue dextran. *Proc. Natl. Acad. Sci. USA* 78,2977–2980.

Smith, W. O., Jr., and Correll, D. L. (1975). Phytochrome: A re-examination of the quarternary structure. *Plant Physiol.* 56,340–343.

Smith, W. O., Jr., and Daniels, S. M. (1981). Purification of phytochrome by affinity chromatography on agarose-immobilized Cibacron Blue 3GA. *Plant Physiol.* 68,443–446.

Song, P.-S. (1983). The molecular basis of phytochrome (Pfr) and its interactions with model receptors. *Symp. Soc. Exp. Biol.* 36,181–206.

Song, P.-S., Chae, Q., and Gardner, J. D. (1979). Spectroscopic properties and chromophore conformations of the photomorphogenic receptor: Phytochrome. *Biochim. Biophys. Acta* 576,479–495.

Song, P.-S., Kim, I. S., and Hahn, T. R. (1981). Purification of phytochrome by Affi-gel Blue chromatography: An effect of lumichrome on purified phytochrome. *Anal. Biochem.* 117,32–39.

Tobin, E. M., and Briggs, W. R. (1973). Studies on the protein conformation of phytochrome. *Photochem. Photobiol.* 18,487–495.

Tokutomi, S., Yamamoto, K. T., and Furuya, M. (1981). Photoreversible changes in hydrophobicity of undegraded pea phytochrome determined by partition in an aqueous two-phase system. *FEBS Lett.* 134,159–162.

CHAPTER **23** _____

Membrane Hypothesis

By 1975, there were "basically two general hypotheses to account for the manifold and rapid effects of phytochrome on cell metabolism and development. The first proposes that one or more 'second messengers' act as intermediaries between the single primary action of phytochrome at the membrane, and its effects on enzyme synthesis and activity. The alternative view is that more than one primary reaction of phytochrome is possible, depending perhaps on its location within the cell'' (Smith, 1976).

NYCTINASTIC PLANTS

Ruth L. Satter (Fig. 23-1) tested the membrane hypothesis using nyctinastic plants at Yale, where she became a postdoctorate with Galston (Fig. 23-2) in 1967.

Previous work had involved whole pinnae of *Albizzia julibrissin,* but Satter floated single pairs of pinnules attached to small pieces of rachilla (Fig. 23-3) on experimental solutions. Each pinnule had at its base a tiny motor organ, the pulvinule, whose vascular core was sandwiched between subepidermal motor cells.

Measuring solute[1] concentrations in freeze-dried sections of tissue with a microprobe, Satter determined that the potassium (K^+) content of ventral motor cells decreased and that of dorsal motor cells increased when she darkened pinnules a few hours into the photoperiod, a treatment that also induced closure. Whereas FR before darkness inhibited K^+ efflux from ventral pulvinar tissue, K^+ influx into dorsal tissue required darkness but not R pretreatment (Satter et al, 1970).

Diurnal variations in leaflet angle paralleled changes in potassium-to-calcium ratio in ventral cells but were a mirror image of those in dorsal

1. Because water moves into and out of plants cells in response to changes in solute concentrations.

312

FIGURE 23-1 Ruth L. Satter.

cells. But fluxes were somewhat out of phase, indicating the involvement of additional cells in storage and shunting (Satter and Galston, 1971).

When pinnules were briefly exposed to R shortly before they normally opened, they opened less and their ventral motor cells acquired less K^+ than after a FR pulse (Satter et al., 1972). Thus Pfr enhanced K^+ efflux from ventral motor cells during leaflet opening as well as during nyctinastic closing, perhaps by altering the permeability of a membrane that was sensitive to its actions just before and during the initial hours of the

FIGURE 23-2 Arthur W. Galston.

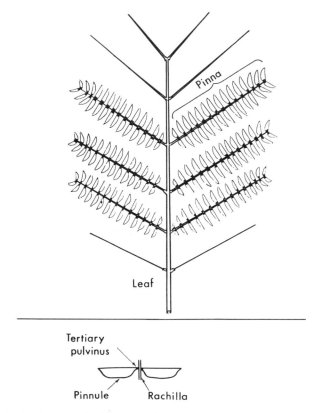

FIGURE 23-3   Leaves of *Albizzia julibrissin*. (From Satter, 1979. Copyright 1979, Springer-Verlag.)

photoperiod (Satter and Galston, 1973)—leaflet closing is not under phytochrome control during the later hours of a photoperiod (Hillman and Koukkari, 1967; Jaffe and Galston, 1967).

Around 1973, Satter began to work also with *Samanea saman*,[2] a leguminous tree whose large, cylindrical pulvinules are amenable to biochemical analysis. Although its leaves close rhythmically, the dorsal motor cells of pulvinules (secondary pulvini) become flaccid during closure, whereas those of tertiary pulvini become turgid (Fig. 23-4). To avoid confusion, Satter adopted the terms *extensors* for cells that swell during opening and shrink during closure and *flexors* for cells that contract during opening and expand during closure, changing in shape more than size. The microprobe detected huge K^+ fluxes during opening and closing in this species (Satter

2. *Samanea* has since been renamed *Pithecolobium*.

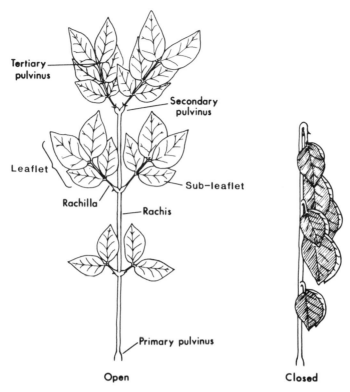

FIGURE 23-4   Open and closed leaves of *Samanea saman*. (From Satter et al., 1974b. Reproduction from the *Journal of General Physiology*, 1974, vol. 64, pp. 413–430 by copyright permission of the Rockefeller University Press.)

et al., 1974b). As in *Albizzia,* phytochrome acted in conjunction with an endogenous rhythm to control movement of this ion,[3] promoting K^+ movement into extensor cells.[4]

Chloride (Cl^-) was the major negative ion accompanying K^+ in the *Albizzia* pulvinus (Schrempf et al., 1976). In *Samanea,* pulvinar angle increased as the combined concentration of these two ions in the mid-extensor region increased relative to that in the mid-flexor region. Movements of other ions helped maintain electrical balance (Satter et al., 1977).

3. Exposure of excised *Samanea* pulvini to R or FR pulses during nearly a week of darkness revealed that the photoreversible pigment could also entrain the circadian rhythm (Simon et al., 1976). A pigment that absorbs B and FR also shifts the rhythm, both in *Samanea* and *Albizzia* (Satter et al., 1981).

4. Several other groups found a link between Pfr formation and potassium fluxes (e.g., Brownlee and Kendrick, 1977, 1979; Brownlee et al., 1979).

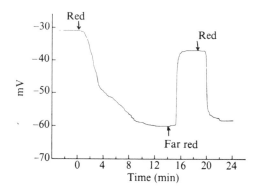

FIGURE 23-5 Phytochrome-controlled changes in transmembrane potential of *Samanea* flexor motor cells. (From Racusen and Satter, 1975. Reprinted by permission from *Nature* vol. 255, pp. 408–410. Copyright © 1975 Macmillan Magazines Ltd.)

Because the plasma membrane is rather impermeable to Cl⁻ in most higher plants, the Yale workers wondered if the ions passed instead through cell interiors and plasmodesmata[5] (Schrempf et al., 1976). But measurement at Cornell University with a scanning electron microscope fitted with an energy-dispersive X-ray analyzer confirmed that the plasma membrane was "an important site of ion fluxes and . . . a locus for phytochrome-clock interaction" (Campbell et al., 1981).

In the mid 1970s, Richard Racusen at the University of Vermont, in collaboration with Satter, measured transmembrane potentials by impaling cells with microelectrodes. Excising *Samanea* pulvinules at successive intervals during an extended dark period, he observed rhythmic changes in both extensor and flexor cells, although the amplitude in the latter was larger. The phase difference was about 8 h, and neither rhythm was quite in phase with leaflet movements.

Flexor cell interiors became increasingly negative over a 10-min period after excised pulvini were exposed to R, a phenomenon known as *hyperpolarization*. FR reversed the R effect within 90 s, and a second R treatment reversed the FR treatment (Fig. 23-5). Thus phytochrome might "regulate rapid, membrane-associated phenomena." K⁺ fluxes were too slow to account for the electrical changes (Racusen and Satter, 1975). Later work at Yale suggested that cotransport of sucrose and protons (H⁺) into the cell by a carrier in the plasma membrane contributed to membrane depolariza-

5. Plasmodesmata run through cell walls. Because they connect the interiors of adjacent cells, substances can pass from one cell to another without crossing plasma membrane. The continuum of living, interconnected cell interiors is named the *symplast*. The combined extracellular spaces, in and between cell walls, is the *apoplast*.

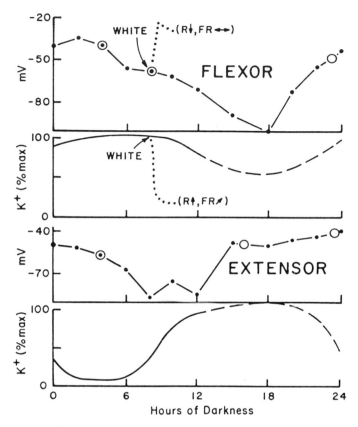

FIGURE 23-6 K^+ content and membrane potential of *Samanea* (—) and *Albizzia* (---) extensor and flexor cells. Plants were darkened for 24 h, except where indicated. Large circles indicate times when cells depolarize in response to external protons and sucrose. Direction of the small arrows after R and FR indicates effect of irradiation. (From Satter and Galston, 1981. Reproduced with permission from the Annual Review of Plant Physiology, vol. 32. © 1981 by Annual Reviews, Inc.)

tion, partly annulling the potentials generated in motor cells (Racusen and Galston, 1977).

Extensor cells from plants darkened for 24 h began to hyperpolarize a few hours before K^+ and Cl^- uptake began, and cells became most negative just as these ions started to accumulate. Flexor cells reached maximal hyperpolarization when concentrations of the two ions was lowest, 8–10 h later (Fig. 23-6). Satter and Galston interpreted "light-regulated and rhythmic changes in potential in terms of electrogenic H^+ secretion and back diffusion" (Satter and Galston, 1981).

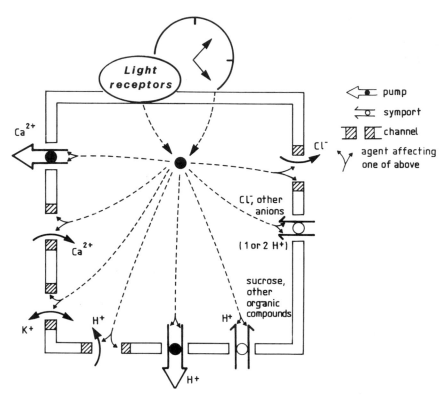

FIGURE 23-7   Satter's 1988 version of Satter and Galston's 1981 model of ion transport through plasma membrane of a pulvinal motor cell. Phytochrome is envisioned as affecting Ca^{2+} entry, which in turn alters other cellular processes. The oscillator may affect diffusion pathways for H^+, K^+, Cl^-, and OH^-, as well as H^+ pumps, H^+-sucrose symports, and Cl^-/OH^- antiports. (From Satter and Galston, 1981; Satter et al., 1988.)

The importance of proton-extruding pumps in ion transport had recently been acknowledged (see Poole, 1978; Spanswick, 1981). In motor cells, such a pump would increase the negativity of the cell interior, creating an electrical gradient along which K^+ could move into the cell, provided specific channels were open. It would also generate a pH gradient that might energize the uptake of Cl^- and organic ions by proton-enhanced carriers. Both light and the clock would regulate its activity (Fig. 23-7).

In 1980, Satter moved to the University of Connecticut in Storrs where she obtained evidence that an energy-requiring system pumped protons out of the cell, opposing their passive inward diffusion. Proton extrusion appeared to be coupled to inward K^+ movement, and protons also moved along pH gradients (Iglesias and Satter, 1983; Lee and Satter, 1987). Weak organic ions appeared to regulate cytoplasmic pH and alter electrical gradients (Satter et al., 1987).

Monitoring apoplastic pH changes with a proton-sensitive liquid membrane electrode (Lee and Satter, 1988), Youngsook Lee found that darkening the pulvinus when phytochrome was largely Pfr increased the pH in the extensor apoplast and decreased that in the flexor apoplast, increasing the pH difference between the two regions from 10- to 100-fold. When pulvini were closed by a light/dark transition and then exposed to continuous WL, they reopened after 20 min, and the pH increase in the flexor apoplast accelerated. The extensor apoplast pH rose slightly after the WL treatment and then decreased rapidly before stabilizing (Lee and Satter, 1989). Thus "the H^+ pump appears to be activated by darkness in flexor cells and by light in extensor cells *in situ* as well as *in vitro*" (Satter et al., 1988).

Cara Zucker Lowen continuously monitored K^+ movements with a K^+-sensitive microelectrode. She detected an increase in K^+ activity in the extensor apoplast 2 min after the beginning of a R pulse; the level stabilized in the dark and then decreased in WL. The opposite occurred in the flexor apoplast (Fig. 23-8), supporting "the hypothesis that K^+ channels in the plasma membranes of *Samanea* motor cells are utilized for K^+ efflux from extensor cells and uptake by flexor cells following a WL → RL → D transition" (Lowen and Satter, 1989). Evidence for the presence in *Samanea* pulvini of membrane-spanning proteins that, through conformational changes, form open or closed pores led to that proposal that "the final steps in a mechanism for cell shrinking [are] membrane depolarization → K^+ channel opening → K^+ efflux cell shrinkage" (Moran et al., 1988).

MEMBRANE POTENTIAL CHANGES

Pfr-induced changes in membrane potential were not unique to *Samanea*, as Jaffe's investigation of the Tanada effect (see Chapter 16) had demonstrated. Racusen, who found that root tips adhered to a negative platinum electrode as well as to glass (Racusen and Miller, 1972), observed that plasmolyzed protoplasts moved around in the cell in response to R, showing that the charge changes underlying the Tanada effect were associated with the plasma membrane and not the cell wall (Racusen and Etherton, 1975).

Visiting Harvard, Ian A. Newman detected Pfr-induced electrical potential changes on the surfaces of oat seedlings within 10 s of R treatment (Newman and Briggs, 1972). In Tasmania, he placed microelectrodes

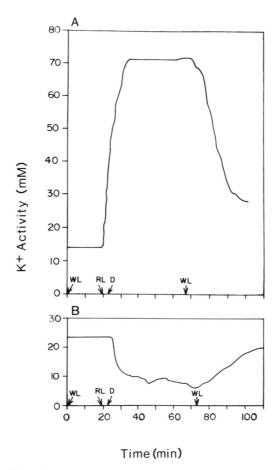

FIGURE 23-8 K^+ activity in extensor (A) and flexor (B) regions of apoplast of a *Samanea* pulvinus irradiated sequentially with WL (20 min), R (3 min), D (40 min), and WL (60 min). The first WL treatment began at hour 4.5 of the L portion of daily L-D cycle. (From Lowen and Satter, 1989. Copyright 1989, Springer-Verlag.)

inside coleoptile cells, which hyperpolarized in R (Newman and Sullivan, 1976), giving the expected, opposite response to that obtained with surface electrodes. But at Yale, Racusen observed R-induced depolarization with the same system (Racusen, 1976). In both labs, the response began within 5 to 10 s, but it was transient in Newman's lab, whereas depolarization persisted in the dark in Racusen's experiments. "The rapid changes in membrane potential are consistent with the idea of a membrane locale for early phytochrome action," Racusen concluded (Racusen, 1976). Extending the work with surface electrodes, Newman detected changes 4.5 s after the onset of light pulses (Newman, 1981). The response was thus the most rapid phytochrome-mediated physiological change recorded—the dark reaction leading to pelletability had a half-time of only 2 s, but it involved a change in the phytochrome molecule itself rather than the induction of a secondary reaction.

INTRINSIC PHYTOCHROME

The membrane hypothesis was also explored through cell fractionation, both in pelletability studies (see Chapter 21) and in attempts to identify phytochrome intrinsic to organelles or plasma membrane.

At Nagoya University, Katsushi Manabe isolated a mitochondrial fraction from etiolated pea epicotyls whose ability to reduce $NADP^+$ was R/FR-reversible. Spectrophotometrically detectable phytochrome remained in the membrane fraction after the mitochondria were lysed. Manabe and Furuya (who moved from Nagoya to the University of Tokyo in 1968) concluded that "the phytochrome appeared quite likely to be bound mainly in or on mitochondrial membranes" (Manabe and Furuya, 1974). Interaction of phytochrome with mitochondrial membrane (Cedel and Roux, 1980a,b; Roux et al., 1981) is still a controversial topic, however (Gross et al., 1979), and immunoelectron microscopy of oat and rye sections has failed to produce any evidence (McCurdy and Pratt, 1986; Speth et al., 1986; Warmbrodt et al., 1989).

There were numerous reports of phytochrome-mediated changes in isolated mitochondria, plastids, plasma membrane, and endoplasmic reticulum (See Marmé, 1977; Quail, 1982). Most turned out to be "false-alarms and fade-outs" (Quail, 1982), although a R response of etioplasts—an increase in the amount of extractable gibberellin-like substances—was observed in several labs (Cooke and Saunders, 1975; Cooke et al., 1975; Cooke and Kendrick, 1976; Evans, 1976; Evans and Smith, 1976a, b; Hilton and Smith, 1980).

Smith's group also detected phytochrome in a crude membrane fraction. Being Mg^{2+}-independent and immediately R/FR-reversible, the association was distinct from the pelletability previously described (Watson and Smith, 1982a,b; Napier and Smith, 1987a,b).

At the Glasshouse Crops Research Institute in Littlehampton, Brian Thomas and co-workers used a sensitive, enzyme-like immunosorbent assay (ELISA) (Thomas et al., 1984) to detect phytochrome in a high-speed, microsomal[6] fraction even in the absence of R and Mg^{2+} (Jordan et

6. Fragments of endoplasmic reticulum and plasma membrane.

al., 1984). Also, very clean plasma membrane preparations, prepared from dark-grown wheat leaves through phase partitioning[7] (Terry et al., 1989a), contained 1–2 ng immunologically detectable native phytochrome per microgram membrane protein (Terry et al., 1989b). The concentration of the membrane-bound pigment failed to relate to that of soluble phytochrome after various *in vivo* light treatments.

SYNTHETIC MEMBRANE

At Yale, Roux and Juan Yguerabide deposited a solution of oxidized cholesterol across a tiny hole in a Teflon cup and applied a small electrical potential. Injection of partly purified small phytochrome into KCl solution immersing the membrane decreased resistance across the membrane. A R pulse caused a further, rapid drop, a response that was almost completely reversed by subsequent FR (Fig. 23-9) (Roux and Yguerabide, 1973). "This suggested," says Roux, "that small Pfr could directly interact with lipid portions of membranes and affect their permeability to ions."

At the University of Pittsburgh, Roux explored phytochrome binding to liposomes, which are more stable than planar lipid bilayers (Georgevich et al., 1976). The addition of large phytochrome to medium suspending salt-loaded liposomes resulted in transitory changes in conductivity. And with lecithin/dicetylphosphate liposomes, Pfr induced a two- to threefold greater change in conductivity than Pr, although not at 4°C, suggesting that the pigment interacted only with fluid membrane (Georgevich, 1980; Georgevich and Roux, 1982).

Song determined that twice as much large Pfr as Pr bound to lecithin/cholesterol spheres, although binding was not R/FR-reversible (Kim and Song, 1981). But when the study was repeated with native phytochrome (see Chapter 28), there was only about a 25% difference between Pr and Pfr binding (Singh et al., 1989).

AN APPRAISAL

Although Pfr certainly promoted ion movements and changed membrane potentials, these events were possibly only secondary consequences of its action. Asking if any of the rapid actions of phytochrome were fast enough to represent *direct* alteration of membrane properties, Quail scrutinized

7. Plasma membrane has strings of sugars on its outer surface. When a mixed membrane preparation is suspended in a mixture of dextran and polyethylene glycol, right side-out vesicles of plasma membrane move into the dextran layer, while other types of membrane— which lack the sugar coating—distribute themselves in the polyethylene glycol.

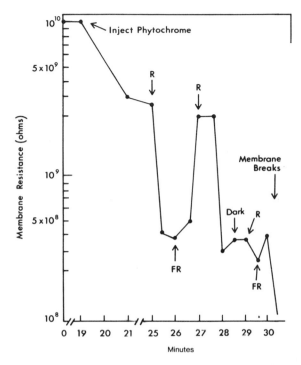

FIGURE 23-9 Resistance changes of an oxidized cholesterol membrane on addition of 6 μg small phytochrome to bathing solution and R or FR irradiation. (From Roux and Yguerabide, 1973.)

Newman's data. The transient change in membrane potential had a lag of at least 4.5 s after 1-s R (Fig. 23-10) (Newman, 1981). A FR pulse immediately after R partly prevented the R-induced depolarization (Newman, 1974). So Pfr had to act for a few seconds to induce a response, and the lag was not solely due to the time required for photon absorption or phytochrome phototransformation—or, presumably, to interaction of phytochrome with a receptor. It was possible, therefore, that "change in potential is indirect and is mediated by one or more second messengers that transmit the signal from phytochrome to the membrane" (Quail, 1983). Among the many examples of phytochrome rapid action, none showed that phytochrome was involved *and* that membrane properties altered *and* that the observed changes resulted from direct physical interaction of phytochrome with that membrane (Quail, 1983).

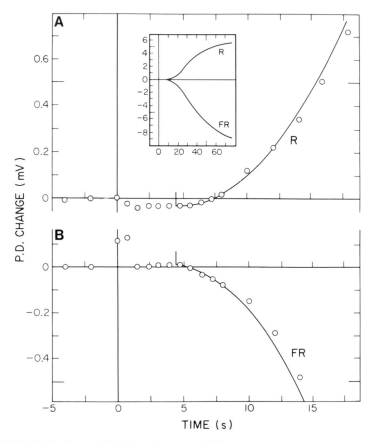

FIGURE 23-10 R- and FR-induced potential difference changes in oat coleoptiles recorded with surface electrodes. Pulses of 1 s R (A) and FR (B) began at time zero in each case.
FR pulse was given 10 min after initial R. Surface potential before each pulse has been
designated "zero" in each case but differs in value. Open circles are data points; solid lines
are hal-Gaussian curves fitted using an assumed lag period of 4.5 s. **Inset:** longer-term
response on lower sensitivity scale. (From Quail, 1983, compiled from Newman, 1981.
Reproduced by permission of Springer-Verlag and the American Society of Plant Physiologists.)

CALCIUM: A SECOND MESSENGER?

There were, however, "strongly suggestive data from Ca^{2+} studies"
(Quail, 1983) in algae. In 1975, Haupt presented a hypothesis, proposed by
his former student Manfred H. Weisenseel, that phytochrome might function as a calcium carrier or transport mediator for Ca^{2+} in the plasma
membrane, which is otherwise rather impermeable to divalent ions. Once

calcium entered the cell, it would trigger further steps in the transduction chain by binding to membrane proteins or contractile proteins or enzymes (Haupt and Weisenseel, 1976).

Weisenseel explored this hypothesis with *Nitella,* a fresh-water alga. Puncturing internodes with microelectrodes, he found that the extent of R-induced depolarization related to the Ca^{2+} content of the medium. The cells repolarized after the light was turned off, becoming internally negative even more rapidly if a FR pulse followed the R treatment. Thus both Pfr and Ca^{2+} appeared to be involved in depolarization (Weisenseel and Ruppert, 1977).

Gottfried Wagner and Karin Klein detected a slow, parallel loss of phytochrome-induced chloroplast movement and insoluble Ca^{2+} from Ca^{2+}-deprived *Mougeotia* cells. Movement was restored when Ca^{2+} once more became available (Wagner and Klein, 1978). The insoluble Ca^{2+} was sequestered in osmiophilic organelles clustered near the chloroplast's edge (Wagner and Rossbacher, 1980)—these structures were isolated and characterized *in vitro* (Grolig and Wagner, 1987, 1989). Moreover, single filaments of actin, a possibly calcium-activated protein widely involved in cellular movement, were detected between the chloroplast's edge and the plasma membrane (Wagner and Klein, 1978; Klein et al., 1980; Haupt and Wagner, 1984; Grolig et al., 1990). Wagner and Klein proposed that Pfr releases Ca^{2+} from internal stores and that the Ca^{2+} activates actin into a form that can fasten to and physically turn the chloroplast. To explain the precision of chloroplast movement, they suggested that Pfr occupies anchorage sites for actin filaments on the plasma membrane. So the creation of a tetrapolar Pfr gradient would cause actin filaments to seek regions of low Pfr, where there were still vacant anchorage sites. As the filaments moved, they would swing the edges of the chloroplast from high-Pfr to low-Pfr regions (Wagner and Klein, 1981; Grolig and Wagner, 1988).

Through the use of a window in a gold-plated microscope slide, Eva M. Dreyer and Weisenseel were able to compare the uptake of radioactive Ca^{2+} into cells of the same filament that had been either R-irradiated or not irradiated and then kept in darkness for several minutes before exposure to the ion. They detected larger amounts of label in R-irradiated cells, and the effect was FR-reversible (Figs. 23-11). Thus Pfr either promoted Ca^{2+} uptake or inhibited its efflux (Dreyer and Weisenseel, 1979)—assuming that the radioactive ions had not merely passed into the cell wall (Grolig, 1986). R also induced Ca^{2+} uptake into *Mougeotia* protoplasts, and there was an accompanying rise in extracellular pH (Weisenseel, 1986).

In 1979, University of Georgia biochemists identified an acidic, heat-stable and highly conserved regulatory protein named calmodulin in plants (Anderson and Cormier, 1978; Anderson et al., 1980). In animal cells,

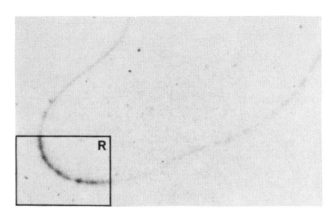

FIGURE 23-11   Autoradiograph from partially irradiated *Mougeotia* threads subsequently exposed to $^{45}Ca^{2+}$, illustrating effect of a 3-s R irradiation. (From Dreyer and Weisenseel, 1979. Copyright 1979, Springer-Verlag.)

Ca^{2+}-activated calmodulin activates certain enzymes or proteins that control key processes, such as contraction or hormone secretion. As Ca^{2+} is sequestered by the endoplasmic reticulum and mitochondria and pumped out through the plasma membrane, calmodulin once again becomes inactive.

Calmodulin was soon isolated from many plant species, including *Mougeotia,* where calmodulin antagonists inhibited phytochrome-mediated chloroplast movement (Wagner et al., 1984). So Wagner's group extended its 1981 model to include calmodulin in the signal transduction chain leading to chloroplast movement (Fig. 23-12).

In 1978, Roux (Fig. 23-13) moved to the University of Texas at Austin, where he also began to design experiments to see whether phytochrome can influence Ca^{2+} fluxes in biological membranes. Using the chromometallic dye murexide to monitor Ca^{2+} efflux continuously from apical tips and isolated protoplasts of etiolated oat coleoptiles, his group detected an R/FR-reversible increase in external Ca^{2+} (Hale and Roux, 1980). This was the first observation of Pfr-induced Ca^{2+} fluxes from higher plant cells, but the direction was opposite to that observed in *Nitella* and *Mougeotia,* where Ca^{2+} influxes were detected. Calvin C. Hale II also observed a Ca^{2+} efflux from *Mougeotia* (Hale, 1981). Current models, Roux

FIGURE 23-12   Model for transduction chain involved in *Mougeotia* chloroplast orientation to light. (From Haupt et al., 1986. Reprinted by permission of Kluwer Academic Publishers.)

LIGHT

$$P_r \longrightarrow P_{fr}$$

(phytochrome)

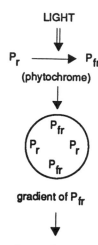

gradient of P_{fr}

change in membrane properties
change in ion fluxes

redistribution of Ca^{2+}
calcium release from internal stores

differential control function in space
of Ca^{2+} and calmodulin

activation or reorganization of actin microfilaments
actin microfilaments connect
chloroplast edge to plasmalemma

chloroplast's edge moves through cytoplasm

FIGURE 23-13   Stanley J. Roux, 1990. Courtesy of
I. Spear, University of Texas.

explains, propose that an initial increase in cytosolic Ca^{2+} activates the
calmodulin system, which turns on pumps in the plasma membrane that
promote Ca^{2+} efflux. "So when the pump is working," he says, "you
might see a net efflux if you were looking from the outside. Looking on the
inside, you might see an equilibrium level because there would be a release
of Ca^{2+} from internal stores."

   Turning to *Mougeotia,* Roux reasoned that localized introduction of
Ca^{2+} should trigger chloroplast rotation if the ion is part of the signal
transduction chain. When the Ca^{2+} ionophore A23187 was applied to
opposite sides of a *Mougeotia* cell, near the chloroplast's edges, part of the
chloroplast slowly turned through 90°. It did not move in a Ca^{2+}-deficient
medium or when a K^+ ionophore was used. Unilateral application of
A23187 also failed to induce turning, however. In pairs of structurally
similar calmodulin antagonists with differing affinities for the protein, the
member with the higher affinity was also most inhibitory. Thus perhaps
"Ca^{2+}, through the activation of a calcium-binding regulatory protein, can
function as a secondary messenger in promoting a phytochrome-
controlled event" (Serlin and Roux, 1984).

   Attractive as calcium-calmodulin hypotheses appear, Ekkehard Schönbohm's group failed to find a dependence on external Ca^{2+} for any *Mougeotia* chloroplast response—which does not rule out the participation

of internal Ca^{2+} stores. But chloroplast anchorage (tested by chloroplast centrifugability) was enhanced, not diminished, by Ca^{2+} starvation (Schönbohm et al., 1990). An alternative hypothesis, involving biophenolic compounds (Schönbohn and Schönbohm, 1984; Schönbohm, 1987) has yet to be tested by other workers.

FERN SPORE GERMINATION

At the University of Massachusetts, Amherst, Randy Wayne and Peter Hepler discovered that Ca^{2+} was required for R-induced germination of unicellular green *Onoclea sensibilis* spores. The small percentage of spores that germinated in the dark did not require Ca^{2+} in the external medium, although the Ca^{2+}-ionophore A23187 increased dark germination if spores were bathed in a Ca^{2+}-containing medium. The latter observation suggested that Pfr acted in part by increasing the internal Ca^{2+} concentration. Cobalt and lanthanum ions, which are Ca^{2+} antagonists, inhibited germination only if they were present before or during irradiation, being ineffective if added immediately after. The transitory requirement for Ca^{2+} could be satisfied many hours after irradiation. Experiments with calmodulin antagonists suggested that calmodulin might be involved in signal transduction (Wayne and Hepler, 1984).

By drying irradiated spores and subjecting them to atomic absorption spectroscopy, Wayne and Hepler showed that R quintupled the rate of Ca^{2+} uptake during the first 5 min of irradiation; this increase was FR-reversible (Wayne and Hepler, 1985a,b). By contrast, nuclear magnetic resonance spectroscopy and conventional polarography failed to detect *any* involvement of intracellular pH or transplasma membrane proton fluxes in the R-induced signal transduction chain (Wayne et al., 1986).

In Erlangen, Robert Scheuerlein had detected a requirement for both Pfr and Ca^{2+} during germination of nongreen *Dryopteris paleacea* spores (Haupt et al., 1986). Collaborating in Roux's lab, Scheuerlein and Wayne monitored chlorophyll formation as an indicator of germination. They found that lanthanum cancelled the effect of R applied at an optimal Ca^{2+} concentration. But Ca^{2+}-free spores responded to Ca^{2+} for at least 40 h after a R treatment, and even at 24 h immediately after a FR treatment. Thus "Ca^{2+} appears to interact with a product of Pfr rather than with Pfr itself" (Scheuerlein et al., 1989). Scheuerlein's group in Erlangen then noted a 15-h gap between the Pfr-requiring phase and the period of Ca^{2+} competence. "Thus the direct interaction of Pfr with external Ca^{2+} as a

first transduction step can be excluded experimentally" (Dürr and Scheuerlein, 1990).

CALCIUM ASSAY

Roux proposed that phytochrome acts on the plasma membrane either
directly or indirectly to release Ca^{2+} (Fig. 23-14) (Roux et al., 1986).

"But before we can accept the phytochrome-Ca^{2+} connection," he
says, "we still need two critical pieces of information. You have to show
that, following R, there is a change in free Ca^{2+} in the cytosol of the
plant—although there are several mechanisms where Ca^{2+} can produce
responses without a major change in concentration. You also have to
demonstrate a target of Ca^{2+} action."

Direct measurements, using fluorescent probes, of phytochrome-
induced changes in Ca^{2+} concentration in higher plant cells have now been
performed (Miyoshi, 1986; Chae et al., 1990). Also, Wagner and co-
workers loaded the calcium-sensitive fluorescent dye indo-1 into *Mougeotia scalaris* cells and protoplasts. When Ca^{2+} was removed from the
external medium, UV-A, B or R pulses increased cytoplasmic free Ca^{2+},
which presumably was released from internal stores. Both A23187 and
the Ca^{2+} channel agonist Bay-K8644 speeded R-mediated chloroplast rotation. "Based on these and other observations," the Giessen workers
concluded, "a Ca^{2+}-induced decrease in cytoplasmic viscosity in *Mougeotia* is presumed to occur" (Russ et al., 1991). Their most recent model
proposed that a light-mediated although not phytochrome-specific change
in cytoplasmic Ca^{2+} levels affects the kinetics rather than orientation of
chloroplast movement (Fig. 23-15) (Wagner et al., 1991).

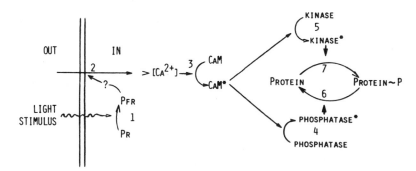

FIGURE 23-14   Roux's 1986 model of phytochrome-mediated activation of nuclear
proteins. Asterisk designates the relatively more active form of a protein or enzyme. (From
Roux, et al., 1986.)

PROTOPLASTS AS MODEL SYSTEMS

In the early 1980s, Thomas and Steve Blakeley discovered a response—protoplast swelling—that at last could be studied easily and reproducibly *in vitro* (Blakely et al., 1983). In Wageningen, Margreet E. Bossen adapted this system. R-irradiated protoplasts were consistently 13–18% larger than dark controls, and the response was FR-reversible. Moreover, Ca^{2+} induced swelling. And A23187 could replace R treatment, whereas lanthanum and the Ca^{2+}-channel blocker Verapamil were inhibitory. Ca^{2+} was not required immediately after phytochrome phototransformation but could be supplied up to 10 min later (Bossen et al., 1988).

Murexide assay of the external medium yielded similar results to Roux's: an R/FR-reversible efflux of Ca^{2+} inhibited by the Ca^{2+}-channel blocker Verapamil and the calmodulin antagonist W_7 but not by W_5. "It is proposed that red light causes the opening of Ca^{2+}-channels in the plasma membrane, and that the resulting enhanced influx of Ca^{2+} is followed by a large release of Ca^{2+} from internal stores and/or from Ca^{2+} bound to membranes," the Wageningen researchers concluded. "This Ca^{2+} is extruded from the protoplasts by means of a plasma membrane located Ca^{2+}-ATPase" (Bossen et al., 1990).

Briggs's group, meanwhile, was detecting Ca^{2+} fluxes through pea plasma membrane fractions, after noticing that the preparations contained immunodetectable phytochrome (Gallagher et al., 1988). "We can dissolve away more than 60% of the total protein with Triton and the phytochrome is still there," Briggs explains. "We can put in 1 M potassium chloride and the phytochrome is still there. So it is our very strong feeling that the phytochrome is associated with plasma membrane."

Samples from different heights on the stem had a rather constant ratio of membrane-associated phytochrome to protein, whereas the ratio of soluble phytochrome to protein declined steeply toward the bottom of the stem (Eisinger et al., 1989).

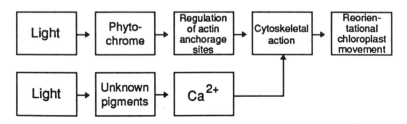

FIGURE 23-15   Wagner's 1991 model for induction of chloroplast movement in *Mougeotia* (From Wagner et al., 1991. Reproduced with permission from CRC Press.)

PHOSPHATIDYLINOSITOL PATHWAY

If Pfr does promote Ca^{2+} movement through the plasma membrane but does not directly interact with the ion, part of the signal transduction pathway must lie between the two. In invertebrate cells, a common link

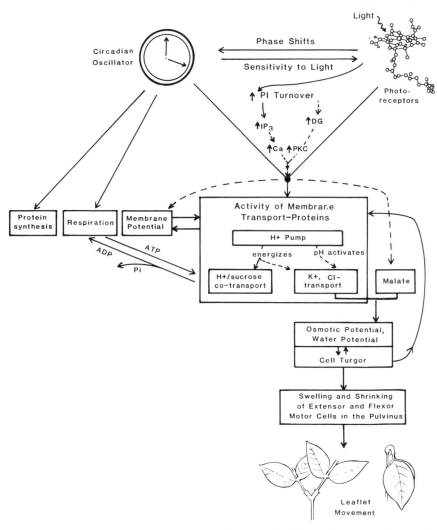

FIGURE 23-16 Scheme depicting biochemical and physiological processes that couple leaflet movement in *Samanea* to light absorption and the clock. Solid line indicates demonstrated control pathway; dotted line indicates postulated control pathway. DG, diacylglycerol; IP_3, inositol 1,4,5-triphosphate; PI, membrane-localized phosphoinositides; PKC, protein kinase C. (From Satter et al., 1988.)

between Ca^{2+} and light or hormonal signals involves the turnover of phosphoinositides, which are minor components of many cellular membranes. A light signal stimulates cleavage of phosphatidylinositol 4,5-bisphosphate, and one of the products, membrane-bound diacylglycerol, activates protein kinase C, which then activates or deactivates certain proteins through phosphorylation. The other product, water-soluble inositol triphosphate, mobilizes intracellular Ca^{2+} by interacting with a receptor on the endoplasmic reticulum, temporarily increasing the ion's cytosolic concentration.

Firm confirmation that polyphosphoinositides occur in plant cells was obtained in the late 1980s (Coté et al., 1989; Irvine et al., 1989). After observing small, light-stimulated changes in levels of phosphatidylinositol phosphates and inositol phosphates in excised pulvini, Satter incorporated the phosphatidylinositol pathway into her model of leaflet movement (Fig. 23-16) (Satter et al., 1988). She was unable to pursue the topic further because she died on August 3, 1989.

MEMBRANE TRANSITIONS

Hendricks continued to gather evidence for the membrane hypothesis after his retirement in 1970, collaborating with Ray B. Taylorson. Taylorson and Hendricks discovered a striking response to reduced temperature in imbided *Amaranthus retroflexus* (pigweed) seeds: a sharp transition between temperatures at which the seeds could or could not respond to the low Pfr level produced by a FR pulse, although subsequent germination required a higher temperature (Fig. 23-17) (Hendricks and Taylorson, 1978). "We interpreted that, "Taylorson says, "as a change in state of the membrane brought about by the drop in temperature. This change would allow phytochrome to associate with the membrane better and therefore facilitate germination better than if the seed had never seen cold." Leakage rates of amino acids also increased sharply between 30 and 35°C in many seeds (Hendricks and Taylorson, (1976).

When pigweed, lettuce, and wintercress seeds were incubated with a fluorescent probe, fluorescence varied logarithmically with the reciprocal of temperature but in the case of pigweed and lettuce, the slopes of the lines changed abruptly at about 30°C. Hendricks also found several examples in the literature of physiological processes that changed markedly near this temperature (Hendricks and Taylorson, 1979).

Taylorson, meanwhile, was working with gibberellins, which stimulated germination in many species (Taylorson and Hendricks, 1976), but only in very large amounts. Suspecting a permeability problem, he dissolved the hormone in ethanol, which was known to perturb membranes. Getting

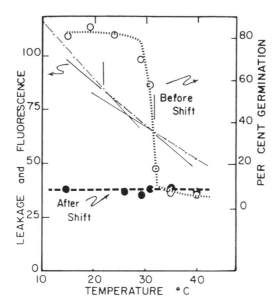

FIGURE 23-17 Germination of *Amaranthus retroflexus* seeds after a shift for 2 h to a lower temperature and a saturating exposure to FR before (open circles) or after (closed circles) the shift. Seeds were imbibed in darkness at 40°C before and after treatments. Diagonal dash-dot line shows fluorescence of *Amaranthus* seed membrane preparation in 1,8-anilino-naphthalene sulfonate as a function of temperature. (From Hendricks and Taylorson, 1978. Reproduced with permission from the American Society of Plant Physiologists.)

huge responses, he found that "it wasn't the gibberellin that was causing the response; it was the alcohol."

Certain other small organic molecules, including some that were metabolically inert (Taylorson and Hendricks, 1981), also promoted germination after R irradiation, even in stubbornly dormant *Panicum dichotomiflorum* (Taylorson and Hendricks, 1979). The only common property seemed to be anesthetic action in animals. And anesthetics were believed to act on membrane.

Knowing that pressure reverses anesthetic effects, the collaborators "were able to show in panicum seeds that pressure could oppose the effects of these compounds" (Hendricks and Taylorson, 1980). Thus "both phytochrome and anesthetic action involve association with membrane" (Taylorson and Hendricks, 1981).

In his last weeks of life, Hendricks started to write, "How phytochrome acts —perspectives on the continuing quest." (Hendricks and VanDerWoude, 1983). "It was suggested in 1967," he said "that phytochrome functions through action on membranes of cells and organelles. This concept has withstood critical examination."

FIGURE 23-18   Hendricks receives National Medal of Science in 1976 from U.S. President Gerald Ford.

Hendricks died at age 78 on January 4, 1981. During his 58 years of research, he had been showered with honors (Fig. 23-18).[8] But his most coveted prize was "freedom to inquire into the nature of things, . . . a rewarding privilege granted to a few by a permissive society" (Hendricks, 1970).

8. In addition to honorary degrees from the University of Arkansas, the University of North Carolina, and Kansas State University, Hendricks was awarded the Hillebrand Prize for chemistry in 1937, the Science Award of the Washington Academy of Sciences in 1942, and the Day Medal for geology in 1952. He was elected to the National Academy of Sciences in 1952, after being considered for nomination by the botanists and geologists as well as the chemists. In 1958, he was one of five government employees who received the first President's Award for Distinguished Civilian Service from President Eisenhower. He obtained the Rockefeller Public Service Award in 1960, and in 1962 he and Borthwick shared both the Hoblitzelle Award in Agricultural Sciences and the Stephen Hales Award from the American Society of Plant Physiologists. The ASPP also gave Hendricks the Charles Reid Barnes Honorary Life Membership Award in 1969. He was elected to the American Philosophical Society in 1967 and received the Distinguished Alumnus Award from the California Institute of Technology in 1968. In 1976, he was presented with the National Medal of Science (See Fig. 23-18). That year, he also received the Finsen Medal from the International Society of Photobiology. He had been president of both the American Society of Plant Physiologists and the American Mineralogical Society. As a member, from 1974, of the Committee for Research and Exploration of the National Geographic Society, he made field trips to Kenya, Tanzania, Jordan, Iran, and Israel.

336

REFERENCES

Anderson, J. M. Charbonneau, H., Jones, H. P., McCann, R. O., and Cormier, M. J. (1980). Characterization of the plant nicotinamide adenine dinucleotide kinase activator protein and its identification as calmodulin. *Biochemistry* 19,3113–3120.

Anderson, J. M., and Cormier, M. J. (1978). Calcium-dependent regulator of NAD kinase. *Biochem. Biophys. Res. Commun.* 84,595–602.

Blakeley, S. D., Thomas, B., Hall, J. L., and Vince-Prue, D. (1983). Regulation of swelling of etiolated-wheat-leaf protoplasts by phytochrome and gibberellic acid. *Planta* 158, 416–421.

Bossen, M. E., Dassen, H. H. A., Kendrick, R. E., and Vredenberg, W. J. (1988). The role of calcium ions in phytochrome-controlled swelling of etiolated wheat (*Triticum aestivum* L.) protoplasts. *Planta* 174,94–100.

Bossen, M. E., Verhoeven-Jaspers, P. A. P. M., Kendrick, R. E., and Vredenberg, W. J. (1990). Phytochrome dependent changes in external Ca^{2+}-concentration in suspensions of etiolated wheat (*Triticum aestivum* L.) protoplasts. *In* "Plant Protoplasts as a Model System to Study Phytochrome-regulated Changes in the Plasma Membrane" (H. Bossen, ed.), pp. 47–56. Wageningen Agricultural University, Wageningen.

Brownlee, C., and Kendrick, R. E. (1977). Phytochrome and potassium uptake by mung bean hypocotyl sections. *Planta* 137,61–64.

Brownlee, C., and Kendrick, R. E. (1979). Ion fluxes and phytochrome protons in mung bean hypocotyl segments. *Plant Physiol.* 64,206–210.

Brownlee, C., Roth-Bejerano, N., and Kendrick, R. E. (1979). The molecular mode of phytochrome action. *Sci. Prog.* 66,217–229.

Campbell, N. A., Satter, R. L., and Garber, R. C. (1981). Apoplastic transport of ions in the motor organ of *Samanea. Proc. Natl. Acad. Sci. USA* 78,2981–2984.

Cedel, T. E., and Roux, S. J. (1980a). Further characterization of the *in vitro* binding of phytochrome to a membrane fraction enriched for mitochondria. *Plant Physiol.* 66,696–703.

Cedel, T. E., and Roux, S. J. (1980b). Modulation of a mitochondrial function by oat phytochrome *in vitro. Plant Physiol.* 66,704–709.

Chae, Q., Park, H. J., and Hong, S. D. (1990). Loading of quin2 into the oat protoplast and measurement of cytosolic calcium ion concentration changes by phytochrome action. *Biochim. Biophys. Acta* 1051,115–122.

Cooke, R. J., and Kendrick, R. E. (1976). Phytochrome controlled metabolism in etioplast envelopes. *Planta* 131,303–307.

Cooke, R. J., and Saunders, P. F. (1975). Phytochrome mediated changes in extractable gibberellin activity in a cell-free system from etiolated wheat leaves. *Planta* 123, 299–302.

Cooke, R. J., Saunders, P. F., and Kendrick, R. E. (1975). Red light induced production of gibberellin-like substances in homogenates of etiolated wheat leaves and in suspensions of intact etioplasts. *Planta* 124,319–328.

Coté, G. G., and Crain, R. C. (1990). Failure to detect an effect of red, blue, or white light on diacylglycerol levels in *Samanea saman* pulvini. *Plant Physiol.* 93(suppl. 152).

Coté, G. G., DePass, A. L., Quarmby, L. M., Tate, B. F., Morse, M. J., Satter, R. L., and Crain, R. C. (1989). Separation and characterization of inositol phospholipids from the pulvini of *Samanea saman. Plant Physiol.* 90,1422–1428.

Datta, N., Chen, Y.-R., and Roux, S. J. (1985). Phytochrome and calcium stimulation of protein phosphorylation in isolated pea nuclei. *Biochem. Biophys. Res. Commun.* 128,1403–1408.

Dreyer, E. M., and Weisenseel, M. H. (1979). Phytochrome-mediated uptake of calcium in *Mougeotia* cells. *Planta* 146,31–39.

Dürr, S., and Scheuerlein, R. (1990). Characterization of a calcium-requiring phase during phytochrome-mediated fern-spore germination. *Photochem. Photobiol.* 52,73–82.

Eisinger, W. R., Short, T. W., and Briggs, W. R. (1989). Light regulation of calcium fluxes in isolated membrane vesicles from pea (*Pisum sativum* L.) seedlings. European Symposium on Photomorphogenesis in Plants, Freiburg. Book of abstracts, p. 33.

Evans, A. (1976). Etioplast phytochrome and its *in vitro* control of gibberellin efflux. *In* "Light and Plant Development" (H. Smith, ed.), pp. 129–142. Butterworths, London.

Evans, A., and Smith, H. (1976a). Localization of phytochrome in etioplasts and its regulation *in vitro* of gibberellin levels. *Proc. Natl. Acad. Sci. USA* 73,138–142.

Evans, A., and Smith, H. (1976b). Spectrophotometric evidence for the presence of phytochrome in the envelope membrane of barley etioplasts. *Nature (Lond.)* 299,323–325.

Gallagher, S., Short, T. W., Ray, P. M., Pratt, L. H., and Briggs, W. R. (1988). Light-mediated changes in two proteins found associated with plasma membrane fractions from pea stem sections. *Proc. Natl. Acad. Sci. USA* 85,8003–8007.

Georgevich, G. (1980). Phytochrome interaction with lipid membranes. Ph.D. thesis. University of Pittsburgh.

Georgevich, G., and Roux, S. J. (1982). Permeability and structural changes induced by phytochrome in lipid vesicles. *Photochem. Photobiol.* 36,663–671.

Georgevich, G., Krauland, J., and Roux, S. J. (1976). Phytochrome interaction with lipid bilayers. *Plant Physiol.* 57(suppl. 20).

Georgevich, G., Cedel, T. E., and Roux, S. J. (1977). Use of ^{125}I-labeled phytochrome to quantitate phytochrome binding to membranes of *Avena sativa*. *Proc. Natl. Acad. Sci. USA* 74,4439–4443.

Grolig, F. (1986). Calcium-Vesikel und lichtabhängige Chloroplastenreorientierung bei *Mougeotia* spec. Doktorarbeit, Universität Gießen.

Grolig, F., and Wagner, G. (1987). Vital staining permits isolation of calcium vesicles from the green alga *Mougeotia*. *Planta* 171,433–437.

Grolig, F., and Wagner, G. (1988). Light-dependent chloroplast reorientation in *Mougeotia* and *Mesotaenium:* Biased by pigment-regulated plasmalemma anchorage sites to actin filaments? *Bot. Acta* 101,2–6.

Grolig, F., and Wagner, G. (1989). Characterization of the isolated calcium-binding vesicles from the green alga *Mougeotia scalaris,* and their relevance to chloroplast movement. *Planta* 177,169–177.

Grolig, F., Weigang-Köhler, K., and Wagner, G. (1990). Different extent of F-actin bundling in walled cells and in protoplasts of *Mougeotia scalaris*. *Protoplasma* 157,225–230.

Gross, J., Ayadi, A., and Marmé, D. (1979). Protochlorophyl (lide)$_{630}$ photosensitizes active Ca^{2+} accumulation in microsomal and mitochondrial fraction isolated from plants. *Photochem. Photobiol.* 30,615–621.

Hale, C. C., II. (1981). Studies on the role of calcium in mediating phytochrome effects. Ph.D. thesis. University of Texas, Austin.

Hale, C. C., II, and Roux, S. J. (1980). Photoreversible calcium fluxes induced by phytochrome in oat coleoptile cells. *Plant Physiol.* 65,658–662.

Haupt, W. (1986). Photomovement. *In* "Photomorphogenesis in Plants" (R. E. Kendrick and G. H. M. Kronenberg, eds.), pp. 415–441. Martinus Nijhoff, Dordrecht.

Haupt, W., and Wagner, G. (1984). Chloroplast movement. *In* "Membranes and Sensory Transduction" (G. Colombetti and F. Lenci, eds.), pp. 331–375. Plenum, New York.

Haupt, W., and Weisenseel, M. H. (1976). Physiological evidence and some thoughts on localised responses, intracellular localisation and action of phytochrome. *In* "Light and Plant Development" (H. Smith, ed.), pp. 63–74. Butterworths, London.

Haupt, W., Scheuerlein, R., Mische, S., and Mader, U. (1986). Control of fern-spore germination by phytochrome and inorganic ions. Proceedings XVI Yamada Conference, Okazaki, p. 119.

Hendricks, S. B. (1970). The passing scene. Quotation reproduced, with permission, from the *Annual Review of Plant Physiology*, vol. 21, pp. 1–10. © 1970 by Annual Reviews Inc.

Hendricks, S. B., and Taylorson, R. B. (1976). Variation in germination and amino acid leakage of seeds with temperature related to membrane phase change. *Plant Physiol.* 58,7–11.

Hendricks, S. B., and Taylorson, R. B. (1978). Dependence of phytochrome action in seeds on membrane organization. *Plant Physiol.* 61,17–19.

Hendricks, S. B., and Taylorson, R. B. (1979). Dependence of thermal responses of seeds on membrane transitions. *Proc. Natl. Acad. Sci. USA* 76,778–781.

Hendricks, S. B., and Taylorson, R. B. (1980). Reversal by pressure of seed germination promoted by anesthetics. *Planta* 149,108–111. ·

Hendricks, S. B., and VanDerWoude, W. J. (1983). How phytochrome acts—The continuing quest. *In* "Encyclopedia of Plant Physiology, New Series" (W. Shropshire, Jr. and H. Mohr, eds.), vol. 16A, pp. 3–23. Springer, Berlin, Heidelberg.

Hillman, W. S., and Koukkari, W. L. (1967). Phytochrome effects in the nyctinastic leaf movements of *Albizzia julibrissin* and some other legumes. *Plant Physiol.* 42,1413–1418.

Hilton, J. R., and Smith, H. (1980). The presence of phytochrome in purified barley etioplasts and its *in vitro* regulation of biologically active gibberellin levels in etioplasts. *Planta* 148,312–318.

Iglesias, A., and Satter, R. L. (1983). H$^+$ fluxes in excised *Samanea* motor tissue. I. Promotion by light. *Plant Physiol.* 72,564–569.

Irvine, R. F., and Letcher, A. J., Drobak, D. J., Dawson, A. P., and Musgrave, A. (1989). Phosphatidylinositol (4,5) bisphosphate and phosphatidylinositol (4) phosphate in plant tissues. *Plant Physiol.* 89,888–892.

Jaffe, M. J., and Galston, A. W. (1967). Phytochrome control of rapid nyctinastic movements and membrane permeability in *Albizzia julibrissin*. *Planta* 77,135–141.

Jordan, B. R., and Partis, M. D., and Thomas, B. (1984). A study of phytochrome-membrane association using an enzyme-linked immunosorbent assay and Western blotting. *Physiol. Plant.* 60,416–421.

Kim, I.-S., and Song, P.-S. (1981). Binding of phytochrome to liposomes and protoplasts. *Biochemistry* 20,5482–5489.

Klein, K., Wagner, G., and Blatt, M. R. (1980). Heavy-meromyosin-decoration of microfilaments from *Mougeotia* protoplasts. *Planta* 150,354–356.

Lee, Y., and Satter, R. L. (1987). H$^+$ uptake and release during circadian rhythmic movements of excised *Samanea* motor organs. *Plant Physiol.* 83,856–862.

Lee, Y., and Satter, R. L. (1988). The apoplast pH of the *Samanea* pulvinus. *Plant Physiol.* 86(suppl. 93).

Lee, Y., and Satter, R. L. (1989). Effects of white, blue, red light and darkness on pH of the apoplast in the *Samanea* pulvinus. *Planta* 178,31–40.

Lowen, C. Z., and Satter, R. L. (1989). Light-promoted changes in apoplastic K$^+$ activity in the *Samanea saman* pulvinus, monitored with liquid membrane microelectrodes. *Planta* 179,421–427.

Manabe, K., and Furuya, M. (1974). Phytochrome-dependent reduction of nicotinamide nucleotides in the mitochondrial fraction isolated from etiolated pea epicotyls. *Plant Physiol.* 53,343–347.

Marmé, D. (1977). Phytochrome: Membranes as possible sites of primary action. *Annu. Rev. Plant Physiol.* 28,173–198.

McCurdy, D. W., and Pratt, L. H. (1986). Immunogold electron microscopy of phytochrome in *Avena:* Identification of intracellular sites responsible for phytochrome sequestering and enhanced pelletability. *J. Cell Biol.* 103,2541–2550.

Miyoshi, Y. (1986). Phytochrome-induced changes of the intracellular Ca^{2+} concentration in protoplasts from higher plants. Proceedings XVI Yamada Conference, Photomorphogenesis in Plants, Okazaki, p. 101.

Moran, N., Ehrenstein, G., Iwasa, K., Mischke, C., Bare, C., and Satter, R. L. (1988). Potassium channels in motor cells of *Samanea saman. Plant Physiol.* 88,643–648.

Morse, M. J., Crain, R. C., and Satter, R. L. (1987). Light-stimulated inositolphospholipid turnover in *Samanea saman* leaf pulvini. *Proc. Natl. Acad. Sci. USA* 84,7075–7078.

Napier, R. M., and Smith, H. (1987a). Photoreversible association of phytochrome with membranes. I. Distinguishing between two light-induced binding responses. *Plant Cell Environ.* 10,383–389.

Napier, R. M., and Smith, H. (1987b). Photoreversible association of phytochrome with membranes. II. Reciprocity tests and a model for the binding reaciton. *Plant Cell Environ.* 10,391–396.

Newman, I. A. (1974). Electric responses of oats to phytochrome phototransformation. *In* "Mechanisms of Regulation of Plant Growth" (R. L. Bieleski, A. R. Ferguson, and M. M. Cresswell, eds.). *R. Soc. N. Z. Bull.* 12, 355–360. Wellington.

Newman, I. A. (1981). Rapid electric responses of oats to phytochrome show membrane processes unrelated to pelletability. *Plant Physiol.* 68,1494–1499.

Newman, I. A., and Briggs, W. R. (1972). Phytochrome-mediated electric potential changes in oat seedlings. *Plant Physiol.* 50,687–693.

Newman, I. A., and Sullivan, J. K. (1976). Auxin transport in oats: A model for the electric changes. *In* "Transport and Transfer Processes in Plants" (I. Wardlaw and J. Passioura, eds.), pp. 153–159. Academic Press, New York.

Poole, R. J. (1978). Energy coupling for membrane transport. *Annu. Rev. Plant Physiol.* 29,437–460.

Quail, P. H. (1982). Intracellular location of phytochrome. *Trends Photobiol. Proc. Int. Congr. Photobiol. 8th,* pp. 485–500.

Quail, P. H. (1983). Rapid action of phytochrome in photomorphogenesis. *In* "Encyclopedia of Plant Physiology, New Series" (W. Shropshire, Jr. and H. Mohr, eds.), vol. 16A, pp. 178–212. Springer, Berlin, Heidelberg.

Racusen, R. H. (1976). Phytochrome control of electrical potentials and intercellular coupling in oat-coleoptile tissue. *Planta* 132,25–29.

Racusen, R. H., and Etherton, B. (1975). Role of membrane-bound, fixed-charge changes in phytochrome-mediated mung bean root tip adherence phenomenon. *Plant Physiol.* 55,491–495.

Racusen, R. H., and Galston, A. W. (1977). Electrical evidence for rhythmic changes in the cotransport of sucrose and hydrogen ions in *Samanea* pulvini. *Planta* 135,57–62.

Rucasen, R. H., and Miller, K. (1972). Phytochrome-induced adhesion of mung bean root tips to platinum electrodes in a direct current field. *Plant Physiol.* 49,654–655.

Racusen, R. H., and Satter, R. L. (1975). Rhythmic and phytochrome-regulated changes in transmembrane potential in *Samanea* pulvini. *Nature (Lond.)* 255,408–410.

Roux, S. J., and Yguerabide, J. (1973). Photoreversible conductance changes induced by phytochome in model lipid membranes. *Proc. Natl. Acad. Sci. USA* 70,762–764.

Roux, S. J., McEntire, K., Slocum, R. D., Cedel, T. E., and Hale, C. C., II. (1981). Phytochrome induces photoreversible calcium fluxes in a purified mitochondrial fraction from oats. *Proc. Natl. Acad. Sci. USA* 78,283–287.

Roux, S. J., Wayne, R. O., and Datta, N. (1986). Role of calcium ions in phytochrome responses: An update. *Physiol. Plant.* 66,344–348.

Russ, U., Grolig, F., and Wagner, G. (1991). Changes of cytoplasmic free Ca^{2+} in the green alga *Mougeotia scalaris* as monitored with indo-1, and their effect on the velocity of chloroplast movements. *Planta* 184,105–112.

Satter, R. L. (1979). Leaf movements and tendril curling. *In* "Encyclopedia of Plant Physiology, New Series" (W. Haupt and M. E. Feinleib, eds.), vol. 7, pp. 442–484. Springer, Berlin, Heidelberg, New York.

Satter, R. L., and Galston, A. W. (1971). Potassium flux: A common feature of *Albizzia* leaflet movement controlled by phytochrome or endogenous rhythm. *Science* 174,518–520.

Satter, R. L., and Galston, A. W. (1973). Leaf movements: Rosetta stone of plant behavior? *BioScience* 23,407–416.

Satter, R. L., and Galston, A. W. (1981). Mechanisms of control of leaf movements. *Annu. Rev. Plant Physiol.* 32,83–110.

Satter, R. L., Marinoff, P., and Galston, A. W. (1970). Phytochrome controlled nyctinasty in *Albizzia julibrissin.* II. Potassium flux as a basis for leaflet movement. *Am. J. Bot.* 57,916–926.

Satter, R. L., Marinoff, P., and Galston, A. W. (1972). Phytochrome controlled nyctinasty in *Albizzia julibrissin.* IV. Auxin effects on leaflet movement and K flux. *Plant Physiol.* 50,235–241.

Satter, R. L., Geballe, G. T., Applewhite, P. B., and Galston, A. W. (1974). Potassium flux and leaf movement in *Samanea saman.* I. Rhythmic movement. *J. Gen. Physiol.* 64, 413–430.

Satter, R. L., Schrempf, M., Chaudri, J., and Galston, A. W. (1977). Phytochrome and circadian clocks in *Samanea. Plant Physiol.* 59,231–235.

Satter, R. L., Guggino, S. E., Lonergan, T. A., and Galston, A. W. (1981). The effects of blue and far red light on rhythmic leaflet movements in *Samanea* and *Albizzia. Plant Physiol.* 67,965–968.

Satter, R. L., Xu, Y., and DePass, A. (1987). Effects of temperature on H^+ secretion and uptake by excised flexor cells during dark-induced closure of *Samanea* leaflets. *Plant Physiol.* 85,850–855.

Satter, R. L., Morse, M. J., Lee, Y., Crain, R. C., Coté, G. G., and Moran, N. (1988). Light- and clock-controlled leaflet movements in *Samanea saman:* A physiological, biophysical and biochemical analysis. *Bot. Acta* 101,205–282.

Scheuerlein, R., Wayne, R., and Roux, S. J. (1989). Calcium requirement of phytochrome-mediated fern-spore germination: No direct phytochrome–calcium interaction in the phytochrome-initiated transduction chain. *Planta* 178,25–30.

Schönbohm, E. (1987). Movement of *Mougeotia* chloroplasts under continuous weak and strong light. *Acta Physiol. Plant.* 9,109–135.

Schönbohm, E., and Schönbohm, E. (1984). Biophenole: Steuernde Faktoren bei der licht-orientierten Chloroplastenbewegung? *Biochem. Physiol. Pflanz.* 179,489–505.

Schönbohm, E., Schönbohm, E., and Meyer-Wegener, J. (1990). On the signal-transduction chains of two Pfr-mediated short-term processes: Increase of anchorage and movement of *Mougeotia* chloroplasts. *Photochem. Photobiol.* 52,203–209.

Schrempf, M., Satter, R. L., and Galston, A. W. (1976). Potassium-linked chloride fluxes during rhythmic leaf movement of *Albizzia julibrissin. Plant Physiol.* 58,190–192.

Serlin, B. S., and Roux, S. J. (1984). Modulation of chloroplast movement in the green alga *Mougeotia* by the Ca^{2+} ionophore A23187 and by calmodulin antagonists. *Proc. Natl. Acad. Sci. USA* 81,6368–6372.

Simon, E., Satter, R. L., and Galston, A. W. (1976). Circadian rhythmicity in excised *Samanea* pulvini. III. Resetting the clock by phytochrome conversion. *Plant Physiol.* 58,421–425.

Singh, B. R., Choi, J.-K., and Song, P.-S. (1989). Binding of 124-kilodalton oat phytochrome to liposomes and chloroplasts. *Physiol. Plant.* 76,319–325.

Smith, H. (1976). The mechanism of action and the function of phytochrome. *In* "Light and Plant Development" (H. Smith, ed.), pp. 493–502. Butterworths, London, Boston.

Spanswick, R. M. (1981). Electrogenic ion pumps. *Annu. Rev. Plant. Physiol.* 32,267–289.

Speth, V., Otto, V., and Schäfer, E. (1986). Intracellular localisation of phytochrome in oat coleoptiles by electron microscopy. *Planta* 168,299–304.

Taylorson, R. B., and Hendricks, S. B. (1976). Interactions of phytochrome and exogenous gibberellic acid on germination of *Lamium amplexicaule* L. seeds. *Planta* 132,65–70.

Taylorson, R. B., and Hendricks, S. B. (1979). Overcoming dormancy in seeds with ethanol and other anesthetics. *Planta* 145,507–510.

Taylorson, R. B., and Hendricks, S. B. (1981). Anesthetic release of seed dormancy—An overview. *Isr. J. Bot.* 29,273–280.

Terry, M. J., Hall, J. L., and Thomas, B. (1989a). Purification of plasma membrane from wheat leaves and characterization of the associated vanadate-sensitive Mg^{2+}ATPase activity. *J. Plant Physiol.* 134,756–761.

Terry, M. J., Thomas, B., and Hall, J. L. (1989b). Analysis of the association of phytochrome with wheat leaf plasma membranes by quantitative immunoassay and Western blotting. European Symposium on Photomorphogenesis in Plants, Freiburg. Book of abstracts, p. 22.

Thomas, B., Crook, N. E., and Penn, S. E. (1984). An enzyme-linked immunosorbent assay for phytochrome. *Physiol. Plant.* 60,409–415.

Wagner, G., and Klein, K. (1978). Differential effect of calcium on chloroplast movement in *Mougeotia. Photochem. Photobiol.* 27,137–140.

Wagner, G., and Klein, K. (1981). Mechanism of chloroplast movement in *Mougeotia. Protoplasma* 109,169–185.

Wagner, G., and Rossbacher, R. (1980). X-ray microanalysis and chlorotetracycline staining of calcium vesicles in the green alga *Mougeotia. Planta* 149,298–305.

Wagner, G., Valentin, P., Dieter, P., and Marmé, D. (1984). Identification of calmodulin in the green alga *Mougeotia* and its possible function in chloroplast reorientational movement. *Planta* 162,62–67.

Wagner, G., Russ, U., and Quader, H. (1991). Calcium, a regulator of cytoskeletal activity and cellular competence. *In* "The Cytoskeleton of the Algae" (D. Menzel, ed.). CRC Press, Boca Raton, Florida (in press).

Warmbrodt, R. D., VanDerWoude, W. J., and Smith, W. O. (1989). Localization of phytochrome in *Secale cereale* L. by immunogold electron microscopy. *Bot. Gaz.* 150, 219–229.

Watson, P. J., and Smith, H. (1982a). Integral association of phytochrome with a membranous fraction from *Avena* shoots: Red/far-red photoreversibility and *in vitro* characterisation. *Planta* 154,121–127.

Watson, P. J., and Smith, H. (1982b). Integral association of phytochrome with a membranous fraction from *Avena* shoots: *In vivo* characterisation and physiological significance. *Planta* 154,128–134.

Wayne, R., and Hepler, P. K. (1984). The role of calcium ions in phytochrome-mediated germination of spores of *Onoclea sensibilis* L. *Planta* 160,12–20.

Wayne, R., and Hepler, P. K. (1985a). Red light stimulates an increase in intracellular calcium in the spores of *Onoclea sensibilis. Plant Physiol.* 77,8–11.

Wayne, R., and Hepler, P. K. (1985b). The atomic composition of *Onoclea sensibilis* spores. *Am. Fern J.* 75,12–18.

Wayne, R., Rice, D., and Hepler, P. K. (1986). Intracellular pH does not change during phytochrome-mediated spore germination in *Onoclea*. *Dev. Biol.* 113,97–103.

Weisenseel, M. H. (1986). Uptake and release of Ca^{2+} in the green algae *Mougeotia* and *Mesotaenium*. *In* "Molecular and Cellular Aspects of Calcium in Plant Development" (A. J. Trewavas, ed.), pp. 193–199. Plenum, New York.

Weisenseel, M. H., and Ruppert, H. K. (1977). Phytochrome and calcium ions are involved in light-induced membrane depolarization in *Nitella*. *Planta* 137,225–229.

CHAPTER 24

High-Irradiance Responses

The high-energy reaction and resulting responses were renamed the *high-intensity reaction* (Hartmann, 1969, Mohr et al., 1971) and then the *high-irradiance reaction and responses* (Shropshire, 1972a,b). *High-irradiance reaction* (HIR) was widely adopted after the Eretria symposium and the publication of Mohr's Amherst lectures (Mohr, 1972).

Despite Hartmann's impressive evidence (see Chapter 17) that phytochrome was the receptor for the HIR FR peak, a convincing explanation for its action in this mode was lacking. Thus HIR models proliferated in the 1970s (Smith, 1970; Gammerman and Fukshansky, 1974; Mancinelli and Rabino, 1975; Jose and Vince-Prue, 1978; Johnson and Tasker, 1979). They attempted to explain the HIR's lack of photoreversibility, its requirement for FR or B, its failure to obey the Bunsen-Roscoe reciprocity law, its immediate cessation when a light source is extinguished, and the necessity for fluences at least 100 times greater than those that saturate the low-energy response.

SCHÄFER'S HIR MODELS

Schäfer's first HIR model grew out of the Freiburg pelletability studies (see Chapter 21). Finding that pelletability was induced by R but not reversed by FR, Quail had proposed that phytochrome becomes pelletable by binding to a receptor, X, which then changes conformation to yield the complex PfrX'. FR generates PrX' from PfrX', leaving phytochrome still tightly bound. PrX regenerates slowly in the dark from PrX' (Quail et al., 1973).

Through mathematical manipulations, Schäfer (Fig. 24-1) determined the probable fates of the various phytochrome–receptor complexes under steady-state conditions. At a given wavelength, predicted PrX and PrX' levels decreased with increasing irradiance, whereas those of Pfr$X$ and

FIGURE 24-1   Eberhard Schäfer, 1989. Photo by Linda Sage.

PrX' increased, up to saturation level. The concentration of PrX' was greatest at 710 nm, corresponding quite well to the FR peaks of many HIRs.

Although the model accounted for wavelength and irradiance dependence, it predicted that the total amount of phytochrome, P_{total}, was independent of irradiance. But Schäfer had determined, both theoretically and experimentally, that this was not so during long-term FR treatments (Schäfer and Mohr, 1974)—in contrast to the situation under inductive conditions (Kendrick and Frankland, 1968; Marmé, 1969; Schäfer, 1971). Thus the model would have to take account of phytochrome destruction and synthesis. Dark reversion could be ignored, because it was minimal in mustard seedlings under these conditions.

Schäfer decided that a fast transition between two unknown states of Pfr preceded phytochrome destruction in *Cucurbita pepo* (Schäfer and Schmidt, 1974; Schmidt and Schäfer, 1974). Drawing on these and other studies, he proposed that Pfr destruction always starts from PfrX'. Synthesis, however, was known to generate unbound Pr. Thus he was able to convert the closed model to an open version (Fig. 24-2).

To make this model mathematically accessible, Schäfer eliminated the faster reactions (Fig. 24-3) and assigned more receptor sites than phytochrome molecules (Schäfer, 1974, 1975a). Solving numerous equations, he determined that "PfrX' is irradiance and wavelength independent. Therefore, PfrX' cannot be the effector of the HIR!" (Schäfer, 1975a). Levels of PrX and Pfr$X$, however, rose with increasing irradiance and were maximal in the FR (Fig. 24-4). "The Pfr$X$ pool, which is a transient pool and is irradiance and wavelength dependent, is the effector element under

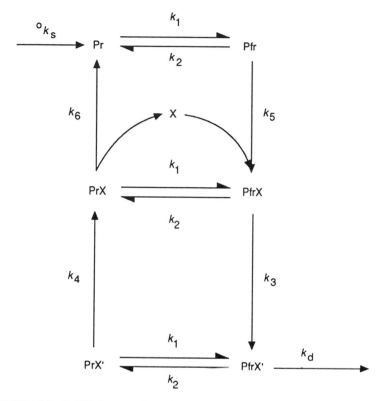

FIGURE 24-2   Schäfer's open phytochrome-receptor model. (From Schäfer, 1975b.)

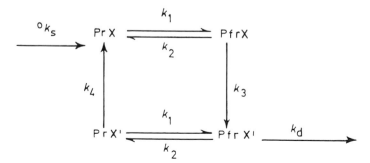

FIGURE 24-3   Abbreviated version of Fig. 25-2. (From Schäfer, 1975b.)

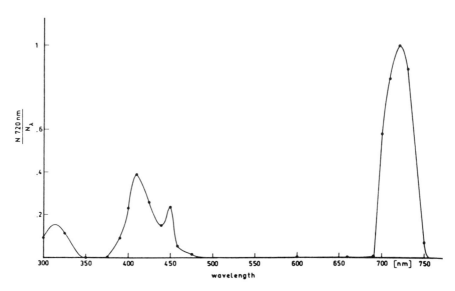

FIGURE 24-4   Calculated action spectrum for amount of Pfr$X$, a proposed HER effector. Relative quantum responsivity is plotted as a function of wavelength. (From Schäfer, 1975a. Copyright 1975, Springer-Verlag.)

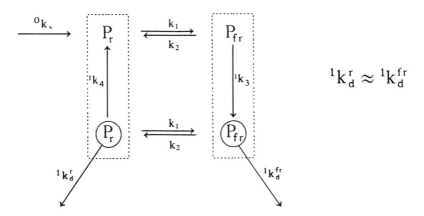

FIGURE 24-5   Schäfer's modified open model of phytochrome action. (From Fukshansky and Schäfer, 1983. Copyright 1983, Springer-Verlag.)

HIR conditions," Schäfer concluded. "Under induction conditions, the regulation will be from the PfrX' pool" (Schäfer, 1975a). Based on kinetic studies of pelletability, he decided that at least two different states of pelletable Pfr (Pfr$X$ and PfrX') must exist (Lehmann and Schäfer, 1978).

This model managed to avoid the concept of overcritically activated Pfr while still equating the HIR effector with an ephemeral species. It did not suggest that Pfr$X$ was the sole HIR effector at shorter wavelengths, for the calculated action spectrum differed significantly from experimentally derived curves in the B and UV (Fig. 24-4). Schäfer attributed the discrepancy not only to the use of data derived from *in vitro* absorption measurements but also to the simultaneous operation of a B photoreceptor (Schäfer, 1975a).

A modified model (Fig. 24-5) (Fukshansky and Schäfer, 1983), was barely tested. "For me," Schäfer says, "modeling failed to be possible because the complexity of the system suddenly became so dramatic."

RECIPROCITY FAILURE

In 1975, Mancinelli (Fig. 24-6) and doctoral student Isaac Rabino published a paper that addressed a key problem of the HIR: reciprocity failure. Ku and Mancinelli (1972) had already observed that cyclic light treatments

FIGURE 24-6   Alberto L. Mancinelli, 1989. Courtesy of James Prince.

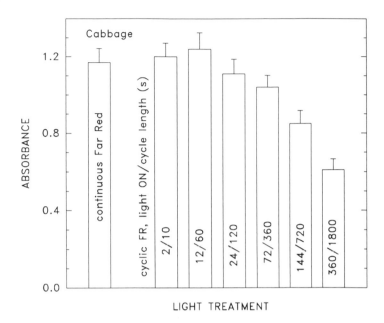

FIGURE 24-7 Effect of continuous and cyclic FR on anthocyanin production in cabbage seedlings. Duration from start to end of light treatments, 48 h. Fluence applied, 180 kJ m^{-2}. (Redrawn from data of Mancinelli and Rabino, 1975. Courtesy of A. L. Mancinelli.)

could be quite effective in eliciting anthocyanin production. Mancinelli and Rabino (1975) then showed that cyclic FR treatments were just as effective as continuous FR, provided they delivered the same total energy, spanned the same length of time, and permitted only short dark intervals between successive FR exposures (Fig. 24-7). Moreover, anthocyanin production under cyclic FR obeyed the reciprocity law (Fig. 24-8). Because cycles lasting from 10 to 120 s were equally effective and 360-s cycles only slightly less effective, the HIR effector appeared to persist for at least a few minutes after the FR was turned off. These observations failed to "support the hypothesis of cycling between Pr and Pfr and formation of Pfr* as an explanation for the irradiance dependence and reciprocity failure of the HIR" (Mancinelli and Rabino, 1975).

SPECTRAL SENSITIVITY

As soon as Mancinelli and his students began to study anthocyanin production, they found that experimental conditions markedly affected the spectral sensitivity of the response under prolonged irradiation. Thus it was possible to obtain almost any desired spectral sensitivity in a single

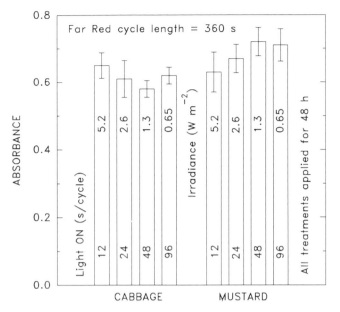

FIGURE 24-8   Reciprocity of anthocyanin production under cyclic FR. Duration of treatments from start to end, 48 h (= 480 cycles). (Redrawn from data of Mancinelli and Rabino, 1975. Courtesy of A. L. Mancinelli.)

species. This was not unexpected because previous workers had reported marked variability in FR effectiveness (Evans et al., 1965; Turner and Vince, 1969).

Mancinelli's group observed significant effects of irradiance and length of exposure on the relative efficiencies of B, R, and FR (Ku and Mancinelli, 1972; Rabino et al., 1977; Mancinelli and Rabino, 1978; Mancinelli and Walsh, 1979). And despite obvious differences in the spectral sensitivity of anthocyanin production in cabbage and tomato, there was a common pattern of change: the relative efficiency of FR increased with increasing exposure durations (Fig. 24-9). So although Mancinelli classified HIR responses into three groups—action in the UV-B-R-FR; action in the UV-B-R; action in the UV-B only (Mancinelli, 1980)—he still believed that some differences in spectral sensitivity result from the varying responses of different systems to a range of experimental conditions.

DE-ETIOLATION

Loss of the FR peak of the HIR action spectrum had been noted in aging or de-etiolated seedlings (Evans et al., 1965; Turner and Vince, 1969; Black and Shuttleworth, 1974; Jose and Vince-Prue, 1977a). Using high-

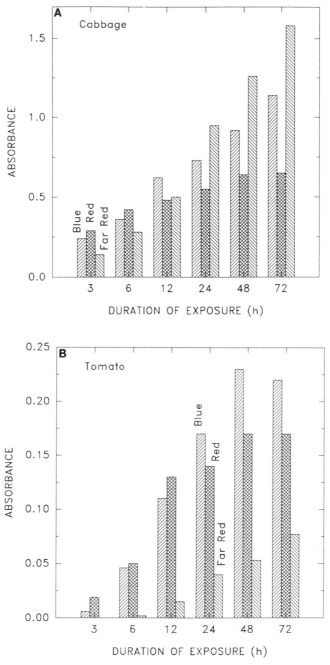

FIGURE 24-9 Effect of exposure duration on anthocyanin production in (A) cabbage seedlings and (B) tomato seedlings under continuous B, R, and FR. (Redrawn from the data of Mancinelli and Walsh, 1979. Courtesy of A. L. Mancinelli.)

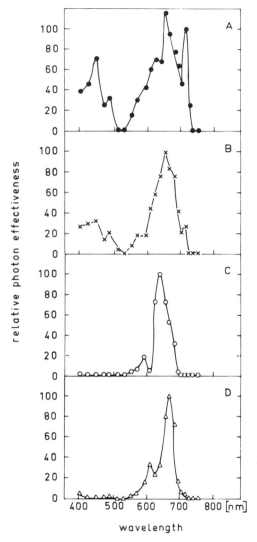

FIGURE 24-10   Action spectra for 50% inhibition of hypocotyl growth in *Sinapis alba* L.
under continuous light. (A) For dark-grown seedlings, 54 h dark + 24 h monochromatic light;
relative photon effectiveness normalized to 716 nm = 100%. (B) For dark-grown seedlings
pretreated with R pulses (52 h dark + 5 min R + 55 min dark + 5 min R + 55 min dark + 24 h
monochromatic light); relative photon effectiveness normalized to 655 nm = 100%. (C) For
WL-grown, green seedlings (54 h W + 24 h monochromatic light); relative photon effectiveness normalized to 640 nm = 100%. (D) For WL-grown herbicide-treated seedlings (54 h WL
+ 24 h monochromatic light, all in the presence of San 9789); relative photon effectiveness
normalized to 667 nm = 100%. (From Beggs et al., 1980. Reproduced with permission from
the American Society of Plant Physiologists.)

intensity, monochromatic light sources, Schäfer's group studied this change in detail (Beggs et al., 1980). With hypocotyls of 54-h dark-grown mustard seedlings, an action spectrum for growth inhibition peaked in the B (448 nm), R (655 nm), and FR (716 nm) (Fig. 24-10A), closely resembling a curve for inhibition of radish hypocotyl growth (Jose and Vince-Prue, 1977b). After R pulses at 52 and 53 h after sowing, mustard hypocotyls were much less responsive to prolonged B and FR, although the R peak of the action spectrum remained (Fig. 24-10B). The level of total phytochrome in these plants was one-quarter that of the fully dark controls.

Plants grown under WL were completely unresponsive to B and FR (Fig. 24-10C), but R was inhibitory, although most effectively at 640 rather than 655 nm, presumably because of screening by chlorophyll. The R peak persisted—at 667 nm—in seedlings grown in the presence of San 9789 (norflurazon) and therefore free of chlorophyll (Fig. 24-10D). Spectrophotometry of these white plants detected only 2.5% as much photoreversible pigment as in fully dark-grown seedlings. Thus exposure to either continuous WL or R pulses simultaneously reduced phytochrome content and decreased the inhibitory effects of B and FR.

The coordinated disappearance of the B and FR peaks suggested that "the blue peak in *Sinapis* (at this age and for this response) may also be due to phytochrome" (Beggs et al., 1980). But later data indicated the involvement of a separate B-absorbing pigment (see Schäfer, 1981). Schäfer attributed the R peak that persisted in the light to phytochrome, which presumably acted in a different manner than the phytochrome associated with the FR peak.

REFERENCES

Beggs, C. J., Holmes, M. G., Jabben, M., and Schäfer, E. (1980). Action spectra for the inhibition of hypocotyl growth by continuous irradiation in light and dark-grown *Sinapis alba* L. seedlings. *Plant Physiol.* 66,615–618.

Black, M., and Shuttleworth, J. (1974). The role of cotyledons in the photocontrol of hypocotyl extension in *Cucumis sativus* L. *Planta* 117,57–66.

Evans, L. T., Hendricks, S. B., and Borthwick, H. A. (1965). The role of light in suppressing hypocotyl elongation in lettuce and *Petunia*. *Planta* 64,201–218.

Fukshansky, L., and Schäfer, E. (1983). Models in photomorphogenesis. *In* "Encyclopedia of Plant Physiology, New Series" (W. Shropshire, Jr. and H. Mohr, eds.), vol. 16A, pp. 69–95. Springer, Berlin, Heidelberg, New York, Tokyo.

Gammerman, A. Ya., and Fukshansky, L. (1974). A mathematical model of phytochrome— The receptor of photomorphogenetic processes in plants. *Ontogenez* 5,122–129.

Hartmann, K. M. (1969). Lecture, University of Marburg, Germany. Unpublished.

Johnson, C. B., and Tasker, R. (1979). A scheme to account quantitatively for the action of phytochrome in etiolated and light-grown plants. *Plant Cell Environ.* 2,259–265.

Jose, A. M., and Vince-Prue, D. (1977a). Light-induced changes in the photoresponses of plant stems; the loss of a high irradiance response to far-red light. *Planta* 135,95–100.

Jose, A. M., and Vince-Prue, D. (1977b). Action spectra for inhibition of growth in radish hypocotyls. *Planta* 136,131–134.

Jose, A. M., and Vince-Prue, D. (1978). Phytochrome action: A reappraisal. *Photochem. Photobiol.* 27,209–216.

Kendrick, R. E., and Frankland, B. (1968). Kinetics of phytochrome decay in *Amaranthus* seedlings. *Planta* 82,317–320.

Ku, P.-K., and Mancinelli, A. L. (1972). Photocontrol of anthocyanin synthesis. *Plant Physiol.* 49,212–217.

Lehmann, U., and Schäfer, E. (1978). Kinetics of phytochrome pelletability. *Photochem. Photobiol.* 27,767–773.

Mancinelli, A. L. (1980). The photoreceptors of the high irradiance responses of plant photomorphogenesis. *Photochem. Photobiol.* 32,853–857.

Mancinelli, A. L., and Rabino, I. (1975). Photocontrol of anthocyanin synthesis. IV. Dose dependence and reciprocity relationships in anthocyanin synthesis. *Plant Physiol.* 56,351–355.

Mancinelli, A. L., and Rabino, I. (1978). The "high irradiance responses" of plant photomorphogenesis. *Bot. Rev.* 44,129–180.

Mancinelli, A. L., and Walsh, L. (1979). Photocontrol of anthocyanin synthesis. VII. Factors affecting the spectral sensitivity of anthocyanin synthesis in young seedlings. *Plant Physiol.* 63,841–846.

Marmé, D. (1969). Photometrische Messungen am Phytochromsystem von Senfkeimlingen (*Sinapis alba* L.). *Planta* 88,43–57.

Mohr, H. (1972). "Lectures on Photomorphogenesis." Springer, New York, Heidelberg, Berlin.

Mohr, H., Bienger, I., and Lange, H. (1971). Primary reaction of phytochrome. *Nature (Lond.)* 230,56–58.

Quail, P. H., Schäfer, E., and Marmé, D. (1973). *De novo* synthesis of phytochrome in pumpkin hooks. *Plant Physiol.* 52,124–127.

Rabino, I., Mancinelli, A. L., and Kuzmanoff, K. M. (1977). Photocontrol of anthocyanin synthesis. VI. Spectral sensitivity, irradiance dependence, and reciprocity relationships. *Plant Physiol.* 59,569–573.

Schäfer, E. (1971). Detaillierte photometrische Messungen *in vivo* am Phytochromsystem von *Sinapis alba* L. und *Cucurbita pepo* L. Doctoral dissertation, University of Freiburg.

Schäfer, E. (1974). Evidence for binding of phytochrome to membranes. *In* "Membrane Transport in Plants" (U. Zimmermann and J. Dainty, eds.), pp. 435–440. Springer, Berlin, Heidelberg, New York.

Schäfer, E. (1975a). A new approach to explain the "high irradiance responses" of photomorphogenesis on the basis of phytochrome. *J. Math. Biol.* 2,41–56.

Schäfer, E. (1975b). The "high irradiance reaction." *In* "Light and Plant Development" (H. Smith, ed.), pp. 45–59. Butterworths, London, Boston.

Schäfer, E. (1981). Phytochrome and daylight. *In* "Plants and the Daylight Spectrum" (H. Smith, ed.), pp. 461–480. Academic Press, London, New York.

Schäfer, E., and Mohr, H. (1974). Irradiance dependency of the phytochrome system in cotyledons of mustard (*Sinapis alba* L.). *J. Math. Biol.* 1,9–15.

Schäfer, E., and Schmidt, W. (1974). Temperature dependence of phytochrome dark reactions. *Planta* 116,257–266.

Schmidt, W., and Schäfer, E. (1974). Dependence of phytochrome dark reactions on the initial photostationary state. *Planta* 116,267–272.

Shropshire, W., Jr. (1972a). Phytochrome, a photochromic sensor. *In* "Photophysiology" (A. C. Giese, ed.), vol VII, pp. 33–72. Academic Press, New York, London.

Shropshire, W., Jr. (1972b). Action spectroscopy. *In* "Phytochrome" (K. Mitrakos and W. Shropshire, Jr., eds.), pp. 162–181. Academic Press, London, New York.

Smith, H. (1970). Phytochrome and photomorphogenesis in plants. *Nature (Lond.)* 227,665–668.

Turner, M. R., and Vince, D. (1969). Photosensory mechanisms in the lettuce seedling hypocotyl. *Planta* 84,368–382.

CHAPTER 25

Very Low Fluence Response;
A Dimer Model

Dina F. Mandoli was a first-year doctoral student at Stanford University in 1978 when she discovered that even green safelights can have physiological effects on plants. "I was young and compulsive and worried that I would leave out a control," she explains, "so I did controls for handling oat mesocotyls and coleoptiles in absolute darkness as well as under the safelight. I consistently saw a dramatic difference—mesocotyls handled under the safelight were shorter than those handled in darkness, while coleoptiles were longer."

Green safelights had been in use for 30 years, after red light was found to be physiologically active (see Chapter 5). Action spectra for phytochrome phototransformation appeared to confirm the safety of such lights, because 500 nm was the least effective wavelength in the visible spectrum (see Fig. 11-4).

There were periodic reports, however, that even minute amounts of light could have physiological effects (Chon and Briggs, 1966; Blaauw et al., 1968; Raven and Shropshire, 1975). "But such sensitive responses," Mandoli says, "had never been documented or explained in a rigorous, statistical fashion."

VLFR

As a predoctoral fellow at the Carnegie Institution, Mandoli (Fig. 25-1) set out to determine the sensitivities of mesocotyls and coleoptiles that had *never* been exposed to light (Mandoli and Briggs, 1981). Enduring 8–10-h shifts in a hot, humid, and pitch-black growth room, she obtained "reagent-grade darkness" by banning entry during experiments and venting the projector without letting out stray light. "But the major source of

FIGURE 25-1   Dina F. Mandoli.

light,'' she recalls, ''was static electricity. I had to label everything in the dark, and I lost many experiments by picking up a piece of tape and having a spark fly out.''

She taped Lodi oat seeds to glass plates lined with absorbent paper. Between 54 and 56 h after the beginning of imbibition, she irradiated the seedlings briefly and then allowed them to grow in darkness for a further 24 h. She could measure only the final lengths of mesocotyls and coleoptiles; initial lengths were determined from a developmental time course plotted for hundreds of dark-grown seedlings, using spot checks in each experiment.

After R, green, and FR treatments, both coleoptiles and mesocotyls exhibited two clearly separated levels of response. In R, one response was observable at 10^{-4} μmol m^{-2} and saturated at 5×10^{-2} μmol m^{-2}. The other began at 10^0 μmol m^{-2} and saturated at 3×10^2 μmol m^{-2}. It thus became clear that responses to extremely low light levels—very low fluence responses (VLFR)—formed a class that was quite distinct from the classic low fluence response (Mandoli and Briggs, 1981).

The thresholds of the VLF and LF responses differed by four orders of magnitude (Fig. 25-2), and the amount of light needed to trigger the former was so low that, when given over 60 s, it was invisible to Mandoli's dark-adapted eyes unless she looked directly into the projector. Because the threshold level for the LFR was so far above the saturation level for the VLFR, there was a fluence range where increasing irradiance failed to enhance response.

Green light was only one-tenth as effective as R, and FR was 1,000 times less effective than R. But because FR could saturate the VLFR, responses to very low doses of R were not FR-reversible. They did obey the reciprocity law, provided a light treatment lasted no longer than 10 min.

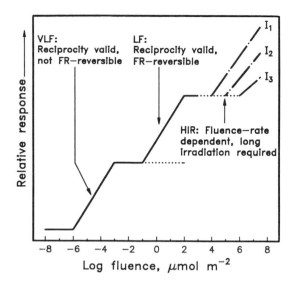

FIGURE 25-2   Three types of phytochrome responses. Relative magnitudes of representa-
tive responses are plotted against increasing R fluence. Short light pulses potentiate VLF and
LF responses. For HIR, three fluence rates given over long periods are shown ($I_1 < I_2 < I_3$).
(From Briggs et al., 1984.)

Estimated Pfr levels at threshold and saturation points were 0.01% and
0.4% for the VLFR and 2% and 87% for the LFR. Thus the conversion of
just one of every 10,000 phytochrome molecules to Pfr partially suppressed mesocotyl growth and partially enhanced the growth of oat coleoptiles (Mandoli and Briggs, 1981).

LIGHT PIPING

Mandoli then determined which parts of the young, dark-grown plant are
most light-sensitive. Securing oat seedlings behind a shield, she directed R
from a diode to discrete, 2-mm areas. Localized application slowed mesocotyl growth and enhanced coleoptile growth in both the VLF (Fig. 25-3A)
and LF (Fig. 25-3B) ranges. Both organs responded maximally when light
was applied 23 to 25 mm from the base of the seedling.

A graded response along the entire length was puzzling. "I went home
for a week and stared at the data, trying to make sense of it," Mandoli
remembers. "I couldn't, since we had assumed there would be a discrete
site of photoperception. So then I said, 'What if there's a discrete site of
photoperception from which a signal travels through the tissue?' But the

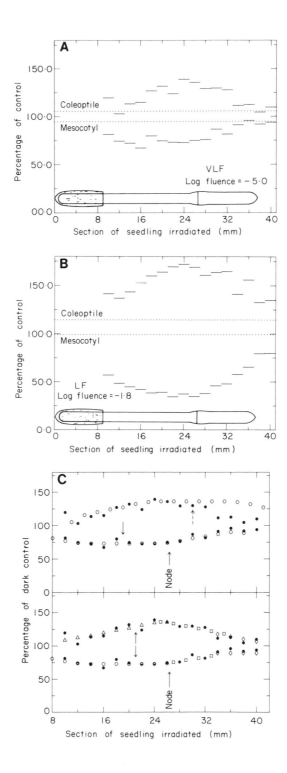

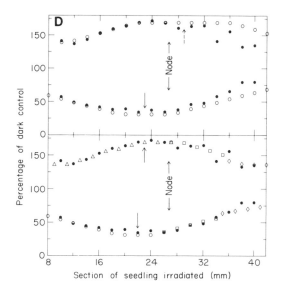

FIGURE 25-3 (A) Response pattern of 72-h-old seedlings to a very low fluence ($10^{-5.0}$ mol m^{-2}) given to 2-mm areas of the seedling. Dotted lines represent controls that were prepared for, but not exposed to, light. Apex of the seedling is on right of figure. (B) Response pattern of 72-h-old seedlings to a low fluence ($10^{-1.8}$ mol m^{-2}). (C) Prediction of response of oat tissues to localized irradiation according to light-piping properties of tissue for VLFR. Closed symbols are derived from (A). Open symbols do not (*above*) or do (*below*) take light attenuation by structures such as the node into account. (D) As (C) but for low-fluence irradiation. Closed symbols are derived from (B). (From Mandoli and Briggs, 1982a.)

theoretical predictions didn't fit the experimental data. So then I began to think about the interesting possibility that the light itself was moving through the tissue—I hadn't done a control for that. So I bicycled back to the lab, very excited, and irradiated one end of a bent seedling. Light came out of the other!''

Cropping roots and seeds from seedlings, she placed cut ends in front of a photomultiplier tube. While irradiating discrete areas, she measured how much light emerged. R was able to travel 25–45 mm through the plant, and mesocotyl and coleoptile had surprisingly similar light-piping properties. For both tissues, the percentage of light transmitted was logarithmically proportional to distance.

Comparing a theoretical response pattern with the experimental data, Mandoli observed a striking fit in both the VLF (Fig. 25-3C) and LF (Fig. 25-3D) ranges. ''It was obvious then that the tissues were piping light,'' Mandoli says. Taking these properties into account, she localized two sites of photoperception: one near the top of the mesocotyl, which controlled the mesocotyl response, and one just above the node. Both sites had to be irradiated before the coleoptile would respond (Mandoli and Briggs, 1982a).

Mandoli continued to explore the fiber optic properties of oat seedlings with a laser rather than a light-emitting diode. She found that mesocotyls were the most effective light guides but that waves could also travel through other parts of plants. Oat mesocotyl was 0.7% as effective as an optical fiber and 10% as effective as a glass rod, and the tissues behaved like multifiber, rather than single-fiber, optic bundles (Mandoli and Briggs, 1982b).

"This optical property of the pre-emergent seedlings," she pointed out, "could allow light transmission from the coleoptilar tip to the sites of perception and thus induce photomorphogenesis well before the majority of the seedling had emerged from the soil" (Mandoli and Briggs, 1982a).

At Michigan State University, Brian M. Parks and Kenneth L. Poff demonstrated that light piped through plant tissues is photochemically effective. Using a spectrophotometer to measure photoconversion of phytochrome just below the coleoptilar node of dark-grown *Zea* seedlings, they observed that transformation of Pr to Pfr decreased log-linearly with increasing distance between the irradiation point and measuring point (Parks and Poff, 1985). Piped light was also physiologically effective. After staining *Zea* coleoptiles with methylene blue to increase the axial R gradient, the researchers transplanted the seedlings and covered them with soil. Emerging under natural lighting, the mesocotyls were shorter than those of unstained seedlings (Parks and Poff, 1986).

SENSITIZED SEEDS

Even before the VLFR was recognized as a separate response class, it was known that seeds may develop an extreme sensitivity to light during burial in the soil (Wesson and Wareing, 1969; Taylorson, 1970, 1972) or, in the lab, after low (see Chapter 13) or high-temperature treatments (Fig. 25-4) (Blaauw-Jansen and Blaauw, 1975).

Vivian Toole, who retired in 1976, continued to explore interactions between light and temperature in collaboration with William J. VanDer-Woude at Beltsville (Fig. 25-5). The researchers imbibed Grand Rapids lettuce seeds on thermal plates that maintained a temperature gradient between the two ends of each row. "Ultimately," says VanDerWoude, "we saw that treatment for hours at low temperature would lead to responses to far-red, which had generally been thought to inhibit germination."

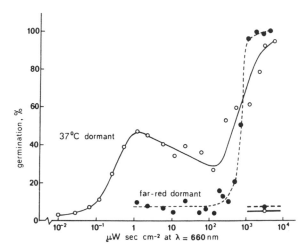

FIGURE 25-4   Dose–response curves for induction of germination of May Queen lettuce seeds exposed either to FR for 22 h or maintained at 37°C for 24 h. (From Blaauw-Jansen and Blaauw, 1975.)

After three FR pulses and 24 h darkness (no Pfr), only 11% of the seeds germinated, regardless of temperature (Fig. 25-6). But subsequent R treatment promoted maximal germination, whose lower temperature limit related to the length of time Pfr remained in the seed before being converted back to Pr. But at 12°C (for the 6-h curve), low temperatures suddenly began to promote germination, and the responses for the high Pfr periods of differing durations coincided closely. The researchers attributed these findings not to "phytochrome potentiation during the high Pfr periods, but rather to subsequent potentiation of germination at 20°C by the low level of Pfr produced by the FR irradiation which terminated these periods. Low temperature treatments apparently enhanced subsequent sensitivity of the germination response at 20°C to low Pfr levels" (VanDerWoude and Toole, 1980).

Omitting the R treatment, they applied the short FR treatment after 6 h on the thermal gradient. Seeds that had been below 13°C had higher germination levels than the dark controls, whereas those kept above that temperature failed to respond to the FR treatment. As little as 1 h at 4°C enhanced the response, which increased linearly with duration of the cold treatment. VanDerWoude had previously determined that 10 h at 4°C enhanced sensitivity to R 2,000-fold. Prechilling failed to alter the escape time from phytochrome control of germination.

"The above findings indicate that prechilling enhances the sensitivity to low levels of Pfr but does not alter the rate of potentiation of germination,"

FIGURE 25-5 William J. VanDerWoude, 1988.
Photo by Linda Sage.

the researchers decided. "This suggests that prechilling modifies the control of potentiation by phytochrome rather than the process of potentiation" (VanDerWoude and Toole, 1980). Prechilling also enhanced germination in seeds that had never been irradiated, presumably by sensitizing them to already present Pfr. Influenced by the studies of Hendricks and Taylorson (see Chapter 23), the Beltsville workers speculated that "a change in membrane order and lipid composition, reflected as a decrease in viscosity, is responsible for the prechilling phenomenon" (VanDerWoude and Toole, 1980).

Kendrick's group later obtained biphasic fluence response curves for light-induced germination of *Arabidopsis thaliana* seeds after high-temperature treatment (Cone et al., 1985). Researchers in Ghent discovered that *Kalanchoë blossfeldiana* seeds would become extremely light-sensitive if they were incubated with gibberellic acid (de Petter et al., 1985).

WEED SEEDS

While the ecological significance of the VLFR is still under debate, studies at the University of Buenos Aires suggest that it is not "some sort of aberrant behavior that can only be demonstrated within the strange world of dark rooms," Carlos L. Ballaré says. His experiments in collaboration with Ana L. Scopel and Rodolfo A. Sánchez suggested that "the VLFR could play a central role in the perpetuation of weed species in arable fields."

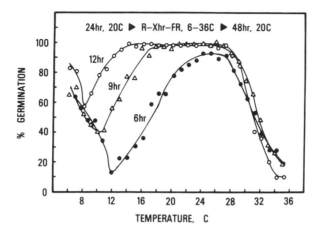

FIGURE 25-6   Effects of temperature on phytochrome potentiation of germination. All seeds were treated with FR pulses during planting and then incubated in the dark at 20°C for 24 h. The temperature gradient was then established and a R pulse given. After 6, 9, or 12 h in darkness, seeds were irradiated for 8 min with FR and returned to 20°C. Germination was scored 48 h after this last irradiation. "Dark" controls received a 6-h exposure to the treatment temperature but no R or subsequent FR light treatments. A green safelight was used only during planting. (From VanDerWoude and Toole, 1980.)

Datura ferox (chinese thornapple), an aggressive arable weed in Argentina, emerges in flushes after soil cultivation. Comparing air-dried seeds with those buried for 2 months at a depth of 7 cm, the researchers observed R/FR-reversible induction of germination. But whereas they obtained a LFR curve for the dry seeds, the majority of the exhumed seeds exhibited a VLFR, germinating even if less than 0.01% of their phytochrome was converted to Pfr. "These results clearly document, for the first time, a massive, natural transition to the typical VLFR state," the researchers concluded (Scopel et al., 1991).

Aiming to simulate natural conditions, Ballaré and Scopel piped sunlight from the soil surface to buried seeds and counted seedlings 2 weeks later. About one-third of the seeds sprouted, and even a fluence equal to 0.1 s of full sunlight triggered some germination. If the seeds were retrieved at midday using a light shield, promptly exposed to sunlight in a shutter box, and then buried again, fluences equal to a few miliseconds full sunlight were sufficient to promote more than half the seeds. Calculations revealed that four-fifths of the seeds that germinated required only between 0.0001% and 0.01% Pfr. Thus burial increased light sensitivity 10,000-fold, allowing seeds to respond to split-second exposures. "Because short pulses of this nature are very likely to be the most important light signals of soil cultivation," the Buenos Aires workers concluded, "the natural switch from the LFR to the VLFR state . . . is probably essential for the perception of soil disturbances by light-requiring weed seeds" (Scopel et al., 1991).

THE DIMER HYPOTHESIS

VanDerWoude investigated the mechanism of sensitization to very low Pfr levels in detailed fluence-response studies of lettuce seeds sensitized by prechilling, brief exposure to high-temperature or preincubation in ethanol. All treatments yielded biphasic fluence-response curves having VLF and LF regions.

For about a year, he tried to understand how phytochrome could induce both types of response. He eventually focused on an unpublished model[1] for the action of monomeric phytochrome in which Hendricks had tried to explain the *Zea* paradox (see Chapter 12). Hendricks suggested that receptors for Pfr are much less abundant than phytochrome molecules, as Hartmann had proposed. R doses too small to produce spectrophotometrically measurable Pfr could produce enough Pfr to fill all the receptors and thus give a physiological response. A FR dose that established a photoequilibrium of about 1% Pfr would reverse rather than promote the response by converting most of the bound Pfr molecules back to Pr, blocking the receptor sites with inactive molecules while Pfr stood idly by in the cell. Although Hendricks's model accounted for much biphasic fluence-response behavior, inhibition by FR was not observed in many VLFR, including germination of prechilled lettuce seeds.

In December 1980, VanDerWoude suddenly visualized Hendricks's model functioning with dimeric, rather than monomeric, phytochrome molecules, and he developed a set of differential equations based on the proposed mechanism. In the literature, he found evidence that the pigment might be a dimer *in vivo*, because light treatments that produced only low levels of Pfr made Pr and Pfr equally pelletable (Pratt and Marmé, 1976). This supported the concept that phytochrome in the plant was a dimeric molecule with two chromophores, only one of which needed to be in the Pfr form for pelletability.

VanDerWoude presented a dimeric model in 1981 (VanDerWoude, 1983) and a revised and expanded version (added to in Fig. 25-7) later

---

1. Described by Vanderhoef, Quail, and Briggs (1979) and, after Hendricks's death, by VanDerWoude (Hendricks and VanDerWoude, 1983).

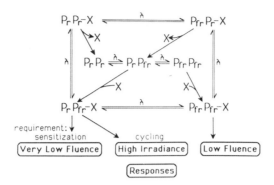

FIGURE 25-7 Dimeric model of phytochrome action adapted to include high-irradiance responses. (From VanDerWoude, 1987.)

(VanDerWoude, 1985). "What this model does," he says, "is to provide an explanation of the *Zea* paradox and other major difficulties in understanding the relationship between photoconversion and action. It shows that phytochrome action is not strictly related to its photoconversion but depends on the sensitization that has developed and on the action of a heterodimer–receptor complex."

He proposed that phytochrome exists in three forms: PrPr (the Pr dimer), PrPfr (the heterodimer), and PfrPfr (the Pfr dimer). These are interconvertible by light, and each chromophore of a dimer is transformed independently of its partner.

Very low fluences produce heterodimer, which reacts with a membrane-associated receptor, X. The resulting highly stable complex, PrPfr-X, elicits a VLFR in sensitized systems. "One possibility," VanDerWoude says, "is that this whole complex freely interacts with the next step in the transduction chain. Let's say a substance Y is involved, which the complex can activate repeatedly, owing to its mobility in a membrane."

Low fluences fully transform phytochrome to the Pfr dimer, which also reacts with X, forming PfrPfr-X. This second complex forms a stable association with Y, provoking the LFR. Interconversion of the various unstable dimer–receptor complexes constantly generates and removes PrPfr-X and PfrPfr-X.

Computer simulations revealed that the proposed mechanism could account for the biphasic fluence responses of sensitized seeds. They suggested that:

1. VLF promote the formation of PrPfr-X

2. Saturation of the VLFR in seeds occurs when all the receptors are occupied by PrPfr molecules

3. PrPfr-X is not converted to PfrPfr-X until fluences reach the LF range

4. LF responses are linearly proportional to the concentration of PfrPfr-X

5. VLF responses are proportional to the logarithm of PrPfr-X concentration

6. Sensitization of seeds increases the physiological effectiveness of PrPfr-X but not PfrPfr-X

7. There are about 1,000 times as many phytochrome molecules as receptor molecules

The last prediction suggested that "the postulated formation of PrPfr-X and PfrPfr-X should not be equated with the phenomena of *in vivo*, light-induced pelletability and sequestering that suggest massive aggregation or association of phytochrome with subcellular components. Rather, the possibility of a very high and specific affinity of the receptor for these two forms of the dimer is supported" (VanDerWoude, 1985).

DIMERS AND HIR

By 1986, VanDerWoude had extended his model to account for high-irradiance responses (Fig. 26-7). Making assumptions about the relative affinities of the heterodimer and Pfr dimer for X, the half-lives of the unstable dimer–receptor complexes, and the invulnerability to destruction of phytochrome attached to X, he simulated the wavelength and fluence dependencies of the various components of the model during a 12-h irradiation.

The most realistic "action spectrum" was obtained for PfrPfr-X in the R and the product of PrPfr-X and its cycling rate[2] in the B and FR—if action was summed over 12 h (Fig. 25-8). The combined curves bore a striking resemblance to that obtained by Schäfer's group for inhibition of hypo-cotyl growth in dark-grown mustard seedlings (see Fig. 24-10A). VanDer-Woude therefore proposed that PrPfr-X is the HIR effector in the B and FR regions, that PfrPfr-X produces the R peak, and that, as turnover of the heterodimer–receptor complex becomes more rapid, the HIR response in the B and FR increases. The model also predicted that preirradiation

---

2. In this context, cycling was defined as the continuous formation and removal of PrPfr-X by other reactions in the model, although presumably other types of cycling would also occur.

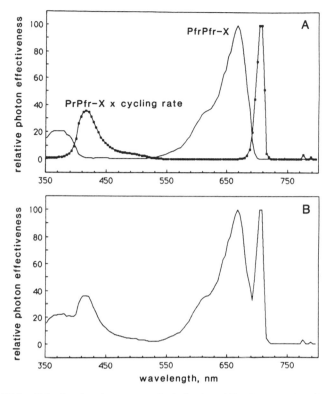

FIGURE 25-8 Calculated action spectra for behavior of VanDerWoude's dimeric model under continuous irradiation. (A) For PfrPfr-X and for PrPfr-X × its cycling rate, each normalized to its maximum. (B) Normalized sum of the two action spectra in (A). (From VanDerWoude, 1987.)

leading to phytochrome destruction would reduce the B and FR peaks, while sparing the peak in the R (VanDerWoude, 1987), as Schäfer's group had observed in hypocotyls of mustard seedlings (see Chapter 24).

VanDerWoude's model was well-received, though with reservations about membrane-bound receptors. But Mohr criticized it in light of studies of NADH-dependent glutamate synthase (Hecht and Mohr, 1990).

A DIMER *IN VIVO?*

Schäfer, meanwhile, had become convinced that phytochrome is a dimer *in vivo*. "I had a brilliant student, Jörg Brockmann," he said, "who suggested that a dimeric phytochrome would explain a lot of very old,

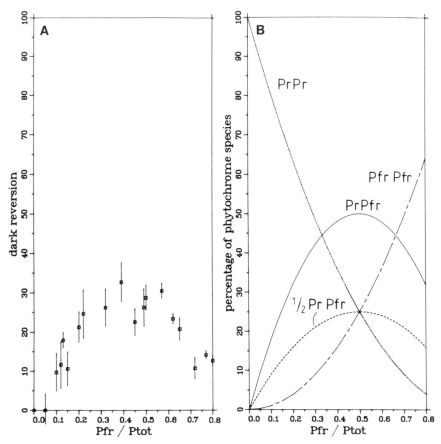

FIGURE 25-9   (A) Dependence of amount of dark-revertible phytochrome in mustard hooks on initial Pfr/P_{total} ratio. Reversion was measured after a 40-min dark period. (B) Theoretical relationship between spectrophotometrically measurable Pfr/P_{total} ratio and the relative concentrations of phytochrome dimers. Lower optimum curve represents half the PrPfr pool (because only the Pfr portion reverts), and it corresponds to the measured amount of dark-revertible Pfr in (A). (From Brockmann et al., 1987.)

problematic data. My first graduate student had determined that the amount of dark reversion you observe is not constant but depends on the photoequilibrium established with the light pulse'' (Schmidt and Schäfer, 1974).

Dark reversion in mustard hooks was complete within 10 min of a R pulse, when about 65% of the total phytochrome was present as Pfr. But with the aid of different light sources, Werner Schmidt had observed maximal reversion of Pfr when about 40% of P_{total} was initially Pfr (Fig. 25-9A). After reversion had ceased, it began again if a second light pulse restored this photoequilibrium. Thus there seemed to be a pool of Pfr that could revert and a pool that could not, and the latter could replenish the former.

Brockmann estimated the concentrations of the Pr dimer, heterodimer, and Pfr dimer at various photoequilibria (Fig. 25-9B). The resulting curves suggested that Schmidt's results could be explained if the Pfr portion of PrPfr reverted, because the curve for the Pfr dimer bore no resemblance to the experimentally obtained curve. Once reversion was complete, a second pulse would induce it again by producing additional heterodimer.

The kinetics of phytochrome destruction in dark-grown oat and *Amaranthus* seedlings also suggested that phytochrome was a dimer in the plant. Pr and Pfr disappeared simultaneously during 2 h of darkness after a 3-ms R flash that established 40% Pfr. Because Pr itself is destroyed more slowly and because the flash was too short to cycle Pr through Pfr, it seemed probable that the Pr had been destroyed because it was attached to Pfr, as part of a heterodimer (Brockmann et al., 1987).

REFERENCES

Blaauw-Jansen, G., and Blaauw, O. H. (1975). A shift of the response threshold to red irradiation in dormant lettuce seeds. *Acta Bot. Neerl.* 24,199–202.

Blaauw, O. H., Blaauw-Jansen, G., and van Leeuwen, W. J. (1968). An irreversible red-light induced growth response in *Avena. Planta* 82,87–104.

Briggs, W. R., Mandoli, D. F., Shinkle, J. R., Kaufman, L. S., Watson, J. C., and Thompson, W. F. (1984). Phytochrome regulation of plant development at the whole plant, physiological, and molecular levels. *In* "Sensory Perception and Transduction in Aneural Organsims" (G. Colombetti, F. Lenci, and P. -S. Song, eds.), pp. 265–280. Plenum, New York.

Brockmann, J., Rieble, S., Kazarinova-Fukshansky, N., Seyfried, M., and Schäfer, E. (1987). Phytochrome behaves as a dimer *in vivo. Plant Cell Environ.* 10,105–111.

Chon, H. P., and Briggs, W. R. (1966). Effects of red light on the phototropic sensitivity of corn coleoptiles. *Plant Physiol.* 41,1715–1724.

Cone, J. W., Jaspers, P. A. P. M., and Kendrick, R. E. (1985). Biphasic fluence-response curves for light induced germination of *Arabidopsis thaliana* seeds. *Plant Cell Environ.* 8,605–612.

Hecht, U., and Mohr, H. (1990). Relationship between phytochrome photoconversion and response. *Photochem. Photobiol.* 51,369–373.

Hendricks, S. B., and VanDerWoude, W. J. (1983). How phytochrome acts—Perspectives on the continuing quest. *In* "Encyclopedia of Plant Physiology, New Series" (W. Shropshire, Jr. and H. Mohr, eds.), vol. 16A, pp. 3–23. Springer, Berlin, Heidelberg.

Mandoli, D. F., and Briggs, W. R. (1981). Phytochrome control of two low-irradiance responses in etiolated oat seedlings. *Plant Physiol.* 67,733–739.

Mandoli, D. F., and Briggs, W. R. (1982a). The photoperceptive sites and the function of tissue light-piping in photomorphogenesis of etiolated oat seedlings. *Plant Cell Environ.* 5,137–145.

Mandoli, D. F., and Briggs, W. R. (1982b). Optical properties of etiolated plant tissues. *Proc. Natl. Acad. Sci. USA* 79,2902–2906.

Parks, B. M., and Poff, K. L. (1985). Phytochrome photoconversion as an *in situ* assay for effective light gradients in etiolated seedlings of *Zea mays*. *Photochem. Photobiol.* 41,317–322.

Parks, B. M., and Poff, K. L. (1986). Altering the axial light gradient affects photomorphogenesis in emerging seedlings of *Zea mays* L. *Plant Physiol.* 81.75–80.

de Petter, E., van Wiemeersch, L., Rethy, R., Dedonder, A., Fredericq, H., de Greef, J., Steyaert, H., and Stevens, H. (1985). Probit analysis of low and very-low fluence-responses of phytochrome-controlled *Kalanchoë blossfeldiana* seed germination. *Photochem. Photobiol.* 42,697–703.

Pratt, L. H., and Marmé, D. (1976). Red light-enhanced phytochrome pelletability. Reexamination and further characterization. *Plant Physiol.* 58,686–692.

Raven, C. W., and Shropshire, W., Jr. (1975). Photoregulation of logarithmic fluence-response curves for phytochrome control of chlorophyll formation in *Pisum sativum*. *Photochem. Photobiol.* 21,423–429.

Schmidt, W., and Schäfer, E. (1974). Dependence of phytochrome dark reactions on the initial photostationary state. *Planta* 116,267–272.

Scopel, A. L., Ballaré, C. L., and Sánchez, R. A. (1991). Induction of extreme light sensitivity in buried weed seeds and its role in the perception of soil cultivations. *Plant Cell Environ.* 14, 501–508.

Taylorson, R. B. (1970). Changes in dormancy and viability of weed seeds in soils. *Weed Sci.* 18,265–269.

Taylorson, R. B. (1972). Phytochrome controlled changes in dormancy and germination of buried weed seeds. *Weed Sci.* 20,417–422.

Vanderhoef, L. N., Quail, P. H., and Briggs, W. R. (1979). Red light-inhibited mesocotyl elongation in maize seedlings. *Plant Physiol.* 63,1062–1067.

VanDerWoude, W. J. (1983). Mechanisms of photothermal interactions in the phytochrome control of seed germination. *In* "Beltsville Symposia in Agricultural Research 6: Strategies of Plant Reproduction" (W. J. Meudt, ed.), pp. 135–143. Allanheld, Osmun & Co., London, Toronto, Sydney.

VanDerWoude, W. J. (1985). A dimeric mechanism for the action of phytochrome: Evidence from photothermal interactions in lettuce seed germination. *Photochem. Photobiol.* 42,655–661.

VanDerWoude, W. J. (1987). Application of the dimeric model of phytochrome action to high irradiance responses. *In* "Phytochrome and Photoregulation in Plants. Proceedings of the XVI Yamada Conference" (M. Furuya, ed.), pp. 249–258. Academic Press, New York.

VanDerWoude, W. J. and Toole, V. K. (1980). Studies of the mechanism of enhancement of phytochrome-dependent lettuce seed germination by prechilling. *Plant Physiol.* 66, 220–224. Quotations reprinted with permission from the American Society of Plant Physiologists.

Wesson, G., and Wareing, P. F. (1969). The induction of light sensitivity in weed seeds by burial. *J. Exp. Bot.* 20,414–425.

CHAPTER **26** _____

Shade Avoidance

"Why do plants have phytochrome?" Harry Smith asked Borthwick in Eretria in 1971. "It hasn't evolved so that photophysiologists can shine 2 or 3 min of red or far-red light on plants grown artificially in the dark." Coming away from that symposium, Smith (Fig. 26-1) says, "I was absolutely convinced that where I had to go in the future was to work on plants in simulated natural environments." Writing a text on phytochrome (Smith, 1975), he realized that an important function of the pigment must be shade avoidance.

LIGHT QUALITY AND SEED GERMINATION

The Beltsville workers knew that shade might affect seed germination. Studying the action of light filtered through fresh crop leaves, Taylorson and Borthwick suggested that alterations by foliage of the R:FR ratio could influence the germination of light-requiring weed seeds (Taylorson and Borthwick, 1969)—light transmitted through leaves has a lower R:FR ratio than sunlight, because chlorophyll selectively removes wavelengths below 700 nm. Michael Black found that *Tilia* leaves act as perfect FR filters, preventing R-induced germination of Grand Rapids lettuce seed (Black, 1969).

Meischke had used leaves as experimental light filters over 30 years earlier, during his investigations of infrared inhibition of photoblastic seeds (see Chapter 3). Much later, Cumming drew attention to the possible ecological significance of this effect, after observing that *Chenopodium* seeds germinated better in light with a R:FR ratio similar to that of sunlight than under ratios resembling those of sunlight filtered through green leaves. "There may be restriction of germination in areas shaded by green plants," he suggested (Cumming, 1963). Thus phytochrome's

FIGURE 26-1   Harry Smith

unique photochromic properties might allow shaded, light-sensitive seeds
to remain dormant until overhanging vegetation disappears.

LIGHT QUALITY AND
VEGETATIVE DEVELOPMENT

Because FR was known to promote stem elongation, Hendricks suggested
that overhanging vegetation might also modify vegetative growth. "Be-
neath the forest canopy . . . the position of the [phytochrome] equilibrium is changed resulting in effects on metabolic activity. Obvious effects
are modification of stem length and leaf size" (Hendricks and Borthwick,
1963).

Plants under forest canopies tolerate shade, however, producing few
internodes during the growing season and photosynthesizing efficiently at
low irradiances. Shade avoiders thrive on grassland and recently cleared
land, where by rapidly growing taller they may overtop neighboring grass
plants or the many other plants that quickly colonize bare soil (Grime,
1965, 1981).

The biological effects of light quality in the field also interested Kasperbauer. Controlled environment experiments at a USDA research station in
Lexington revealed that R or FR at the end of 8-h WL photoperiods
affected tobacco plant development, especially at the seedling stage. End-
of-day 5-min FR exposures produced plants with much taller stems and
smaller roots than end-of-day 5-min R exposures. Moreover, the FR-

treated plants failed to branch from the axils of the lower leaves. And although plants in both treatment groups had the same number of leaves, those of the FR-treated plants were narrower, thinner, and lighter in color (Kasperbauer, 1971), more photosynthetically efficient (Kasperbauer and Peaslee, 1973), and had higher chlorophyll a/b ratios and altered chloroplast polypeptide content (Jones and Kasperbauer, 1985). The FR-treated plants also had lower concentrations of free amino acids and higher concentrations of free sugars and organic acids (Kasperbauer et al., 1970). And their chloroplasts had more numerous, smaller grana and fewer, smaller starch grains (Kasperbauer and Hamilton, 1984). The R and FR effects on free-sugar concentration and starch grain size were evidence that phytochrome regulated the partitioning of photosynthate to sugar or starch at the cellular level. Phytochrome regulation of sugar uptake in tobacco callus cultures was demonstrated in the 1960s (Kasperbauer and Reinert, 1966).

Soybean showed a similar responsiveness to phytochrome, and end-of-day FR treatment caused both soybean and tobacco to partition relatively more photosynthate to shoots than to roots (Kasperbauer et al., 1984; Kasperbauer and Jones, 1985; Kasperbauer, 1987, 1988). Conversely, end-of-day R treatment produced soybean plants with lower shoot:root biomass ratios, larger roots, and more nitrogen-fixing root nodules (Kasperbauer et al., 1984).

The controlled environment studies of the mid 1960s led Kasperbauer to investigate tobacco canopies, where he suspected that alterations in light quality might also affect development. Using a 16-waveband portable spectroradiometer he detected FR:R ratios[1] seven times higher at soil level under the canopy than on a roadway some distance from growing plants. The ratio was 6.5 times higher within the canopy, and 4.2 times higher below a single leaf. The differences appeared to be significant in adaptation to competition because plants at the west ends of east–west rows were shorter than plants within rows, even though they had the same number of leaves (Kasperbauer, 1971). "Taller plants within higher density populations," says Kasperbauer, "would have a greater probability of keeping some leaves in sunlight above the competition, an adaptation favoring survival among other plants."

ENVIRONMENTAL RADIATION

Returning from Eretria to Sutton Bonington, Smith consulted with John L. Monteith, an environmental physicist. He soon became convinced that he

1. Kasperbauer expresses his data as FR:R ratios (730 nm:645 nm) rather than R:FR ratios.

should *quantitate* the relationship between natural R:FR ratios and phytochrome photoequilibria to determine whether the latter was markedly affected by light-quality changes that occur as plants are shaded. "Quantitative relationships are essential to the construction of any theory," he stresses.

Phytochrome appeared to be the perfect shade detector because it detects light quality rather than irradiance, which fluctuates constantly as the weather changes. With its capacity to provide color vision, it would allow plants to compare the proportions of light at two different wavebands whose ratio is altered in shade.

During the growing seasons of 1973 and 1974, M. Geoffrey Holmes transported ice-cooled cuvettes of phytochrome-rich bean hook sections[2] into croplands near Sutton Bonington. After scanning the spectrum with a spectroradiometer that fit into canopies without disturbing plants (Holmes and Smith, 1977a), he exactly replaced the meter with a cuvette, allowing time for the phytochrome in the sections to reach photoequilibrium. He then placed the cuvette back on ice, put the ice bath back in its lightproof black box, and raced back to Smith's new Perkin-Elmer dual wavelength spectrophotometer.

R:FR ratios[3] varied widely in the natural environment, although they were largely unaffected by weather, in contrast to total radiation, which fluctuated markedly and unpredictably. In the open, when the sun was high in the sky, the mean R:FR value for total global radiation (direct radiation from the sun plus scattered radiation from the atmosphere) was 1.15 on both clear and overcast days. It was 44% lower in twilight, when the sun was low in the sky and its rays had to pass through a much larger air mass (Holmes and Smith, 1977a). Within a wheat (*Triticum aestivum*) canopy, composed of closely spaced, erect plants with narrow leaves, light was depleted of B and R but not FR, giving a mean value at ground level one June day of 0.21. The ratio was even reduced in sunflecks, which had altered proportions of direct, diffuse, and reflected (from neighboring plants) radiation (Holmes and Smith, 1977b).

In the dense shade of a sugar beet canopy, where plants with broad leaves were widely spaced, R:FR ratio was only about 0.04. Because weeds grow within crop canopies, "terrestrial plants normally exist under conditions in which [R:FR] ranges from about 1.2 in full sunlight to about 0.05 in dense vegetation" (Holmes and Smith, 1975).

2. These were dark-grown seedlings, because there was no way of measuring phytochrome levels in light-grown material.

3. Smith compares total photon irradiances between 655 and 665 nm with those between 725 and 735 nm.

PHYTOCHROME PHOTOEQUILIBRIUM

To determine whether phytochrome would be a suitable shade detector within this range, Holmes plotted the R:FR values against the Pfr/P_{total} values obtained with the spectrophotometer, securing additional points under artificial light sources. The curve (Fig. 26-2), which approximated a rectangular hyperbola, revealed that large changes in the phytochrome photoequilibrium resulted from small changes in R:FR ratio in exactly the range plants experience in the natural environment. Phytochrome was most sensitive between R:FR values of 0 and 1.0. Once the ratio reached about 1.5, large increases in the proportion of R produced only small shifts in the photoequilibrium as the curve approached the asymptote (Holmes and Smith, 1977c).

Pfr/P_{total} never reaches 0.8 in nature as it does when laboratory plants are exposed to R. The proportion of total phytochrome as Pfr ranged from 0.04 in the dense shade of sugar beet to 0.20 within a wheat canopy to 0.54 above both canopies. Because the pigment reached equilibrium in 5 s in midday sunlight and about 30 s in conditions mimicking dense canopy

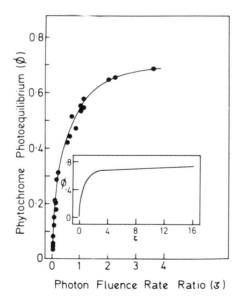

FIGURE 26-2   Relationship between phytochrome photoequilibrium and R:FR photon fluence rate ratio for a series of natural and artificial light regimes. (From Holmes and Smith, 1977c.)

shade, Smith and Holmes concluded that "phytochrome is capable of responding to changes in spectral energy distribution which occur as sunflecks progressively move across the canopy floor due to changes in solar zenith and azimuth angles" (Holmes and Smith, 1977c). Phototransformation was too slow to cope with sunflecks that danced across the ground on windy days. So plants would ignore rapid fluctuations of light lacking physiological significance (Holmes and Smith, 1977c).

EFFECTS ON GROWTH

It remained to be shown that light-quality changes of the magnitude found in nature influence vegetative growth. David C. Morgan established a quantitative relationship between plant growth and Pfr/P_{total}, by using cabinets where light quality could be varied independently of light quantity. Mixing FR and fluorescent sources, he established phytochrome photoequilibria ranging between 0.29 and 0.71 in dark-grown bean hooks. During 21 16-h days under these sources, *Chenopodium* plants differed strikingly in growth rates, exhibiting a linear relationship between the logarithm of stem extension rate and estimated Pfr/P_{total} (Fig. 26-3). The

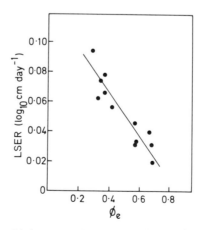

FIGURE 26-3   Relationship between phytochrome photoequilibrium and logarithmic rate constant of stem extension for *Chenopodium album* seedlings. (From Morgan and Smith, 1976. Reprinted by permission from *Nature* vol. 262, pp. 210–212. Copyright © 1976 Macmillan Magazines Ltd.)

data supported the proposal "that the principal function of phytochrome is to detect the quality of light in the red and far red wavelength regions, and thus to perceive mutual shading" (Morgan and Smith, 1976).

Because end-of-day FR also promotes internode elongation, Morgan compared the effects of all-day (16 h W+FR), daytime ($15\frac{1}{2}$ h W+FR, $\frac{1}{2}$ h W), and end-of-day ($15\frac{1}{2}$ h W, $\frac{1}{2}$ h W+FR) regimes on logarithmic stem extension rates. End-of-day response was only 36% as great as the all-day response, whereas the daytime response was 87% as great. Thus "the daytime spectral photon distribution is more important than end-of-day photon distribution in determining growth response." The researchers pointed out that "(a) daytime detection offers more complete information; and (b) the non-vegetational change in [R/FR] at dusk would complicate shade detection" (Morgan and Smith, 1978).

Morgan reasoned that habitat should relate to responsiveness to R:FR ratios. Logarithmic stem extension rates were linearly proportional to Pfr/P$_{total}$ in several weed species, and the order of the lines correlated with relative shade avoidance or tolerance (Fig. 26-4) (Morgan and Smith, 1979).

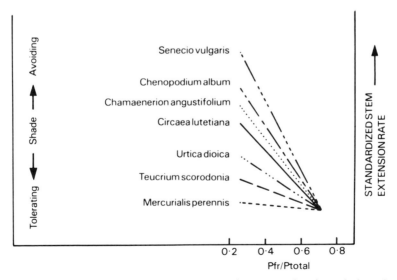

FIGURE 26-4 Relationship between stem extension rate and estimated phytochrome photoequilibrium for a range of herbaceous plants. (From Smith, 1982, based on data from Morgan and Smith, 1979. Reproduced with permission from the Annual Review of Plant Physiology, vol. 33. © 1982 by Annual Reviews Inc.)

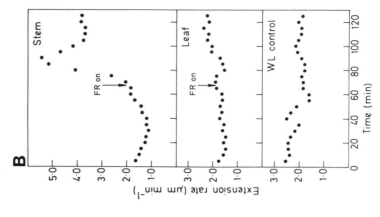

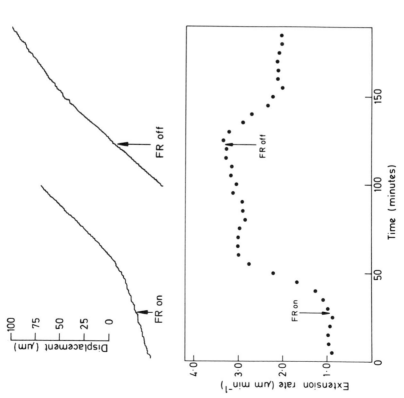

FIGURE 26-5 (A) Effect on extension rate of irradiating a whole mustard seedling with supplementary FR (719 nm) for 90 min against a WL background; (B) either stem or leaf with supplementary FR for 60 min against a whole-plant WL background. (From Morgan et al., 1980. Copyright 1980, Springer-Verlag.)

RAPIDITY OF GROWTH RESPONSES

After Smith became chair of botany at the University of Leicester in 1978, Morgan and Richard Child began to investigate the rapidity of the stem extension response, using a transducer that could reliably record increments of only a ten-thousandth of a millimeter (0.1 μm) per minute.

FR added to WL markedly increased the rate of mustard seedling stem extension after a 13-min lag. When the FR source was extinguished after 90 min, stem growth gradually slowed (Fig. 26-5A). Irradiation of the first internode with FR from an optical fiber produced a similar response under background WL. Again there was a 10–15-min lag before a large but transient increase in stem extension rate (phase I). After half an hour, extension decreased to a constant rate that was twice that in WL alone (phase II) (Fig. 26-5B). If FR irradiation continued, a further gradual increase occurred after 4–5 h. Supplementary FR to the primary leaves provoked only a gradual enhancement of stem extension growth rate after about 3 h. When the FR source was turned off after 1 or several hours, stem extension rate decreased after 15 min if the supplementary light had been applied to the internode but not if it had irradiated the leaf. So it appeared that the growing internode itself was the primary receptive site for the perception of the environmental signal.

When Morgan applied FR from a fiber optic probe directly to the first internode of a mustard plant growing in WL, the addition of R during phase II of the response decreased extension rate after 10–15-min lag. Removal of R an hour later increased extension rate, after a similar lag. Thus the *response*—rather than the induction of the response—was FR/R-reversible (Morgan et al., 1980), and phytochrome was acting as a dimmer rather than an on/off switch.

These experiments, performed with completely de-etiolated plants, strongly suggested that the development of maturing, light-grown plants responds rapidly to minute changes in light quality in the natural environment and that phytochrome is the responsible photoreceptor.

REFLECTED LIGHT

Because shade avoidance involves rapid stem elongation, which requires photosynthate, Peter Hayward tested mustard seedlings' ability to perceive changes in overhead R:FR ratios at different fluence rates of background WL. Between values of 60 and 138 μmol m^{-2} s^{-1}, R:FR ratio perception was unaffected by fluence rate (Fig. 26-6). But at lower fluence rates, this ability was seriously impaired (Smith and Hayward, 1985). Thus

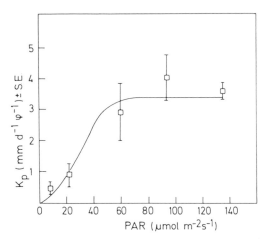

FIGURE 26-6   Effect of photosynthetically active radiation (PAR) on mustard seedling responsivity to R:FR ratio. K_p, slope of the relationship between extension growth rate and photoequilibrium that a mixture of light sources would establish in purified phytochrome. (From Smith and Hayward, 1985.)

the capacity to respond to shade might require a supply of photosynthetic products, which clearly would be limiting in already shaded plants. Shade avoidance would therefore be most effective if plants used R:FR perception as an early warning system.

At the University of Buenos Aires in 1983, Jorge J. Casal countered the decrease in branching that normally accompanies increasing population density by placing R-emitting diodes around the bases of potted grass plants (Casal et al., 1986). He was surprised to observe these effects at low plant densities, where there was little mutual shading.

Ballaré and Scopel (see Chapter 25), meanwhile, were investigating the population dynamics of *Datura ferox* (chinese thornapple) seedlings. Making detailed measurement of microclimate and observing responses to stand density early in the development of the crop, they concluded that the seedlings were not responding to mutual shading (Ballaré et al., 1987, 1988). Thus plants could perhaps respond to the quality of *reflected* light and thus detect neighbors before becoming directly shaded. Green leaves reflect mainly light above 700 nm and absorb much of the B and R in sunlight (Fig. 26-7). The amount of reflected FR received by a plant depends on the size, nearness, and number of competing green plants.

In 1985, Caludio M. Ghersa and Ballaré positioned an integrating cylinder in *D. ferox* canopies as if it were the stem of a seedling and observed

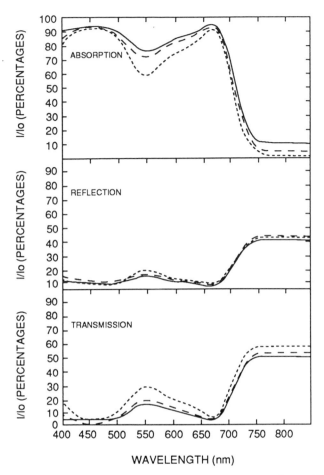

FIGURE 26-7   Optical properties of soybean leaves. (.....), expanding unifoliate; (----), expanded first trifoliate; (———), expanded trifoliate from full-sun, field-grown plant. (From Kasperbauer, 1988.)

reductions in R:FR ratios at very low leaf area index[4] values, in the absence of a reduction in photosynthetic light energy. Conventional sensors placed horizontally did not perceive these changes. The plants responded to increased density by elongating their internodes before there was mutual shading (Ballaré et al., 1987).

4. $LAI = \dfrac{\text{area of leaves}}{\text{area of ground surface}}$.

In the winter of 1986, Ballaré and Scopel placed pots of white mustard on the northern sides of grass hedges, which are never shaded in the southern hemisphere. The first internodes of plants in front of green hedges elongated 54% more than those of plants set in front of bleached hedges. Thus plants in full sunlight could distinguish between growing and dead neighbors, although the only detectable difference was in spectral properties. Proximity to mirrors that selectively reflected the FR wavelengths of sunlight also enhanced stem elongation in *D. ferox, Sinapis alba,* and *Chenopodium album,* whereas proximity to a mirror that selectively reflected R did not. This supported the hypothesis "that through the perception of the R:FR light quality via phytochrome, green plants can detect not only the actual degree of vegetation shade but also an early warning signal of oncoming competition" (Ballaré et al., 1987).

When Child detected transitory changes in mustard stem extension rates in WL after applying even brief FR pulses from a fiber optic light guide to the first internode, Smith concluded that "the rapid, and repeatable, increases in extension rate in response to transient simulated shading means that, in nature, shade-avoiding species such as mustard would be capable of reacting sensitively to the first signals of competition from neighbors" (Child and Smith, 1987).

Wanting to know how plants in the field perceive such competition, Ballaré and colleagues showed with an optic fiber that small changes in canopy density generate large changes in the amount of FR reaching the stem interior (Ballaré et al., 1989). When they placed *D. ferox* and *S. alba* seedlings into stands of similar stature, the seedlings' stems began to grow more quickly, even though they were not directly shaded by other plants. But extension rates were much less stimulated in plants whose stems wore Plexiglas collars containing a FR-absorbing solution of copper sulfate that maintained the R:FR ratio equal to that outside the canopy (Ballaré et al., 1990). Such observations could be explained in terms of the differential sensitivity of growth to light presented by fiber optic probe to leaves or internodes, shown previously in Morgan's experiments (Morgan et al., 1980). Thus early detection of neighbors "is triggered by low R:FR ratios received at the stem level and . . . this localized drop in the R:FR balance is mainly a consequence of the FR light reflected from nearby leaves" (Ballaré et al., 1990).

USDA STUDIES

At the USDA Coastal Plains Soil and Water Conservation Research Center in Florence, South Carolina, in 1983, Kasperbauer found that the upper

TABLE 26-1

FR:R Photon Fluence Rate Ratios at the Outer Surface of the Upper Canopy of Soybean Plants Grown in North–South (N–S) or East–west (E–W) rows [a][b]

Row orientation	Side of row	Far-red/red photon fluence rate ratio	
		Parallel to soil	45° Angle from soil
N–S	East	1.13 d	1.72 d
	West	3.29 c	6.14 a
E–W	North	1.51 d	3.10 c
	South	1.90 d	4.84 b

[a] From Kasperbauer et al. 1984.

[b] Measurements taken between 0852 and 0944 on July 27, 1983, near Florence, S C. FR:R ratio of incoming sunlight above the canopy was 1.05. Values followed by the same letter do not differ significantly.

parts of plants in north–south rows received higher FR:R ratios on the west side in the morning, when the sun was in the east (Table 26-1), and on the east side later in the day, when the sun was in the west. Thus the increased FR:R ratio appeared to be due mainly to FR reflected back from plants in adjacent rows (Kasperbauer et al., 1984).

Averages of FR:R readings at upper leaf surfaces on both sides of rows throughout the day were about 1.15 for north–south rows and 1.05 for east–west rows. Kasperbauer concluded that the heliotropic movements of soybean leaves, which are broad and have long petioles, might contribute to this difference, especially at the beginning and end of day (Kasperbauer, 1987). "As the sun goes to lower and lower angles," he explains, "it has less B and a greater FR:R ratio, and if much of that FR is reflected off leaves that are facing the sun and acting like directional FR reflectors, there will be a large amount of FR reflected to nearby plants. So row orientation in plants like soybean will affect the amount of FR, especially toward the end of day, which could influence the action of phytochrome during darkness."

Kasperbauer also studied the effects of row orientation on soybean plant development. Plants in north–south rows received higher FR:R ratios and were taller and produced heavier shoots and a greater yield of beans than identically spaced plants in east–west rows. In seedlings, stem length increased as distance between rows decreased and FR:R ratios rose (Kasperbauer, 1987). These results suggested that FR:R ratios affected photosynthate partitioning in the field as well as in controlled environments. Similar results were obtained with bushbeans grown on the darker soil of Kentucky (Kaul and Kasperbauer, 1988).

In wheat stands in the field, FR:R ratios in sunflecks on the soil were greatest when seedlings were closely spaced (Fig. 26-8), and the ratios increased toward sunset. Moreover, closely spaced plants developed fewer tillers than those farther apart. Because end-of-day FR treatment in controlled environments also photoreversibly reduced the number of tillers per wheat seedling, Kasperbauer suggested that "natural bioregulation of tillering (branching) in wheat is controlled by the relative amounts of FR and R acting through a sensing mechanism which signals the plant how to adapt to improve its chances for survival and for seed production under the sensed growth environment" (Kasperbauer and Karlen, 1986).

"From a survival standpoint," Kasperbauer pointed out, "partitioning more photosynthate to stem growth rather than to branches and root growth seems realistic if the plant is growing in a high population. However, if there is no competition from other plants, survival of the species could best be served by developing more branches (with more flowers and seed) and a larger root system to support development of this seed" (Kasperbauer, 1987).

END-OF-DAY VERSUS DAYTIME R:FR RATIOS

In 1987, Casal arrived in Leicester where he studied both the rapid, transitory response to lowered daytime R:FR ratio and the response to

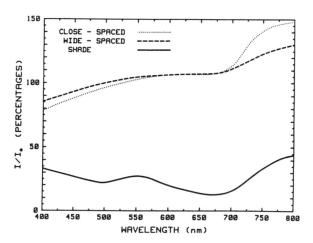

FIGURE 26-8 Spectral distribution of light in sunflecks (dashed lines) and shade (solid line) at soil level in field plots of wheat seedlings. Early afternoon, March 14, 1984, near Florence, SC. Values are expressed as percentages of incoming sunlight measured 2 m above canopy. (From Kasperbauer and Karlen, 1986.)

end-of-day FR, which accelerates growth rate during the following hours of darkness. "He wanted to know," Smith explains, "why plants do not respond to twilight, when the R:FR ratio falls from 1.1 to 0.7."

The rapid response, although caused by FR perception by the internode, occurred only if leaves were also irradiated with broad-band WL. Thus WL + FR directed at the internodes when the rest of the plant was in darkness did not accelerate stem extension, whereas FR irradiation of only the first internode followed by WL irradiation of only one primary leaf did. Moreover, the overhead source had to contain B, which, perceived by the leaf, appeared to set the sensitivity of the internode to FR and R. "This does not seem to be a requirement for blue light for photosynthesis," Smith explains. "The response appears to require transport of something from growing leaves to internodes." Because the irradiance of B is low at twilight, "the B-modulation of internode responsivity to Pfr/P_{total} could

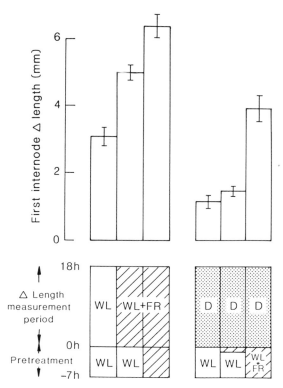

FIGURE 26-9 Internode elongation as affected by light conditions during growth measurement period (0–18 h) and preceding 7 h. WL, WL + FR, or WL terminated by a 10-min W + FR pulse. (From Casal and Smith, 1989.)

prevent internodes in mustard from exerting a full rapid response to the twilight drop in the R:FR ratio. On the other hand, during daytime when B irradiance is higher, a dramatic response to small reductions of R:FR caused by neighboring plants would be possible" (Casal and Smith, 1988). Because the drop in B at twilight probably parallels the drop in R:FR ratio, this interaction between two photoreceptors would allow the plant to ignore biologically uninformative environmental signals. "That," says Smith, "is an important feature of any sensing system."

Unlike the rapid response, the end-of-day response resulted from a lowering of Pfr/P$_{total}$ in the leaves, not the internode (Casal and Smith, 1988). But overhead FR for the last 10 min of the day was much less effective than FR applied during the last 7 h of the photoperiod (Fig. 26-9). This observation, combined with the discovery that internode growth rate in completely de-etiolated mustard seedlings declines after the transition from WL to darkness, suggested that perhaps "under natural conditions, a seedling could have a high internode extension rate during the night if the R:FR ratio was reduced by neighbor shade during the day. In contrast, a similar reduction of the R:FR ratio restricted to the end of the photoperiod . . . would be much less effective" (Casal and Smith, 1989).

ACTIVE FORM OF PHYTOCHROME

The discovery that light-grown plants, which have very little phytochrome, are sensitive to small changes in the phytochrome photoequilibrium, led Smith to question the dogma that Pfr is the active form of phytochrome—or, at least, the only active form (Smith, 1983). He proposed that Pfr/P$_{total}$ rather than the absolute amount of Pfr may determine stem elongation rate in light-grown plants, and he pointed to Fig. 26-3 as evidence. This can be interpreted as a linear relationship between Pfr concentration and extension growth only if plants under different R:FR ratios (Fig. 26-10) have the same amount of phytochrome. "Since Pfr destruction is a first order reaction," he said, applying data from dark-grown plants, "the rate of loss of total phytochrome is greater under light sources which maintain a high Pfr/P$_{total}$, than under those establishing low Pfr/P$_{total}$" (Smith, 1986). Thus under a R:FR ratio of 2.28, a plant would be expected to have a large percentage of a small amount of phytochrome in the Pfr form, whereas a ratio of 0.18 would produce a small percentage of a much larger amount of phytochrome as Pfr. Such a situation would be unlikely to yield a linear relationship when Pfr/P$_{total}$ was plotted against the log of extension growth rate.

FIGURE 26-10 *Chenopodium album* seedlings grown in four cabinets that provided equal
amounts of photosynthetically active radiation but differing R:FR ratios. (From Smith, 1986.
Reproduced by permission of Kluwer Academic Publishers.) (See also Fig. 26-3.)

Actual phytochrome levels in green plants could not be determined, but
they could be assayed in light-grown seedlings cultivated in the presence of
the herbicide norflurazon, which prevents chlorophyll accumulation (see
Chapter 31). When 7-day-old maize seedlings were transferred from continuous WL (with a R:FR ratio resembling daylight) to light cabinets, the
total amount of phytochrome in the second leaf changed substantially,
accumulating under sources that established low percentages of Pfr and
decreasing when Pfr percentages were higher.[5] Daily increments in leaf
length were inversely proportional to estimated phytochrome photoequilibria but inversely proportional to the amount of Pfr only during the first
24 h. "These data show that growth rate is not a function of the Pfr amount
or concentration," Smith concluded. "To account for phytochrome action
in light-grown plants, therefore, we must abandon the present doctrinaire
adherence to the view that Pfr is the only active form of phytochrome"

5. Proportions of Pfr could not be measured directly; the data were obtained from dark-
grown maize leaves, which had about 50 times more phytochrome than the light-grown
leaves.

(Smith, 1981). A decade later, it is known (see Chaper 31) that the pool of phytochrome measured in these experiments is almost certainly not the pool responsible for R:FR ratio perception and modulation of extension growth (Smith and Whitelam, 1990).

The Freiburg workers had shown that anthocyanin synthesis correlates quantitatively with the amount of Pfr and that lipoxygenase synthesis is controlled by a Pfr threshold (Oelze-Karow and Mohr, 1973; Drumm and Mohr, 1974; Steinitz et al., 1979). When Reinhard Schmidt managed to enhance phytochrome-mediated anthocyanin synthesis by irradiating mustard seedlings *before* their development came under phytochrome control (Schmidt and Mohr, 1981), Mohr realized that a comparison of R and FR pretreatments might distinguish between response to Pfr and response to Pfr/P_{total}.

Both R and FR irradiation for 21 or 24 h after sowing, terminated by a saturating FR pulse, strongly increased the amounts of anthocyanin induced by a subsequent R pulse given before about 30 h after sowing. But the R pretreatment drastically decreased the total phytochrome content of cotyledons, whereas the FR pretreatment did not. When anthocyanin levels were related to Pfr concentration, the kinetics of the response were very similar for the two pretreatments (Fig. 26-11). But when they were related to the phytochrome equilibrium established by the R pulse (0.8), the curve after FR pretreatment remained above that for the dark controls,

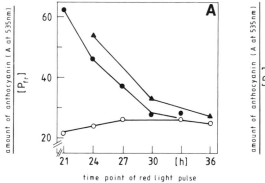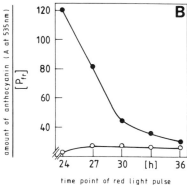

FIGURE 26-11   Time course of responsiveness to a saturating R pulse applied at points indicated. Response is referred to amount of Pfr established by R pulse. (A) FR pretreatment; (B) R pretreatment. Closed circles, 21-h pretreatment; closed triangles, 24-h pretreatment; open circles, no light pretreatment. (From Schmidt and Mohr, 1982.)

whereas that after the R pretreatment crossed the control curve (Fig. 26-12). Thus the data were meaningful only if "anthocyanin synthesis of the mustard seedling, dark-grown or light-pretreated, responds to the amount of Pfr" (Schmidt and Mohr, 1982).

At Reading University, Chris A. H. Kilsby and Chris Johnson also determined that total measurable phytochrome varied according to light quality, when they studied 4-day-old light-grown mustard seedlings (Kilsby and Johnson, 1981). But in partially norflurazon-bleached *mature* mustard and impatiens plants, they detected similar levels of phytochrome under different R:FR ratios and even under monochromatic R or FR sources. Thus there seemed to be only a stable pool of the pigment, making Pfr content entirely dependent on light quality and equal to Pfr/P_{total} (Kilsby and Johnson, 1982). But transgenic plants with abnormally high phytochrome levels (see Fig. 29-5) are much shorter in WL than plants with normal, low levels of the pigment (Boylan and Quail, 1989; Keller et al., 1989), suggesting that total phytochrome content does affect extension growth.

HIGH FLUENCE RATE EXPERIMENTS

Smith hoped to settle this argument by testing perception of R:FR ratio during abrupt changes in fluence rate. He reasoned that, if a large proportion of phytochrome exists as intermediate(s) at high fluence rates, as

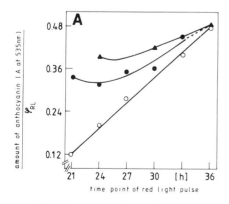
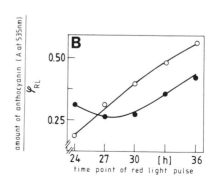

FIGURE 26-12 As Fig. 26-8 but with response referred to Pfr/P_{total} established by R pulse. (From Schmidt and Mohr, 1982.)

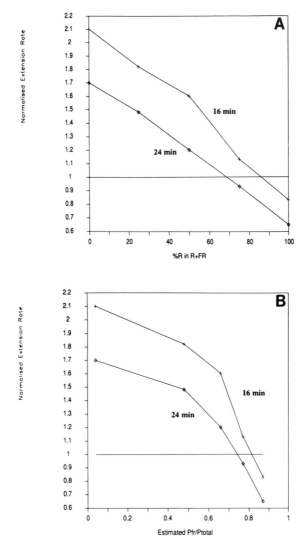

FIGURE 26-13   Mustard seedling extension rate responses to mixed R and FR expressed as (A) function of R:FR ratio and (B) estimated Pfr/P$_{total}$. (From Smith, 1990.)

indirect measurements suggested (Smith et al., 1988), Pfr concentration would fall as fluence rate rose. If the stem elongation response were a function of Pfr concentration, therefore, a step-up to a higher fluence rate should trigger a long-term acceleration of growth rate, whereas growth rate would decrease if fluence rate was stepped down.

The data provided no clear-cut answer. A sudden, 100-fold decrease in fluence rate (from ~1,400 to ~14 μmol m^{-2} s^{-1}) temporarily decreased stem extension rates, but the rate then increased and decreased again, only to equal the original after about 60 min. A step-up transition between the two fluence rates triggered a rapid increase in stem extension rate followed by a decline. Thus the data did not suggest that the concentration of Pfr controls stem extension rates, but they were "not amenable to facile interpretation" (Smith, 1990).

Because of the very high irradiances used, Smith could determine, for the first time, whether phytochrome controls stem extension rates at fluence rates approaching those of sunlight "as opposed to the twilight that is characteristic of most growth rooms." Child had previously observed a strict, inverse linearity between response and Pfr/P_{total} values between about 0.14 and about 0.65. But total fluence rates in those experiments never exceeded 250 μmol m^{-2} s^{-1} (equivalent to light shade) (Child and Smith, 1987).

When R + FR mixtures were supplied at a total fluence rate of 1,400 μmol m^{-2} s^{-1}, response was inversely proportional to Pfr/P_{total} values between 0.87 and 0.5, but not at lower values (Fig. 26-13B). A computer simulation provided one possible explanation: At high fluence rates, the actual Pfr/P_{total} (or [Pfr]) may be much lower than that measured after the light is turned off, progressively reducing values for Pfr/P_{total} with increasing Pfr/P_{total} (Smith, 1990). Response was, however, always proportional to the percentage of R in the R + FR mixture (Fig. 26-13A), suggesting that phytochrome was in some manner mediating the response to light-quality changes in mustard seedlings. Smith concluded "that phytochrome is able to control extension growth at fluence rates approaching those of summer sunlight, thereby providing the capacity to sense the presence of neighbouring vegetation *before* shading seriously compromises photosynthesis" (Smith, 1990).

REFERENCES

Ballaré, C. L., Sánchez, R. A., Scopel, A. L., Casal, J. J., and Ghersa, C. M. (1987). Early detection of neighbour plants by phytochrome perception of spectral changes in reflected sunlight. *Plant Cell Environ.* 10, 551–557.

Ballaré, C. L., Sánchez, R. A., Scopel, A. L., and Ghersa, C. M. (1988). Morphological responses of *Datura ferox* seedlings to the presence of neighbors. Their relations to canopy microclimate. *Oecologia* 76,288–293.

Ballaré, C. L., Scopel, A. L., and Sánchez, R. A. (1989). Photomodulation of axis extension in sparse canopies. Role of the stem in the perception of light quality signals of stand density. *Plant Physiol.* 89,1324–1330.

Ballaré, C. L., Scopel, A. L., and Sánchez, R. A. (1990). Far-red radiation reflected from adjacent leaves: An early signal of competition in plant canopies. *Science* 247,329–332.

Black, M. (1969). Light-controlled germination of seeds. *Symp. Soc. Exp. Biol.* 23,193–217.

Boylan, M. T., and Quail, P. H. (1989). Oat phytochrome is biologically active in transgenic tomatoes. *Plant Cell* 1,765–773.

Casal, J. J., and Smith, H. (1988). The loci of perception for phytochrome control of internode growth in light-grown mustard: Promotion by low phytochrome photoequilibria in the internode is enhanced by blue light perceived by the leaves. *Planta* 176,277–282. Copyright 1988, Springer-Verlag.

Casal, J. J., and Smith, H. (1989). The "end-of-day" phytochrome control of internode elongation in mustard: Kinetics, interaction with the previous fluence rate, and ecological implications. *Plant Cell Environ.* 12,511–520.

Casal, J. J., Sánchez, R. A., and Deregibus, V. A. (1986). The effect of plant density on tillering: The involvement of R/FR ratio and the proportion of radiation intercepted per plant. *Environ. Exp. Bot.* 26,365–371.

Child, R., and Smith, H. (1987). Phytochrome action in light-grown mustard: Kinetics, fluence-rate compensation and ecological significance. *Planta* 172,219–229.

Cumming, B. G. (1963). The dependence of germination on photoperiod, light quality, and temperature in *Chenopodium* spp. *Can. J. Bot.* 41,1211–1233.

Drumm, H., and Mohr, H. (1974). The dose response curve in phytochrome-mediated anthocyanin synthesis in the mustard seedling. *Photochem. Photobiol.* 20,151–157.

Grime, J. P. (1965). Shade avoidance and shade tolerance in flowering plants. *In* "Light as an Ecological Factor" (R. Bainbridge, G. C. Evans, and O. Rackham, eds.), pp. 187–207. Wiley, New York.

Grime, J. P. (1981). Plant strategies in shade. *In* "Plants and the Daylight spectrum" (H. Smith, ed.), pp. 159–186. Academic Press, London.

Hendricks, S. B., and Borthwick, H. A. (1963). Control of plant growth by light. *In* "Environment Control of Plant Growth" (L. T. Evans, ed.), pp. 233–263. Academic Press, New York, London.

Holmes, M. G., and Smith, H. (1975). The function of phytochrome in plants growing in the natural environment. *Nature (Lond.)* 254,512–514.

Holmes, M. G., and Smith, H. (1977a). The function of phytochrome in the natural environment—I. Characterization of daylight for studies in photomorphogenesis and photoperiodism. *Photochem. Photobiol.* 25,533–538.

Holmes, M. G., and Smith, H. (1977b). The function of phytochrome in the natural environment—II. The influence of vegetation canopies on the spectral energy distribution of natural daylight. *Photochem. Photobiol.* 25,539–545.

Holmes, M. G., and Smith, H. (1977c). The function of phytochrome in the natural environment—III. Measurement and calculation of phytochrome photoequilibria. *Photochem. Photobiol.* 25,547–550.

Jones, L. H., and Kasperbauer, M. J. (1985). Red and far-red light bioregulation of chlorophyll *a/b* ratio and chloroplast polypeptides. *Plant Physiol.* 77(suppl. 28).

Kasperbauer, M. J. (1971). Spectral distribution of light in a tobacco canopy and effects of end-of-day light quality on growth and development. *Plant Physiol.* 47,775–778.

Kasperbauer, M. J. (1987). Far-red light reflection from green leaves and effects on phytochrome-mediated assimilate partitioning under field conditions. *Plant Physiol.* 85,350–354. Quotation reprinted with permission from the American Society of Plant Physiologists.

Kasperbauer, M. J. (1988). Phytochrome involvement in regulation of the photosynthetic apparatus and plant adaptation. *Plant Physiol. Biochem.* 26,519–524.

Kasperbauer, M. J., and Hamilton, J. L. (1984). Chloroplast structure and starch grain accumulation in leaves that received different red and far-red levels during development. *Plant Physiol.* 74,967–970.

Kasperbauer, M. J., and Jones, L. H. (1985). Phytochrome regulation of chloroplast structure, polypeptides and pigments. *Int. Congr. Plant Mol. Biol 1st Abst.,* p. 84.

Kasperbauer, M. J., and Karlen, D. L. (1986). Light-mediated bioregulation of tillering and photosynthate partitioning in wheat. *Physiol. Plant.* 66,159–163.

Kasperbauer, M. J., and Peaslee, D. E. (1973). Morphology and photosynthetic efficiency of tobacco leaves that received end-of-day red or far red light during development. *Plant Physiol.* 52,440–442.

Kasperbauer, M. J., and Reinert, R. A. (1966). Biological detection of phytochrome in callus tissue of *Nicotiana tabacum* L. *Nature (Lond.)* 211,744–745.

Kasperbauer, M. J., Tso, T. C., and Sorokin, T. P. (1970). Effects of end-of-day red and far-red radiation on free sugars, organic acids and amino acids of tobacco. *Phytochemistry* 9,2091–2095.

Kasperbauer, M. J., Hunt, P. G., and Sojka, R. E. (1984). Photosynthate partitioning and nodule formation in soybean plants that received red or far-red light at the end of the photosynthetic period. *Physiol. Plant.* 61,549–554.

Kaul, K., and Kasperbauer, M. J. (1988). Row orientation effects on FR/R light ratio, growth and development of field-grown bush bean. *Physiol. Plant.* 74,415–417.

Keller, J. M., Shanklin, J., Vierstra, R. D., and Hershey, H. P. (1989). Expression of a functional monocotyledonous phytochrome in transgenic tobacco. *EMBO J.* 8,1005–1012.

Kilsby, C. A. H., and Johnson, C. B. (1981). The influence of light quality on the phytochrome content of light grown *Sinapis alba* L. and *Phaseolus aureus* Roxb. *Planta* 153,109–114.

Kilsby, C. A. H., and Johnson, C. B. (1982). The *in vivo* spectrophotometric assay of phytochrome in two mature dicotyledonous plants. *Photochem. Photobiol.* 35,255–260.

Morgan, D. C., and Smith, H. (1976). Linear relationship between phytochrome photoequilibrium and growth in plants under simulated natural radiation. *Nature (Lond.)* 262,210–212.

Morgan, D. C., and Smith, H. (1978). The relationship between phytochrome photoequilibrium and development in light grown *Chenopodium album* L. *Planta* 142,187–193.

Morgan, D. C., and Smith, H. (1979). A systematic relationship between phytochrome-controlled development and species habitat, for plants grown in simulated natural radiation. *Planta* 145,253–258.

Morgan, D. C., O'Brien, T., and Smith, H. (1980). Rapid photomodulation of stem extension in light-grown *Sinapis alba* L. *Planta* 150,95–101.

Oelze-Karow, H., and Mohr, H. (1973). Quantitative correlation between spectrophotometric phytochrome assay and physiological response. *Photochem. Photobiol.* 18,319–330.

Schmidt, R., and Mohr, H. (1981). Time-dependent changes in the responsiveness to light of phytochrome-mediated anthocyanin synthesis. *Plant Cell Environ.* 4,433–437.

Schmidt, R., and Mohr, H. (1982). Evidence that a mustard seedling responds to the amount of Pfr and not the Pfr/P$_{tot}$ ratio. *Plant Cell Environ.* 5,495–499.

Smith, H. (1975). "Phytochrome and Photomorphogenesis." McGraw-Hill, London.

Smith, H. (1981). Evidence that Pfr is not the active form of phytochrome in light-grown maize. *Nature (Lond.)* 293,163–165.

Smith, H. (1982). Light quality, photoperception, and plant strategy. *Annu. Rev. Plant Physiol.* 33,481–518.

Smith, H. (1983). Is Pfr the active form of phytochrome? *Philos. Trans. R. Soc. Lond. B* 303,443–452.

Smith, H. (1986). The perception of light quality. *In* "Photomorphogenesis in Plants" (R. E. Kendrick and G. H. M. Kronenberg, eds.), pp. 187–217. Martinus Nijhoff, Dordrecht.

Smith, H. (1990). Phytochrome action at high photon fluence rates: Rapid extension rate responses of light-grown mustard to variations in fluence rate and red:far-red ratio. *Photochem. Photobiol.* 52,131–142.

Smith, H., and Hayward, P. (1985). Fluence rate compensation of the perception of red:far-red ratio by phytochrome in light-grown seedlings. *Photochem. Photobiol.* 42,685–688.

Smith, H., and Whitelam, G. C. (1990). Phytochrome, a family of photoreceptors with multiple physiological roles. *Plant Cell Environ.* 13,695–707.

Smith, H., Jackson, G. M., and Whitelam, G. C. (1988). Photoprotection of phytochrome. *Planta* 175,471–477.

Steinitz, B., Schäfer, E., Drumm, H., and Mohr, H. (1979). Correlation between far-red absorbing phytochrome and response in phytochrome-mediated anthocyanin synthesis. *Plant Cell Environ.* 2,159–163.

Taylorson, R. B., and Borthwick, H. A. (1969). Light filtration by foliar canopies: Significance for light-controlled weed seed germination. *Weed Sci.* 17,48–51.

27 _____

The Chromophore

At the University of Saarlandes, Saarbrucken in 1968, chemist Wolfhart Rüdiger (Fig. 27-1), who had deduced the structures of several bile pigments, including phycocyanobilin and phycoerythrobilin[1], began to characterize the phytochrome chromophore. Since the phytochrome he obtained from Correll was no longer photoreversible but turned green-yellow when the pH was raised and became blue again when acidified, Rüdiger decided to use the two forms as models for Pr and Pfr.

Siegelman had cleaved the chromophore with boiling methanol, but the yield was very poor. Rüdiger therefore analyzed the chromophore while it was still attached to the protein, oxidizing it with chromic acid. Because he identified the degradation products by thin-layer chromatography, only 1 nmol of starting material was required (Rüdiger, 1968).

Chromate quickly discolored the blue and yellow samples, identifying the chromophore as an open-chain bilin, because porphyrins and chlorophylls are much more resistant to chromate oxidation. The degradation products confirmed the pyrrolic nature of the chromophore and established the structures of rings B and C (Fig. 27-2) (Rüdiger and Correll, 1969). But the data from rings A and D proved to be misleading.

The products of both the acid and alkaline forms had identical β substituents, which cast doubt on Siegelman's hypothesis that the A and D rings of Pr and Pfr differed in these positions (see Fig. 17-1). Rüdiger did believe, however, that the two chromophores differed in chemical structure (Rüdiger and Correll, 1969; Grombein et al., 1975) rather than in mere conformation (Burke et al., 1972).

---

1. Phycocyanobilin is the isolated chromophore of phycocyanin. Phycoerythrobilin is the isolated chromophore of phycoerythrin.

FIGURE 27-1   Wolfhart Rüdiger, 1989. Photo by
Linda Sage.

CHEMICAL STUDIES

At the University of Munich in 1971, Rüdiger began to purify small oat
phytochrome. But when he oxidized the pigment with chromic acid, he
obtained no product from ring A, and that from ring D differed from the
oxidation product of denatured phytochrome. Its identification as methyl-
vinyl maleimide indicated that ring D had a vinyl (-CH=CH$_2$) group in
place of the ethyl (-CH$_2$-CH$_3$) group in phycocyanin (Rüdiger, 1972).

The A ring was identified much later, after Rüdiger managed to cleave it
from small phytochrome with a combination of chromic acid and ammonia. The product—from both Pr and Pfr—was methylethylidene succinimide, which could only have formed if ring A was hydrogenated (Fig.
27-2). Ring A must also have been linked to the apoprotein by a thioether
linkage at carbon-3' (Klein et al., 1977), because Rüdiger's group had
previously determined that methylethylidene succinimide formed under
the experimental conditions only if the sulfur substituent was on that
particular carbon (Schoch et al., 1974).

The α substituents, although destroyed by microdegradation, could be
tentatively identified through spectroscopic measurements after unfolding
the polypeptide chain. A large number of semi-synthetic tetrapyrroles
prepared from bilirubin were used as reference samples. The studies confirmed the close similarities between the chromophore in phytochrome and
phycocyanobilin and A-dihydrobiliverdin and supported the assignment of
a vinyl group to ring D. The free base, cation, and zinc complex of the Pr

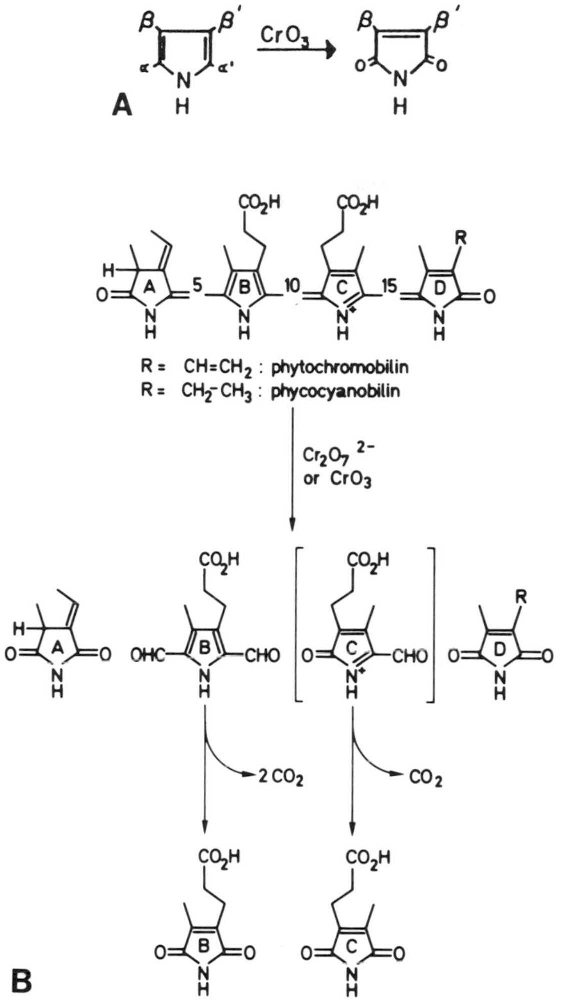

FIGURE 27-2 (A) Chromic acid oxidizes α substituents of pyrrole rings.(From Rüdiger, 1972.) (B) Chromic acid and chromate degradation products of phytochrome chromophore (phytochromobilin) and phycocyanobilin. (From Rüdiger, 1986. Reproduced by permission of Kluwer Academic Publishers.)

chromophore had absorption maxima 10 nm longer than those of phycocyanin in the absence of oxygen (Grombein et al., 1975). The same red shift was observed when bile pigments known to contain a vinyl group were compared with ethyl- bile pigments (Rüdiger and O'Carra, 1969).

The proposed structure for phytochromobilin (Fig. 27-2) was confirmed directly in 1980. In 1979, University of Berlin chemists Albert Gossauer and Jens-Peter Weller synthesized a racemic mixture of a bile pigment with that formula (Weller and Gossauer, 1980). At the same time, Rüdiger's group managed to obtain free phytochrome chromophore—phytochromobilin—using hydrogen bromide in trifluoroacetic acid, a procedure developed by Kroes for cleavage of phycocyanobilin from phycocyanin (Kroes, 1970b). This approach failed to release the chromophore from phytochrome, but it was successfully applied to phytochrome chromopeptides. Comparing the spectral properties of phytochromobilin dimethyl ester with those of the synthetic phytochromobilin dimethyl ester prepared by Gossauer's group, Rüdiger concluded that the hypothetical formula for phytochromobilin (Fig. 28-2) was correct (Rüdiger et al., 1980).

NMR SPECTRA FOR PR

At the University of California, Berkeley, J. Clark Lagarias (Fig. 27-3) decided to characterize the phytochrome chromophore using NMR spectroscopy. Because phytochrome itself is much too large to yield interpretable spectra, he digested the chromoprotein with pepsin. Fry and Mumford at du Pont had isolated an 11-residue chromopeptide in this fashion. Its sequence was Leu-Arg-Ala-Pro-His-(Ser,Cys)-His-Leu-Gln-Tyr, and the chromophore was possibly linked to the cysteine residue (Fry and Mumford, 1971).

Model compounds were also required, and Lagarias obtained a seven-residue chromophore-containing peptide from C-phycocyanin from Alexander N. Glazer. C-phycocyanin has two distinct polypeptide chains, α and β. The α chain bears one chromophore, the β chain two, and the α and β1 chromophores have similar NMR spectra, whereas that of the β2 chromophore is different. Fortunately for Lagarias, Glazer had prepared a β1-derived chromopeptide, which was more similar to the chromopeptide derived from phytochrome.

Lagarias analyzed the samples with a 270-MHz NMR spectrometer. Over the next year and a half, he also synthesized the seven-residue phycocyanin peptide, obtained its NMR spectrum, and subtracted that from the spectrum of the chromopeptide. This produced data for the C-phycocyanin chromophore and its linkage to the apoprotein (Lagarias et al., 1979).

After stockpiling 614 mg of phytochrome while finishing his doctoral thesis, Lagarias recorded in his lab notebook:

FIGURE 27-3   J. Clark Lagarias, 1989. Photo by
Linda Sage.

Monday, September 10, 1979: Precipitated all 614 mg with ammonium
sulfate, then brought it up in 5% formic acid overnight. *Tuesday, September 11:* Digested the protein with pepsin. Applied the digest to a Bio-Gel
column. There were four major peptide fractions. *Monday, September 17:*
Decided to shorten the 11-residue peptide to simplify the NMR. *Wednesday, September 26:* Generated an octapeptide with thermolysin. *Thursday, September 27:* Applied the thermolysin digest to a Sephadex G–50
column and collected the major peak. *Saturday, September 29:* Ran this
peak on high-performance liquid chromatography and collected two major
fractions. *Monday, October 8:* Sent these fractions to Al Smith at the
University of California-Davis, who sequenced them on October 24. One
was the 11-residue peptide and the other the octapeptide. The undecapeptide's sequence [Fig. 27-4] was as Fry and Mumford had suggested. The
octapeptide lacked the three last residues. Because the blue color could be
extracted from the preparation only after the first seven residues had been
removed, cysteine was the chromophore-bearing residue. *Week of October 8:* Took the rest of the peptide preparation to Willy Shih, who helped
me collect 270 MHz ^{1}H NMR spectra. The analysis worked beautifully.
The phytochrome chromopeptide was perfectly analogous to that of the
β1 chromopeptide from phycocyanin, so we could assign every band.
These analyses established the structure of the A ring and its linkage
to the peptide [Fig. 27-4], and double irradiation experiments confirmed
the presence of a vinyl group on the D ring (Lagarias and Rapoport,
1980).

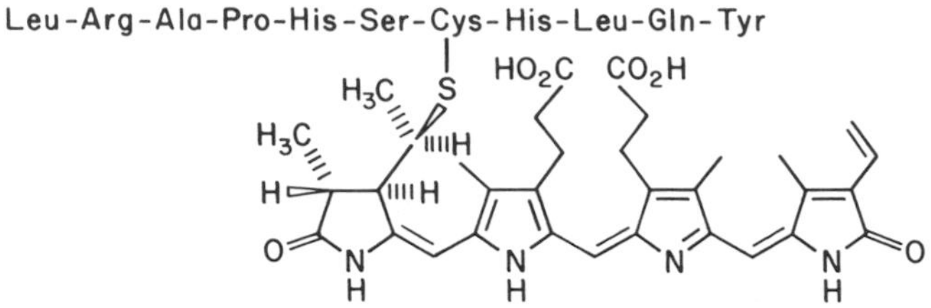

Leu-Arg-Ala-Pro-His-Ser-Cys-His-Leu-Gln-Tyr

FIGURE 27-4 Structure of phytochrome chromophore attached to an undecapeptide, deduced from NMR spectra. Pyrrole rings A to D from left to right. (From Lagarias and Rapoport, 1980. Reprinted with permission from the *Journal of the American Chemical Society* 102, 4821–4828. Copyright 1980 American Chemical Society.)

STRUCTURE OF PFR

In Rüdiger's lab, Siegbert Grombein obtained different light absorption spectra from unfolded Pr and Pfr—if he worked in the dark. In the light, denatured Pfr mimicked denatured Pr, presumably because of photoconversion. The denatured chromoprotein was not photoreversible, Pr being stable in the light under acid conditions. This suggested a chemical difference between the two forms in the α substituents.

Unfolding the apoprotein under acid conditions moved the absorption maximum of Pr from 662 to 675 nm but of Pfr from 718 (small phytochrome) to 615 nm. Thus denatured Pfr absorbed maximally at a lower rather than a higher wavelength than Pr. This curious shift led the group to conclude, as Kroes (1970a,b) had done, that the long-wavelength absorption of Pfr arises from interactions between chromophore and apoprotein (Grombein et al., 1975).

In 1981, the group characterized the chromopeptide prepared en route to free phytochromobilin; it was spectrally identical to denatured Pr (Brandlmeier et al., 1981). They also digested Pfr with pepsin, obtaining a photosensitive chromopeptide with absorption maxima very close to those of denatured Pfr. The Pfr and Pr peptides separated from each other on Bio-Gel plus silica gel columns, differing in chromatographic behavior as well as spectral properties.

Thinking about chemical reactions that might shift the absorption maxima of linear tetrapyrroles, Rüdiger considered photoaddition/elimination, protonation/deprotonation, and *cis-trans* isomerization at a methine bridge. The differing chromatographic behaviors of Pr and Pfr peptides favored the first possibility. "Our presumption, which was wrong," Rüdiger says, "was that something was sticking to Pfr to change the mobility of the chromopeptide. But we found nothing." Moreover, photoxidation or photoaddition of model compounds yielded products that absorbed at inappropriate wavelengths (Thümmler et al., 1981).

During pioneering studies of phytochrome phototransformation at Brandeis University, Henry Linschitz had attributed spectral changes during flash photolysis to chromophore isomerization (Linschitz et al., 1966). Hendricks and Sieglman had proposed that "*cis* and *trans* configurations are formally possible for each of the neighboring pyrrole rings of a bilatriene although none has been reported" (Hendricks and Siegelman, 1967). Rüdiger was able to investigate this possibility after University of Vienna chemist Heinz Falk prepared the first synthetic Z,E isomers[2] of bile pigments, including a bilatriene (Falk and Grubmayr, 1977).

In collaboration with Hugo Scheer, Rüdiger's group used a procedure Falk had applied to biliverdins (Falk et al., 1980) to convert (ZZZ)-A-dihydrobilindione to its (ZZE)-isomer (Fig. 27-5), thus obtaining model compounds with the same ring structure as phytochrome (Kufer et al., 1982). They also isomerized phycocyanin peptide and Pr and Pfr peptides, although they were unable to purify the phytochrome (E)-chromopeptide.[3]

The spectral properties of the Pr and Pfr peptides were strikingly similar to those of the model compounds. At low pH, the (E)-isomers of A-dihydrobilindione, phycocyanin peptide, and the phytochrome peptides all had long-wavelength absorption maxima shorter than those of the ZZZ isomers. Moreover, they absorbed more intensely in this part of the spectrum. (E)- and ZZZ isomers had almost identical short-wavelength maxima, but the (E)-isomers absorbed less strongly than the ZZZs in this region. Corresponding exactly, Pfr peptide had a shorter but more intense maximum than Pr peptide in the long wavelengths and an identical but less intense maximum in the B region.

The (E)-isomers and Pfr peptide were also more polar than the ZZZ isomers and Pr peptide. Moreover, Pfr peptide and the (E)-chromopeptides were rapidly converted to Pr peptide or ZZZ-like chromopeptides by light, even reverting in the dark refrigerator. Thus "the Pfr chromopeptide behaves like an E isomer of the Pr chromopeptide, which itself behaves like 'normal' Z models" (Thümmler and Rüdiger, 1983).

---

2. Prefaces Z and E refer to the configurations of the methine bridges. Z indicates that the two pyrrole rings on either side of a bridge are in the *cis* configuration; E indicates that they are in the *trans* configuration. Because there are three bridges in bilatrienes, three prefaces are necessary, and they refer to C-5, C-10, and C-15 in that order. ZZZ and ZZE bilatrienes are shown in Fig. 27-6 with the A ring at the upper right in each case. The conversion of a Z bridge to an E bridge involves rotation around a double bond. This cannot normally happen at room temperature but is possible when the pigment is excited by light. The isomeric product is then stable in the ground state.

3. In which one of the methine bridges has an E configuration.

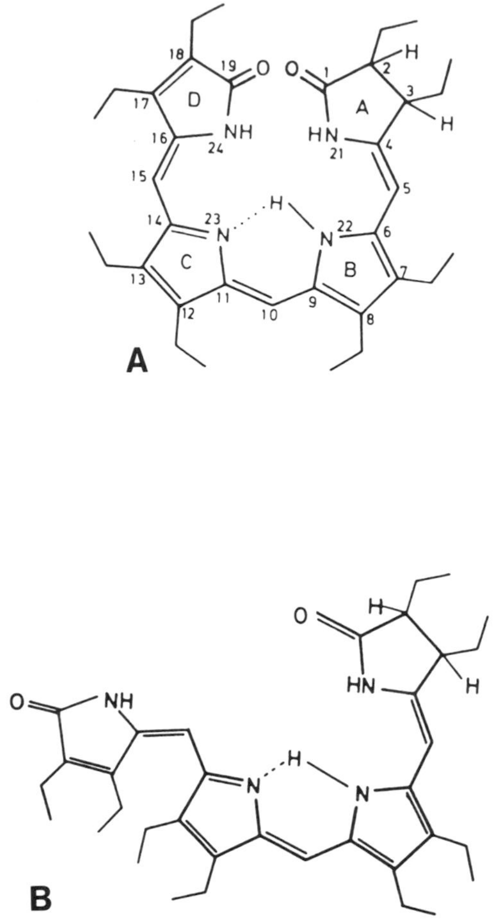

FIGURE 27-5   ZZZ (*above*) and ZZE (*below*) isomers of a bilatriene. (From Kufer et al., 1982.)

The Munich workers were unable to identify the isomerization site, but the experimental procedure was known to produce bilindiones with the E configuration at C-15. And because biliproteins—unlike free bilins—have typical double bonds only at the 4,5 and 15,16 positions, Pfr would have to be an EZZ or a ZZE bilatriene.

NMR SPECTRA FOR PFR

After 6 months of phytochrome preparation, doctoral candidate Fritz Thümmler and physical chemist Edmund Cmiel were able to obtain NMR spectra from both Pr and Pfr peptides, using a high-resolution instrument at Firma Bruker in Karlsruhe. During further work in Munich, which took another half year, the paper by Lagarias and Rapoport was published. "So it was very easy for us to attribute our Pr data to what they saw,"Rüdiger explains. "But our Pfr data were new."

Comparing the NMR spectra of the phytochrome peptides with those from phycocyanin peptide and other model compounds, the group identified the resonances of the methine bridge protons. By comparing Pfr peptide with Pr peptide, they identified features specific to the (E)- and (Z)-configurations (Thümmler et al., 1983). Comparison with 15E, 15Z, 4E, and 4Z model chromophores revealed that Pr peptide had a 15Z configuration and Pfr peptide a 15E configuration (Rüdiger et al., 1983). Thus the Pr peptide was a ZZZ isomer and the Pfr peptide a ZZE isomer (Fig. 27-6), requiring the 15,16 double bond to rotate during photoconversion. Dark relaxation of the protein presumably followed this isomerization, generating the long-wavelength absorption of Pfr.

CHROMOPHORE BIOSYNTHESIS AND HOLOPHYTOCHROME ASSEMBLY

Until the late 1980s, there was little information about the synthesis and attachment of the phytochrome chromophore. But in 1984, Gardner and Holly Gorton, at a Shell research center in California, heard that *gabaculine*[4] inhibited chlorophyll formation. This inhibition was ameliorated by 5-aminolevulinic acid (ALA) (Flint, 1984), a precursor of the chlorophyll pyrrole rings. To test whether ALA is also a precursor of phytochrome, Gardner and Gorton germinated pea seeds in the presence of gabaculine. The resulting seedlings had only about one-tenth as much spectrophotometrically detectable phytochrome as untreated seedlings (Gardner and Gorton, 1985). The inhibitor did not affect synthesis of immunodetectable phytochrome, however (Jones et al., 1986; Konomi and Furuya, 1986), suggesting that the pea phytochrome apoprotein is stable and that its synthesis is not coordinated with that of the chromophore. "This meant there could be chromophoreless phytochrome pools," says Alan M. Jones, who performed his study in Quail's lab, "and it served as a warning to those using only spectrophotometric or immunoblot analysis alone for quantitating phytochrome levels."

When Lagarias and Tedd D. Elich fed radiolabeled ALA to explants of gabaculine-treated oat seedlings, the label appeared in a chromoprotein

4. Gabaculine, secreted by *Streptomyces toyocaensis*, is 5-amino-1,3-cyclohexadienyl-carboxylic acid.

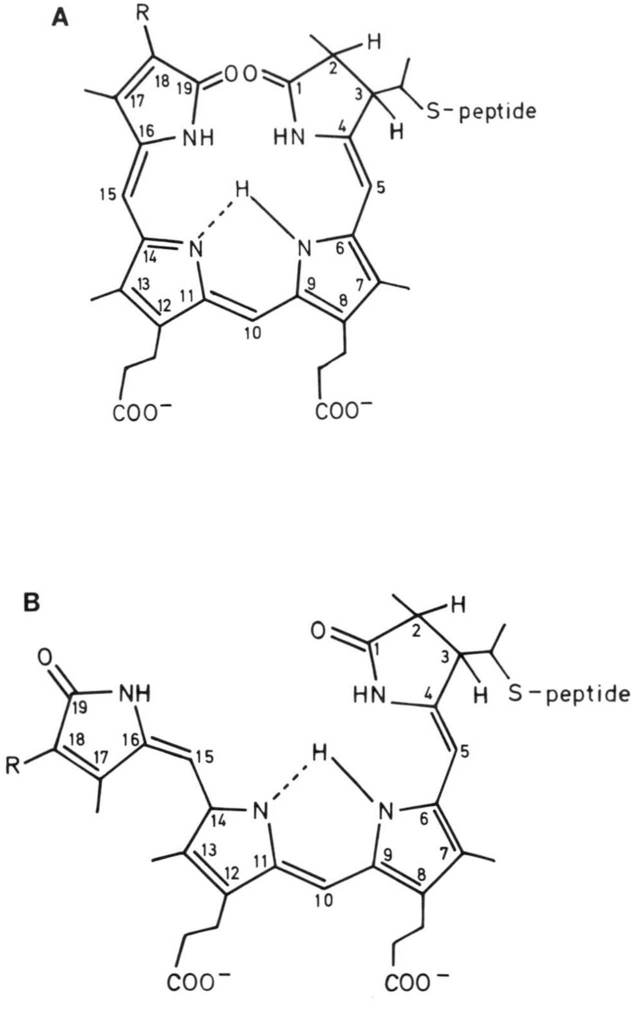

FIGURE 27-6 (A) Pr chromophore and (B) Pfr chromophore of phytochrome. R = CH=CH$_2$. (From Rüdiger, et al., 1985.)

the size of phytochrome that was precipitated by antiphytochrome antiserum. Moreover, its incorporation paralleled the increase in spectrophotometrically detectable phytochrome. Biliverdin, known to be a precursor of phycocyanobilin, also overcame gabaculine inhibition. Because mammals convert ALA to heme, the colored part of hemoglobin, and excrete heme by first breaking it open to biliverdin, Lagarias proposed that a pathway involving ALA, heme, and biliverdin might enable plants to synthesize the phytochrome chromophore (Elich and Lagarias, 1987).

Using a more potent inhibitor of ALA synthesis, aminohexynoic acid (AHA) (Elich and Lagarias, 1988), Elich determined that radiolabeled biliverdin IXa (Fig. 27-7), one of the four isomers that can arise from heme, is incorporated into phytochrome and confers normal spectral properties. Because the next bilatriene in the proposed pathway, phytochromobilin, was unavailable, he substituted phycocyanobilin (PCB). Explants incorporated PCB into a photoreversible phytochrome with a unique, blue-shifted spectrum, supporting the deduction that phytochromobilin is normally the immediate precursor of the phytochrome chromophore (Elich et al., 1989).

Wanting to see if the PCB–apophytochrome complex could form *in vitro,* Elich incubated the bilatriene with soluble extracts of AHA-grown seedlings. Within a couple of minutes he saw the adduct's tell-tale, blue-shifted spectral signature. The magnitude of the absorbance increase revealed that all the apophytochrome in the extract had combined with bilin. When the product was electrophoresed under conditions that destroy noncovalent linkages, it fluoresced orange in the presence of zinc ions (Berkelman and Lagarias, 1986), showing that it was a covalently linked biliprotein. Biliverdin failed to yield a spectrally active or zinc-fluorescent chromoprotein, revealing that conversion to phytochromobilin (or PCB) preceded attachment.

Surprisingly, PCB attached to apophytochrome adsorbed onto antiphytochrome immunoglobulin-coated Sepharose beads. Thus "phytochrome assembly *in vivo* appears to be mediated by the apoprotein itself and requires no additional enzymes or cofactors" (Elich and Lagarias, 1989).

To test this remarkable conclusion, biochemist Donna M. Lagarias spent a 2-month sabbatical in her husband's lab. "Since it's always possible to have a copurifying ligation enzyme that comes down with the protein," Lagarias says, "we had to address the question through cloning. Quail sent us two incomplete clones that we cut and pasted into a full-length coding sequence."

When combined with a piece of regulatory DNA, the sequence could be transcribed into mRNA and then translated into a small amount of polypeptide. The synthetic apophytochrome, entirely free of other plant proteins, was still able to combine with PCB, and no cofactors were necessary, although the yield was less than 10%.

There was insufficient adduct for spectral studies, but knowing that Pr and Pfr disintegrate into different fragments when attacked by proteolytic enzymes (See Chapter 28), Donna Lagarias digested a portion of PCB–apophytochrome reaction mixture after R treatment and another after FR treatment. Electrophoresis yielded Pr and Pfr peptide patterns, whereas electrophoresis after a R/FR treatment yielded only the Pr pattern, indicating photoreversibility of PCB-apophytochrome (Lagarias and Lagarias, 1989). "This shows very conclusively," Lagarias says, "that phytochrome apoprotein, given the chromophore, will self-assemble."

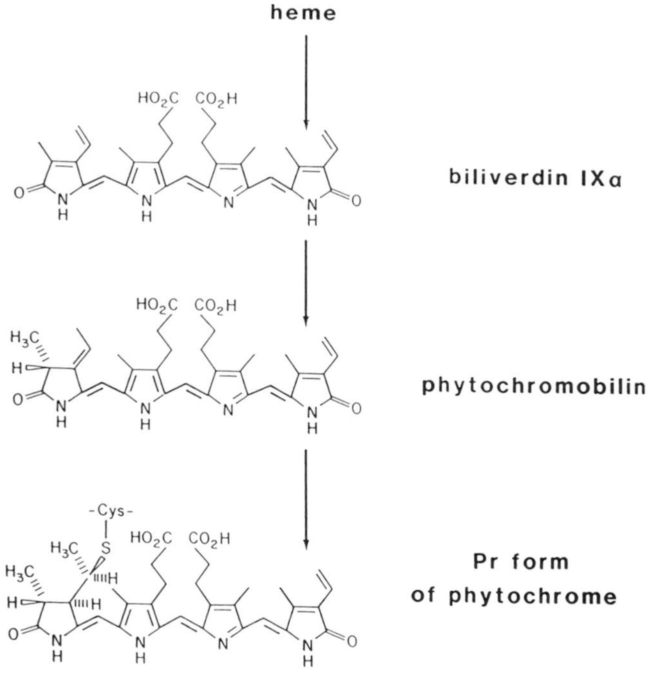

FIGURE 27-7   Proposed latter steps of pathway of phytochrome chromophore biosynthesis. (From Elich et al., 1989.)

REFERENCES

Aramendia, P. F., Ruzsicska, B. P., Braslavsky, S. E., and Schaffner, K. (1987). Laser flash photolysis of 124-kdalton oat phytochrome in H_2O and D_2O solutions: Formation and decay of I_{700} intermediates. *Biochemistry* 26,1418–1422.

Berkelman, T. R., and Lagarias, J. C. (1986). Visualization of bilin-linked peptides and proteins in polyacrylamide gels. *Anal. Biochem.* 156,194–201.

Brandlmeier, T., Scheer, H., and Rüdiger, W. (1981). Chromophore content and molar absorptivity of phytochrome in the Pr form. *Z. Naturforsch.* 36c,431–439.

Braslavsky, S. E. (1984). The photophysics and photochemistry of the plant photosensor pigment phytochrome. *Pure Appl. Chem.* 56,1153–1165.

Brock, H., Ruzsicka, B. P., Arai, T., Schlamann, W., Holzwarth, A. R., Braslavsky, S. E., and Schaffner, K. (1987). Fluorescence lifetimes and relative quantum yields of 124-kDa oat phytochrome in H_2O and D_2O solutions. *Biochemistry* 26,1412–1417.

Burke, M. J., Pratt, D. C., and Moscowitz, A. (1972). Low-temperature absorption and circular dichroism studies of phytochrome. *Biochemistry* 11,4025–4031.

Eilfeld, P., Eilfeld, P., and Rüdiger, R. (1986). On the primary photoprocess of 124-kdalton phytochrome. *Photochem. Photobiol.* 44,761–769.

Elich, T. D., and Lagarias, J. C. (1987). Phytochrome chromophore biosynthesis. Both 5-aminolevulinic acid and biliverdin overcome inhibition by gabaculine in etiolated *Avena sativa* L. seedlings. *Plant Physiol.* 84,304–310.

Elich, T. D., and Lagarias, J. C. (1988). 4-Amino-5-hexynoic acid—A potent inhibitor of tetrapyrrole biosynthesis in plants. *Plant Physiol.* 88,747–751.

Elich, T. D. and Lagarias, J. C. (1989). Formation of a photoreversible phycocyanobilin-apophytochrome adduct *in vitro*. *J. Biol. Chem.* 264,12902–12908.

Elich, T. D., McDonagh, A. F., Palma, L. A., and Lagarias, J. C. (1989). Phytochrome chromophore biosynthesis. Treatment of tetrapyrrole-deficient *Avena* explants with natural and non-natural bilatrienes leads to formation of spectrally active holoproteins. *J. Biol. Chem.* 264,183–189.

Falk, H., and Grubmayr, K. (1977). A geometrically isomeric bilatriene-abc. *Angew. Chem. Int. Ed. Eng.* 16,470–471.

Falk, H., Müller, N., and Schlederer, T. (1980). Beiträge zur Chemie der Pyrrolpigmente, 35. Mitt.: Eine regioselektive, reversible Addition an Bilatriene-abc. *Monatsh. Chem.* 111,159–175.

Flint, D. H. (1984). Gabaculine inhibits δ-ALA synthesis in chloroplasts. *Plant Physiol.* 75(suppl. 965).

Fry, K. T., and Mumford, F. E. (1971). Isolation and partial characterization of a chromophore-peptide fragment from pepsin digests of phytochrome. *Biochem. Biophys. Res. Commun.* 45,1466–1473.

Gardner, G., and Gorton, H. L. (1985). Inhibition of phytochrome synthesis by gabaculine. *Plant Physiol.* 77,540–543.

Grombein, S., Rüdiger, W., and Zimmermann, H. (1975). The structures of the phytochrome chromophore in both photoreversible forms. *Hoppe Seylers Z. Physiol. Chem.* 356,1709–1714.

Hendricks, S. B., and Siegelman, H. W. (1967). Phytochrome and photoperiodism in plants. *Compr. Biochem.* 27,211–235.

Jones, A. M., Allen, C. D., Gardner, G., and Quail, P. H. (1986). Synthesis of phytochrome apoprotein and chromophore are not coupled obligatorily. *Plant Physiol.* 81,1014–1016.

Kendrick, R. E., and Spruit, C. J. P. (1976). Phototransformations of phytochrome. *In* "Light and Plant Development" (H. Smith, ed.), pp. 31–43. Butterworths, London.

Klein, G., Grombein, S., and Rüdiger, W. (1977). On the linkages between chromophore and protein in biliproteins. VI. Structure and protein linkage of the phytochrome chromophore. *Hoppe Seylers Z. Physiol. Chem.* 358,1077–1079.

Konomi, K. and Furuya, M. (1986). Effects of gabaculine on phytochrome synthesis during imbibition in embryonic axes of *Pisum sativum* L. *Plant Cell Physiol.* 27,1507–1512.

Kroes, H. H. (1970a). The structure of the pigment phytochrome. *Physiol. Veg.* 8,533–549.

Kroes, H. H. (1970b). A study of phytochrome, its isolation, structure and photochemical transformations. *Meded. Landbouwhogesch. Wageningen* 70(18),1–112.

Kufer, W., Cmiel, E., Thümmler, F., Rüdiger, W., Schneider, S., and Scheer, H. (1982). Studies on plant bile pigments. II. Regioselective photochemical and acid catalyzed Z,E isomerization of dihydrobilindione as phytochrome model. *Photochem. Photobiol.* 36,603–607.

Lagarias, J. C., and Lagarias, D. M. (1989). Self-assembly of synthetic phytochrome holoprotein *in vitro*. *Proc. Natl. Acad. Sci. USA* 86,5778–5780.

Lagarias, J. C., and Rapoport, H. (1980). Chromopeptides from phytochrome. The structure and linkage of the Pr form of the phytochrome chromophore. *J. Am. Chem. Soc.* 102,4821–4828.

Lagarias, J. C., Glazer, A. N., and Rapoport, H. (1979). Chromopeptides from C-phycocyanin. Structure and linkage of a phycocyanobilin bound to the β-subunit. *J. Am. Chem. Soc.* 101,5030–5037.

Linschitz, H., Kasche, V., Butler, W. L., and Siegelman, H. W. (1966). The kinetics of phytochrome conversion. *J. Biol. Chem.* 241,3395–3403.

Rüdiger, W. (1968). Bile pigments: A new degradation technique and its application. *In* "Porphyrins and Related Compounds" (T. W. Goodwin, ed.), pp. 121–130. Academic Press, London, New York.

Rüdiger, W. (1972). Isolation and purification of phytochrome. Chemistry of the phytochrome chromophore. *In* "Phytochrome" (K. Mitrakos and W. Shropshire, Jr., eds.), pp. 105–141. Academic Press, London, New York.

Rüdiger, W. (1986). The chromophore. *In* "Photomorphogenesis in Plants" (R. E. Kendrick and G. H. M. Kronenberg, eds.), pp. 17–33. Martinus Nijhoff, Dordrecht.

Rüdiger, W., and Correll, D. L. (1969). Über die Struktur des Phytochrom-Chromophors und seine Protein-Bindung. *Liebigs Ann. Chem.* 723,208–212.

Rüdiger, W., and O'Carra, P. (1969). Studies on the structure and apoprotein linkages of the phycobilins. *Eur. J. Biochem.* 7,509–516.

Rüdiger, W., Brandlmeier, T., Blos, I., Gossauer, A., and Weller, J. P. (1980). Isolation of the phytochrome chromophore. The cleavage reaction with hydrogen bromide. *Z. Naturforsch.* 35c,763–769.

Rüdiger, W., Thümmler, F., Cmiel, E., and Schneider, S. (1983). Chromophore structure of the physiologically active form (Pfr) of phytochrome. *Proc. Natl. Acad. Sci. USA* 80,6244–6248.

Rüdiger, W., Eilfeld, P., and Thümmler, F. (1985). Phytochrome, the visual pigment of plants: Chromophore structure and chemistry of photoconversion. *In* "Optical Properties and Structure of Tetrapyrroles" (G. Blauer and H. Sund, eds.), pp. 349–366. Walter de Gruyter, Berlin, New York.

Schoch, S., Klein, G., Linsenmeier, U., and Rüdiger, W. (1974). Synthese und Reaktionen von Äthyliden-methylsuccinimid. *Tetrahedron Lett.* 1974,2465–2468.

Siegelman, H. W., Chapman, D. J., and Cole, W. J. (1968). The bile pigments of plants. *In* "Porphyrins and Related Compounds" (T. W. Goodwin, ed.), pp. 107–120. Academic Press, London, New York.

Thümmler, F., and Rüdiger, W. (1983). Models for the photoreversibility of phytochrome. Z,E isomerization of chromopeptides from phycocyanin and phytochrome. *Tetrahedron* 39,1943–1951.

Thümmler, F., Brandlmeier, T., and Rüdiger, W. (1981). Preparation and properties of chromopeptides from the Pfr form of phytochrome. *Z. Naturforsch.* 36c,440–449.

Thümmler, F., Rüdiger, W., Cmiel, E., and Schneider, S. (1983). Chromopeptides from phytochrome and phycocyanin. NMR studies of the Pfr and Pr chromophore of phytochrome and E,Z isomeric chromophores of phycocyanin. *Z. Naturforsch.* 38c, 359–368.

Weller, J. P., and Gossauer, A. (1980). Synthese und Photoisomerisierung des racem. Phytochromobilin-dimethylesters. *Chem. Ber.* 113,1603–1611.

CHAPTER 28

Native Phytochrome

At a plant physiology meeting in Sainte-Foy, Quebec, in June 1981, Richard Vierstra announced that large phytochrome was not intact; a small piece of the molecule was still missing.

The in-house skeptic of Quail's group in Madison, George Bolton, grilled Vierstra, convinced he was wrong. "So here were two guys from the same lab having a heated discussion," laughs Quail. "Other people were also very skeptical because the initial data were messy.[1] But to me it was as clear as night and day."

Quail had moved to the University of Wisconsin in 1979 to become associate professor of botany. There Bolton began to isolate and translate phytochrome mRNA, while Vierstra hooked phytochrome antibodies to polyacrylamide beads and bacterial cells, hoping to tag a phytochrome-binding fraction in crude pellets (Vierstra et al., 1981).

"MODIFIED" PHYTOCHROME

Both Bolton and Vierstra had observed two phytochrome bands on SDS-polyacrylamide gels, indicating two molecular sizes. And they had noticed that extraction of freeze-dried seedlings into boiling detergent to inactivate any proteolytic enzymes before immunoprecipitation lowered the chromoprotein's electrophoretic mobility (Quail, 1982).

Pratt and Maury L. Boeshore had observed that Pr obtained from pelleted Pfr electrophoresed more slowly than Pr extracted from dark tissue. It also had greater micro complement fixation activity and behaved like a larger molecule during Sephadex chromatography (Boeshoe and Pratt, 1980). They concluded that phytochrome extracted as Pfr but not Pr is enlarged *in vitro* (Boeshore and Pratt, 1981).

---

1. Because they had been obtained only in May. Also, a photographer dropped a camera on the gels.

In 1977, Epel and Benjamin Horwitz reported from Tel Aviv that Pfr in crude extracts of R-irradiated dark-grown oat seedlings absorbed maximally at 735 nm—like Pfr *in vivo*—rather than at the usual 725 nm (Horwitz and Epel, 1977). Because the 735-nm peak was generally attributed to receptor-bound Pfr, the researchers thought they had extracted phytochrome in a modified, active form. This "long-wavelength" Pfr later proved to be stable during storage and partial purification and could be formed *in vitro* in crude but not centrifuged extracts of dark-grown tissue (Epel, 1981).

124-KILODALTON PHYTOCHROME

On April 15, 1981, "it suddenly occurred to me that everything was upside down," Quail says. "I thought backwards and said that the real molecule is larger and has the spectral properties of the *in vivo* molecule and then undergoes changes after extraction to give it a small size and different spectral properties."

Because earlier studies had stressed Pfr's instability (see Chapters 11 and 18), phytochrome was routinely extracted as Pr. But Quail bet Vierstra that extracted Pfr would be larger than 120 kDa but would reduce to that size on standing. Vierstra confirmed this 2 days later.

Pr purified from dark-grown oats using Hunt and Pratt's immunoaffinity procedure (see Chapter 22) gave three closely spaced bands on SDS-polyacrylamide gels at the 118-, 114-, and 112-kDa positions.[2] But when seedlings were irradiated with R and Pfr was rapidly immunoprecipitated from the crude extract, there was only a single phytochrome band at 124 kDa, which also appeared when Pr was extracted into hot SDS-containing buffer. The extra 6–10 kDa corresponded to 50–90 amino acids.

When Pfr was extracted into SDS-free buffer, photoconverted to Pr, and then incubated before immunoprecipitation, the 124-kDa band was progressively replaced by a 118-kDa band and then a 114-kDa band, even in the cold room. PMSF, the serine protease inhibitor, prevented this size reduction, strongly suggesting the action of an endoprotease that preferentially attacked Pr (Vierstra and Quail, 1982a,b).

As the pigment diminished in size with a half-time at 20°C of only 40 min, its difference spectrum minimum shifted from 730 to 722 nm. Maintaining the pigment as Pfr and/or protecting it with PMSF prevented both the spectral shift and the size decrease (Vierstra and Quail, 1982c). Thus the labile polypeptide portion of Pfr contributed to spectral integrity.

2. The 112-kDa band formed only when phytochrome was heated in the electrophoresis buffer (Lagarias and Mercurio, 1985).

At the Max-Planck-Institute for Biochemistry in Martinsried, Lorenz Kerscher and Susanne Nowitzki independently demonstrated that native phytochrome was bigger than large phytochrome, detecting PMSF-inhibited Pr degradation in crude extracts of rye, oat, and maize seedlings. Degradation was especially rapid in preparations of frozen seedlings, a common starting material (Kerscher and Nowitzki, 1982).

Confirmation of the larger size of native phytochrome was, in fact, already in the literature. Both Bolton and Schäfer's student Klaus Gottmann had translated phytochrome mRNA by incubating poly(A) RNA[3] from dark-grown oats with radiolabeled amino acids and a cell-free protein-synthesizing system. The immunoprecipitated product had the same electrophoretic mobility as phytochrome extracted from freeze-dried tissue into boiling SDS buffer. But it migrated more slowly than immunoaffinity-purified phytochrome.

Bolton was unsure of the relationship between large phytochrome and this larger relative (Bolton and Quail, 1981). Gottmann and Schäfer attributed the increased size to posttranslational modification (Gottmann and Schäfer, 1982). But in January 1982, Bolton and Quail submitted a paper that equated the 124-kDa polypeptide with native phytochrome and the "120"-kDa species with degradation products (Bolton and Quail, 1982). So phytochrome was now intact, 23 years after it was first extracted.

PURIFICATION

Because immunoaffinity purification yielded degraded oat phytochrome, Vierstra combined the traditional hydroxylapatite column with a modification of Smith's Affi-gel Blue procedure (see Chapter 22), maintaining the pigment as Pfr and including PMSF until proteolytic enzymes were eliminated (Vierstra and Quail, 1983b). Before chromatography, he added poly(ethylenimine) (PEI) to precipitate nucleic acids and other macro-molecules (Bolton, 1979; Bolton and Pratt, 1980).

In 2 days, he could obtain a 220-fold purification with a 10%–15% yield. The product was stable for several hours at room temperature and had absorbance ratios (A_{666}/A_{280}) up to 0.97. Because it yielded no detectable amino acid derivatives on end group analysis, its amino terminus was blocked.[4] So large phytochrome, with its free amino termini, obviously lacked the amino terminus of the native molecule.

---

3. Eukaryotic messenger RNA, poly(A) RNA, has a tail of about 200 adenine nucleotides attached to the 3' end. This allows it to separate from other types of RNA during chromatography on cellulose modified with adenine's complementary nucleotide, thymine.

4. The amino group that normally lies at the beginning of a polypeptide chain is modified.

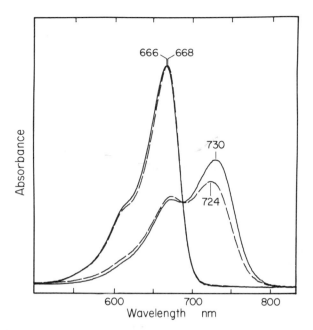

FIGURE 28-1 Absorbance spectrum for 124-kDa phytochrome. (From Vierstra and Quail, 1983a. Reproduced with permission from the American Society of Plant Physiologists.)

The amino acid composition was similar to that of large phytochrome and typical for a water-soluble molecule. Absorbance spectra had peaks at 666–668 nm for Pr and 730 nm for Pfr (Fig. 28-1). Like monocot phytochrome *in vivo*, the native oat molecule did not revert *in vitro*, even in the presence of sodium dithionite (Vierstra and Quail, 1983b). Native phytochrome from *Cucurbita pepo* also failed to revert, even though dicot phytochrome does so *in vivo* (Vierstra and Quail, 1985).

PHOTOCHEMICAL PROPERTIES

Photochemical properties of Vierstra's preparation differed significantly from those of large phytochrome, because the quantum yield ratio ($\Phi r/\Phi fr$), proportion of phytochrome in the Pfr form after saturating R irradiation ($[\mathrm{Pfr}]^{665}/\infty$), and quantum yield for the conversion of Pr to Pfr (Φr) were all higher (Table 28-1). Thus 86.2% rather than 79.1% (or 75%) of the pigment was in the Pfr form after saturating R irradiation. Molar absorption coefficients were 73,000 $\mathrm{dm}^3 \mathrm{\ mol}^{-1} \mathrm{\ cm}^{-1}$ for Pr at 666 nm and 49,000 $\mathrm{dm}^3 \mathrm{\ mol}^{-1} \mathrm{\ cm}^{-1}$ for Pfr at 730 nm (Vierstra and Quail, 1983a).

TABLE 28-1

Photochemical Characteristics of 124- and 118/114-kDa *Avena* Phytochrome[a]

Phytochrome preparation	$\Phi r/\Phi fr$	$[Pfr]_\infty^{665}$	Φr	Φfr
124 kD$_a$	1.76 ± 0.10	0.862 ± 0.005	0.17 ± 0.02	0.10 ± 0.02
118/114 kD$_a$	0.98 ± 0.02	0.791 ± 0.021	0.11 ± 0.02	0.12 ± 0.02

[a] From Vierstra and Quail, 1983a. Reproduced with permission from the American Society of Plant Physiologists.

Lagarias's group, which obtained a highly enriched native prepavation, despite purification as Pr, calculated values of 121,000 dm^3 mol^{-1} cm^{-1} for oat Pr at 668 nm and 68,000 for Pfr at 730 nm (Litts et al. 1983). To determine quantum yields, they adopted an approach-to-equilibrium method (Johns, 1969), which required actinometric measurements of light absorption by phytochrome. Butler's use of initial photoconversion rates "is only valid if Pr concentration does not change during the course of the photoreaction," Lagarias says. "But that assumption doesn't always hold under practical conditions."

The quantum yield (Φr) for the conversion of Pr to Pfr under continuous R was 0.152 ± 0.008 mol Einstein^{-1}. The value of Φfr under continuous FR was 0.069 ± 0.004 mol Einstein^{-1}. Thus the quantum yield ratio ($\Phi r/\Phi fr$), at 2.20 ± 0.01, was considerably higher than that (Table 28-1) obtained by Vierstra and Quail, who also used a less accurate method of protein determination. The mole fraction of Pfr under continuous R was very close to Vierstra and Quail's estimate—0.876 (Kelly and Largarias, 1985).

Comparative analyses on highly purified rye (prepared at the Smithsonian) and oat phytochromes led to revised molar absorption coefficients for Pr and Pfr (Lagarias et al., 1987)—both were slightly larger than previous estimates (Table 28-2). But neither these nor Vierstra and Quail's estimates were close to those (mole fraction Pfr at 664 nm = 0.76; $\Phi r/\Phi fr = 1.38$) calculated from *in vivo* measurements of dark-grown zucchini seedlings in Freiburg (Seyfried and Schäfer, 1985).

Kelly and Lagarias's value for the proportion of Pfr *in vitro* after an R irradiation was later confirmed at Leicester by Garry C. Whitelam and Mary L. Holdsworth. They obtained the first pure solution of Pfr by immunoprecipitating Pr from a Pr/Pfr equilibrium mixture with bacterial cells coated with the monoclonal antibody LAS 41, which reacted only with Pr. The Pfr solution had a larger peak at 730 nm and a smaller peak at 670 nm than a Pfr/Pr mixture. Comparing the Δ (Δ A)665:730 for the two

TABLE 28-2

Photochemical Characteristics of *In Vitro* Native Oat and Rye Phytochrome[a]

Plant source	M_r (kDa)	SAR	[P], μM	$\epsilon^{P.r.}_{\lambda max}$ ($mM^{-1} cm^{-1}$)	Φ_2	Φ_{f2}	Φ_r/Φ_{fr}	$X^{fr}_{rcd\ cq}$
Oat	124	0.98–1.0	2–3	132 ± 10	0.152 ± 0.004	0.060 ± 0.002	2.515 ± 0.046	0.876 ± 0.002
Rye	124	1.0 –1.1	2–3	132 ± 3	0.174 ± 0.001	0.074 ± 0.0004	2.343 ± 0.019	0.890 ± 0.001
Oat	124	0.98–1.0	2–3	132 ± 10	0.156 ± 0.006	0.065 ± 0.001	2.378 ± 0.073	0.874 ± 0.004
Rye	124	1.0 –1.1	2–3	132 ± 3	0.172 ± 0.004	0.078 ± 0.002	2.217 ± 0.032	0.884 ± 0.002

[a] Lagarias et al., 1987.

preparations, the researchers directly measured the mole fraction of Pfr at photoequilibrium. The value was 0.874 ± 0.0015 (Holdsworth and White-lam, 1987).

MONOCLONAL ANTIBODIES

Structural studies of native phytochrome relied heavily on the use of monoclonal antibodies (MAbs), which, unlike polyclonals, recognize only one epitope on a protein's surface. In 1981/82, Marie-Michèle Cordonnier collaborated with Pratt's group to obtain nine different MAbs against pea large phytochrome and 20 against oat large phytochrome (Cordonnier et al., 1983). Furuya's group, which was now at the National Institute for Basic Biology in Okazaki, obtained six MAbs against rye large phytochrome. All six recognized both large and small rye phytochrome, but none cross-reacted with pea phytochrome, supporting the major antigenic distinction between monocots and dicots. Only two cross-reacted with oat phytochrome, pointing to antigenic differences even between different monocot proteins (Nagatani, et al., 1983). Six MAbs to pea large phytochrome also failed to cross-react with monocot phytochromes (Furuya, 1983; Nagatani et al., 1984).

The first lab to prepare MAbs to native phytochrome was Quail's, where Susan Daniels obtained 46 MAbs against Vierstra's 124-kDa oat preparation. She classified them according to their affinity with peptides derived by degradation with endogenous proteases. Type-1 MAbs recognized an epitope that was missing from 118-kDa large phytochrome and therefore seemed to lie in the 6-kDa amino terminus. Because a 74-kDa chromopeptide from Pfr also contained this epitope, the 6-kDa terminus appeared to begin the chromophore-containing part of the molecule. Daniels assigned type-2 MAbs to an adjacent region and type-3 MAbs to a site in the carboxyl-terminal end, 88–97 kDa from the amino terminus (Daniels and Quail, 1984). Type-1 MAbs were unequivocally mapped when they were later shown to bind to synthetic peptides matching the amino-terminal sequence (Quail et al., 1987; Plumpton et al., 1988).

PRIMARY SEQUENCE

The nucleotide sequence of phytochrome mRNA was deduced by postdoctorate Howard P. Hershey, who joined Quail's group in September 1981 (Hershey et al., 1984, 1985). After enriching poly(A) RNA for phytochrome mRNA by size fractionation, Hershey synthesized complementary DNA (cDNA)[5] with the aid of reverse transcriptase. After converting the single strands into double helices with DNA polymerase, he placed the cDNA into bacterial cells, obtaining a library of copied oat messages.

To select cells from the library that contained phytochrome cDNA, the Madison workers radiolabeled single-stranded DNA copies of poly(A) RNA isolated from oat seedlings that either had never seen light or had received a phytochrome-depleting R treatment. They then noted which colonies hybridized more strongly with the "enriched probe" than with the "depleted probe."

To identify the cDNA subgroup specific for phytochrome, they cultured cells from the selected colonies, obtaining enough plasmid DNA for analysis. Translating the fraction of etiolated oat poly(A) RNA to which this DNA hybridized, they obtained protein products. Colonies contained phytochrome cDNA if they yielded an *in vitro*-translated protein that was precipitable by antiphytochrome and had the same electrophoretic mobility as the native chromoprotein.

There were two such colonies, and one contained a long—4.1 kbp[6]— insert. But this hybridized to a 4.2-kbp band of electrophoresed poly(A) RNA, revealing that a small piece of the mRNA sequence had been lost during cDNA preparation (Hershey et al., 1984).

After screening an additional 125,000 transformants, the researchers identified 47 more positive colonies. But all the inserts in their plasmids were incomplete, and they could not be combined into a complete sequence. They fell into four classes—3, 4, 5, and 6—of closely related cDNAs, suggesting that at least four closely related phytochrome genes are transcribed in oat, a hexaploid species.

The missing information for classes 3 and 4 was eventually obtained from the phytochrome gene itself. The researchers isolated genomic DNA (gDNA) from etiolated oats, cloned fragments into a bacteriophage, and identified plaques[7] containing parts of the phytochrome sequence through hybridization with a radiolabeled piece of phytochrome cDNA. Sequencing the inserts from the positive clones revealed that two contained the required pieces.

From the complete nucleotide sequences, they derived the amino acid sequences of two related phytochrome apoproteins that were 97.8% identical. Both were 1,128-amino-acid residues long, with molecular masses of

5. cDNA (complementary DNA) is an *in vitro* double-stranded DNA copy of mRNA (messenger RNA).
6. kbp = 1,000 nucleotide base pairs.
7. Bacteriophages can be located on agar plates because they make clear areas or plaques where they lyse bacterial cells.

```
   1  SSSRPASSSS  SRNRQSSQAR  VLAQTTLDAE  LNAEYEESGD  SFDYSKLVEA  QRDGPPVQQG
  61  RSEKVIAYLQ  HIQKGKLIQT  FGCLLALDEK  SFNVIAFSEN  APEMLTTVSH  AVPSVDDPPR
 121  LGIGTNVRSL  FSDQGATALH  KALGFADVSL  LNPILVQCKT  SGKPFYAIVH  RATGCLVVDF
 181  EPVKPTEFPA  TAAGALQSYK  LAAKAISKIQ  SLPGGSMEVL  CNTVVKEVFD  LTGYDRVMAY
 241  KFHEDDHGEV  FSEITKPGLE  PYLGLHYPAT  DIPQAARLLF  MKNKVRMICD  CRARSIKVIE
 301  AEALPFDISL  CGSALRAPHS  [C]HLQYMENMN  SIASLVMAVV  VNENEEDDEA  ESEQPAQQQK
 361  KKKLWGLLVC  HHESPRYVPF  PLRYACEFLA  QVFAVHVNRE  FELEKQLREK  NILKMQTMLS
 421  DMLFREASPL  TIVSGTPNIM  DLVKCDGAAL  LYGGKVWRLR  NAPTESQIHD  IAFWLSDVHR
 481  DSTGLSTDSL  HDAGYPGAAA  LGDMICGMAV  AKINSKDILF  WFRSHTAAEI  RWGGAKNDPS
 541  DMDDSRRMHP  RLSFKAFLEV  VKMKSLPWSD  YEMDAIHSLQ  LILRGTLNDA  SKPKREASLD
 601  NQIGDLKLDG  LAELQAVTSE  MVRLMETATV  PILAVDGNGL  VNGWNQKAAE  LTGLRVDDAI
 661  GRHILTLVED  SSVPVVQRML  YLALQGKEEK  EVRFEVKTHG  PKRDDGPVIL  VVNACASRDL
 721  HDHVVGVCFV  AQDMTVHKLV  MDKFTRVEGD  YKAIIHNPNP  LIPPIFGADE  FGWCSEWNAA
 781  MTKLTGWNRD  EVLDKMLLGE  VFDSSNASCP  LKNRDAFVSL  CVLINSALAG  EETEKAPFGF
 841  FDRSGKYIEC  LLSANRKENE  GGLITGVFCF  IHVASHELQH  ALQVQQASEQ  TSLKRLKAFS
 901  YMRHAINNPL  SGMLYSRKAL  KNTDLNEEQM  KQIHVGDNCH  HQINKILADL  DQDSITEKSS
 961  CLDLEMAEFL  LQDVVVAAVS  QVLITCQGKG  IRISCNLPER  FMKQSVYGDG  VRLQQILSDF
1021  LFISVKFSPV  GGSVEISSKL  TKNSIGENLH  LIDLELRIKH  QGLGVPAELM  AQMFEEDNKE
1081  QSEEGLSLLV  SRNLLRLMNG  DVRHLREAGV  STFIITAELA  SAPTAMGQ
```

FIGURE 28-2   Amino acid sequence of 124-kDa phytochrome deduced from the nucleotide sequence of cDNA and genomic clones designated type 3. Residue number is indicated
on the left Cysteine-321 is boxed. R, Arginine; K, Lysine; N, Asparagine; D, Aspartate; Q,
Glutamine; E, Glutamate; H, Histidine; P, Proline; Y, Tyrosine; W, Tryptophan; S, Serine;
T, Threonine; G, Glycine; A, Alanine; M, Methionine; C, Cysteine; F, Phenylalanine; L,
Leucine; V, Valine; I, Isoleucine. (From Vierstra and Quail, 1986. Reproduced by permission
of Kluwer Academic Publishers.)

124.870 kDa and 124.949, respectively. Each contained only one copy of
the 11-amino acid sequence that constituted the chromophore-binding
site (see Fig. 27-4); this lay between amino acids 315 and 325, with the
chromophore-anchoring cysteine residue at position 321 (Fig. 28-2). Thus
the phytochrome monomer had only one chromophore, attached to the
amino-terminal half of the molecule.

Distribution of hydrophobic and hydrophilic regions revealed that phytochrome was unlikely to be a membrane-spanning protein. The 6-kDa
amino-terminal segment was predominantly hydrophilic and therefore
likely to be at the molecule's surface; other hydrophilic segments centered
around residues 347, 691, and 1079. The chromophore attachment site was
located in a mildly hydrophilic region between two strongly hydrophobic
regions, one of which extended from about residue 80 to residue 315
(Hershey et al., 1985).

When Rudolf Grimm in Rüdiger's lab proteolyzed the protein and micro-
sequenced the resulting fragments, his data perfectly matched Hershey's
derived amino acid sequence (Grimm et al., 1986, 1988a). Subjecting a
12-residue fragment from the amino-terminus to amino acid analysis and
mass spectrometry, Grimm detected a serine residue at the beginning of
the polypeptide chain. This was acetylated to N-acetylserine, accounting
for the blocked amino terminus. Thus the codon for methionine, which

begins the nucleotide sequence, is only a start signal for polypeptide synthesis, and an initial methionine is lacking in the mature protein (Grimm et al., 1988b).

Grimm also deduced that more than one variant of the phytochrome gene must be expressed in etiolated oat, because a fragment beginning at residue 212 contained either valine or methionine at position 220 (Grimm et al., 1987).

Hershey, meanwhile, had sequenced the gene for type 3 oat phytochrome, using a cDNA probe to identify genomic fragments. The gene spanned about 5.9 kbp with six exons and five introns[8] (Hershey et al., 1987).

COMPARATIVE STUDIES

The Madison workers turned next to a dicot phytochrome, from zucchini (*Cucurbita pepo*). Using an *Avena* probe to screen a library, graduate student James L. Lissemore immediately isolated a full-length cDNA (Lissemore et al., 1987), which Robert A. Sharrock sequenced. The derived amino acid sequence—corresponding to a single-copy gene—was 1,123 residues long, and its calculated molecular mass was 124.950 kDa (Sharrock et al., 1986), rather than the 120-kDa value obtained on SDS-PAGE (Vierstra et al., 1984). The zucchini and oat sequences were only 65% homologous, diverging most in the carboxyl-terminal region.

Twenty-nine residues around the chromophore attachment site in oat phytochrome and around cysteine-322 in zucchini phytochrome were identical, suggesting that cysteine-322 holds the chromophore in zucchini. The hydrophobic sequence between residues 150 and 300 was also highly conserved, supporting the notion of a chromophore-containing pocket (Sharrock et al., 1986).

The Davis workers obtained immunochemical evidence for conservation of the chromophore attachment site between monocots and dicots. Polyclonal antibodies to a synthetic 11-residue peptide at the chromophore-binding site cross-reacted in an ELISA assay against amino-terminal pepsin digests of partially purified[9] oat, maize, and pea phytochromes (Mercurio et al., 1986).

8. Eukaryotic genes consist of exons that code for portions of polypeptides interspersed with introns, which are deleted from the primary RNA transcription product before translation occurs.

9. Crude phytochrome preparations were partly purified on SDS-polyacrylamide gels using a buffer system that contained zinc acetate. This made any phytochrome bands fluoresce orange under UV (Berkelman and Lagarias, 1986). The fluorescent bands were then eluted and prepared for assay.

Phytochrome cDNAs from several other species have now been sequenced (Sato, 1988; Kay et al., 1989; Christensen and Quail, 1989).

CHEMICAL STUDIES

The chromophore proved to be chemically less reactive in native than large phytochrome, suggesting that the labile amino terminus shielded the tetrapyrrole. In Rüdiger's lab, postdoctorate Peter Eilfeld probed the molecule with 8-anilinonaphthalene-1-sulfonate (ANS), which bleaches the chromophore of large Pfr. During 14 h, peak absorbance of native Pr decreased by only about 20% and that of Pfr by about 45%. So a hydrophobic site was more exposed in native Pfr than Pr, but it was much less exposed than in partially degraded phytochrome (Eilfeld and Rüdiger, 1984).

Hahn oxidized native phytochrome with tetranitromethane, a gentler and more specific oxidizing agent than the permanganate he had used previously (see Chapter 22). It reacted with native Pfr eight times faster than with native Pr, but there was a 40-fold difference between the two forms of large phytochrome, whose Pfr form was highly susceptible to oxidation (Hahn et al., 1984). "Obviously," Song says, "the chromophore is not as fully exposed in 124-kDa phytochrome as in "120"-kDa phytochrome, where you do not have the initial 6-kDa peptide chain at the amino-terminus."

The differences between the two native forms appeared to be largely caused by chromophore accessibility because the rate constant for oxidation with ozone was only 10% greater for Pfr than Pr (Thümmler et al., 1985). Unlike ANS and tetranitromethane, ozone is small enough to reach internal nooks and crannies.

PHYTOCHROME DIGESTS

As well as comparing native with large phytochrome, researchers deliberately dismantled the native molecule. Vierstra retrieved 74- and 55-kDa peptides after digesting Pfr with an endoprotease in crude extracts lacking PMSF (Daniels and Quail, 1984; Vierstra et al., 1984). Because cleavage occurred with ease, the two major domains appeared to be tenuously linked together.

The 74-kDa fragment proved to be spectrally similar to native phytochrome, suggesting that its interaction with the 55-kDa carboxyl-terminal domain did not involve the chromophore. Like native phytochrome, it showed minimal Pfr to Pr dark reversion, reinforcing this conclusion (Jones et al., 1985).

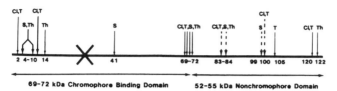

FIGURE 28-3 Structural model for *Avena* phytochrome. Arrows indicate proteolytic cleavage sites by trypsin (T), thermolysin (Th), subtilisin (S), and clostripain (Cl). X, chromophore binding site. Bold lines (large arrows) in the 4-10 kDa region indicate sites that are cleaved more rapidly in the Pr form. Dashed lines indicate cleavage sites that are favored in the Pfr form. (From Lagarias and Mercurio, 1985.)

When Hershey sequenced the phytochrome gene, he found the nucleotide sequence for the 74-kDa polypeptide fragment in exon 2, along with part of the 5' untranslated region of the mRNA. The coding sequence for the 55-kDa carboxyl-domain was divided among the other exons. So perhaps "the various functional domains of multidomain proteins have arisen over evolutionary time by the assembly of multiple exons into the contiguously transcribed sequence of single genes (Gilbert, 1985)" (Hershey et al., 1987).

Using commercial proteases, Lagarias and Frank Mercurio also deduced that 60–72-kDa and 52–55-kDa domains were connected by a protease-sensitive region (Fig. 28-3). Antiserum to the chromophore-binding site recognized the larger domain and its 62/61-kDa degradation product. Once severed, the chromophoric domain was relatively insensitive to further degradation, except for its terminal, Pfr-labile portion.

The 52–55-kDa domain was easily proteolyzed and thus likely to be less compact than the chromophoric domain. Because it was cleaved more easily from Pfr than Pr, there appeared to be "interdomain allosteric interaction in the isolated phytochrome molecule" (Lagarias and Mercurio, 1985).

Grimm digested native oat phytochrome (Grimm and Rüdiger, 1986) with a variety of proteases, locating cleavage sites by microsequencing the fragments and comparing his data with the known primary sequence (Grimm et al., 1986, 1988a). Because digestion was very slow, he determined which sites were readily cleaved and therefore superficial and which were protected from cleavage and therefore buried inside the protein. The regions around residues 54–66, 426, and 595–596 were quickly cleaved by trypsin, which attacks polypeptide chains at arginine and lysine residues. But four lysine residues at 361–364 remained uncleaved at the end of the experiment.

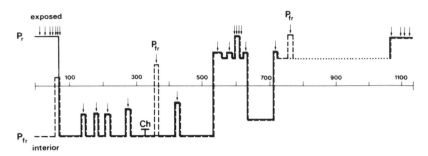

FIGURE 28-4   Exposed and interior parts of the peptide chain of phytochrome. Degree of exposure or protection is estimated from number and amount of fragments arising from a particular proteolytic cleavage site. Solid line, Pr; dashed line, Pfr; dotted line, no individual early cleavage sites observed in this region; Ch, chromophore. Arrows indicate localized sites of proteolytic cleavage. (From Grimm et al., 1988a. Copyright 1988, Springer-Verlag.)

The data confirmed that the chromophoric half of phytochrome is much more compact than the carboxyl-terminal half and placed the proteolytically sensitive link between residues 595–600 (Fig. 28-4). The end of the carboxyl domain was also very labile, and several large fragments with a common amino-terminal sequence differed in size. Rüdiger therefore proposed that 118-kDa large phytochrome lacks 6 kDa from the amino terminus but that 114-kDa large phytochrome lacks both this segment and a 3.5-kDa portion from the carboxyl terminus (Grimm et al., 1986).

PHYTOCHROME DIMER

Under nondenaturing conditions, the 55-kDa domain eluted from a Bio-Gel column with a molecular mass of 250–260 kDa (Jones et al., 1985). Thus it possibly contained a dimerization site. Steric exclusion chromatography suggested that native phytochrome was a dimer, giving a molecular mass of about 350 kDa (Lagarias and Mercurio, 1985).

In Madison, Alan M. Jones provided unequivocal evidence for the dimeric nature of native phytochrome. First, he observed that highly pure, native phytochrome behaved as a single population with a molecular mass of 253 kDa during equilibrium centrifugation under conditions precluding artifactual aggregation. Centrifugation in a strong salt solution reduced the molecular mass to 155 kDa, suggesting that ionic bonds are involved in holding two subunits together. These conclusions were confirmed by cross-linking studies.

Through MAb analysis of tryptic peptides in conjunction with cross-linking analysis, Jones mapped the dimerization site(s) within a 37-kDa domain in the carboxyl-terminal region. He identified each tryptic fragment using MAbs.

Those fragments lacking the 37-kDa domain within the carboxyl terminus behaved like monomers during size exclusion chromatography, whereas the 55- and 37-kDa fragments, which contained this domain, migrated as 160- and 74-kDa species. So the chromophoric half of the molecule appeared to be both globular and monomeric in isolation, whereas the carboxyl-terminal half was most likely a nonglobular dimer.

Noting glutaraldehyde levels at which non–cross-linked peptides disappeared, Jones deduced that readily cross-linked peptides contained or came from the 37-kDa carboxyl-terminal fragment.

Because dithiothreitol, which breaks the disulfide bridges that internally cross-link polypeptides, failed to alter the position of the phytochrome peak during size exclusion chromatography, Jones concluded that disulfide bridges were not solely responsible for linking the two subunits. Phytochrome behaved as a 300–350-kDa protein, giving a Stokes radius of 56 Å. This was identical to that reported by Lagarias and Mercurio but smaller than the 81 Å derived for large Pfr from quasi-electric light-scattering (Sarkar et al., 1984). The frictional ratio, which indicates the deviation of molecular shape from a perfect sphere (ratio=1.0), was estimated to be 1.37 for the native dimer and 1.34 for the monomers produced at high salt concentrations. The dimerized 55-kDa peptides had a frictional ratio of 1.44 (Jones and Quail, 1986). Thus the dimer seemed to be composed ot two elongated monomers, each shaped like a tadpole, with a globular "head"—the chromophoric domin—and an elongated carboxyl-terminal "tail."

Holdsworth and Whitelam used LAS 41 to determine whether the phytochrome dimer can be a PrPfr heterodimer, as had previously been proposed (see Chapter 25). They pointed out that an R-irradiated solution containing 87.4% Pfr would theoretically be 76.4% PfrPfr, 22% PrPfr, and 1.6% PrPr. Thus an antibody that bound exclusively to Pr should react with 23.6% of the molecules in such a mixture. Titrating phytochrome solutions with LAS 41, they obtained a curve that saturated at 100% for Pr and 24.5% for the photoequilibrium mixture. These data were "quantitatively consistent with the hypothesis that phytochrome does exist as stable heterodimers *in vitro*" (Holdsworth and Whitelam, 1987).

MOLECULAR SHAPE

After moving to the University of North Carolina in Chapel Hill, Jones collaborated with Harold P. Erickson, an electron microscopist at Duke University, to obtain negatively stained and rotary shadowed images of native oat phytochrome. The two observed Y-shaped particles with three globular domains, each 7–8 nm in diameter (Fig. 28-5A). The angle of the Y, although most commonly 90°, was variable. The arms of the Y did not touch; their centers were separated by 15 nm. After the hinges were cleaved by trypsin and the resulting particles centrifuged on glycerol gradients, the amino-terminal domains—identified by Quail's MAbs—appeared as single globules.

Jones proposed that each Y had a globular amino-terminal domain for each arm and two laterally juxtaposed carboxyl-terminal domains for its stem (Fig. 28-5B). Between the two domains of each monomer was a flexible—and proteolytically sensitive—hinge region (Jones and Erickson, 1989). "Since the rotary shadow micrographs look like Mickey Mouse, we affectionately call this the Mickey Mouse model," Jones says. "The globular chromophore domains are Mickey's ears." Using a novel method to map putative dimerization regions, Jones and graduate student Mike Edgerton have recently identified a protein–protein interaction between the two subunits that resides in a small region very close to the hinge region (unpublished data).

Researchers at the National Institute for Basic Biology at Okazaki and at Tohoku University were separately studying the three-dimensional structure of etiolated pea phytochrome by steric exclusion column chromatography and small-angle X-ray scattering (SAXS) (Tokutomi et al., 1987). Related to X-ray diffraction, SAXS measures scattering through a small rather than a wide angle and consequently has a resolution of only 1–100 nm. X-ray diffraction can pinpoint individual atoms, but it would require a crystal of phytochrome, which has not yet been obtained. And crystallization may change the conformations of proteins, whereas SAXS analyzes proteins in solution.

From scattering data obtained at the National Laboratory for High Energy Physics in Tsukuba, the collaborators calculated gyration radii[10] of phytochrome samples to determine spatial extension. The labile amino terminus appeared to contribute little to overall dimension, because average values were 54.0 Å for native pea P_r and 53.8 Å for large pea Pr. The maximum distance of an electron pair in 114-kDa pea Pr was 140 Å, suggesting that the shape of phytochrome deviated greatly from a sphere. The calculated molecular mass was 228 kDa for the 114-kDa chromopeptide, confirming its dimeric nature. Values of 314–318 kDa were obtained by steric exclusion chromatography, indicating a nonspherical shape. These values increased to 321–324 kDa after R irradiation, but SAXS analysis of Pfr, which precipitates at high concentrations, was unsuccessful.

10. The root-mean square of the distances of all electrons from their center of gravity.

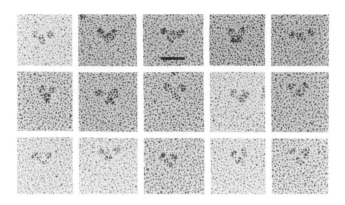

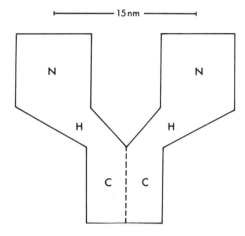

FIGURE 28-5   (*Top*) Electron micrographs of some *Avena* phytochrome molecules after rotary shadowing. Bar represents 24 nm. (*Bottom*) Schematic diagram of phytochrome dimer. N, amino-terminal domain; H, hinge region; C, carboxy-terminal domain. (From Jones and Erickson, 1989.)

Because a sphere with a 53.8-Å radius of gyration has an actual radius of 69.5 Å, the molecular mass of the large pea phytochrome dimer would be 1,140 kDa, according to Jones and Quail's value for the partial specific volume of oat phytochrome. This was five times as great as 228 kDa, indicating that phytochrome's shape deviates greatly from the spherical and/or that its electrons are unevenly distributed (Tokutomi et al., 1988).

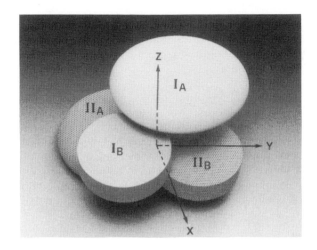

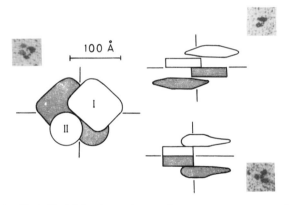

FIGURE 28-6 (*Top*) Model for the tertiary and quarternary structure of dimeric pea Pr (subunit molecular mass, 114 kDa). (From Tokutomi el al., 1989.) (*Bottom*) Comparison of three projection views of the phytochrome model with typical images of phytochrome I 114-kDa chromopeptide dimer obtained with low-angle, rotary-shadowing electron microscopy (×225,000). (From Nakasako et al., 1990.)

The possibilities were examined with the SAXS contrast variation method, which showed that pea Pr did indeed have an unusual shape (Nakasako et al., 1990).

Electron microscopy yielded bewildering two-, three-, and four-part images, foiling attempts to construct a model of phytochrome's three-dimensional structure. But Nakasako eventually realized that the images might represent different views of the same molecule. Comparing the scattering data with curves from known shapes, he proposed that the phytochrome monomer consists of an oblate ellipsoid—the amino-terminal 59-kDa chromophoric domain—and a plate-shaped carboxyl domain. The two subunits are juxtaposed at the lower flat plane of one chromophoric domain and the upper flat plane of the other to form a "four-leaved clover" (Fig. 28-6A). Afer rotary shadowing (Fig. 28-6B), such a structure would generate four-part images if viewed from the z axis and two-part images if viewed in the x-y plane. Three-part images would result if the platinum deposits used for shadowing clogged the space between the two carboxyl domains (Furuya, 1989; Tokutomi et al., 1989; Nakasako et al., 1990).

REFERENCES

Baron, O., and Epel., B. L. (1982). Studies on the capacity of Pr *in vitro* to photoconvert to the long-wavelength Pfr-form. A survey of ten plant species. *Photochem. Photobiol.* 36,79–82.

Berkelman, T. R., and Lagarias, J. C. (1986). Visualization of bilin-linked peptides and proteins in polyacrylamide gels. *Anal. Biochem.* 156,194–201.

Boeshore, M. L., and Pratt, L. H. (1980). Phytochrome modification and light-enhanced, *in vivo*-induced phytochrome pelletability. *Plant Physiol.* 66,500–504.

Boeshore, M. L., and Pratt, L. H. (1981). Characterization of a molecular modification of phytochrome that is associated with its conversion to the far-red-absorbing form. *Plant Physiol.* 68,789–797.

Bolton, G. W. (1979). Phytochrome: Aspects of its protein and photochemical properties. Ph.D. dissertation, University of Minnesota.

Bolton, G. W., and Pratt, D. C. (1980). A new method for the isolation of phytochrome. *Plant Physiol.* 65,(suppl. 2).

Bolton, G. W., and Quail, P. H. (1981). Cell-free synthesis of *Avena* phytochrome. *Plant Physiol.* 67(suppl. 130).

Bolton, G. W., and Quail, P. H. (1982). Cell-free synthesis of phytochrome apoprotein. *Planta* 155,212–217.

Chou, P. Y., and Fasman, G. D. (1978). Empirical predictions of protein conformation. *Annu. Rev. Biochem.* 47,251–276.

Christensen, A. H., and Quail, P. H. (1989). Sequence and expression of a phytochrome gene from maize. *Gene* 85,381–390.

Cordonnier, M.-M., Smith, C., Greppin, H., and Pratt, L. H. (1983). Production and purification of monoclonal antibodies to *Pisum* and *Avena* phytochrome. *Planta* 158,369–376.

Daniels, S. M., and Quail, P. H. (1984). Monoclonal antibodies to three separate domains on 124 kilodalton phytochrome from *Avena*. *Plant Physiol* 76,622–626.

Eilfeld, P., and Rüdiger, W. (1984). On the reactivity of native phytochrome. *Z. Naturforsch.* 39c,742–745.

Epel, B. L. (1981). A partial characterization of the long-wavelength "activated" far-red absorbing form of phytochrome. *Planta* 151,1–5.

Furuya, M. (1983). Molecular properties of phytochrome. *Philos. Trans. R. Soc. Lond. B* 303,361–375.

Furuya, M. (1989). Molecular properties and biogenesis of phytochrome I and II. *Adv. Biophys.* 25,133–167.

Gilbert, W. (1985). Genes-in-pieces revisited. *Science* 228,823–828.

Gottmann, K., and Schäfer, E. (1982). *In vitro* synthesis of phytochrome apoprotein directed by mRNA from light and dark grown *Avena* seedlings. *Photochem. Photobiol.* 35,521–525.

Grimm, R., and Rüdiger, W. (1986). A simple and rapid method for isolation of 124 kDa oat phytochrome. *Z. Naturforsch.* 41c,988–992.

Grimm, R., Lottspeich, F., Schneider, H. A. W., and Rüdiger, W. (1986). Investigation of the peptide chain of 124 kDa phytochrome: Localization of proteolytic fragments and epitopes for monoclonal antibodies. *Z. Naturforsch.* 41c,993–1000.

Grimm, R., Lottspeich, F., and Rüdiger, W. (1987). Heterogeneity of the amino acid sequence of phytochrome from etiolated oat seedlings. *FEBS Lett.* 225,215–217.

Grimm, R., Eckerskorn, C., Lottspeich, F., Zenger, C., and Rüdiger, W. (1988a). Sequence analysis of proteolytic fragments of 124-kilodalton phytochrome from etiolated *Avena sativa* L.: Conclusions on the conformation of the native protein. *Planta* 174,396–401.

Grimm, R., Kellermann, J., Schäfer, W., and Rüdiger, W. (1988b). The amino-terminal structure of oat phytochrome. *FEBS Lett.* 234,497–499.

Hahn, T.-R., Song. P.-S., Quail, P. H., and Vierstra, R. D. (1984). Tetranitromethane oxidation of phytochrome chromophore as a function of spectral form and molecular weight. *Plant Physiol.* 74,755–758.

Hershey, H. P., Colbert, J. T., Lissemore, J. L., Barker, R. F., and Quail, P. H. (1984). Molecular cloning of cDNA for *Avena* phytochrome. *Proc. Natl. Acad. Sci. USA* 81,2332–2336.

Hershey, H. P., Barker, R. F., Idler, K. B., Lissemore, J. L., and Quail P. H. (1985). Analysis of cloned cDNA and genomic sequences for phytochrome: Complete amino acid sequences for two gene products expressed in etiolated *Avena*. *Nucleic Acids Res.* 13,8543–8559.

Hershey, H. P., Barker, R. F., Idler, K. B., Michael, M. G., and Quail, P. H. (1987). Nucleotide sequence and characterization of a gene encoding the phytochrome polypeptide from *Avena*. *Gene* 61,339–348.

Holdsworth, M. L., and Whitelam, G. C. (1987). A monoclonal antibody specific for the red-absorbing form of phytochrome. *Planta* 172,539–547.

Horwitz, B. A., and Epel, B. L. (1977). A far-red form of phytochrome exhibiting *in vivo* spectral properties: Studies with crude extracts of oats and squash. *Plant Sci. Lett.* 9,205–210.

Johns, H. E. (1969). Photochemical reactions in nucleic acids. *Methods Enzymol.* 16,253–316.

Jones, A. M., and Erickson, H. P. (1989). Domain structure of phytochrome from *Avena sativa* visualized by electron microscopy. *Photochem. Photobiol.* 49,479–483.

Jones, A. M., and Quail, P. H. (1986). Quarternary structure of 124-kilodalton phytochrome from *Avena sativa* L. *Biochemistry* 25,2987–2995.

Jones, A. M., Vierstra, R. D., Daniels, S. M., and Quail, P. H. (1985). The roles of separate molecular domains in the structure of phytochrome from etiolated *Avena sativa* L. *Planta* 164,501–506.

Kay, S. A., Keith, B., Shinozaki, K., and Chua, N.-H. (1989). The sequence of the rice phytochrome gene. *Nucleic Acids Res.* 17,2865–2866.

Kelly, J. M., and Lagarias, J. C. (1985). Photochemistry of 124-kilodalton phytochrome under constant illumination *in vitro*. *Biochemistry* 24,6002–6010.

Kerscher, L., and Nowitzki, S. (1982). Western blot analysis of a lytic process *in vitro* specific for the red light absorbing form of phytochrome. *FEBS Lett.* 146,173–176.

Lagarias, J. C. and Mercurio, F. M. (1985). Structure function studies on phytochrome. *J. Biol. Chem.* 260,2415–2423.

Lagarias, J. C., Kelly, J. M., Cyr, K. L., and Smith, W. O., Jr. (1987). Comparative photochemical analysis of highly purified 124 kilodalton oat and rye phytochromes *in vitro*. *Photochem. Photobiol.* 46,5–13.

Lissemore, J. L, Colbert, J. T., and Quail, P. H. (1987). Cloning of cDNA for phytochrome from etiolated *Cucurbita* and coordinate photoregulation of the abundance of two distinct phytochrome transcripts. *Plant Mol. Biol.* 8,485–496.

Litts, J. C., Kelly, J. M., and Lagarias, J. C. (1983). Structure–function studies on phytochrome. *J. Biol. Chem.* 258,11025–11031.

Mercurio, F. M., Houghten, R. A., and Lagarias, J. C. (1986). Site-directed antisera to the chromophore binding site of phytochrome: Characterization and cross-reactivity. *Arch. Biochem. Biophys.* 248,35–42.

Nagatani, A., Yamamoto, K. T., Furuya, M., Fukumoto, T., and Yamashita, A. (1983). Production and characterization of monoclonal antibodies to rye (*Secale cereale*) phytochrome. *Plant Cell Physiol.* 24,1143–1149.

Nagatani, A., Yamamoto, K. T., Furuya, M., Fukumoto, T., and Yamashita, A. (1984). Production and characterization of monoclonal antibodies which distinguish different surface structures of pea (*Pisum sativum* cv Alaska) phytochrome. *Plant Cell Physiol.* 25,1059–1068.

Nakasako, M., Wada, M., Tokutomi, S., Yamamoto, K, T., Sakai, J., Kataoka, M., Tokunaga, F., and Furuya, M. (1990). Quarternary structure of pea phytochrome I dimer studied with small-angle X-ray scattering and rotary shadowing electron microscopy. *Photochem. Photobiol.* 52,3–12.

Plumpton, C., Partis, M. D., Thomas, B., Huskisson, N. S., and Butcher, G. W. (1988). Site-directed monoclonal antibodies against the amino-terminus of 124-kDa phytochrome from *Avena sativa* L. *Plant Cell Environ.* 11,487–491.

Quail, P. H. (1982). Intracellular location of phytochrome. *Trends Photobiol. Proc. Int. Congr. Photobiol. 8th,* pp. 485–500.

Quail, P. H., Gatz, C., Hershey, H. P., Jones, A. L., Lissemore, J. L., Parks, B. M., Sharrock, R. A., Barker, R. F., Idler, K., Murray, M. G., Koornneef, M., and Kendrick, R. E. (1987). Molecular biology of phytochrome. *In* "Phytochrome and Photoregulation of Plants" (M. Furuya, ed.), pp. 23–37. Academic Press, Tokyo.

Sarkar, H. K., Moon, D.-K., Song, P.-S., Chang, T., and Yu, H. (1984). Tertiary structure of phytochrome probed by quasi-elastic light scattering and rotational relaxation time measurements. *Biochemistry* 23,1882–1888.

Sato, N. (1988). Nucleotide sequence and expression of the phytochrome gene in *Pisum sativum:* Differential regulation by light of multiple transcripts. *Plant Mol. Biol.* 11,697–710.

Seyfried, M., and Schäfer, E. (1985). Action spectra of phytochrome *in vivo*. *Photochem. Photobiol.* 42,319–326.

Sharrock, R. A., Lissemore, J. L., and Quail, P. H. (1986). Nucleotide and amino acid sequence of a *Cucurbita* phytochrome cDNA clone: Identification of conserved features by comparison with *Avena* phytochrome. *Gene* 47,287–295.

Thümmler, F., Eilfeld, P.. Rüdiger, W., Moon, D.-K., and Song, P.-S. (1985). On the chemical reactivity of the phytochrome chromophore in the Pr and Pfr form. *Z. Naturforsch.* 40c,215–218.

Tokutomi, S., Kataoka, M., and Tokunaga, F. (1987). Small-angle X-ray scattering, a useful tool for studying the structure of phytochrome. *In* "Phytochrome and Photoregulation in Plants" (M. Furuya, ed.), pp. 167–177. Academic Press, Tokyo.

Tokutomi, S., Kataoka, M., Sakai, J., Nakasako, M., Tokunaga, F., Tasumi, M., and Furuya, M. (1988). Small-angle X-ray scattering studies on the macromolecular structure of the red-light-absorbing form of 121 kdalton pea phytochrome and its 114 kDa chromopeptide. *Biochim. Biophys. Acta* 953,297–305.

Tokutomi, S., Nakasako, M., Sakai, J., Kataoka, M., Yamamoto, K. T., Wada, M., Tokunaga, F., and Furuya, M. (1989). A model for the dimeric molecular structure of phytochrome based on small-angle X-ray scattering. *FEBS Lett.* 247,139–142.

Vierstra, R. D., and Quail, P. H. (1982a). Native phytochrome: Inhibition of proteolysis yields a homogeneous monomer of 124 kdalton from *Avena*. *Plant Physiol.* 69(suppl. 85).

Vierstra, R. D., and Quail, P. H. (1982b). Native phytochrome: Inhibition of proteolysis yields a homogeneous monomer of 124 kilodaltons from *Avena*. *Proc. Natl. Acad. Sci. USA* 79,5272–5276.

Vierstra, R. D., and Quail, P. H. (1982c). Proteolysis alters the spectral properties of 124 kdalton phytochrome from *Avena*. *Planta* 156,158–165.

Vierstra, R. D., and Quail, P. H. (1983a). Photochemistry of 124 kilodalton *Avena* phytochrome *in vitro*. *Plant Physiol.* 72,264–267.

Vierstra, R. D., and Quail, P. H. (1983b). Purification and initial characterization of 124-kilodalton phytochrome. *Biochemistry* 22,2498–2505.

Viersta, R. D., and Quail, P. H. (1985). Spectral characterization and proteolytic mapping of native 120-kilodalton phytochrome from *Cucurbita pepo L*. *Plant Physiol.* 77,990–998.

Vierstra, R. D., and Quail, P. H. (1986). The protein. *In* "Photomorphogenesis in Plants" (R. E. Kendrick and G. H. M. Kronenberg, eds.), pp. 35–60. Martinus Nijhoff, Dordrecht.

Vierstra, R. D., Tokuhisa, J. G., Newcomb, E. H., and Quail, P. H. (1981). A solid-phase antibody approach to identifying phytochrome-bearing structures in particulate fractions. *Plant Physiol.* 67(suppl. 130).

Vierstra, R. D., Cordonnier, M.-M., Pratt, L. H., and Quail, P. H. (1984). Native phytochrome: Immunoblot analysis of relative molecular mass and *in-vitro* proteolytic degradation for several plant species. *Planta* 160,521–528.

CHAPTER **29** _____

Native Pr and Pfr

"The physiologically active Pfr form of phytochrome must have a particular surface structure—'active center'—which is not present in Pr and which is able to interact with reaction partners in the cell in order to start the phytochrome transduction chain," Rüdiger explains. "Regions of the peptide chain which are exposed only in Pfr are good candidates for constituents of such an active center."

CIRCULAR DICHROISM SPECTRA

The first CD spectra from native phytochrome had negative bands at 284 and 668 nm for Pr and at 302 and 670 for Pfr. Pr had a positive band at 370 nm, whereas Pfr had positive bands at 358, 430, and 720 nm. Because the 430-nm band was much smaller in large phytochrome samples (Bolton, 1979), the Davis researchers attributed it to "the influence of the proteolytically labile domain on the Pfr chromophoric region" (Litts et al., 1983).

Visiting Song's lab in 1983, Vierstra obtained similar data in the visible region of the spectrum, also observing more pronounced peaks in native than large Pfr (Fig. 29-1A). Below 250 nm, "to our surprise and pleasure, there was a reproducible, though small difference between 124 kDa Pr and Pfr," says Song (Fig. 29-1B). Such a difference, which translated into a 3% change in α-helical content, had not been observed with small or large phytochrome. Thus "the N-terminus domain appears to be at least one of the protein domains where the light-induced conformational change occurs" (Vierstra et al., 1987).

MONOCLONAL ANTIBODIES

Initial immunoassays with native phytochrome failed to discriminate between Pr and Pfr (Saunders and Pratt, 1983; Cordonnier et al., 1984; Nagatani et al., 1983, 1984). But when Thomas and virologist Norman

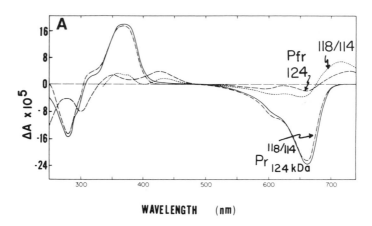

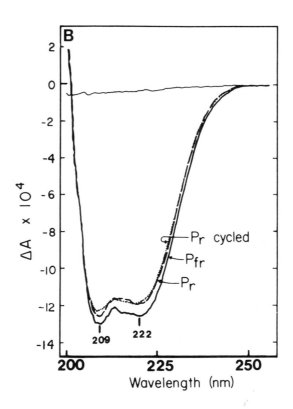

FIGURE 29-1   (A) Circular dichroic spectra of native and large Pr and Pfr. (B) UV-circular dichroic spectra of native Pr and Pfr. (From Vierstra, et al., 1987.)

Crook developed an ELISA in which they sandwiched the phytochrome between rabbit antiphytochrome immunoglobulins and an MAb and then added rabbit antirat alkaline phosphatase conjugate (Thomas et al., 1984a), three MAbs reacted differentially with the pigment's two forms. MAC 49 and 52 bound preferentially to Pr, and MAC 50 bound more strongly to Pfr, even after alternating FR/R irradiation (Thomas et al., 1984b; Thomas and Penn, 1986).

Loading ELISA wells with rabbit antibodies to mouse immunoglobulins, an MAb, phytochrome, and then rabbit anitbodies conjugated to alkaline phosphatase, Cordonnier detected four- to fivefold greater affinity for Pr than Pfr among three previously tested MAbs, which appeared to correspond to Quail's type-1 antibodies. Binding of the representative Oat-25 stabilized the amino terminus of Pr (Cordonnier et al., 1985). Later analyses confirmed that the epitope recognized by Oat-25 is "wholly located near the N terminus of phytochrome" (Pratt et al., 1988).

Even though Oat-25 had less affinity for Pfr than Pr, it stimulated reversion and decreased and blue-shifted the FR absorption peak. This suggested that its epitope normally interacted with the chromophore and took part in conformational changes. In the cell, perhaps binding partners "could provide the plant with a means to regulate the disappearance of the active Pfr form during the night. Variable availability of an appropriate binding partner could also explain why reversion, when it is observed in situ, occurs in a limited and variable population of the phytochrome that is present" (Cordonnier et al., 1985).

Several MAbs in Furuya's lab affected the reversion rate of native pea phytochrome, a 121-kDa molecule obtained through immunoaffinity purification. The preparation reverted slowly *in vitro,* but an MAb that bound to native but not large phytochrome increased the rate and also blue-shifted the Pfr peak to 723 nm, behaving similarly to Oat-25. Four MAbs to large pea phytochrome—which could not have been type-1 antibodies—either altered rates of Pfr disappearance or Pr reappearance or affected both rates. Because their epitopes mapped to different sites (Fig. 29-2), there seemed to be multiple points on Pfr's surface where reversion could be regulated. So "depending on the distribution and type of binding sites, reversion may occur at different rates in different cellular locations" (Lumsden et al., 1985).

Young-Gya Chai discovered in Song's lab that Oat-25 abolished the difference between Pr and Pfr in the UV region of CD spectra (Chai et al., 1987). "So it was very clear to us," Song says, "that the increase in α-helicity is coming from the 6-kDa amino-terminal sequence." Rüdiger, however, believes that Siebert's infrared data (see Chapter 27) "exclude a

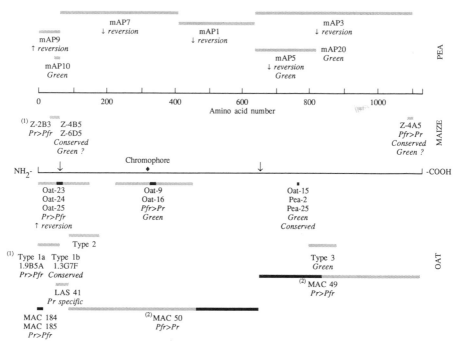

FIGURE 29-2   Cordonnier's summary of locations and effects of MAbs to pea, maize, and oat phytochrome. Large arrows mark sites of unusual susceptibility to proteolysis. (2) Tentative assignations. (From Cordonnier, 1989.)

real difference in α-helix content between Pr and Pfr which is larger than 0.2%.'' Chai and Song reported a 3% difference.

By attaching a synthetic peptide corresponding to the first 18 residues of the phytochrome amino terminus to tuberculin-purified protein derivative, which stimulates the immune system without eliciting antibodies to itself, Thomas and colleagues obtained MAC 184 and MAC 185. Both MAbs bound four times more tightly to Pr than to Pfr (Plumpton et al., 1988), providing ''absolute proof that the amino terminus is undergoing conformational change.''

LAS 41 (see Chapter 28) was the first MAb that bound exclusively to one form of phytochrome (Pr). Through peptide mapping, Whitelam and Holdsworth located its epitope in a 4-kDa region immediately adjacent to the 6-kDa amino-terminal fragment. Thus at least part of this region was exposed in Pr but hidden in Pfr. LAS 41 binding shifted maximal Pr absorption from 665 to 667 nm and induced dithionite-dependent reversion of Pfr. Thus this site also interacted with the chromophore and contributed to conformational changes (Holdsworth and Whitelam, 1987).

Like the epitopes for type 1a[1] MAbs and LAS 41, part of the carboxyl-terminal domain is also more exposed in Pr, because Thomas's MAC 49 and 52 mapped to this region (Fig. 29-2) (Thomas et al., 1986). An epitope more exposed in Pfr than Pr lies between residues 200 and 450, where the epitopes for Oat-9 and Oat-16 are located (Shimazaki et al., 1986). A second lies between 1088 and 1106 (Schneider-Poetsch et al., 1988), the epitope for Z-4A5 (Grimm et al., 1986) and several other MAbs (Schneider-Poetsch et al., 1989) obtained by Hansjörg Schneider's group in Cologne (Schwarz and Schneider, 1987). All this group's MAbs exhibited slight preferences for Pr or Pfr (Schneider-Poetsch et al., 1989).

CONSERVED SITES

Of the detectable differences between Pr and Pfr, the most significant may be those that have been retained during evolution, because they may be indispensable to biological function (Cordonnier et al., 1983).

There were scattered early reports of MAbs with some monocot–dicot cross-reactivity (Cordonnier et al., 1984; Whitelam et al., 1985). Pea-25, an MAb against native phytochrome from etiolated pea, recognized a phytochrome-sized polypeptide in immunoblots[2] from a wide range of dark-grown and light-grown monocots, dicots, algae, and a moss (Cordonnier et al., 1986). It did not discriminate between Pr and Pfr, and its epitope appeared to be about 85 kDa from the amino terminus (Pratt et al., 1988).

Using an oat phytochrome cDNA library prepared by Cordonnier, Laura K. Thompson identified a nucleotide sequence that coded from residue 464 to the carboxyl terminus. Inserting this into a bacterial plasmid, Lyle Crossland at Ciba-Geigy cut it into defined fragments. "We made six subclones from one end and six from the other," Pratt explains, "so we had two nested sets of subclones that overlapped one another [upper half of Fig. 29-3]. We then grew the plasmids in *Escherichia coli,* asking the bacteria to make whatever protein was coded for by these fused pieces of DNA. Then we asked whether Pea-25 or other antibodies detected these proteins."

Pea-25 failed to recognize proteins lacking residues 747–830 (Fig. 29-4). Crossland therefore cut the DNA insert near the codons for these residues

1. Quail's group identified two types of type-1 antibodies. Type 1a bound to one or more closely spaced sites, altered spectral properties, had greater affinity for Pr than Pfr, and induced reversion. Type 1b, typified by MAb 1.3G7F, bound to a spatially separate site in the 6-kDa amino-terminal domain, recognized Pr and Pfr equally, and did not alter phytochrome's properties (Quail et al., 1987).

2. Electrophoresed proteins are visualized immunochemically after electrophoretic transfer from the polyacrylamide gel to a strip of nitrocellulose, a technique known as immunoblotting or Western blotting.

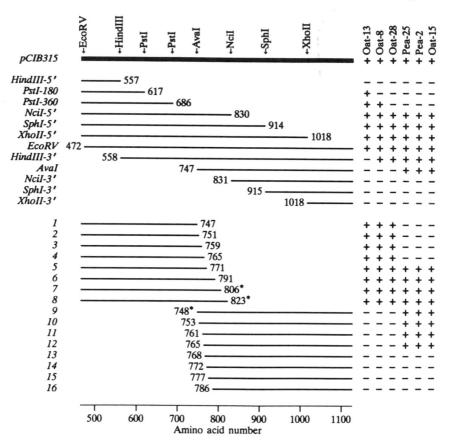

FIGURE 29-3   Diagram of cDNA inserts into plasmid UC18 used by Pratt's group for epitope mapping. Restriction endonuclease sites are indicated at top, corresponding amino acids at bottom. Designations of subclones are on left; reactivity of expressed fusion proteins with MAbs is on right. Terminating or initiating amino acids, which were verified by sequencing except for those marked by an asterisk, are indicated for each construct. (From Thompson, et al., 1989.)

and added an enzyme that slowly removed nucleotides from the cut ends. Thompson found that Pea-25 recognized only the resulting fusion proteins that contained residues 765–771 (Fig. 29-3, subclones 5–12). Two other MAbs, Pea-2 and Oat-15, also bound to this epitope (Thompson et al., 1989). Thus the site for three MAbs was precisely delineated to residues 765–771, in the carboxyl domain, through the first use of systematically created, nested sets of subclones.

Several other MAbs that recognize highly conserved epitopes are now available, including Z-3B1, against *Zea* phytochrome (Schneider-Poetsch et al., 1988; Lindemann et al., 1989; López-Figueroa et al., 1989) and Z-4A5, which binds to a 1088–1106-residue fragment from the extreme carboxyl end of the molecule and prefers Pfr to Pr (Schneider-Poetsch et al., 1988).

TRANSGENIC PLANTS

Evidence that the active site of phytochrome is conserved between distantly related species came from transgenic plants containing phytochrome DNA. In 1988, Hershey transferred a chimeric oat phytochrome structural gene with an uninterrupted phytochrome (*phyA*)[3] coding region into tobacco. The resulting transgenic plants overexpressed photoreversible phytochrome, were shorter than wild-type plants, and had smaller but darker green leaves, indicating that the foreign protein was active (Keller et al., 1989).

After Quail moved to Albany, California, in 1987 to become director of a newly created USDA Plant Gene Expression Center and a professor at the University of California, Berkeley, Margaret T. Boylan obtained transgenic tomato plants containing a cDNA encoding a full-length PHYA polypeptide. These also had an altered phenotype (Fig. 29-4) (Boylan and Quail, 1989). At the Rockefeller University in New York, Steve A. Kay and co-workers introduced rice *phyA* cDNA into tobacco (Kay et al., 1989).

Identification of mutant sequences that fail to alter phenotype may reveal which parts of the molecule are necessary for activity. Hershey, Vierstra, and Joel R. Cherry progressively deleted oat *phyA* sequences and transferred them into tobacco. As well as noting effects on phenotype, they extracted the resulting foreign proteins and determined their photoreversibility, spectral properties, rates of dark reversion, and ability to dimerize. Carboxyl-terminal deletions were biologically inactive and unable to dimerize and, when the deletion was large, to attach chromophore *in vitro*. "We also have phytochrome that is missing amino acids 7 though 69 (Cherry et al., 1991)," Vierstra says. "By all criteria it is identical to large phytochrome—it's still a dimer in solution, but its spectral properties are shifted and its Pfr form reverts to Pr. When you express this construct in plants, you get the same phenotype as if the plant didn't have any foreign phytochrome—the plants are tall. This says that the small

3. *phyA* is the gene for the predominant phytochrome in dark-grown seedlings. It encodes the polypeptide PHYA.

region that there was so much fuss over in the middle 1980s is critical—not for chromophore attachment or dimerization—but for biological activity.''

CHEMICAL STUDIES

After his circular dichroism studies, Song decided that ''if interaction with the Pfr chromophore causes the amino-terminal sequence to fold into a helix, we should be able to abolish the conformational change by modify-

FIGURE 29-4   Transgenic tomato plants containing oat phytochrome gene. Plant on *left* is not expressing the foreign gene, whereas the one on *right* is overexpressing it (From Boylan and Quail, 1989. Reproduced by permission from the American Society of Plant Physiologists.)

ing the chromophore from an extended to cyclic conformation." When Debbie Sommer mixed phytochrome with an equal ratio of zinc ions, the 730-nm absorption band of Pfr disappeared, while the spectral properties of Pr remained unchanged. Thus the chromophore was certainly more exposed in the Pfr form. A CD spectrum of the Pfr–zinc complex indicated that β-sheet folding had increased at the expense of α-helical folding (Sommer and Song, 1990). "This means," Song says, "that when the chromophore is sitting in its pocket, it is obviously interacting with the protein to maintain its extended structure. But when you add zinc, the chromophore becomes cyclic and that interaction is disrupted, which causes the conformation of the protein around the chromophore binding site to undergo dramatic change. Because we have this very strong indication that the chromophore is very efficient at stabilizing the α-helical conformation, we think that when the Pfr chromophore becomes exposed and begins to interact with the amino terminus, that terminus folds into an α-helix" (Fig. 29-5).

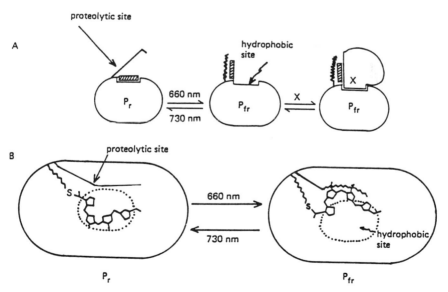

FIGURE 29-5 Schematic model for topographic changes involved in Pr to Pfr photoconversion. (A) side view; (B) top view. X represents a putative receptor. (From Song and Yamazaki, 1987.)

SUSCEPTIBILITY TO PROTEOLYSIS

The proteolytic studies at Munich (see Chapter 28) also revealed that, while the amino terminus was more proteolytically sensitive in Pr than Pfr, two regions were more susceptible in Pfr (Grimm et al., 1988). One included residue 354, a glutamate very near the chromophore and the epitope for Oat-9 and Oat-16 (see Fig. 28-4). The other included residue 753, in the carboxyl half of the molecule. Because its cleavage produced an 83-kDa peptide, it corresponded to one of the Pfr sites reported by Lagarias (see Fig. 28-3).

The proteolytic sensitivity of rice phytochrome, isolated by Rüdiger's group, was very similar. The first 70 amino acid residues at the amino terminus were exposed in Pr, whereas residues 340–350 and 740–750 were exposed in Pfr (Schendel et al., 1989). Pfr also proved to have more acidic groups at its surface than Pr, and this difference was not seen with denatured phytochrome (Schendel and Rüdiger, 1989).

Just before the Munich peptide map (see Fig. 28-4) was published, Jones and Quail submitted a map of the amino-terminal domain obtained through progressive digestion of Pr and Pfr with the serine protease subtilisin (Fig. 29-6). Although not as precise as that obtained by microsequencing, it was in general agreement, except that cleavage site d, near residue 354, was not Pfr-specific. Jones used three markers to deduce cleavage sites: the chromophore; the type-1 epitope; and the type-2 epitope, which he mapped to a position between 10 and 20 kDa from the amino terminus.

After extensive digestion, Jones obtained a mixture of peptides that resisted further attack. This migrated as a single 35–40-kDa peak during nondenaturing size exclusion chromatography and was photoreversible between a form with maximal absorbance at 656 nm and a bleached form. SDS-polyacrylamide gel electrophoresis separated several peptides from this 35-kDa complex, including a 16-kDa chromopeptide and a 14-kDa fragment that reacted with a type-2 MAb.

Because immunoprecipitating the 14-kDa species failed to affect the photoreversibility of the remaining mixture, the remaining peptides, or probably just the 16-kDa species, contained all the necessary structure for photoreversibility. The 16-kDa fragment was hydrophobic except for 10 residues around residue 245; thus it was possibly part of a chromophore-containing hydrophobic cavity (Jones and Quail, 1989).

Furuya's group had previously isolated a 40.3-kDa photoreversible peptide from pea through limited proteolysis (Yamamoto and Furuya, 1983), while Rüdiger's group had obtained a corresponding 39-kDa oat peptide (Reiff et al., 1985). Because trypsin excised the latter near cleavage sites a

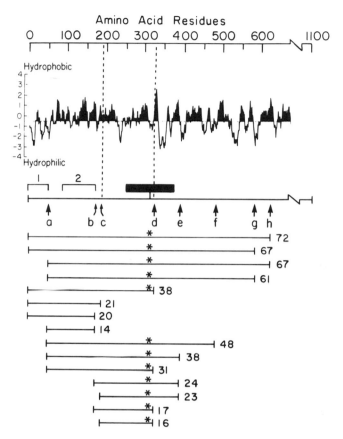

FIGURE 29-6   Peptide map of *Avena* phytochrome obtained through subtilisin digestion. Brackets 1 and 2 indicate location of type-1 and type-2 epitopes. Black box and asterisks indicate position of chromophore. Peptides with molecular masses in kilodalton on right are derived from cleavage sites denoted by letters *a–h*. Dotted vertical lines delineate 16-kDa chromopeptide. (From Jones and Quail, 1989. Copyright 1989, Springer-Verlag.)

and *e* (Fig. 29-6) (Grimm et al., 1986), the structure required for photo-reversibility lay in the *a-e* region. And because the superfluous 14-kDa peptide could be removed at *b* or *c* from a 31-kDa chromopeptide beginning at *a*, the structure could be delimited further to within the 22-kDa portion between sites *c* and *e*, residues 200–400.

Irradiation of the Pr (P_{660}) form of the 39-kDa fragment generated intermediates lumi-R and *meta*-R_a. "This result supports the assumption," the Munich workers concluded, "that the non-covalent chromophore binding site of the protein is different in Pfr from that in Pr (or P_{660}), lumi-R and *meta*-Ra" (Reiff et al., 1985).

Because small (60-kDa) phytochrome was fully photoreversible whereas the 39-kDa chromopeptide photoconverted between only between a R-absorbing and a bleached form (possibly *meta*-R_b), Rüdiger deduced that part of or all a 21-kDa sequence was required for the bleached intermediate to convert to Pfr and acquire its characteristic long-wavelength absorbance.

PHOSPHORYLATION

Knowing that phytochrome is a phosphoprotein (see Chapter 22), Lagarias wondered if phosphorylatable sites were equally available on Pr and Pfr. Yum-Shing Wong therefore phosphorylated oat phytochrome with radioactive ATP and each of three mammalian kinases. The discovery of one site in the 10-kDa amino terminus of Pr, another in the 69–84 kDa region of Pfr and a third, less readily phosphorylatable site in both Pr and Pfr, within the chromophore-binding domain, raised "the possibility that light-dependent phosphorylation (or dephosphorylation) may play an important role in the regulation of phytochrome-mediated biochemical changes in plant cells" (Wong et al., 1986).

After reaction with radioactive ATP, Pr yielded two radioactive peptides whose identical amino acid sequence corresponded to residues 15–20 of the known primary sequence (see Fig. 28-2). A synthetic peptide containing this sequence was phosphorylatable at Ser_{16}. Pfr yielded two phosphopeptides containing residues 596–607. The corresponding synthetic peptide was phosphorylatable at Ser_{598}. Thus a kinase A-phosphorylatable site was exposed near the amino terminus in Pr. But an available site in Pfr lay in the proteolytically sensitive hinge region.

Wong included an unusual set of controls in the mammalian kinase experiments. To determine the activity of these enzymes, he provided histone as a substrate, as the standard assay required. But one control contained only phytochrome, radiolabeled ATP, and histone. Much to his surprise, both the phytochrome and histone became labeled (Fig. 29-7). In the absence of histone, phytochrome was not labeled by ATP.

Substituting a nonprotein polycation, synthetic polylysine, Wong obtained the same result except that "the polylysine itself couldn't be phosphorylated. But Pr was phosphorylated and, to a lesser degree, Pfr," Lagarias explains. "So Pr turned out to be a preferred substrate, as with kinase activity in the phytochrome preparations that was stimulated by polycations." Using labeled ATP analogues, which bind to kinases but are

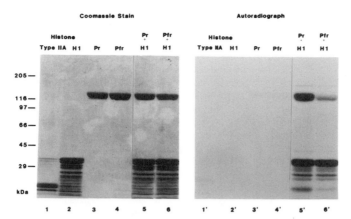

FIGURE 29-7   Phosphorylation of phytochrome by the endogenous polycation-dependent protein kinase associated with purified *Avena* phytochrome preparations. *Left:* Stained SDS-polyacrylamide gel. *Right:* Corresponding autoradiograph. Pr or Pfr was incubated with radioactive ATP in absence (third and fourth lanes) or presence (fifth and sixth lanes) of histone. Samples run in first and second lanes contained histone and ATP but no phytochrome. (From Wong et al., 1986.)

metabolically inactive, Wong deduced that polycations expose an ATP-binding site on phytochrome (Wong and Lagarias, 1989).

Rüdiger's group contested this proposal, because kinase activity in Grimm's phytochrome preparations decreased as purity increased and was associated with a 450-kDa protein that separated from phytochrome during electrophoresis (Grimm et al., 1989). Song's group also detected a protein kinase in highly purified phytochrome preparations, but repeated Bio-Gel filtration freed such solutions from kinase activity (Kim et al., 1989). But this issue has not been resolved because phytochrome purified as Pfr at Davis, using the Munich protocol, also lacked kinase activity (Wong et al., 1989).

In Schäfer's lab, meanwhile, Veit Otto detected rapid phytochrome-mediated changes in protein phosphorylation *in vivo*. When he labeled coleoptile tips with radioactive phosphate and pulsed them with R, the amount of label changed in three small proteins. A and C lost label (dephosphorylation) very rapidly, while B gained it (Fig. 29-8). The changes were partly FR-reversible, and their half-lives at 0°C were 1 s for A and 3 s for B and C, making this the most rapid Pfr-mediated physiological response reported. Thus "this reaction may be an integral part of a primary molecular mechanism of phytochrome action. Another plausible alternative is that the phosphorylation and dephosphorylation of proteins may be

time from the onset of irradiation [s]

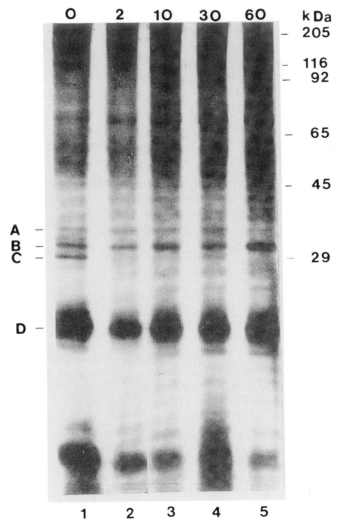

FIGURE 29-8   Autoradiograms of SDS-polyacrylamide gels on which supernatants of boiled homogenates were electrophoresed. *Lane 1:* Unirradiated coleoptile tips. *Lanes 2–5:* 2 s R pulse + darkness. Time between irradiation and addition of stop solution was 2s (lane 2), 10s (lane 3), 20s (lane 4), or 60 s (lane 5). (From Otto and Schäfer, 1988.)

a signal leading to the initiation of Pfr destruction'' (Otto and Schäfer, 1988).

REFERENCES

Bolton, G. W. (1979). Phytochrome: Aspects of its protein and photochemical properties. Ph.D. dissertation, University of Minnesota.

Boylan, M. T., and Quail, P. H. (1989). Oat phytochrome is biologically active in transgenic tomatoes. *Plant Cell* 1,765–773.

Chai, Y. -G., Song, P. -S., Cordonnier, M. -M., and Pratt, L. H. (1987). A photoreversible circular dichroism spectral change in oat phytochrome is suppressed by a monoclonal antibody that binds near its N-terminus by chromophore modification. *Biochemistry* 26,4947–4952.

Cherry, J. R., Hondred, D., Hershey, H., and Vierstra, R. (1991). Functional analysis of oat phytochrome by deletion mutagenesis. *J. Cell. Biochem.* (suppl.15A),112.

Cordonnier, M. -M. (1989). Monoclonal antibodies: Molecular probes for the study of phytochrome. *Photochem. Photobiol.* 49,821–831.

Cordonnier, M. -M., Smith, C., Greppin, H., and Pratt, L. H. (1983). Production and purification of monoclonal antibodies to *Pisum* and *Avena* phytochrome. *Planta* 158,369–376.

Cordonnier, M. -M., Greppin, H., and Pratt, L. H. (1984). Characterization by enzyme-linked immunosorbent assay of monoclonal antibodies to *Pisum* and *Avena* phytochrome. *Plant Physiol.* 74,123–127.

Cordonnier, M. -M., Greppin, H., and Pratt, L. H. (1985). Monoclonal antibodies with differing affinities to the red-absorbing and far-red-absorbing forms of phytochrome. *Biochemistry* 24,3246–3253.

Cordonnier, M. -M., Greppin, H., and Pratt, L. H. (1986). Identification of a highly conserved domain on phytochrome from angiosperms to algae. *Plant Physiol.* 80,982–987.

Grimm, R., Lottspeich, F., Schneider, H. A. W., and Rüdiger, W. (1986). Investigation of the peptide chain of 124 kDa phytochrome: Localization of proteolytic fragments and epitopes for monoclonal antibodies. *Z. Naturforsch.* 41c,933–1000.

Grimm, R., Eckerskorn, C., Lottspeich, F., Zenger, C., and Rüdiger, W. (1988). Sequence analysis of proteolytic fragments of 124-kilodalton phytochrome from etiolated *Avena sativa* L.: Conclusions on the conformation of the native protein. *Planta* 174,396–401.

Grimm, R., Gast, D., and Rüdiger, W. (1989). Characterization of a protein-kinase activity associated with phytochrome from etiolated oat (*Avena sativa* L.) seedlings. *Planta* 178,199–206.

Holdsworth, M. L., and Whitelam, G. C. (1987). A monoclonal antibody specific for the red-absorbing form of phytochrome. *Planta* 172,539–547.

Jones, A. M., and Quail, P. H. (1989). Phytochrome structure: Peptide fragments from the amino-terminal domain involved in protein–chromophore interactions. *Planta* 178, 147–156.

Kay, S. A., Nagatani, A., Keith, B., Deak, M., Furuya, M., and Chua, N. -H. (1989). Rice phytochrome is biologically active in transgenic tobacco. *Plant Cell* 1,775–782.

Keller, J. M., Shanklin, J., Vierstra, R. D., and Hershey, H. P. (1989). Expression of a functional monocotyledonous phytochrome in transgenic tobacco. *EMBO J.* 8,1005–1012.

Kim, I. S., Bai, U., and Song, P. -S. (1989). A purified 124-kDa oat phytochrome does not possess a protein kinase activity. *Photochem. Photobiol.* 49,319–323.

Lindemann, P., Braslavsky, S. E., Hartmann, E., and Schaffner, K. (1989). Partial purification and initial characterization of phytochrome from the moss *Atrichum undulatum* P. Beauv. grown in the light. *Planta* 178,436–442.

Litts, J. C., Kelly, J. M., and Lagarias, J. C. (1983). Structure–function studies on phytochrome. *J. Biol. Chem.* 258,11025–11031.

López-Figueroa, F., Lindemann, P., Braslavsky, S. E., Schaffner, K., Schneider-Poetsch, H. A. W., and Rüdiger, W. (1989). Detection of a phytochrome-like protein in macroalgae. *Bot. Acta* 102,117–180.

Lumsden, P. J., Yamamoto, K. T., Nagatani, A., and Furuya, M. (1985). Effect of monoclonal antibodies on the *in vitro* Pfr dark reversion of pea phytochrome. *Plant Cell Physiol.* 26,1313–1322.

Nagatani, A., Yamamoto, K. T., Furuya, M., Fukumoto, T., and Yamashita, A. (1983). Production and characterization of monoclonal antibodies to rye (*Secale cereale*) phytochrome. *Plant Cell Physiol.* 24,1143–1149.

Nagatani, A., Yamamoto, K. T., Furuya, M., Fukumoto, T., and Yamashita, A. (1984). Production and characterization of monoclonal antibodies which distinguish different surface structures of pea (*Pisum sativum* cv Alaska) phytochrome. *Plant Cell Physiol.* 25,1059–1068.

Otto, V., and Schäfer, E. (1988). Rapid phytochrome-controlled protein phosphorylation and dephosphorylation in *Avena sativa* L. *Plant Cell Physiol.* 29,1115–1121.

Plumpton, C., Partis, M. D., Thomas, B., Huskisson, N. S., and Butcher, G. W. (1988). Site-directed monoclonal antibodies against the amino-terminus of 124-kDa phytochrome from *Avena sativa* L. *Plant Cell Environ.* 11,487–491.

Pratt, L. H., Cordonnier, M. -M., and Lagarias, J. C. (1988). Mapping of antigenic domains on phytochrome from etiolated *Avena sativa* L. by immunoblot analysis of proteolytically derived peptides. *Arch Biochem. Biophys.* 267,723–735.

Quail, P. H., Barker, R. F., Colbert, J. T., Daniels, S. M., Hershey, H. P., Idler, K. B., Jones, A. M., and Lissemore, J. L. (1987). Structural features of the phytochrome molecule and feedback regulation of the expression of its genes in *Avena. In* "Molecular Biology of Plant Growth Control" (J. E. Fox and M. Jacobs, eds.), pp. 425–439. Alan R. Liss, New York.

Reiff, U., Eilfeld, P., and Rüdiger, W. (1985). A photoreversible 39 kdalton fragment from the Pfr form of 124 kdalton oat phytochrome. *Z. Naturforsch.* 40c,693–698.

Saunders, M. J., and Pratt, L. H. (1983). Immunocytochemical localization of phytochrome in pea (*Pisum sativum* L.) using monoclonal antibodies. *Plant Physiol.* 72(suppl. 84).

Schendel, R., and Rüdiger, W. (1989). Electrophoresis and electrofocusing of phytochrome from etiolated *Avena sativa* L. *Z. Naturwiss.* 44c,12–18.

Schendel, R., Tong, Z., and Rüdiger, R. (1989). Partial proteolysis of rice phytochrome: Comparison with oat phytochrome. *Z. Naturforsch.* 44c,757–764.

Schneider-Poetsch, H. A. W., Schwarz, H., Grimm, R., and Rüdiger, W. (1988). Cross-reactivity of monoclonal antibodies against phytochrome from *Zea* and *Avena*. *Planta* 173,61–72.

Schneider-Poetsch, H. A. W., Braun, B., and Rüdiger, W. (1989). Phytochrome—All regions marked by a set of monoclonal antibodies reflect conformational changes. *Planta* 177,511–514.

Schwarz, H., and Schneider, H. A. W. (1987). Immunological assay of phytochrome in small sections of roots and other organs of maize (*Zea mays* L.) seedlings. *Planta* 170,152–160.

Shimazaki, Y., Cordonnier, M. -M., and Pratt, L. H. (1983). Phytochrome quantitation in crude extracts of *Avena* by enzyme-linked immunosorbent assay with monoclonal antibodies. *Planta* 159,534–544.

Shimazaki, Y., Cordonnier, M. -M., and Pratt, L. H. (1986). Identification with monoclonal antibodies of a second antigenic domain on *Avena* phytochrome that changes upon its photoconversion. *Plant Physiol.* 82,109–113.

Sommer, D., and Song, P. -S. (1990). Chromophore topography and secondary structure of 124-kdalton *Avena* phytochrome probed by Zn^{2+}-induced chromophore modification. *Biochemistry* 29,1943–1948.

Song, P. -S., and Yamazaki, I. (1987). Structure–function relationship of the phytochrome chromophore. *In* "Phytochrome and Photoregulation in Plants" (M. Furuya, ed.), pp. 139–156. Academic Press, Tokyo.

Thomas, B., and Penn, S. E. (1986). Monoclonal antibody ARC MAC 50.1 binds to a site on the phytochrome molecule which undergoes a photoreversible conformational change. *FEBS Lett.* 195,174–178.

Thomas, B., Crook, N. E., and Penn, S. E. (1984a). An enzyme-linked immunosorbent assay for phytochrome. *Plant Physiol.* 60,409–415.

Thomas, B., Penn, S. E., Butcher, G. W., and Galfre, G. (1984b). Discrimination between the red- and far-red-absorbing forms of phytochrome from *Avena sativa* L. by monoclonal antibodies. *Planta 160,382–384*.

Thomas, B., Partis, M. D., and Jordan, B. R. (1986). Immunological approaches to the study of phytochrome. *In* "Immunology in Plant Science" (T. L. Wang, ed.), pp. 171–195. Cambridge University Press, Cambridge.

Thompson, L. K., Pratt, L. H., Cordonnier, M. -M., Kadwell, S., Darlix, J. -L., and Crossland, L. (1989). Fusion protein-based epitope mapping of phytochrome. *J. Biol. Chem.* 264,12425–12431.

Vierstra, R. D., Quail, P. H., Hahn, T. R., and Song, P. -S. (1987). Comparison of the protein conformations between different forms (Pr vs. Pfr) of native (124 kDa) and degraded (118/114 kDa) phytochromes from *Avena*. *Photochem. Photobiol.* 45,429–432.

Whitelam, G. C., Anderson, M. L., Billett, E. E., and Smith, H. (1985). Immunoblot analysis of cross-reactivity of monoclonal antibodies against oat phytochrome with phytochrome from several plant species. *Photochem. Photobiol.* 42,793–796.

Wong, Y. -S., and Lagarias, J. C. (1989). Affinity labeling of *Avena* phytochrome with ATP analogs. *Proc. Natl. Acad. Sci. USA* 86,3469–3473.

Wong, Y. -S., Cheng, H. -C., Walsh, D. A., and Lagarias, J. C. (1986). Phosphorylation of *Avena* phytochrome *in vitro* as a probe of light-induced conformational changes. *J. Biol. Chem.* 261,12089–12097.

Wong, Y. -S., McMichael, R. W., and Lagarias, J. C. (1989). Properties of a polycation-stimulated protein kinase associated with purified *Avena* phytochrome. *Plant Physiol.* 91,709–718.

Yamamoto, K. T., and Furuya, M. (1983). Spectral properties of chromophore-containing fragments prepared from pea phytochrome by limited proteolysis. *Plant Cell Physiol.* 24,713–718.

CHAPTER **30** _____

Sequestered Phytochrome

"This whole pelletability and sequestering phenomenon," says L. H. Pratt, "was taken by some people as evidence that phytochrome was becoming membrane-associated as Pfr. But observations in the last few years indicate that neither sequestered phytochrome nor pelleted phytochrome involve an association of phytochrome with membrane-containing structures."

Sequestering had been observed in etiolated oat, rice, and maize (see Chapter 21). Jane Saunders then detected it in pea, a dicot, locating phytochrome under the light microscope through immunocytochemical fluorescence (Saunders et al., 1983).

KINETICS OF SEQUESTERING

"A long-standing question, which John Mackenzie had dealt with in his doctoral dissertation," Pratt says, "was to find out how quickly phytochrome is sequestered in the cell. If sequestering and pelletability are the same, you would expect them to have the same time course."

When 4 s FR immediately followed 1 s R, most of the phytochrome in oat coleoptile cells remained diffusely distributed, but some sequestered areas were visible. When 1 s of darkness separated the pulses, sequestering was more marked, and it was complete by 10 s. During 3 min, sequestered areas tripled in size (McCurdy and Pratt, 1986a). Sequestering was thus very rapid in oat, with similar kinetics to pelletability, which had a half-time of 2 s (see Chapter 21). Both pelletability (Pratt and Marmé, 1976) and sequestering (Mackenzie et al., 1975) were reversed within 2 h after a FR pulse, and both were temperature-dependent (McCurdy and Pratt, 1986a; Pratt and Marmé, 1976; Quail and Briggs, 1978). Thus they appeared to result from the same event—which seemed not to involve microtubules or microfilaments (McCurdy and Pratt, 1986a).

448

PHYTOCHROME-ASSOCIATED PARTICLES

To explore the fine structure of sequestered areas, Volker Speth and Otto embedded oat coleoptile sections in a polar acrylic plastic, treated them with antiphytochrome rabbit serum and then with antirabbit serum tagged with gold particles. The microscopically immense particles labeled electron-dense areas that were absent in unirradiated tissue. In conventionally fixed tissue, the areas were amorphous, contained no ribosomes and other organelles, and were certainly not bounded by membrane (Speth et al., 1986).

David W. McCurdy and Pratt reported the presence of granular, membrane-free structures 200–400 nm in diameter in R-irradiated oats (Fig. 30-1). They named these entities *phytochrome-associated structures* (PASs). Gold-tagged antibodies labeled similar objects in pelleted phytochrome, indicating "that the PASs seen *in vivo* are the same structures as those seen *in situ*. These observations provide direct evidence that both enhanced pelletability and intracellular sequestering are different manifestations of the same intracellular event(s)" (McCurdy and Pratt, 1986b).

PASs were most likely made of protein—possibly aggregates of Pfr or Pfr plus unidentified protein(s). Although no MAb epitopes seemed to be masked, "both sequestering and pelletability exhibit characteristics consistent with a meaningful association between Pfr and a receptor. . . .

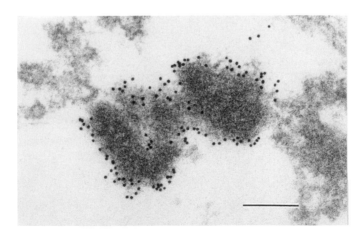

FIGURE 30-1   Pre-embedding immunogold localization of phytochrome as Pfr in dark-grown, R/FR-irradiated *Avena* coleoptile tissue. *Bar,* 0.2 μm. (From McCurdy and Pratt, 1986b. Reproduced from the *Journal of Cell Biology* (1986) vol. 103, pp. 2541–2550 by copyright permission of the Rockefeller University Press.)

However . . . the PASs might represent intracellular sites where the Pfr-dependent dark destruction of phytochrome occurs. Alternatively, the PASs might act as vehicles to transport Pfr to intracellular sites, possibly to vacuoles, where dark destruction occurs'' (McCurdy and Pratt, 1986b). Presumably, Pfr might also function during sequestering and then be destroyed.

SAP CONTENT

Schäfer's group explored the possible link with destruction by searching for ubiquitin in PASs—which they renamed *sequestered areas of phytochrome* (SAPs). Ubiquitin is a highly conserved, small protein that attaches covalently to short-lived animal and yeast proteins, marking them for destruction. Schäfer knew that Vierstra, who first detected ubiquitin in plants (Vierstra et al., 1985), had made a connection between ubiquitin and phytochrome.

Gold-coupled polyclonal antibodies against phytochrome or ubiquitin labeled the same structures in alternate sections of oat coleoptile (Fig. 30-2). But labeling for ubiquitin was less intense than for phytochrome, and it began only 30–120 min after R irradiation. Because destruction in *Avena* began after at least a 30-min lag (Schäfer et al., 1975), perhaps "phytochrome sequestering and the association of the electron-dense areas with ubiquitin may be a prerequisite for this destruction" (Speth et al., 1987).

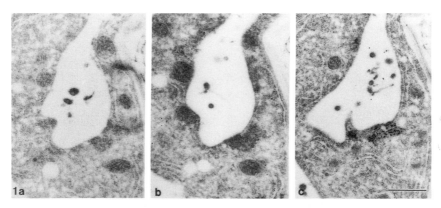

FIGURE 30-2   Serial sections of an *Avena* coleoptile cell from material fixed 30 min after a 5-min R pulse. Electron-dense areas show a positive reaction with both antiubiquitin (B) and antiphytochrome (C) as compared with control serum (A). ×20,000. (From Speth et al., 1987. Copyright 1987, Springer-Verlag.)

Speth and Ellen Hofmann detected SAPs 30 min after converting only 8%–13% (8–13 units) of total Pr to Pfr. But when a nonsaturating R preirradiation reduced P_{total}, at least 38–41 units were required. A saturating R pulse/FR pulse pretreatment, which made SAPs form and then disperse, changed the requirement to 19–23 units. Thus "an adaptation process exists and . . . our data indicate that SAP formation is not a simple self-aggregation of newly formed Pfr" (Hofmann et al., 1990).

Several groups are isolating SAPs in the hope of retrieving a substance that interacts with phytochrome. But partly purified SAPs that were the same shape as SAPs *in vivo* appeared to contain no other major protein (Hofmann et al., 1991). "But the migration of phytochrome purified from those particles is slower than that of classically purified phytochrome that has undergone the same treatment," Schäfer says.

SEQUESTERING IN OTHER SPECIES

In Beltsville, VanDerWoude's group obtained electron micrographs of SAPs in rye coleoptilar tissue, where Mackenzie had failed to detect sequestering under the light microscope (Mackenzie et al., 1978). As in oat, the structures were amorphous and not membrane-associated, but they were less granular and less numerous. Complete sequestering required 30 min, and the particles did not enlarge with time (Warmbrodt et al., 1989).

In soybean, sequestering also occurred slowly, but it was immediately FR-reversible even 10 min after a R pulse (Cope and Pratt, 1989; Cope, 1991). Thus both monocots and dicots sequester phytochrome, but there are interspecific differences.

REFERENCES

Cope, M. (1991). Comparative analysis of the intracellular redistribution of phytochrome in etiolated soybean seedlings. Ph.D. dissertation. University of Georgia, Athens.

Cope, M., and Pratt, L. H. (1989). Subcellular localization of phytochrome in soybean (*Glycine max* L.). European Symposium on Photomorphogenesis in Plants, Freiburg. Book of abstracts, S5.

Hofmann, E., Speth, V., and Schäfer, E. (1990). Intracellular localisation of phytochrome in oat coleoptiles by electron microscopy. Dependence on light pretreatments and the amount of the active, far-red-absorbing form. *Planta* 180,372–377.

Hofmann, E., Grimm, R., Harter K., Speth, V., and Schäfer, E. (1991). Partial purification of sequestered particles of phytochrome from oat (*Avena sativa* L.) seedlings. *Planta* 183,265–273.

Mackenzie, J. M., Jr., Coleman, R. A., Briggs, W. R., and Pratt, L. H. (1975). Reversible redistribution of phytochrome within the cell upon conversion to its physiologically active form. *Proc. Natl. Acad. Sci. USA* 72,799–803.

Mackenzie, J. M., Jr., Briggs, W. R., and Pratt, L. H. (1978). Intracellular phytochrome distribution as a function of its molecular form and of its destruction. *Am. J. Bot.* 65,671–676.

McCurdy, D. W., and Pratt, L. H. (1986a). Kinetics of intracellular redistribution of phytochrome in *Avena* coleoptiles after its photoconversion to the active, far-red-absorbing form. *Planta* 167,330–336.

McCurdy, D. W., and Pratt, L. H. (1986b). Immunogold electron microscopy of phytochrome in *Avena:* Identification of intracellular sites responsible for phytochrome sequestering and enhanced pelletability. Quotation reproduced from the *Journal of Cell Biology* (1986) vol. 103, pp. 2541–2550 by copyright permission of the Rockefeller University Press.

Pratt, L. H., and Marmé, D. (1976). Red-light enhanced phytochrome pelletability: Reexamination and further characterization. *Plant Physiol.* 58,686–692.

Quail, P. H., and Briggs, W. R. (1978). Irradiation-enhanced phytochrome pelletability: Requirement for phosphorylative energy *in vivo. Plant Physiol.* 62,773–778.

Saunders, M. J., Cordonnier, M-M., Palevitz, B. A., and Pratt, L. H. (1983). Immunofluorescence visualization of phytochrome in *Pisum sativum* L. epicotyls using monoclonal antibodies. *Planta* 159,545–553.

Schäfer, E., Lassig, T. U., and Schopfer, P. (1975). Photocontrol of phytochrome destruction in grass seedlings. The influence of wavelength and irradiance. *Photochem. Photobiol.* 22,193–202.

Speth, V., Otto, V., and Schäfer E. (1986). Intracellular localisation of phytochrome in oat coleoptiles by electron microscopy. *Planta* 168,299–304.

Speth, V., Otto, V., and Schäfer, E. (1987). Intracellular localisation of phytochrome and ubiquitin in red-light-irradiated oat coleoptiles by electron microscopy. *Planta* 171, 332–338.

Warmbrodt, R. D., VanDerWoude, W. J., and Smith, W. O. (1989). Localization of phytochrome in *Secale cereale* L. by immunogold electron microscopy. *Bot. Gaz.* 150, 219–229.

Multiple Phytochromes

"The first experiments at Beltsville were with light-grown plants—flowering in *Xanthium* and Downs's work with Pinto beans," Harry Smith stresses. "Downs' paper was cited time and time again for end-of-day, far-red effects on stem extension. But until the mid 1970s, nobody mentioned the experiments that pointed out that Pfr in the light-grown plant is stable."

Downs found that FR-rich light stimulated internode elongation when applied at the end of a photoperiod, although not if an R pulse followed (see Chapter 7). Because R was also effective after several hours of darkness, he deduced that "the active pigment is present and functional in the plant at all times. . . . The pigment system does not deteriorate during an ordinary dark period" (Downs et al., 1957).

The infamous trip to Montreal (see Chapter 9) and consequent discovery of Pfr destruction (see Chapter 11) turned attention to labile phytochrome, however. Phytochrome in cauliflower curd and green-leaf extracts was assumed to be a stabilized version of this unstable molecule.

NORFLURAZON-TREATED PLANTS

In 1975, Gerald F. Deitzer at the Smithsonian Institution Radiation Biology Laboratory began to study photoperiodism in barley. To assess the role of phytochrome, he wanted to assay the pigment in light-grown plants, which had hitherto proved impossible.[1] After use of albino mutants proved unsatisfactory, he began to grow seedlings in solutions of the herbicide norflurazon (Sandoz 9789),[2] which Klein obtained from Beltsville. Norflurazon blocked the synthesis of the carotenoids that normally protect

---

1. Despite attempts to remove chlorophyll from seedlings prior to assay (Akulovich, et al., 1966).

2. Norflurazon (Sandoz 9789) = 4-chloro-5-(methylamino)-2-(α,α,α,-trifluoro-*m*-tolyl)-3(2H)-pyridazinone.

TABLE 31-1

In Vivo Phytochrome Concentrations ($\Delta\Delta$ A$_{730}^{660}$) in
Light- and Dark-grown Seedlings

Treatment	0.2 mM San 9789	Water control
5d D	8.5×10^{-2}	8.0×10^{-2}
5d D + 6 h DF	1.2×10^{-2}	1.0×10^{-2}
6d DF	1.0×10^{-3}	—

[a] From Jabben and Deitzer, 1978a.
[b] DF, daylight fluorescent light; D, darkness.

chlorophyll from photooxidation, generating completely white oat seedlings (Jabben and Deitzer, 1977). It appeared not to affect Pr synthesis (Jabben and Deitzer, 1979) or photomorphogenic responses (Jabben and Deitzer, 1979; Gorton and Briggs, 1980).

Merten Jabben determined that norflurazon-treated seedlings had much less phytochrome in the light than in darkness (Table 31-1), the level in extracts being about 2% (Jabben and Deitzer, 1978a). Values ranging from 2% to 9% were obtained from light-grown seedlings of several dicot species (Jabben et al., 1980).

Jabben and Deitzer performed the first kinetic study on light-grown plants since Butler's cauliflower curd assay. When they transferred 4-day-old, dark-grown, norflurazon-treated oat seedlings to WL, total phytochrome decreased with a half-life of 2 h, reaching a low but steady level of 2.5% after 36 h (Jabben and Deitzer, 1978b). The collaborators attributed the persistent level to a balance between destruction and synthesis. "But that," says Deitzer, "turned out to be completely wrong."

RADIOIMMUNOASSAY

Hunt and Pratt measured phytochrome levels in green oat seedlings with a radioimmunoassay, which was sensitive, unaffected by chlorophyll, and did not require plants "to subsist off stored food like starving animals." To a crude extract, Hunt added a known amount of tritium[3]-labeled large oat phytochrome. Rabbit antibodies to large oat phytochrome complexed with both the added and endogenous phytochrome, whereas goat antibodies to the rabbit immunoglobulins precipitated the complex, allowing the level of radioactivity to be measured. Because the labeled pigment had to compete

3. The radioactive isotope of hydrogen.

with the phytochrome for rabbit antibodies, radiolabeling of the pellet decreased as phytochrome became more abundant in the sample. Assays with standard phytochrome solutions then converted counts into concentrations. The assay was sensitive to about 2 ng of phytochrome, whereas the spectrophotometric assay had a lower limit of 350 ng (Hunt and Pratt, 1979).

Crude extracts of light-grown oat seedlings contained 6.4 ng phytochrome/mg protein when assayed by this method. When seedlings were transferred to darkness, the level increased linearly and sixfold over the first 24 h and then more rapidly to reach 50 times the light-grown level over the next 24 h. During 12-h light/12-h dark cycles, the level increased threefold during darkness, returning each day to the lower, light-grown level (Hunt and Pratt, 1980).

TWO POOLS

Physiological evidence for a light-stable phytochrome pool had surfaced with some regularity (Borthwick et al., 1964; Wagner and Mohr, 1966; Kendrick et al., 1969; Klein, 1969; Wetherell and Koukkari, 1970; Steinitz et al., 1976; Tong and Schopfer, 1978). PAL accumulation in mustard seedlings correlated with photostationary state rather than total amount of pigment, for example. Apparently "phytochrome molecules involved in the regulation of this response by light pulses comprise a small fraction of the total phytochrome of the cotyledons. In contrast to bulk phytochrome, this fraction appears to be not subject to Pfr destruction" (Tong and Schopfer, 1978).

Brigitte Heim in Freiburg observed the familiar initial fast destruction followed by a much slower rate in *Amaranthus* seedlings, and the transition from the fast to the slow phase always occurred when the Pfr level reached 3.5% of that in etiolated seedlings (Fig. 31-1). "It is unclear at the moment whether there is one pool of Pfr showing fast destruction in the beginning and slow destruction afterwards," Schäfer said. "It is very tempting to speculate about two pools of phytochrome, especially since the transition from fast to slow is rather abrupt and the critical Pfr levels are always in the order of 3–4% of the dark Pfr level" (Heim et al., 1981).

When Brockmann established different initial Pfr levels in dark-grown *Amaranthus* seedlings with R pulses, destruction always showed biphasic kinetics, although transitions did not occur at the same Pfr level. Additional Pfr created by a second light pulse during the slow phase was also rapidly destroyed. But because a small amount formed by the first light pulse always escaped destruction, "the experiment suggests that not all

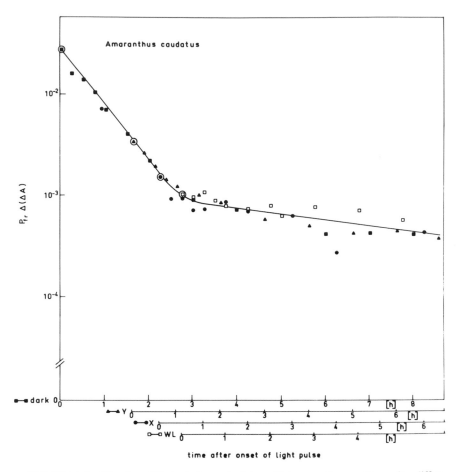

FIGURE 31-1 Kinetics of Pfr in darkness in 4-day-old *Amaranthus* seedlings after different pretreatments: (■) 96 h D + 5 min R; (▲) 94 h D + 2 × (5 min R + 55 min D) + 5 min R;
(●) 93 h D + 3 × (5 min R + 55 min D) + 5 min R; WL (□) 96 h WL + 5 min R. Initial Pfr levels
after treatment are circled. All seedlings were treated with norflurazon. (From Heim et al.,
1981.)

Pfr molecules are accessible to the same destruction mechanism, i.e.,
there are two populations of Pfr'' (Brockmann and Schäfer, 1982).

Phytochrome content changes as seedlings age, but Brockmann detected a constant amount of Pfr 4 h after a R pulse, regardless of the age of
seedlings (Fig. 31-2). The data contradicted "the hypothesis that the more
stable Pfr separates from a common pool" and could "be explained
by assuming the existence of two separate pools of phytochrome"
(Brockmann and Schäfer, 1982).

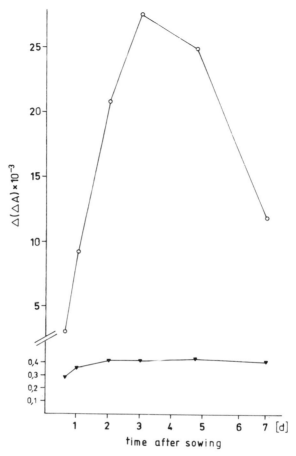

FIGURE 31-2   Time course of P_{total} accumulation (circles) and the amount of "stable" Pfr remaining 4 h after a saturating R pulse (triangles). (From Brockmann and Schäfer, 1982.)

Klaus Gottmann in Schäfer's lab and Jim Colbert in Quail's lab then discovered that phytochrome synthesis decreases drastically in the light (see Chapter 32). "So clearly you couldn't attribute the small but constant pool of phytochrome in light-grown plants to a balance between synthesis and destruction," Deitzer says. "And you couldn't therefore explain it on the basis of intermediates, either. So the only rational possibility became that the phytochrome in light-grown plants was somehow different from that in plants grown in the dark."

The difference between stable and labile phytochrome was generally attributed to changes in intracellular environment—possibly to light-promoted changes in membrane composition that protected phytochrome from destruction (Satter and Galston, 1976) or to spatial separation of stable phytochrome "due to the association with specific 'sites' " (Jabben and Holmes, 1983). There was, however, a much more radical explanation. Perhaps there were two quite different phytochromes: a stable species in light-grown plants and a labile one that predominated in dark-grown seedlings. "That," says Deitzer, "was a nasty suggestion no one wanted to deal with."

MULTIPLE GENE PRODUCTS?

Both Quail and Schäfer, however, had failed to immunoprecipitate phytochrome from the translation products of green-tissue poly(A) mRNA with serum raised against phytochrome from dark-grown seedlings (Gottmann and Schäfer, 1982; Quail et al., 1983). "One potential explanation . . . is that phytochrome detected in green tissue is a second gene product, immunologically distinct from that predominating in etiolated tissue," Quail boldly suggested (Quail et al., 1983). In August 1983, James G. Tokuhisa reported that antibodies against phytochrome from dark-grown oat failed to recognize 70% of the phytochrome in green oat and reacted poorly with the remaining 30% (Tokuhisa and Quail, 1983).

Hearing of this observation previously, Pratt had "just assumed there was some methodological problem." But after he and Cordonnier developed an ELISA with MAbs to phytochrome from dark-grown oats (see Chapter 29), Yukio Shimazaki observed a discrepancy between spectrophotometric and immunochemical assays of light-induced phytochrome disappearance (Fig. 31-3), which "indicates that phytochrome in light-treated plants may be antigenically distinct from that found in fully etiolated plants" (Shimazaki et al., 1983). Thomas made a similar observation while developing his ELISA for phytochrome (Thomas et al., 1984).

By precipitating chlorophyll and other contaminants with poly(ethylenimine) and fractionating with ammonium sulfate, Tokuhisa obtained a clear enough solution for spectral assay. In difference spectra, phytochrome from green oat had a peak of 652 nm, in contrast to the usual 666 nm (Fig. 31-4). Isosbestic points differed by 8 nm (Tokuhisa et al., 1985). This spectrum closely matched one obtained *in vivo* by Deitzer, using light-grown norflurazon-treated maize tissue (Jabben and Deitzer, 1978b).

By immunoblotting SDS-polyacrylamide gels, Tokuhisa was able to detect as little as 50 pg phytochrome. Polyclonal antibodies recognized a minor band at the 124-kDa position and a major band at 118 kDa. The smaller molecule appeared not to be a proteolytic product.

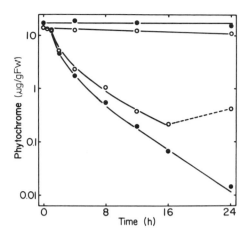

FIGURE 31-3 Phytochrome content in crude extracts of etiolated oat as determined by ELISA (closed circles) and spectrophotometric assay (open circles) after transfer of 5-day-old dark-grown oat seedlings to continuous illumination. Upper two curves are controls, using seedlings kept in darkness. (From Shimazaki et al., 1983. Copyright 1983, Springer-Verlag.)

The polyclonals, which were raised against phytochrome from etiolated oat, immunoprecipitated a 124-kDa polypeptide; this accounted for 30% of the green-tissue phytochrome. But the 118-kDa species stayed in the supernatant, suggesting that green oat shoots contained a small amount of the familiar chromoprotein and a larger amount of a smaller species. The latter was recognized by type-3 but not type-1 or type-2 MAbs.

Staphylococcus aureus V8 protease digests yielded different peptide patterns, suggesting extensive differences in primary sequence. Thus the two proteins were most likely different gene products, although one could have been a posttranslational modification of the other (Tokuhisa et al., 1985).

Cordonnier also obtained different peptide maps from green- and etiolated-oat phytochromes after digestion with *S. aureus* V8 protease (Cordonnier et al., 1986). "But we looked only at those peptides stained by a monoclonal antibody, Pea-25," Pratt explains. "So by definition we were looking at peptides that carry the same sequence."

PARTIAL PURIFICATION

Phytochrome from green oat was rapidly degraded, even in the presence of PMSF, by an endoprotease that did not attack phytochrome from etiolated oat (Cordonnier et al., 1986). But Hiroshi Abe in Okazaki managed to

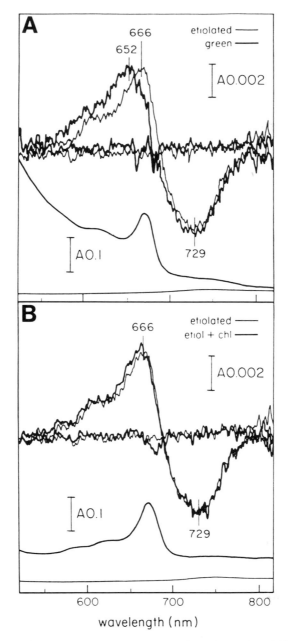

FIGURE 31-4 Absorbance spectra (lower curves in each panel) and difference spectra (upper curves in each panel) of etiolated-oat extract (thin lines) compared with those of (A) green-oat extract (thick lines) and (B) etiolated oat extract colored with chlorophyll at the concentration found in green-oat extract (thick lines). (From Tokuhisa et al., 1985. Copyright 1985, Springer-Verlag.)

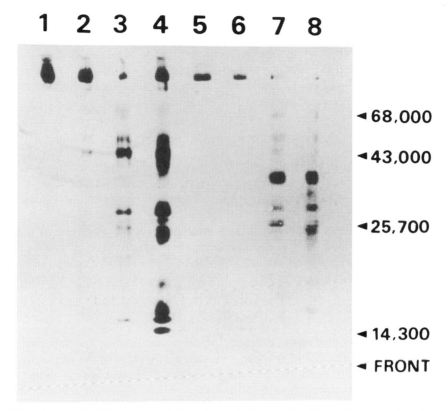

FIGURE 31-5   Peptide mapping of phytochrome I from etiolated tissue (lanes 1 to 4) and phytochrome II from light-grown tissue (lanes 5 to 8). Chromoproteins were digested with 0 ng (lanes 1, 5), 12.5 ng (lanes 2, 6), 50 ng (lanes 3, 7), and 200 ng (lanes 4, 8) of *S. aureus* V8 protease. Peptide fragments were stained with rabbit polyclonal antipea phytochrome antibodies. (From Abe et al., 1985.)

prepare a light-grown-pea pigment to 65% purity by combining ammonium sulfate fractionation, brushite chromatography, and an immunoaffinity procedure.

After brushite chromatography, 40% of the phytochrome complexed with mAP3 (an MAb against etiolated-pea large phytochrome) bound to agarose. The level was 80% for a sample from etiolated pea. Abe and Furuya named the binding and nonbinding fractions phytochrome I and phytochrome II, respectively. Abe further purified phytochrome II on a column containing mAP10, a MAb to native phytochrome from etiolated pea whose epitope lay near the amino or carboxyl terminus.

Phytochrome I from etiolated and green tissues yielded identical *S. aureus* V8 protease digests, but phytochrome II gave a different pattern

(Fig. 31-5). Thus "the phytochrome bound to mAP3 had the same amino acid sequence, no matter whether it was prepared from light-grown or etiolated tissue," while "the primary structure of phytochrome II was different from that of the phytochrome which bound to mAP3" (Abe et al., 1985). Unlike phytochromes I and II from oat, those from pea had similar spectral characteristics (Abe et al., 1985).

PHYTOCHROME I IN LIGHT-GROWN PLANTS

Oat-22, which strongly recognized phytochrome from etiolated oat, failed to react with phytochrome from green oat. Cordonnier thus concluded that light-grown seedlings contain at most a trace of phytochrome I (Cordonnier et al., 1986).

Tokuhisa determined that seeds imbibed over ice (to prevent germination) contained low but detectable levels of phytochrome, which either doubled in amount over 24 h at 25°C in the light and then remained stable or increased 250-fold during 72 h in darkness. By periodic immunoprecipitation with antiserum against oat phytochrome, he separated embryo extracts into a pelletable fraction containing 124-kDa phytochrome and a supernatant containing a 118-kDa species.

Immunoblot analysis of extracts, immunoprecipitates, and supernatants revealed that the 124-kDa species accounted for most of the phytochrome that appeared in darkness. But the 118-kDa species was always detectable in sequential supernatant fractions (Fig. 31-6). It slowly became more abundant in WL, as it did in darkness, whereas the 124-kDa species increased for 24 h and then slowly decreased. Thus both light- and dark-grown seedlings appeared to contain two types of phytochrome: one that was light-regulated and another that slowly became more abundant in both darkness and light to match the growth of the seedling (Tokuhisa and Quail, 1987).

Furuya's group confirmed this deduction, using MAbs. Koji Konomi found that part of the phytochrome extracted from pea embryo axes adsorbed to mAP1-agarose and that the remaining fraction adsorbed to mAP10-agarose. Peptide maps of this second fraction resembled those of the predominant phytochrome in light-grown seedlings. mAP10 was directed against native pea phytochrome, whereas mAP1 was an MAb to dark-grown, large pea phytochrome.

Immunoblots of the mAP1-agarose fraction stained with mAP14, a phytochrome I–specific MAb, whereas the mAP10-agarose fraction stained with phytochrome II–specific mAP11. Both MAbs faintly stained bands derived from dry embryonic axes. After 12 h of dark inhibition, the intensity of the phytochrome I band increased much more than that of the

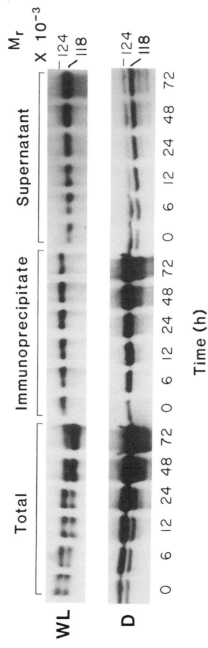

FIGURE 31-6 Immunoblot analysis of fractions obtained from immunoprecipitation of oat embryo/seedling extracts. (From Tokuhisa and Quail, 1987. Copyright 1987, Springer-Verlag.)

phytochrome II band. In the light, the phytochrome I band persisted, whereas the phytochrome II band increased only as much as in darkness.

ELISAs using mAP12, another phytochrome I–specific MAb, and mAP13, an MAb against large etiolated-pea phytochrome that recognized both phytochrome types, quantitated the changes. Dry seed embryos contained 0.007 μg of phytochrome I and 0.18 μg of phytochrome II per axis, whereas each dark-imbibed axis contained 0.2 μg of phytochrome I and 0.05 μg of phytochrome II. Levels of both phytochrome I and phytochrome II in light-imbibed embryos were 0.05 μg per axis (Konomi et al., 1987).

The phenotypes of transgenic plants containing the *phyA* sequence suggest that phytochrome I can operate in light-grown plants, at least under these abnormal conditions. Thus tobacco plants that overexpressed oat phytochrome were shorter than wild-type tobacco and had smaller but darker green leaves (Keller et al., 1989). Tobacco plants containing the rice *phyA* sequence had altered circadian oscillation of phytochrome-regulated gene transcription (Kay et al., 1989). Transgenic tomato plants containing oat *phyA* had short, dark hypocotyls as seedlings and very short stems (see Fig. 29-5) and overpigmented fruits as adults (Boylan and Quail, 1989).

DEFINITIVE EVIDENCE

In 1987, Furuya (Fig. 31-7) became head of the Laboratory for Plant Biological Regulation, Frontier Research Program at RIKEN (**Ri**kagaku **Ken**kyusho = Institute of Physical and Chemical Research), a prestigious

FIGURE 31-7   Masaki Furuya, 1989. Photo by Linda Sage.

governmental research institute in Wako. There he encountered protein chemist Koji Takio, who sequenced parts of the phytochrome II apoprotein, which the group obtained by separating *S. aureus* V8 protease or trypsin digests of partially purified pea phytochrome by SDS-PAGE. When the amino-terminal sequences of two peptides from each digest were aligned against that of pea phytochrome I (Sato, 1988), 37% of the residues differed (Fig. 31-8). The fragment containing residues 922–947 was the least homologous. This was part of the carboxyl-terminal region, which, in phytochrome I, is poorly conserved among different plant species. The RIKEN data provided "the first unequivocal evidence that phytochrome I and phytochrome II are encoded by different genes rather than posttranslationally modified forms of one or the other" (Abe et al., 1989). Microsequencing of a 19-residue degradation fragment of phytochrome immunopurified from green oat verified that "green- and etiolated-plant phytochrome apoproteins are encoded by different genes in monocotyledons as well as dicotyledons" (Pratt et al., 1991a).

ARABIDOPSIS GENES

In Albany, Quail's group began to study *Arabidopsis thaliana,* a mustard with a small genome, hoping to determine the minimal number of phytochrome genes in a higher plant. Using a zucchini probe, Robert A. Sharrock screened an *A. thaliana* library and isolated a single genomic clone. From this he prepared a single-stranded, 0.9-kb DNA probe that hybridized with multiple bands in blots of *A. thaliana* DNA digests, indicating the possible existence of at least five structurally related genes.

The probe also hybridized to several clones in a cDNA library assembled from 3-week-old, green *A. thaliana* leaves. One contained the complete coding sequence for a 124-kDa polypeptide—PHYA. Two that formed less stable hybrids respectively contained the sequence for a 129-kDa polypeptide, PHYB, and that for a second 124-kDa polypeptide, PHYC.

Sharrock subcloned and sequenced these coding inserts and deduced amino acid sequences. "When we then compared A and B," Quail explains, "they were only about 50% identical. When we compared B with C, we got the same story—about 50% or less (Fig. 31-9). So within the same plant, you have three sequences for phytochrome that are only 50% related to each other."

The sequences were largely homologous around residue 361, the chromophore-attaching cysteine. They also had similar hydropathy profiles that suggested solubility in cytosol rather than membrane. The most strik-

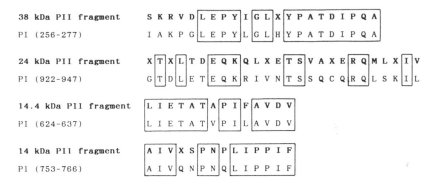

FIGURE 31-8   Amino-terminal amino acid sequences of pea phytochrome II fragments and their homology with pea phytochrome I. Identical residues are enclosed in boxes. Numbers reveal the position of the amino acid residue of PI from the initiator methionine. X indicates an unidentified residue. (From Abe et al., 1989.)

ing difference was in PHYB, which had a 35-residue extension at its amino terminus and an 11-residue extension at its carboxyl terminus.

The *phyA* sequence was most homologous to those determined previously; thus it was likely to represent phytochrome I. *PhyB* and *phyC* were much less similar, presumably because they had undergone long periods of independent evolution after the duplication of an ancestral gene. "We think that at some point before the divergence of the monocots and dicots, which happened 100 million years ago, some more primitive form of phytochrome diverged in at least three different directions and gave rise to *phyA*, *phyB* and *phyC*," says Quail (Fig. 31-10).[4]

4. After moving to Montana State University, Sharrock used the polymerase chain reaction and oligonucleotide primers corresponding to highly conserved regions of phytochrome to clone five ~300-bp fragments of phytochrome-related genomic DNA. Three came from *phyA*, *phyB*, and *phyC*, but the other two proved to be new sequences, *D* and *E*. The *phyD* fragment was highly homologous to *phyB*, suggesting that the two diverged more recently from each other than from *phyA*, *phyC*, and *phyE*. The *phyD* and *phyE* fragments hybridized to bands in Northern blots of *Arabidopsis* mRNA. But the signals were weak, suggesting that *Arabidopsis* grown in continuous WL produces only low levels of *phyD* and *phyE* mRNAs (R. A. Sharrock, unpublished data). Quail's group has now overexpressed *Arabidopsis phyA*, *phyB*, and *phyC* sequences in *E. coli*, showing for the first time that chromophore (phycocyanobilin) can covalently link to each of the three proteins (B. M. Parks, unpublished data).

· The relationships of the various angiosperm *phy* genes to those of lower plants is currently unclear. Rüdiger's group has sequenced part of a phytochrome gene from the moss *Ceratodon purpureus*. The 1474-bp sequence, which included the chromophore attachment site, was respectively 63%, 64%, and 63% identical to the corresponding region in *Arabidopsis phyA*, *phyB*, and *phyC* (Thümmler et al., 1990). A fern gene (M. Wada, unpublished data) and a pine gene (J. Silverthorne, unpublished data) are also being sequenced.

| | *phyA* phytochromes | | | | | | | |
| | monocot | | | dicot | | | | |
	oat	rice	corn	zucchini	pea	*phyA*	*phyB*	*phyC*
Oat	—	89	88	64	65	64	48	49
Rice	89	—	88	64	65	64	49	49
Corn	88	88	—	64	65	64	48	49
Zucchini	64	63	63	—	78	79	52	51
Pea	64	64	64	78	—	79	51	52
phyA	63	63	63	79	79	—	52	52
phyB	46	46	46	49	48	49	—	51
phyC	48	48	48	51	51	52	49	—

FIGURE 31-9   Percent amino acid sequence identity among phytochromes from various plant species and *Arabidopsis thaliana* PHYA, PHYB and PHYC. Because sequences were different lengths, the number of identical residues for each pair of aligned sequences was divided by the total number of positions in the alignment, including gaps and extensions (below the diagonal) or by the number of amino acids in the shorter of the two sequences (above the diagonal) and expressed as a percent. Solid lines box comparisons within monocot or dicot PHYA sequences; dashed lines box those between monocot and dicot sequences. (From Sharrock and Quail, 1989. Copyright © 1989 by Cold Spring Harbor Laboratory Press.)

Sharrock copied the 3' end of each clone to make single-stranded, gene-specific probes. In dark-grown tissue, these detected high levels of *phyA* mRNA (4.0 kb), which decreased drastically several hours after seedlings were illuminated to reveal a larger (4.4 kb) *phyA* mRNA. This suggested that the *phyA* gene was both light-regulated and could be transcribed from several start sites, as was known to happen in pea (see Chapter 33). *phyB* and *phyC* mRNAs were much less abundant in dark-grown tissue, and their levels were not significantly altered by light. "Tokuhisa's data from oats say that the protein we are calling phytochrome II is less abundant than phytochrome I and is also constitutively expressed," says Quail. "So by those two criteria, *B* and *C* genes fit the general characteristics of the type II protein."

The derived amino acid sequence suggests that *phyA* encodes phytochrome I. That from *phyB* resembled the amino-terminal portion of pea phytochrome II. *phyC* corresponded to an unknown phytochrome protein (Sharrock and Quail, 1989).

Because the relationships between *phyA/B/C,* phytochromes I and II, and labile and stable pools are unknown in most instances, Furuya warns

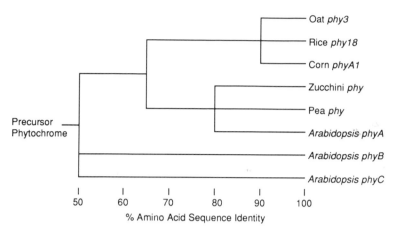

FIGURE 31-10 Speculative evolutionary tree for phytochrome polypeptides. (From Sharrock and Quail, 1989. Copyright © 1989 by Cold Spring Harbor Laboratory Press.)

against unwarranted extrapolations (Furuya, 1989). "Until we have the sequence of a gene or peptide," he says, "we should not say what molecular species it is—PHYA, PHYB, or PHYC. Although if a phytochrome cross-reacts with an antibody that has been produced using a known gene sequence, that is O.K. Otherwise, we should talk about type I and type II phytochromes, which are defined immunochemically. For physiological studies, it's safest to say 'labile' or 'stable' phytochrome. And stable and labile pools might be different depending on the response."

MULTIPLE GENES IN OAT?

Long before the *Arabidopsis* data were obtained, Pratt proposed that at least three different phytochromes might exist in oat, because Oat-9 and Oat-16, which seemed to bind to the same epitope, precipitated only about one-third of the photoreversible phytochrome from green tissue. "It appears," he said, "that phytochrome from green oat shoots may itself be subdivided into at least two immunochemically distinct pools" (Shimazaki and Pratt, 1985).

Pea-25, which stained phytochrome-like proteins from a wide range of species (see Chapter 29), also recognized phytochrome from green oat, and it reacted with the green-oat fraction that Oat-9 failed to precipitate. It failed to react with the Oat-9-precipitated fraction, supporting the notion of two immunologically distinct green-oat phytochromes (Cordonnier et al., 1986).

The use of seven newly obtained MAbs to green-oat phytochrome (Pratt et al., 1991b) confirmed this conclusion, because four recognized a 123-kDa[5] polypeptide that predominated in extracts of light-grown oat leaves, whereas the other three detected a 125-kDa polypeptide that was also present. The two green-oat polypeptides were also differentially sensitive to Zn^{2+}, which altered the electrophoretic mobility of only the 123-kDa species (Wang et al., 1991). "While these differences do not prove that they are encoded by different genes," Pratt concluded, "this hypothesis appears reasonable in view of the recent data of Sharrock and Quail (1989), who documented the presence of at least three markedly different phytochrome genes in *Arabidopsis*" (Pratt et al., 1991a). Sequencing could reveal whether the 123-kDa and 125-kDa species are truly different polypeptides or result from differential modification of a single gene product.

PHOTOMORPHOGENIC MUTANTS

The discovery of multiple phytochromes raised the possibility that different molecular species control different physiological responses (Sharrock and Quail, 1989; Smith and Whitelam, 1990), possibly through different primary actions. "For about 25 years," Smith says, "we argued about how a single phytochrome could operate in so many physiologically distinct modes. With only one phytochrome, the two possibilities were (1) a single primary action leading to branched pathways via a second messenger, or (2) much less plausibly, a large number of distinct primary actions. This logical problem has been solved by the molecular biologists, since each of the different phytochromes could have similar photochemical properties but different biochemical actions."

ARABIDOPSIS HY MUTANTS

Attempts to match phytochrome species with response began with studies of photomorphogenic mutants of higher plants. Geneticist Maarten Koornneef initiated artificial induction of *A. thaliana* mutants in Wageningen in the late 1970s (Koornneef and van der Veen, 1980; Koornneef et al., 1980).

5. Discrepancies between size estimates for green-oat phytochrome in Pratt's and Quail's labs (Cordonnier et al., 1986; Tokuhisa and Quail, 1989) appear to be due to methodological differences.

A. thaliana hy mutants have long hypocotyls in the light as well as in darkness. Despite a severe shortage of spectrophotometrically detectable phytochrome (Koornneef et al., 1980), *hy*1 and *hy*2 mature into almost normal plants, and their seed germination is under phytochrome control (Cone and Kendrick, 1985).

Extracts of *hy*1, *hy*2, and wild-type seedlings contain identical levels of apoprotein—a 116-kDa molecule. But peptide maps of the protein are abnormal and largely unaltered by R or FR irradiation of the samples, suggesting that most of the molecules lack a functional chromophore (Parks et al., 1989). Thus *hy*1 and *hy*2 may be unable to synthesize or attach normal amounts of phytochromobilin but may somehow produce enough functional phytochrome for survival.

A new complementation group, *hy*6, also lacks spectrophotometrically detectable but not immunochemically detectable phytochrome. Its proplastids differentiate into chloroplasts, albeit with lower-than-normal amounts of chlorophyll, high chlorophyll *a/b* ratios, and reduced granal stacking. Levels of several genes known to be under phytochrome control in dark-grown plants (including that for light-harvesting chlorophyll *a/b*-binding protein—CAB—a major component of chloroplast thylakoid membrane) also equal or exceed wild-type levels. Thus the phytochrome absent from *hy*6 appears to regulate photogene transcription in etiolated plants but to be largely dispensable in bright light (Chory et al., 1989a). All *hy* mutants respond to reduced daytime R : FR ratios, a response that must involve a photoreversible chromophore (G. C. Whitelam and H. Smith, unpublished data).

TOMATO *AUREA* MUTANTS

A mutant selected at Wageningen because its seeds were unable to germinate without added gibberellic acid was allelic with already known *aurea* (*au*) mutants, whose leaves are golden. In the greenhouse, *au* has elongated hypocotyls, and these lack spectrophotometrically detectable phytochrome even in the dark (Koornneef et al., 1981). The mutant also fails to respond to light during seed germination, anthocyanin synthesis, and hypocotyl elongation (Koornneef et al., 1985).

In contrast to the *Arabidopsis hy* mutants, *au* lacks phytochrome apoprotein (Parks et al., 1987). But *au* plants make normal levels of phytochrome mRNA, which can be translated *in vitro* (Sharrock et al., 1988). Moreover, the *au* locus resides on chromosome 1 (Khush and Rick, 1968), whereas the coding sequence for phytochrome maps to chromsome 10 (Tanksley et al., 1988). Thus "the *aurea* phytochrome deficiency re-

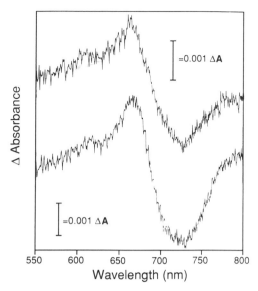

FIGURE 31-11   Phytochrome difference spectra for extracts of light-grown *aurea* and wild-type tomato. (From López-Juez et al., 1990.)

flects, not an inhibition of phytochrome biosynthesis, but a reduced stability of the polypeptide *in vivo*'' (Sharrock et al., 1988). ''The effect clearly involves the product of the *A* gene,'' Quail adds, ''because that is the most abundant phytochrome in wild-type plants. But it is not the *A* gene itself that is mutated here.''

After a R pulse, *cab* mRNA production was much reduced in etiolated *au* seedlings, an observation that provided the first rigorous proof that physiological responses of dark-grown plants to R result from the action of Pfr rather than the easing of inhibition by Pr. If Pr inhibited *cab* mRNA production, its presence in *au* at less than 5% of the wild-type level would result in a high level of *cab* expression in darkness (Sharrock et al., 1988).

Phytochrome can be extracted from light-grown *au* tissues (Fig. 31-11) at 66% wild-type levels, which suggests that perhaps phytochrome I normally accounts for about one-third of the spectrophotometrically detectable phytochrome in illuminated plants (López-Juez et al., 1990). The *au* plants responded to end-of-day FR, moreover, indicating the participation of light-stable pigment in this response (Adamse et al., 1988b; López-Juez et al., 1990). They also show qualitatively normal responses to changes in daytime R : FR ratios (G. C. Whitelam and H. Smith, unpublished data).

Crossing *au* with a spontaneous high-pigment (*hp*) mutant, which over-produced anthocyanin, gave a double *au/hp* mutant. After a B pretreatment and subsequent R pulse, *au/hp* plants synthesized only 3% of the anthocyanin produced by *hp* plants, suggesting "that it is the 'bulk' labile phytochrome pool in etiolated seedlings which regulates anthocyanin synthesis at this stage" (Adamse et al., 1989).

CUCUMBER *1h* MUTANTS

De-etiolated seedlings of a long hypocotyl (*1h*) mutant of cucumber (Koornneef and van der Knaap, 1983) lack the end-of-day FR response (Adamse et al., 1988a) and have long hypocotyls and poorly expanded cotyledons in continuous WL (Fig. 31-12) and, especially, R. Phytochrome levels in flower petals and norflurazon-bleached leaves are only one-third to one-half wild-type levels (Adamse et al., 1988a). Dark-grown *1h* seedlings behave much like etiolated wild-type plants and contain similar levels of spectrophotometrically detectable phytochrome (Adamse et al., 1987), suggesting that phytochrome II is nonfunctional in this mutant (Adamse et al., 1988a).

Immunoblots from extracts of de-etiolated shoots of both wild-type and *1h* cucumber had bands at the 116-kDa position that stained with mAP5, a MAb against pea phytochrome I (Nagatani et al., 1989). An MAb against

FIGURE 31-12   Seedlings of cucumber wild-type (*left*) and *1h* mutant (*right*) germinated in darkness and transferred, when the hypocotyls were in both cases 4 cm long, to continuous WL for 7 days at 25°C. Bar represents 5 cm. (From Adamse et al., 1987.)

tobacco *phyB*, mAT1, which reacts strongly with stable phytochrome from light-grown cucumber plants (Tomizawa et al., 1991) detected phytochrome bands in wild-type but not *lh* extracts, even after extraction into hot SDS-buffer. Thus the *lh* mutant lacks a mAT1-reactable protein—presumably PHYB—and is thus a photoreceptor mutant rather than a signal transduction mutant (E. López, A. Nagatani, K. Tomizawa, M. Deak, R. E. Kendrick, M. Koornneef, and M. Furuya, unpublished data).

The end-of-day stem-elongation response cannot be attributed entirely to PHYB, because *lh* exhibits very small responses to changes in end-of-day and daytime R : FR ratio (G. C. Whitelam, unpublished data). So more than one phytochrome—PHYC or residual PHYA, for example—may regulate hypocotyl growth in the light-grown plant, although PHYB appears to be the main effector. And transgenic tobacco plants that overexpress oat *phyA* (Keller et al., 1989) show a much larger end-of-day FR effect than wild-type tobacco seedlings. "That has to be mediated by phytochrome I," Whitelam says, "although I suspect that under normal conditions the contribution of PHYA is very small."

PEA *lv* MUTANTS

The pea *lv* mutant overelongates in WL, R and B but not in FR or darkness. Light-grown seedlings lack the end-of-day response to FR and are slow to de-etiolate. But both etiolated and green tissues contain normal levels of spectrophotometrically and immunochemically detectable phytochrome—immunoblotting produced positive reactions with both mAP5 and mAP11, a MAb to pea phytochrome II (Nagatani et al., 1990). "So *lv* could lack a third or fourth phytochrome," says Akiro Nagatani, "or it might be a signal transduction mutant."

ARABIDOPSIS DET MUTANTS

The *det* (de-etiolated) mutants, isolated by Joanne Chory in 1987, resemble light-grown plants in darkness, where they have short hypocotyls, expanded cotyledons, and leaves containing anthocyanin (Fig. 31-13). They also express many genes that are normally induced by light. In WL, *det* plants are smaller and paler than wild-type *A. thaliana,* and their roots contain chloroplasts in normally nongreen tissues. Thus DET, the product of the missing gene, may suppress a developmental program in certain tissues of light-grown plants as well as in darkness (Chory et al., 1989b;

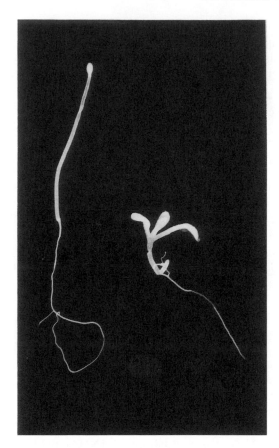

FIGURE 31-13   Wild-type *Arabidopsis* seedling (*left*) and *det*1 (*right*) after 7 days of growth in the dark. Note that *det*1 seedling has a short hypocotyl, expanded cotyledons, and leaves, unlike the developmentally arrested wild-type. Courtesy of Marc Lieberman, Salk Institute Photolab.

Chory and Peto, 1990). *det*2 is small and bushy in the light, and its leaves fail to senesce after plants have flowered (Chory et al., 1991).

When *hy*1 (or *hy*2 or *hy*6) was crossed with *det*1, the resulting double mutant looked very similar to *det*1; the *hy*1/*det*2 mutant appeared similar to *det*2. But when *det*1 was crossed with *det*2, the phenotype of the double mutant was additive, suggesting, Chory says, "that DET1 and DET2 do not interact. The data are most easily explained by an order of gene action that places *hy*1, *hy*2, or *hy*6 before *det*1 and *det*2 on a regulatory pathway and *det*1 and *det*2 on separate pathways. Our current thinking is that DET1 and DET2 are molecular switches that are turned off by phytochrome. If the switch is on, you have high levels of DET activity, and the light developmental program is repressed. Based on our analysis of the pheno-types of the *det*1 and *det*2 mutants, we would predict that high levels of DET1 activity would cause repression of the light developmental program within a given cell type. A second prediction is that DET2 levels would be low during early stages of development but would increase when it is time to turn off light-regulated genes. We don't know if DET1 and DET2 act downstream from the same or different phytochromes."

REFERENCES

Abe, H., Yamamoto, K. T., Nagatani, A., and Furuya, M. (1985). Characterization of green tissue-specific phytochrome isolated immunochemically from pea seedlings. *Plant Cell Physiol.* 26,1387–1399.

Abe, H., Takio, K., Titani, K., and Furuya, M. (1989). Amino-terminal amino acid sequences of pea phytochrome II fragments obtained by limited proteolysis. *Plant Cell Physiol.* 30,1089–1097.

Adamse, P., Jaspers, P. A. P. M., Kendrick, R. E., and Koornneef, M. (1987). Photomor-phogenetic responses of a long hypocotyl mutant of *Cucumis sativus* L. *J. Plant Physiol.* 127,481–491.

Adamse, P., Jaspers, P. A. P. M., Bakker, J. A., Kendrick, R. E., and Koornneef, M. (1988a). Photophysiology and phytochrome content of long-hypocotyl mutant and wild-type cucumber seedlings. *Plant Physiol.* 87,264–268.

Adamse, P., Jaspers, P. A. P. M., Bakker, J. A., Wesselius, J. C., Heeringa, G. H., Kendrick, R. E., and Koornneef, M. (1988b). Photophysiology of a tomato mutant deficient in labile phytochrome. *J. Plant Physiol.* 133,436–440.

Adamse, P., Peters, J. L., Jaspers, P. A. P. M., van Tuinen, A., Koornneef, M., and Kendrick, R. E. (1989). Photocontrol of anthocyanin synthesis in tomato seedlings: A genetic approach. *Photochem. Photobiol.* 50,107–111.

Akulovich, N. K., Godnev, T. N., and Domash, V. I. (1966). A method for determining phytochrome in green plants. *Dokl. Akad. Nauk BSSR* 10,601–603.

Borthwick, H. A., Toole, E. H., and Toole, V. K. (1964). Phytochrome control of *Paulownia* seed germination. *Isr. J. Bot.* 13,122–133.

Boylan, M. T., and Quail, P. H. (1989). Oat phytochrome is biologically active in transgenic tomatoes. *Plant Cell* 1,765–773.

Brockmann, J., and Schäfer, E. (1982). Analysis of Pfr destruction in *Amaranthus caudatus* L.—Evidence for two pools of phytochrome. *Photochem. Photobiol.* 35,555–558.

Chory, J., and Peto, C. A. (1990). Mutations in the *det*1 gene affect cell-type-specific expression of light-regulated genes and chloroplast development in *Arabidopsis. Proc. Natl. Acad. Sci. USA* 87,8776–8780.

Chory J., Peto, C. A., Ashbaugh, M., Saganich, R., Pratt, L., and Ausubel, F. (1989a). Different roles for phytochrome in etiolated and green plants deduced from characteri-zation of *Arabidopsis thaliana* mutants. *Plant Cell* 1,867–880.

Chory, J., Peto, C. A., Feinbaum, R., Pratt, L., and Ausubel, F. (1989b). *Arabidopsis thaliana* mutant that develops as a light-grown plant in the absence of light. *Cell* 58, 991–999.

Chory, J., Nagpal, P., and Peto, C. A. (1991). Phenotypic and genetic analysis of *det2*, a new mutant that affects light-regulated seedling development in *Arabidopsis*. *Plant Cell* 3,445–459.

Cone, J. W., and Kendrick, R. E. (1985). Fluence–response curves and action spectra for promotion and inhibition of seed germination in wildtype and long-hypocotyl mutants of *Arabidopsis thaliana* L. *Planta* 163,43–54.

Cordonnier, M.-M., and Pratt, L. H. (1991). Phytochrome from green *Avena* characterized with monoclonal antibodies directed to it. *In* "Photobiology: The Science and Its Applications" (E. Riklis, ed.), pp. 469–477. Plenum, New York.

Cordonnier, M.-M., Greppin, H., and Pratt, L. H. (1986). Phytochrome from green *Avena* shoots characterized with a monoclonal antibody to phytochrome from etiolated *Pisum* shoots. *Biochemistry* 25,7657–7666.

Downs, R. J., Hendricks, S. B., and Borthwick H. A. (1957). Photoreversible control of elongation of Pinto beans and other plants under normal conditions of growth. *Bot. Gaz.* 118,199–208.

Furuya, M. (1989). Molecular properties and biogenesis of phytochrome I and II. *Adv. Biophys.* 25,133–167.

Gorton, H. L., and Briggs, W. R. (1980). Phytochrome responses to end-of-day irradiations in light-grown corn grown in the presence and absence of Sandoz 9789. *Plant Physiol.* 66,1024–1026.

Gottmann, K., and Schäfer, E. (1982). *In vitro* synthesis of phytochrome apoprotein directed by mRNA from light and dark grown *Avena* seedlings. *Photochem. Photobiol.* 35, 521–525.

Hunt, R. E., and Pratt, L. H. (1979). Phytochrome radioimmunoassay. *Plant Physiol.* 64,327–331.

Hunt, R. E., and Pratt, L. H. (1980). Radioimmunoassay of phytochrome content in green, light-grown oats. *Plant Cell Environ.* 3,91–95.

Jabben, M. (1980). The phytochrome system in light-grown *Zea mays* L. *Planta* 149,91–96.

Jabben, M., and Deitzer, G. (1977). Spectrophotometric measurements of phytochrome in light grown plants. *Plant Physiol.* 59 (suppl. 100).

Jabben, M., and Deitzer, G. (1978a). A method for measuring phytochrome in plants grown in white light. *Photochem. Photobiol.* 27,799–802.

Jabben, M., and Deitzer, G. (1978b). Spectrophotometric measurements in light-grown *Avena sativa* L. *Planta* 143,309–313.

Jabben, M., and Deitzer, G. (1979). Effects of the herbicide San 9789 on photomorphogenic responses. *Plant Physiol.* 63,481–485.

Jabben, M., and Holmes, M. G. (1983). Phytochrome in light-grown plants. *In* "Encyclopedia of Plant Physiology" (Shropshire W. J. Jr., Mohr, H. eds.), New Series, Vol. 16B, pp. 704–722. Springer, Berlin.

Kay, S. A., Nagatani, A., Keith, B., Deak, M., Furuya, M., and Chua, N.-H. (1989). Rice phytochrome is biologically active in transgenic tobacco. *Plant Cell* 1,775–782.

Keller, J. M., Shanklin, J., Vierstra, R. D., and Hershey, H. P. (1989). Expression of a functional monocotyledonous phytochrome in transgenic tobacco. *EMBO J.* 8,1005–1012.

Kendrick, R. E., Spruit, C. J. P., and Frankland, B. (1969). Phytochrome in seeds of *Amaranthus caudatus*. *Planta* 88,293–302.

Khush, G. S., and Rick, C. M. (1968). Cytogenetic analysis of the tomato genome by means of induced deficiencies. *Chromosoma* 23,452–484.

Klein, A. O. (1969). Persistent photoreversibility of leaf development. *Plant Physiol.* 44,897–902.

Konomi, K., Abe, H., and Furuya, M. (1987). Changes in the content of phytochrome I and II apoproteins in embryonic axes of pea seeds during imbibition. *Plant Cell Physiol.* 28,1443–1451.

Koornneef, M., and van der Knaap, B. J. (1983). Another long hypocotyl mutant at the *lh* locus. *Cucurbit. Genet. Coop. Rpt.* 6,13.

Koornneef, M., and van der Veen, J. H. (1980). Induction and analysis of gibberellin-sensitive mutants in *Arabidopsis thaliana* (L.) Heynh. *Theor. Appl. Genet.* 58,257–263.

Koornneef, M., Rolff, E., and Spruit, C. J. P. (1980). Genetic control of light-inhibited hypocotyl elongation in *Arabidopsis thaliana* (L.) Heynh. *Z. Pflanzenphysiol.* 100, 147–160.

Koornneef, M., van der Veen, J. H., Spruit, C. J. P., and Karssen, C. M. (1981). Isolation and use of mutants with an altered germination behaviour in *Arabidopsis thaliana* and tomato. *In* "Induced Mutations—A Tool in Plant Research," pp. 227–232. International Atomic Energy Agency, Vienna.

Koornneef, M., Cone, J. W., Dekens, R. G., O'Herne-Robers, E. G., Spruit, C. J. P., and Kendrick, R. E. (1985). Photomorphogenic responses of long hypocotyl mutants of tomato. *J. Plant Physiol.* 120,153–165.

López-Juez, E., Nagatani, A., Buurmeijer, W. F., Peters, J. L., Furuya, M., Kendrick, R. E., and Wesselius, J. C. (1990). Response of light-grown wild-type and *aurea*-mutant tomato plants to end-of-day far-red light. *J. Photochem. Photobiol. B* 4,391–405.

Nagatani, A., Kendrick, R. E., Koornneef, M., and Furuya, M. (1989). Partial characterization of phytochrome I and II in etiolated and de-etiolated tissues of a photomorphogenetic mutant (*lh*) of cucumber (*Cucumis sativus* L.) and its isogenic wild type. *Plant Cell Physiol.* 30,685–690.

Nagatani, A., Reid, J. B., Ross, J. J., Dunnewijk, A., and Furuya, M. (1990). Internode length in *Pisum*. The response to light quality, and phytochrome type I and II levels in *lv* plants. *J. Plant Physiol.* 135,667–674.

Parks, B. M., Jones, A. M., Adamse, P., Koornneef, M., Kendrick, R. E., and Quail, P. H. (1987). The *aurea* mutant of tomato is deficient in spectrophotometrically and immunochemically detectable phytochrome. *Plant Mol. Biol.* 9,97–107.

Parks, B. M., Shanklin, J., Koornneef, M., Kendrick, R. E., and Quail, P. H. (1989). Immunochemically detectable phytochrome is present at normal levels but is photochemically nonfunctional in *hy-1* and *hy-2* hypocotyl mutants of *Arabidopsis*. *Plant Mol. Biol.* 12,425–437.

Pratt, L. H., Cordonnier, M.-M., Wang, Y.-C., Stewart, S. J., and Moyer, M. (1991a). Evidence for three phytochromes in *Avena*. *In* "Phytochrome Properties and Biological Action" (B. Thomas and C. B. Johnson, eds.), pp. 39–56. Springer, Berlin, Heidelberg, New York.

Pratt, L. H., Stewart, S. J., Shimazaki, Y., Wang, Y.-C., and Cordonnier, M. M. (1991b). Monoclonal antibodies directed to phytochrome from green leaves of *Avena sativa* L. cross-react weakly or not at all with the phytochrome that is most abundant in etiolated shoots of the same species. *Planta* 184,87–95.

Quail, P. H., Bolton, G. W., Hershey, H. P., and Vierstra, R. D. (1983). Phytochrome: Molecular weight, *in vitro* translation and cDNA cloning. *In* "Current Topics in Plant Biochemistry—Physiology" (D. Randall, D. G. Blevins, and R. Larson, eds.), pp. 25–36. University of Missouri, Columbia.

Sato, N. (1988). Nucleotide sequence and expression of the phytochrome gene in *Pisum sativum:* Differential regulation by light of multiple transcripts. *Plant Mol. Biol.* 11, 697–710.

Satter, R. L., and Galston, A. W. (1976). The physiological functions of phytochrome. *In* "Chemistry and Biochemistry of Plant Pigments" (T. W. Goodwin, ed.), vol. 1, pp. 680–735. Academic Press, London.

Sharrock, R. A., and Quail, P. H. (1989). Novel phytochrome sequences in *Arabidopsis thaliana:* Structure, evolution, and differential expression of a plant regulatory photoreceptor family. *Genes Dev.* 3,1745–1757.

Sharrock, R. A., Parks, B. M., Koornneef, M., and Quail, P. H. (1988). Molecular analysis of the phytochrome deficiency in an *aurea* mutant of tomato. *Mol. Gen. Genet.* 213,9–14.

Shimazaki, Y., and Pratt, L. H. (1985). Immunochemical detection with rabbit polyclonal and mouse monoclonal antibodies of different pools of phytochrome from etiolated and green *Avena* shoots. *Planta* 164,333–344.

Shimazaki, Y., Cordonnier, M-M., and Pratt, L. H. (1983). Phytochrome quantitation in crude extracts of *Avena* by enzyme-linked immunosorbent assay with monoclonal antibodies. *Planta* 159,534–544.

Smith, H., and Whitelam, G. C. (1990). Phytochrome, a family of photoreceptors with multiple physiological roles. *Plant Cell Environ.* 13,695–707.

Steinitz, B., Drumm, H., and Mohr, H. (1976). The appearance of competence for phytochrome-mediated anthocyanin synthesis in the cotyledons of *Sinapis alba* L. *Planta* 130,23–31.

Stewart, S. J., Pratt, L. H., and Cordonnier, M.-M. (1989). Effects of end-of-day irradiations on phytochrome levels in light-grown *Avena*. European Symposium on Photomorphogenesis in Plants, Freiburg. Book of abstracts, 21.

Tanksley, S. D., Miller, J., Paterson, A., and Bernatzky, R. (1988). Molecular mapping of plant chromosomes. *In* "Chromosome Structure and Function: Impact of New Concepts" (J. P. Gustafson, ed.), pp. 157–173. Plenum Press, New York.

Thomas, B., Crook, N. E., and Penn, S. E. (1984). An enzyme-linked immunosorbent assay for phytochrome. *Physiol. Plant.* 60,409–415.

Thümmler, F., Beetz, A., and Rüdger, W. (1990). Phytochrome in lower plants. Detection and partial sequence of a phytochrome gene in the moss *Ceratodon purpureus* using the polymerase chain reaction. *FEBS Lett.* 275,125–129.

Tokuhisa, J. G., and Quail, P. H. (1983). Spectral and immunochemical characterization of phytochrome isolated from light grown *Avena sativa*. *Plant Physiol.* 72 (suppl. 85).

Tokuhisa, J. G., and Quail, P. H. (1987). The levels of two distinct species of phytochrome are regulated differently during germination in *Avena sativa* L. *Planta* 172,371–377.

Tokuhisa, J. G., and Quail, P. H. (1989). Phytochrome in green-tissue: Partial purification and characterization of the 118-kilodalton phytochrome species from light-grown *Avena sativa* L. *Photochem. Photobiol.* 50,143–152.

Tokuhisa, J. G., Daniels, S. M., and Quail, P. H. (1985). Phytochrome in green tissue: Spectral and immunochemical evidence for two distinct molecular species of phytochrome in light-grown *Avena sativa* L. *Planta* 164,321–332.

Tomizawa, K.-I., Ito, N., Komeda, Y., Uyeda, T. Q. P., Takio, K., and Furuya, M. (1991). Characterization and intracellular distribution of pea phytochrome I polypeptides expressed in *E. coli*. *Plant Cell Physiol.* 32,95–102.

Tong, W-F., and Schopfer, P. (1978). Absence of Pfr destruction in the modulation of phenylalanine ammonia-lyase synthesis of mustard cotyledons. *Plant Physiol.* 61, 59–61.

Wagner, E., and Mohr, H. (1966). Kinetic studies to interpret "high-energy phenomena" of photomorphogenesis on the basis of phytochrome. *Photochem. Photobiol.* 5,397–406.

Wang, Y.-C., Stewart, S. J., Cordonnier, M.-M., and Pratt, L. H. (1991). *Avena sativa* L. contains three phytochromes, only one of which is abundant in etiolated tissue. *Planta* 184,96–104.

Wetherell, D. F., and Koukkari, W. L. (1970). Phytochrome in cultured wild carrot tissue. *Plant Physiol.* 46,350–351.

CHAPTER **32** _____

Gene Regulation

While Mohr and Smith were debating gene regulation versus enzyme activation (see Chapter 19), "it was discovered," Elaine M. Tobin says, "that mRNA has a polyadenylated tail. So I decided to try to isolate poly(A) RNA from plants, thinking this would be a way to see whether phytochrome changed gene expression."

rbcS

After visiting the Weizmann Institute, where Brian E. Roberts was translating mRNA in a cell-free system (Roberts and Paterson, 1973). Tobin (Fig. 32-1) began to isolate poly(A) RNA from duckweed, *Lemna gibba* G-3, which requires light for growth but can exist in darkness for several days. Exposing slab gels to X-ray film, she found that some mRNAs decreased in the light while others became more abundant (Fig. 32-2) (Tobin and Klein, 1975). She suspected that one of the light-stimulated messages coded for ribulose-1,5-bisphospate carboxylase/oxygenase (RUBISCO), a key enzyme in carbon dioxide assimilation and photo-respiration that is dissolved in the chloroplast stroma and was known to be under phytochrome control (Graham et al., 1971; Brüning et al. 1975). The protein was later found to have eight large (rbcL) and eight small (rbcS) subunits (Ellis, 1979).

In 1975, Tobin moved from Brandeis University to the University of California, Los Angeles, where she purified *Lemna* rbcS, raised rabbit antiserum against it, and repeated the dark/light treatments. Dark-adapted plants placed back into WL light for 18 h had dramatically higher translation levels of two poly(A) RNAs than those harvested after darkness. Size and immunoreactivity identified a 20-kDa polypeptide as a precursor of rbcS. The unidentified protein had an apparent molecular mass of 32 kDa.

The data (Tobin, 1978) remained unpublished for several years because

FIGURE 32-1   Elaine M. Tobin.

Tobin was puzzled. "What I could do with white light, I could do with a brief red illumination," she says. "So I knew that it looked like phytochrome. But I couldn't reverse it with far red." Adopting Hillman's protocol for *Lemna,* she had applied 10 min R followed by 12 min FR.

LIGHT-HARVESTING CHLOROPHYLL a/b-BINDING PROTEIN

In 1976, Klaus Apel (Fig. 32-3) moved back to Freiburg, where he had been a student in the mid 1960s. Fascinated by chloroplast development, he eventually began to work with barley.

At the Max-Planck-Institute for Cell Biology in Wilhelmshaven, Klaus Kloppstech was translating poly(A) RNAs using the wheat-germ system. He and Apel discovered that plastid membranes of dark-grown barley seedlings acquired several proteins after illumination, the most prominent being the size (25 kDa) of chlorophyll *a/b*-binding protein (CAB or LHCP). Through immunoprecipitation and peptide mapping, they identified a prominent translation product as a 29-kDa CAB precursor. This reached a constant high level after 6 h. There was no change in *rbcS* mRNA activity, which is high in darkness in barley and other large-seeded cereals.

Because plastid membranes of a chlorophyll-*b*-deficient mutant contained no detectable CAB protein but made normal levels of *cab* mRNA in the light, "a light-absorbing inductor other than chlorophyll may be responsible for the induction" (Apel and Kloppstech, 1978).

Reading this paper, Tobin realized that the other light-induced *Lemna* polypeptide might also be the CAB precursor, and she confirmed this immunochemically (Tobin, 1981a,c).

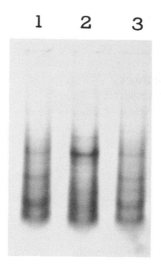

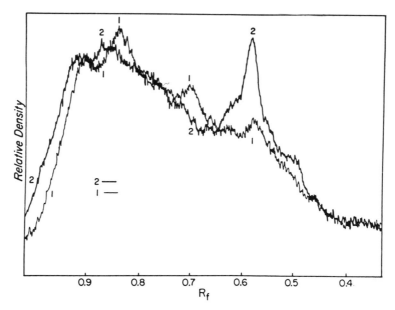

FIGURE 32-2   (*Top*) Autoradiograph of dried slab gels of ^{35}S-methionine-labeled translation products of mRNA isolated from light-grown *Lemna* after (2) 59 h darkness or (1 and 3) 59 h darkness followed by 24 h WL. (*Bottom*) Densitometric tracings of the autoradiograph. Curves 1 and 2 correspond to banding patterns 1 (light) and 2 (dark). (From Tobin and Klein, 1975. Reproduced with permission of the American Society of Plant Physiologists.)

In 1977, Apel presented his findings at Freiburg, where Mohr "immediately suggested that white light was no good and that I should use a defined wavelength. He was sure the photoreceptor was phytochrome."

To obtain constant temperature, Apel had to grow seedlings in a room where there was continuous FR. Even though he used a thick, dark box under a table, he could detect no difference between the R- and FR-treated seedlings. And the dark controls had the same high level of *cab* mRNA activity as the light-treated plants.

Even a safelight in a darkroom induced substantial increases in *cab* mRNA activity. Only when Apel worked in complete darkness did he finally see a clear-cut difference between R- and FR-treated seedlings. Advised by Schäfer and Schopfer, he used only a 30-s R irradiation.

Apel detected translatable *cab* mRNA just 1 h after a R pulse and observed maximal levels at 4 h. Then the relative concentration declined, presumably because Pfr was degraded. No decline occurred in continuous WL.

The R pulse failed to promote CAB insertion into plastid membranes. "The assembly of the light-harvesting chlorophyll *a/b* protein takes place only under continuous illumination which allows chlorophyll synthesis," Apel explained. "Thus, the synthesis and ultimate assembly of this chlorophyll-binding protein depends on the cooperation of two distinct light reactions" (Apel, 1979). When Tobin shortened her R treatment to 1 min, she obtained R/FR-reversible induction of both *rbcS* and *cab* mRNA activities in *Lemna* (Tobin, 1981a,b).

FIGURE 32-3   Klaus Apel, 1989. Photo by Linda Sage.

DOWN-REGULATION OF
PROTOCHLOROPHYLLIDE REDUCTASE

Etioplasts, the photosynthetically inactive plastids of dark-grown angiosperms, accumulate protochlorophyllide. When seedlings see light,
NADPH : protochlorophyllide oxidoreductase (PCR) reduces protochlorophyllide to chlorophyllide, which is esterified to chlorophyll *a*. An
hour or two later, chlorophylls begin to accumulate and protein-packed
thylakoid membrane forms. Activation of a small amount of phytochrome
several hours before WL illumination eliminates the lag (Withrow et al.,
1956; Price and Klein, 1961; Sisler and Klein, 1963; Kasemir et al., 1973).

After isolating PCR (Apel et al., 1980) and raising antibodies, Apel's
group "got a big surprise," he says, "because the PCR activity in the
plastid compartment dropped rapidly after illumination (Santel and Apel,
1981). So there is very little there by the time rapid chlorophyll synthesis
begins 3–4 hours later." Down-regulation of this enzyme's activity by light
had in fact already been reported from Bristol University by W. Trevor
Griffiths (Mapelston and Griffiths, 1979, 1980), who first characterized PCR.

The precursor of PCR proved to be a 44-kDa polypeptide made on
cytoplasmic ribosomes. *pcr* mRNA activity declined to less than 5% of its
initial level if plants were placed in light for a few hours. Thus phytochrome, shown to be the photoreceptor (Apel, 1981), elevated *cab* and
rbcS mRNA activities while simultaneously decreasing that for *pcr*. "This
was the first example of negative control of gene expression by phytochrome," says Apel. It would soon be shown that mRNA activity for
phytochrome itself also decreases after phytochrome photoactivation (see
Chapter 33).

MESSENGER-RNA CONCENTRATIONS

Other workers had discovered that the large subunit of RUBISCO is
encoded in chloroplast DNA (Coen et al., 1977) and synthesized on
chloroplast ribosomes (Blair and Ellis, 1973), whereas a precursor of the
small subunit, like many chloroplast proteins, is encoded on nuclear DNA
(Kung, 1976) and made on cytoplasmic ribosomes (Dobberstein et al.,
1977; Highfield and Ellis, 1978; Chua and Schmidt, 1979). This implicated a
complex control mechanism for enzyme assembly, especially because a
cell may contain several thousand more copies of chloroplast DNA than
nuclear DNA.

In the course of this work, John Ellis and colleagues at the University of
Warwick in Coventry obtained a cDNA from pea *rbcS* mRNA (Bedbrook
et al., 1980) and showed that *rbcS* mRNA is less abundant in dark-grown

than light-grown pea leaves (Smith and Ellis, 1981). Thus light controlled not just mRNA activity but also its steady-state level, which reflects the balance between gene transcription and mRNA degradation. Using cDNA probes, William F. Thompson and colleagues determined the abundance of several nuclear and chloroplast transcripts after etiolated pea was exposed to WL. A R pretreatment strongly increased *rbcS* mRNA abundance, and the effect was FR-reversible. A R pulse itself was sufficient for *cab* mRNA induction, and the effect was partially FR-reversible (Thompson et al., 1983). Gareth Jenkins and colleagues made similar observations (Jenkins, et al., 1983).

By 1981, Apel's group had also begun to clone barley *cab* and *pcr* mRNAs. Of more than 10,000 bacteria containing recombinant plasmids, two proved to be positive for *cab* and one for *pcr*. R increased the amount of mRNA sequences detected with the *cab* cDNA probe (Collmer and Apel, 1983) and decreased levels of sequences detected with the *pcr* cDNA probe (Batschauer and Apel, 1984), suggesting that phytochrome acted at the transcriptional level.

RUN-ON TRANSCRIPTION

Ellis, meanwhile, was anxious to measure transcription more directly. When two Oregon State University researchers prepared nuclei that would finish synthesizing mRNA molecules begun *in vivo* (Luthe and Quatrano, 1980a,b), he realized he could explore chloroplast gene transcription using this run-on assay.

Jane Silverthorne (Fig. 32-4) realized she could look at phytochrome's effects on genes. Moving from Ellis's lab to UCLA in 1981, she entered Tobin's group, which showed that *rbcS* and *cab* mRNA levels fell in darkness and were rapidly—but reversibly—restored by a single R pulse. Thus "phytochrome action must change either the transcription rate or rate of degradation of these mRNAs" (Stiekema et al., 1983).

Silverthorne adapted the run-on transcription assay to *Lemna*. "But nuclei are very fragile," she says. "We used a detergent to strip away the nuclear envelope, so the nuclei were technically lysed but factors needed to continue transcription were still present. It's not the best way to prove transcription is being regulated—but it was all we had."

In March 1982, Tobin left for a 6-month sabbatical that began in Warwick University. At a symposium en route, she reported that phytochrome elicited newly synthesized *rbcS* mRNA in *Lemna*—that it increased gene transcription.

Thomas F. Gallagher was performing similar experiments with pea nuclei, using only WL. In August 1982, he submitted the first published report "of the transcription of specific polypeptide-encoding genes in

FIGURE 32-4   Jane Silverthorne, 1989. Photo by
Linda Sage.

nuclei isolated from a higher plant tissue (Gallagher and Ellis, 1982). WL
increased the level of labeled *rbcS* transcripts 18-fold and *cab* transcripts
ninefold. The increases appeared to be "mediated by an increase in transcription rather than by a decrease in RNA degradation" (Gallagher and
Ellis, 1982). Gallagher later reported that light stimulates transcription of
rbcS and *cab* genes within 20 min—much more rapidly than had been
thought possible (Gallagher et al., 1985).

At UCLA, nuclei from plants treated with intermittent R (2 min every
8 h for 7 weeks) transcribed substantial amounts of *rbcS* and *cab* sequences. Those from plants kept in complete darkness for 7 days before
harvesting made considerably less, but a 1-min R pulse 2 h before harvesting substantially increased synthesis of *rbcS* sequence and dramatically
increased *cab* sequence synthesis (Fig. 32-5). Subsequent FR prevented
the *rbcS* response but only partly prevented the *cab* response, for which
FR alone was a low-level stimulus (Silverthorne and Tobin, 1984).

MULTIPLE *rbcS* GENES

The UCLA workers detected multiple copies of both the *rbcS* and *cab*
genes (Wimpee et al., 1983; Tobin et al., 1984b), as had workers in several
other laboratories. "The coding regions of all these genes are the same,"
Silverthorne explains, "so our clones pulled out any member of the family." Specific probes were obtained for seven of the 12–14 *Lemna rbcS*
genes by cloning part of the 3'-noncoding region of each gene. Use of these
probes revealed that only six of the seven genes produced detectable

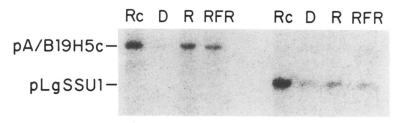

FIGURE 32-5 Autoradiographs showing hybridization of DNA probes for *rbcS* (pLgSSU1) and *cab* (pA/B19H5c) sequences to *in vitro* labeled nuclear transcripts. Rc, plants given 2 min of R every 8 h for 7 weeks. D, as Rc but plants kept in darkness for last 7 days. R, as D but plants given 1 min R 2 h before harvesting. RFR, as R but plants given 10 min FR immediately after the R treatment. (From Silverthorne and Tobin, 1984.)

amounts of mRNA and that they were differentially regulated by phytochrome, with one gene predominating (Tobin et al., 1985; Silverthorne et al., 1990).

Regulation of light-regulated genes according to cell type was reported by researchers at The Rockefeller University in New York. They found that *rbcS* mRNA was a major transcript in photosynthetic tissues such as leaves, pods, and pea seeds but appeared only in small amounts in roots (Coruzzi et al., 1984). And even within leaves, it appeared only in certain cell types (Broglie et al., 1984b). Gene-specific probes revealed that levels of mRNA for individual members of this gene family varied independently in different organs and at different stages of development (Fluhr et al., 1986b).

cab AND *pcr* mRNA INDUCTION

In Schäfer's lab, postdoctorate Egon Mösinger had developed a run-on transcription assay with barley nuclei. Using Apel's *cab* probe in Kiel, he observed that the very low concentration of *cab* transcripts in dark-grown plants increased 10-fold within 3 h after a single R pulse and then declined to a constant, sixfold level (Fig. 32-6). Conversely, the concentration of transcripts for *pcr* fell by one-third during the first 6 h and then rose, by 12 h, to the level of the dark sample. A subsequent FR pulse partly reversed the effects of R but still altered transcript production, as did FR alone (Mosinger et al., 1985). "So the phytochrome effects which previously were found at the level of mRNA accumulation were primarily occurring at the transcriptional level," Apel says. "This told us that phytochrome can both turn on and turn off gene transcription." Because

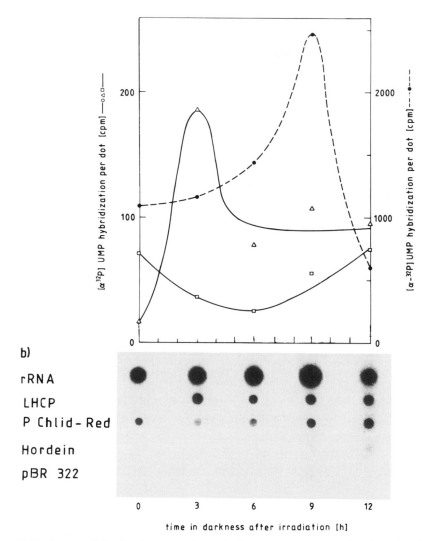

b)

rRNA

LHCP

P Chlid-Red

Hordein

pBR 322

time in darkness after irradiation [h]

FIGURE 32-6   Hybridization of newly synthesized RNA to DNA probes for *cab* (trian-gles), *pcr* (squares), and ribosomal RNA (closed circles). RNA was transcription product of nuclei isolated at 0, 3, 6, 9, and 12 h from dark-grown barley plants that received 5 min R at 0 h. (From Mösinger et al., 1985. Copyright 1985, Springer-Verlag.)

the stimulation did not start from zero or return to zero, the "mechanism cannot be a simple on/off switch as suggested by an early hypothesis of differential gene activation (Hock and Mohr, 1964)" (Mösinger et al., 1985).

TRANSGENIC TISSUE

Prokaryotic and animal genes were known to be regulated by DNA sequences that lie before (upstream from; at the 5' end of) the beginning of the coding region. Such promoters contain a sequence that indicates where transcription will begin plus sequences farther upstream that determine the efficiency of transcription.

At the Rockefeller University in New York, Nam-Hai Chua (Fig. 32-7) wondered how alterations in the promoter would affect the light responsiveness of the pea *rbcS* gene. A test tube approach was impossible because the light signal must be transduced. The solution was to place the altered gene into a different species of plant, where a pea cDNA probe could distinguish it from the host plant's *rbcS* mRNA.

A method for transferring genes between plants made use of *Agrobacterium tumefaciens,* a bacterium that induces crown gall in dicots. This genetic engineer carries gall-inducing genes on circles of DNA or plasmids. When it enters a wound, the genes are spliced into the plant's chromosomal DNA, where they direct cells to make opines, which the bacterium uses as food (see Chilton, 1983). It can also transfer other foreign genes to plants if these are correctly spliced into its plasmid.

Researchers at Monsanto Company in St. Louis used this system to transfer a pea *rbcS* gene into cultured petunia cells. The transformed cells, which grew into an undifferentiated, unruly callus, made mRNA for pea rbcS. Moreover, MAbs detected the pea polypeptide, proving that this mRNA had been translated. MAbs against the large subunit selected pea rbcS, petunia rbcS and petunia rbcL, revealing that pea small subunits were incorporated into the host's RUBISCO.

Darkened calli had 50 times less *rbcS* mRNA than calli kept in the light. Thus the pea gene contained a nucleotide sequence that made it responsive to light (Broglie et al., 1984a).

While this work was in progress, Anthony R. Cashmore at the Rockefeller University initiated bacterial transformation studies with Luis Herrera-Estrella in the laboratory of Marc van Montagu and Jeff Schell in Ghent. Instead of transferring the whole pea *rbcS* gene (family member 3.6) to host cells, they attached its promoter to a bacterial gene and inserted the resulting chimera into cultured tobacco cells. They measured

FIGURE 32-7   Nam-Hai Chua, 1987. Courtesy of
The Rockefeller University.

the *rbcS* promoter's activity in the transgenic cells by assaying for the
bacterial enzyme, whose activity was shown to correlate with changes in
mRNA levels. The bacterial gene thus functioned as a reporter gene.
Because it was expressed more abundantly in WL than in darkness
(Herrera-Estrella et al., 1984), "the sequences in the small subunit gene
that mediate light regulation reside within the promoter fragment,"
Cashmore says. This promoter extended from 973 bp to 4 bp before the
beginning of the coding region ($-973/-4$).

Using a similarly sized promoter ($-1052/-2$) from *rbcS-E9*, Chua's
group also obtained a light-regulated chimeric gene. They further determined that a deletion mutant with only 33 bp in its promoter was subject to
light regulation. Thus the $-35/-2$ region responded to the transduced light
signal (Morelli et al., 1985). This region includes the TATA box
($-35/-24$), an evolutionarily conserved sequence rich in adenine and
thymine that allows one of the RNA polymerase enzymes that transcribe
DNA into mRNA to engage at the right position.

Looking to see if regions of the promoter farther upstream from the
TATA box might enhance light regulation, Mike P. Timko and Albert P.
Kausch removed about 200 bp from the 5' end of the *rbcS*3.6 promoter.
This dramatically reduced the expression of a bacterial reporter gene.
Reversing the $-973/-90$ sequence in the promoter did not destroy its
effectiveness, but placing it at the wrong end (the 3' end) of the bacterial
gene did. This sequence—in either orientation—also conferred responsiveness to light on a truncated (-145) nopaline synthase promoter (Timko
et al., 1985). "So you took the sequences from one promoter and fused

them to a nopaline synthase promoter and they worked," Cashmore says. "Furthermore, if you inverted them, they still worked, demonstrating that they had the characteristic of enhancer sequences, which were just being described for various animal genes."

TRANSGENIC PLANTS

Regulation of foreign genes in calli was not identical to that in whole plants, for rbcS was no longer under phytochrome control (see Fluhr et al., 1986a) and its induction in WL required several days. The next important step was production of transgenic plants, which was accomplished by transforming plant cells with A. tumefaciens plasmids deprived of their tumor-causing genes (deBlock et al., 1984; Fraley et al., 1985; Horsch et al., 1985; Klee et al., 1985). Remarkably, pea rbcS and wheat cab genes were regulated normally by light in transgenic plants and even exhibited correct tissue specificities (Lamppa et al., 1985a,b; Nagy et al., 1985, 1986; Fluhr and Chua, 1986).

Wanting to define the upstream light-responsive region of a promoter more precisely, Chua's group trimmed rbcS pea genes before placing them in tobacco plants. The 3A member of the gene family with its promoter cut at -410 showed identical WL inducibility and organ specificity to tobacco rbcS. And about the same amount of mRNA was transcribed from it as from the unaltered 3A gene in pea plants. When the $-327/-48$ sequence was fused to a -46 light-insensitive viral promoter (cauliflower mosaic virus 35S promoter) attached to a reporter gene (3A-35S-reporter), the latter was expressed in tobacco plants during normal light/dark cycles but not in extended darkness. Sequence $-317/-82$ from E9 gave similar results. Thus light responsiveness was conferred by sequences that lay among 240–280 bp in the upstream region of the rbcS promoter, beyond the TATA box (Fluhr et al., 1986a).

PHYTOCHROME RESPONSIVENESS

These WL experiments were performed with green transgenic plants. But studies of the phytochrome responsiveness of promoter regions required transgenic etiolated tissue, because rbcS induction in mature plants requires continuous B or WL (Fluhr and Chua, 1986; Fluhr et al., 1986b). Because transgenic plants are produced in the light, etiolated tobacco seedlings can be obtained only by self-pollinating a transgenic plant, collecting the seeds, and growing seedlings, which takes about 6 months in all.

FIGURE 32-8   Steve A. Kay, 1989. Photo by Linda Sage.

Most of the phytochrome studies have been performed in Chua's lab, by Steve A. Kay (Fig. 32-8) and Ferenc Nagy. Initial studies revealed that 2.5 kbp of 5' flanking DNA from a pea *cab* gene (Simpson et al., 1986) and 1.8 kbp from a wheat *cab* gene was sufficient to confer phytochrome responsiveness on reporter genes in transgenic, etiolated tobacco seedlings. *cab* itself was also regulated normally by its own promoter in these transgenic plants (Nagy et al., 1986).

In the case of the 3A and 3C pea *rbcS* genes, 0.4 and 2.0 kbp conferred phytochrome responsiveness in transgenic petunia plants (Fluhr and Chua, 1986). The 3A-35S-reporter construct was also responsive to phytochrome, as was the E9-35S-reporter construct. Thus the same 240–280 bp that conferred WL responsiveness made genes reversibly responsive to R/FR light. When these upstream sequences of 3A and E9 were compared, they were found to be almost identical between −196 and the beginning of the coding region (Fluhr et al., 1986a).

Continuing the work with deletion mutants and chimeric genes, Nagy and Marc Boutry identified the sequence −357/−124 as the phytochrome-responsive region for wheat *cab-1* in etiolated transgenic tobacco seedlings. Although the sequence was about the same distance from the coding region as the phytochrome-responsive region in the pea *rbcS* promoters, there was no obvious homology between the two enhancers. "We note that the two genes differ with respect to their fluence responses;" the researchers pointed out, "the *cab-1* gene is more sensitive to red light than the *rbcS* gene. It is possible that the expression of the two genes is mediated by different *cis*-acting elements" (Nagy et al., 1987). Regulatory sequences in 5' flanking regions are named *cis*-acting elements because of their alignment with the coding sequences of genes.

CIRCADIAN RHYTHMS

In Hannover, meanwhile, Kloppstech had measured levels of specific poly(A) RNAs in peas grown on light/dark cycles. Although such plants were not greening, they were making new chloroplasts during growth. To investigate why mRNA for ELIP (early light-inducible protein) seemed to be absent from mature plants when ELIP was abundant in thylakoid membrane, Kloppstech took samples at different times of day. He found that *elip* mRNA, as well as *cab* and *rbcS* mRNAs, declined during the day, reaching minimal levels 2 h after the beginning of the dark period. The *elip* and *cab* mRNAs began to rise again 2 h *before* the next light period. Because these oscillations persisted for at least 3 days of continuous illumination, the rhythm was circadian (Kloppstech, 1985).

Unaware of this work, Nagy and Kay were exploring phytochrome control of *cab* in mature wheat plants. Early one morning, Kay took a control sample of wheat leaves from the greenhouse. At lunchtime, Nagy did the same thing. When the two compared data, they were shocked to discover a huge discrepancy.

Measurements around the clock (Nagy et al., 1988a) revealed that levels of *cab* mRNA—but not, in wheat, *rbcS* mRNA—oscillated in leaves during daily light/dark cycles, rising before dawn each day and even in extended darkness (Fig. 32-9). "This was a heresy," Kay says. "We had said that the *cab* gene was induced by light, but here we were seeing an induction of the gene in the dark."

When Kay and Nagy transferred wheat from 12-h light/12-h dark cycles to extended darkness, the rhythm dampened. But a R pulse before the transfer prevented the dampening and *cab* mRNA was formed at about the same level as in light/dark cycles. "To our joy," says Kay, "a far-red flash at the end of the day abolished *cab* expression in subsequent darkness. There was a faint amount, still in the same phase, peaking at about 10:00 A.M., which is when the *cab* gene wakes up. And this far-red inhibition was relieved by red (Fig. 32-10). Bingo. We had finally seen phytochrome affecting *cab* gene expression in mature plants, but only within the context of a circadian rhythm (Nagy et al., 1988a)."

The researchers propose that "the circadian clock can be considered as a molecular gate that is opening and closing. Behind that gate is the *cab* gene, but phytochrome is on the other side. When the gate is open, as in the morning, Pfr still present from the previous day switches the *cab* gene

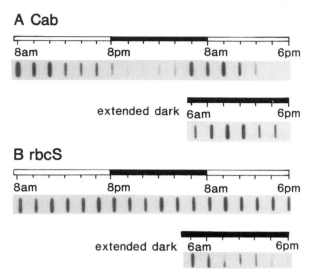

FIGURE 32-9   (A) Circadian oscillations in wheat *cab* gene family mRNA detected by hybridization with a *cab-1* coding sequence probe. (B) Lack of oscillation in *rbcS* gene family mRNA detected with an *rbcS* coding sequence probe. (From Nagy et al., 1988a. Copyright © 1988 by Cold Spring Harbor Laboratory.)

on—you can see this because the gene starts to come back before the light is ever switched on at 8:00 A.M. In the afternoon, the gate is closed, so regardless of how much phytochrome is present, the clock prevents it from inducing this gene.''

Placing the wheat *cab-1* gene into transgenic tobacco plants, Nagy and Kay observed the same circadian rhythm as in wheat, showing that the clock is conserved between monocots and dicots. ''But the thing that really excites people,'' Kay says, ''is that we can attach the *cab* upstream region to a bacterial reporter gene and get the bacterial mRNA cycling in a circadian rhythm inside a transgenic plant (Nagy et al., 1988a). This shows that the *cab* upstream region not only contains phytochrome-responsive elements for inducing the gene in etiolated tissue . . . but also contains an element that is sensitive to the circadian clock of both wheat and tobacco.'' This *cab-1* promoter was 1.8 kb long, and the clock element resided upstream from the −124 position.

On leaving the Rockefeller University, Nagy continued to collaborate with Kay, showing that *cab* transcription as well as steady-state levels of *cab* mRNA oscillate with a circadian rhythm. When a series of wheat *cab-1* deletions were placed in tobacco, truncation at −244 failed to diminish expression, whereas further deletion to −230 decreased expression 10-fold. Removal of an additional 19 bp had no further effect, but deletions

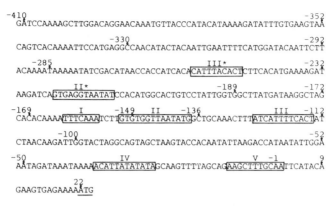

```
-410                                                           -352
GATCCAAAAGCTTGGACAGGAACAAATGTTACCCATACATAAAAGATATTTGTGAAGTAA

                   -330                                       -292
CAGTCACAAAATTCCATGAGGCCAACATACTACAATTGAATTTTCATGGATACAATTCTT

        -285              III*                                -232
ACAAAATAAAAAATATCGACATAACCACCATCACACATTTACACTCTTCACATGAAAAGAT

           II*                           -189                 -172
AAGATCAGTGAGGTAATATCCACATGGCACTGTCCTATTGGTGGCTTATGATAAGGCTAG

-169       I         -149    II    -136              III    -112
CACACAAAATTTCAAATCTTGTGTGGTTAATATGGCTGCAAACTTTATCATTTTCACTAT

         -100                                                 -52
CTAACAAGATTGGTACTAGGCAGTAGCTAAGTACCACAATATTAAGACCATAATATTGGA

-50                  IV                         V    -1        9
AATAGATAAATAAAAACATTATATATAGCAAGTTTTAGCAGAAGCTTTGCAATTCATACA

              22
GAAGTGAGAAAATG
```

FIGURE 32-11   Nucleotide sequence of the pea *rbcS* upstream region showing regions of sequence homology between members of the *rbcS* gene family (boxes I–V). (From Kuhlemeier et al., 1988.)

family (Fluhr et al., 1987) and resemble certain mammalian enhancers. Both boxes II and III are required for light responsiveness of the −166 *rbcS* promoter. And replacement of just two guanines in box II abolishes effectiveness (Kuhlemeier et al., 1988). But the boxes are dispensable in a −410 deletion mutant, because homologous sequences, boxes II* and III* (Fig. 31-11), exist upstream from −166 (Green et al., 1987).

All four (II, III, II*, and III*) boxes are required for full *rbcS* expression in very young leaves and seedlings (Kuhlemeier et al., 1988). Thus permutations of LRE activity might provide differential expression of the gene during development and perhaps also under different fluences (Nagy et al., 1988b).

The boxes may also act as—or overlap with—negative regulatory elements, because reporter gene expression is inhibited in the dark when the −169/−112 sequence is inserted into a constitutive promoter (Kuhlemeier et al., 1987).

TRANS-ACTING FACTORS

In 1985, Kay and Pamela J. Green realized says Kay, "that we were both interested not so much in the DNA elements as in the proteins—*trans*-acting factors—that bind to them."

Green and Kay searched for a nuclear protein that would bind to one of the LREs in the *rbcS* promoter. To detect binding, they used a technique called gel retardation (Fried and Crothers, 1981; Garner and Revzin, 1981).

"If you radiolabel a fragment of the DNA promoter and electrophorese it, you get a band on the gel," Kay explains. "But if you add nuclear extract before you run the gel, proteins that bind to that fragment retard its mobility, so you get two bands—a little bit of the DNA fragment that hasn't bound protein and a band higher up that has."

Labeled −330/−50 *rbcS* DNA was retarded by a pea nuclear protein that proved to have affinity for boxes III, II*, and III* and, especially, II. Because each box contains a GT pair, Green and Kay named this factor GT-1. The factor was present in nuclear extracts from both greenhouse and dark-adapted plants, so it did not appear to be synthesized *de novo* in response to light (Green et al., 1987).

Because mutations within the pea *rbcS* promoter that prevented the gene from being transcribed in transgenic tobacco (Kuhlemeier et al., 1988) also prevented binding of tobacco GT-1 *in vitro,* the factor was possibly an activator of *rbcS* transcription. Footprinting experiments, in which DNA-nuclear protein complexes were digested with DNase to see which regions of the promoter were protected (covered) by the protein, showed that GT-1 from dark-adapted and light-green plants bound to the same sequence.

GT-1 also bound to −410/−330 *rbcS* DNA, and footprinting identified two sequences within this stretch of DNA—boxes II** and III**—that were considerably homologous to the type II and III boxes. Thus, even more permutations would be possible.

In the absence of other type II and III boxes, a mutation in box II not only made box II vulnerable to DNase in the presence of nuclear extract but also lessened protection of box III. Thus factors binding to box II presumably interact with factors binding to box III (Green et al., 1988).

To investigate this interaction, Philip M. Gilmartin changed the spacing between boxes II and III in a −166/−50 *rbcS-3A* promoter fragment. In the wild-type there are 33 bp between the two critical guanines in box II and a critical guanine in box III. This was the maximal distance for normal function. The boxes could be pushed closer together, however, until 10 bp—almost a whole turn of the DNA double helix—were removed.

Deleting 10 bp from between the boxes abolished GT-1 binding *in vitro,* presumably because of steric hindrance, but increasing the spacing had no effect. Therefore the *in vivo* effects of increased spacing must be on GT-1 activity rather than binding. "If there are two molecules of GT-1, perhaps they have to be close enough to interact with each other or even with another factor that may bind to a site between boxes II and III," Gilmartin suggests.

Although GT-1 binding is not wholly sufficient for light-responsive transcription, it appears to be necessary, because the light responsiveness conferred by the −166 *rbcS* promoter on a cropped nopaline synthase promoter is affected by altering the spacing between boxes II and III in just the same way as that of the *rbcS* promoter itself. "These observations," the Rockefeller workers concluded, "strongly suggest that boxes II and III are critical components of the molecular light switch" (Gilmartin and Chua, 1990b).

Four tandem copies of box II conferred light responsiveness on a −90 35S promoter/reporter construct (Lam and Chua, 1990) but not on a −46 35S/reporter construct (Cuozzo-Davis et al., 1990), suggesting that GT-1 must interact with a factor that binds to the 35S promoter between −90 and −46. Chua's group has identified an activation sequence factor, ASF-1, which binds between −83 and −65. Fusion of a tetramer of this sequence to the −90 35S promoter caused constitutive expression of the reporter gene (Lam and Chua, 1990). Thus box II, although unable to promote light regulation alone, would appear to confer responsiveness to light.

Whether GT-1 lies at the end of the phytochrome transduction pathway—or interacts with a protein activated by the phytochrome signal—is unclear, because these experiments were performed with WL. "Our present hypothesis assumes," Chua explained, "that, at least for the *rbcS* and *cab* genes, the cis-acting sequences mediating the white light and phytochrome effects either overlap substantially or may even be identical" (Nagy et al., 1988b).

Whether this proves to be correct, the GT-1 system provides a model of how phytochrome, through its transduction chain, might regulate genes. Chua proposed (Fig. 32-12) that GT-1 binds to box II in both darkness and light and that (1) it is inactivated in the dark by interacting with a negative factor, or (2) it is inactive in the dark but is activated by a positive factor in the light, or (3) the protein itself exists in an active form in the light and an inactive form in the dark (Lam and Chua, 1990).

OTHER REGULATORY SEQUENCES

Other sequences that interact with protein factors have been identified in the upstream regions of light-regulated genes (see Gilmartin et al., 1990). These include the G box (containing CACGTG) (Giuliano et al., 1988; Staiger et al., 1989; Schulze-Lefert et al., 1989a,b) and the AT-1 binding site (AATATTTTATT) (Datta and Cashmore, 1989). "The activity of the AT-1 factor is strongly modulated by phosphorylation," Cashmore says. The GATA motif, found as multiple (Castresana et al., 1987) or single (Castresana et al., 1987; Manzara and Gruissem, 1988; Gidoni et al., 1989)

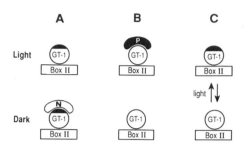

FIGURE 32-12   Models for light activation of transcription mediated by GT-1 binding to box II. (From Lam and Chua, 1990. Copyright 1990 by the American Association for the Advancement of Science.)

copies in pratically all *rbcS* and *cab* genes, may be essential for high-level transcription in the light. A computer search has identified it in the promoters of many phytochrome-responsive genes (Grob and Stuber, 1987). Factor GAF-1 binds specifically to this sequence (Lam et al., 1990).

All the factor-binding sequences—or homologous sequences—appear in the pea *rbcS-3A* promoter. Experiments with defined wavelengths are unraveling their respective roles and interactions.

Using a −255/+20 fragment from the *Lemna rbcS-5B* gene in gel retardation assays, Tobin's group detected the factor LRF-1 (light-regulated nuclear factor) in nuclear extracts. Its binding activity was 10 times higher in light-grown than in dark-adapted plants and increased after plants were returned to the light.

When dark-adapted plants received a R pulse 2 h before harvesting, LRF-1 activity doubled transiently, as phytochrome-induced transcription of *Lemna rbcS* genes is known to do. Moreover, there was a direct correlation between LRF-1 binding to probes for different members of the *rbcS* gene family and phytochrome-induced expression levels *in vivo*.

LRF-1 binds to a GATA sequence at −149 in the 5A gene. But because all the gene-specific probes contained this motif, including one that did not bind to nuclear extract, additional nucleotides are obviously necessary for LRF-1 binding (Buzby et al., 1990).

Sequences around this region are important for phytochrome regulation of *Lemna rbcS in vivo*. Inserting constructs (part of the *rbcS* promoter/CAT/nopaline synthase 3′ untranslated region) into etiolated *Lemna* fronds with a particle acceleration device, Stephen A. Rolfe determined that all constructs in which the promoter extended to −205 showed R/FR-reversible regulation, whereas deletions to positions −83 and −74 resulted in a low level of expression that was not regulated by phytochrome (Rolfe and Tobin, 1991).

KINETIC STUDIES

Kinetic studies of gene transcription suggest that interactions between *cis*-acting elements and *trans*-acting factors must be complex.

In Briggs's lab, Lon S. Kaufman determined fluence requirements for *rbcS* and *cab* transcription. Using cDNA probes, he found that R with a fluence of 10^{-4} μmol m^{-2} elicited *cab* mRNA production in dark-grown pea buds. Because the response was triggered by FR also, it was a VLFR. But 10,000 times more R was required before *rbcS* mRNA was detectable, and this LF response that was fully reversible by FR (Kaufman et al., 1984). Eleven unidentified transcripts also differed in fluence requirements, kinetics (Kaufman et al., 1985), and escape times. Thus induction of one transcript became insensitive to FR before the transcript began to accumulate, whereas that of another was still completely FR-reversible when the maximum amount had formed (Kaufman et al., 1986). "So you may find common elements upstream in the promoter regions of phytochrome-regulated genes," Briggs says, "but you have to assume that factors that regulate, for example, the rate of mRNA appearance once a gene is turned on, are different from those that determine the light requirement."

PLASTIDIC FACTOR

Experiments with photobleached plants revealed that light-regulated nuclear genes also respond to a signal from the chloroplast.

Carotene-deficient maize plants with only rudimentary plastids were found to have abnormally low levels of *cab* mRNA. This suggested that "the accumulation of [*cab*] mRNA is not controlled exclusively by phytochrome" (Harpster et al., 1984) and that "events in the developing plastid affect the accumulation of nuclear-encoded mRNA" (Mayfield and Taylor, 1984).

The effect was not simply due to photooxidation of mRNA, and genes for rbcS and PCR were relatively unaffected. Thus "in addition to phytochrome (Pfr) other plastid-dependent factors are required for a continuous light-dependent transcription of nuclear genes encoding [CAB]" (Batschauer et al., 1986).

In mustard, where rbcS is light-modulated, almost no protein precursor or translatable mRNA for rbcS or CAB was detectable in the cotyledons of photobleached seedlings, although phytochrome-mediated synthesis of representative cytoplasmic, mitochrondrial, and glyoxysomal enzymes proceeded normally (Oelmüller and Mohr, 1986). Experiments with chloramphenicol, which inhibits protein synthesis inside plastids, suggested that the signal from the plastid was short-lived and therefore required continuously and that it was released during a certain stage of plastid development (Oelmüller et al., 1986b). Kinetic studies of *rbcS* mRNA appearance revealed that *rbcS* gene expression is turned on by an endogenous control factor about 12 h before phytochrome can act. Thus, "three factors seem to be involved: an endogenous factor which turns on gene expression, phytochrome which modulates gene expression, and the plastidic factor which is an indispensable prerequisite for the appearance of translatable [*rbcS*] mRNA" (Oelmüller et al., 1986a).

HIERARCHICAL CONTROL OF TRANSCRIPTION

Mohr (Fig. 32-13) is exploring interactions between these regulatory factors by studying enzymes that assimilate the nitrogen used for plastidic amino acid synthesis.

Whereas nitrate reductase (NR), which converts nitrate to nitrite, operates in the cytoplasm, nitrite reductase (NiR), which reduces nitrite to ammonia, acts in plastids, as do the subsequent enzymes in the pathway. Both NR and NiR are induced by nitrate, NR's substrate, but R further increases activity, even when applied several hours before nitrate (Sharma and Sopory, 1984). Phytochrome cannot act alone, however, and neither it nor nitrate can act in the absence of intact plastids (Rajasekhar and Mohr, 1986). Thus NR and NiR levels are controlled first by an endogenous factor, then nitrate, then phytochrome.

FIGURE 32-13   Hans Mohr, 1989. Photo by Linda Sage.

NR production, though cytoplasmic, was much more dependent on the presence of plastidic factor than was production of NiR. So NR seemed to be regulated as if it were a plastidic enzyme (Oelmüller et al., 1988), although one isozyme, which appeared only when ammonia was the nitrogen source, was independent of the factor (Schuster et al., 1989). Plastidic factor decreased not only NR activity but also its protein and mRNA levels (Schuster and Mohr, 1990a).

The phytochrome signal that induces NR and NiR could be stored for up to 12 h. But whereas nitrate and phytochrome induced both NR and NiR in mustard cotyledons between 24 and 96 h after sowing, the ability to store the phytochrome signal for NiR induction declined when the seedlings became 72 h old, whereas ability to store the signal for NR induction remained unimpaired. But although the stored light signal seemed to differ in each case, neither signal appeared to be part of the direct phytochrome transduction chain, because FR induced the enzymes in dark-grown seedlings with a lag phase of 1 h, whereas signal storage could not be detected for 3 h (Fig. 32-14). The signals were stored in the presence of nitrate as well as in its absence, suggesting that they perhaps allow plants to adjust NR and NiR levels during the night to the light conditions of the previous day (Schuster et al., 1987).

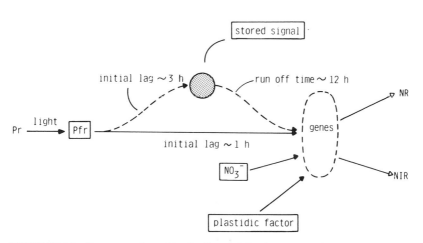

FIGURE 32-14 Summary of combined effects of R, nitrate, and plastidic factor on NiR and NR genes made competent by endogenous factor(s). (From Schuster et al., 1987. Copyright 1987, Springer-Verlag.)

FINE TUNING OF GENE EXPRESSION

The studies of nuclear-encoded chloroplast enzymes have clearly shown that phytochrome regulates gene transcription, as Mohr proposed over a quarter of a century ago. They do not suggest that the phytochrome signal acts only on genes, for changes in mRNA levels, although much more rapid[1] than once thought possible, cannot account for responses that take only seconds (see Chapter 23). Neither do they imply that phytochrome regulates gene expression only at the transcriptional level. "Full gene expression," Mohr pointed out, "means the appearance of a final direct gene product—a protein—active at its physiological site of action. . . . In principle, there are many steps between the initiation of transcription and the accumulation of the gene product at its functional location where gene expression could be regulated" (Mohr, 1988; Schuster et al., 1988).

In the case of nitrite reductase, "phytochrome produces the mRNA, whereas in order to make the enzyme out of the mRNA you need nitrate (Schuster and Mohr, 1990b)," Mohr says. "So you have a beautiful two-step control, which you can take apart completely."

In the case of *rbcS,* phytochrome controls both transcription and a subsequent step in gene expression. Christa Schuster and Ralf Oelmüller observed no correlation between the rate of RUBISCO synthesis (Fig. 32-15A) and the level of *rbcS* mRNA (Fig. 32-15B). Although the enzyme level increased at a constant rate in FR between 54 and 84 h after sowing (i.e., in the absence of turnover, the rate of synthesis of the holoenzyme was constant), the amount of *rbcS* mRNA doubled. Holoenzyme synthesis, which was under phytochrome control throughout, also failed to correlate with the level of *rbcL* mRNA (Oelmüller et al., 1986a). "So there is at least one additional post-translational regulatory step which is also phytochrome-dependent," Mohr concluded. The data suggested "that coarse control of appearance of translatable [*rbcS*]-mRNA by phytochrome is essential for ribulose-1,5-bisphosphate carboxylase to appear at a high rate, but that fine tuning by phytochrome of actual appearance of the holoenzyme is not a matter of transcriptional control" (Mohr, 1988).

Posttranscriptional regulation may also negate the action of phytochrome. Using sequence-specific probes, Silverthorne detected transcription of six members of the *rbcS* gene family in both fronds and roots of *Lemna,* whose roots are normally exposed to light in the plant's aquatic environment. mRNA from one of the genes, SSU5B, was barely detectable, however. "Localization of individual gene transcripts by *in situ*

---

1. Down-regulation of mRNA levels can be observed within 5 min (see Chapter 33).

hybridization showed that SSU1 and SSU5B are expressed in the same cells in fronds," Silverthorne reported. "Thus the mechanism of differential expression is likely to involve an organ-specific post-transcriptional mechanism" (Silverthorne and Tobin, 1990).

Phytochrome appears to control the expression of the gene for PCR at the posttranscriptional as well as transcriptional level. In Schäfer's lab, Briggs examined the effects of various R fluences on barley *pcr* mRNA. Very-low-fluence R strongly decreased the production of that transcript but barely affected its abundance. Low-fluence R had the opposite effects, suggesting that it was affecting mRNA stability (Mösinger et al., 1988).

Tobin first suggested that *cab* gene expression might be controlled posttranscriptionally, after experiments with *Lemna* (Slovin and Tobin, 1982). The observation that chlorophyll-deficient plants have normal levels of *cab* mRNA but no CAB protein suggested that chlorophyll is required for complete *cab* expression. Because phytochrome controls chlorophyll synthesis separately from *cab* mRNA production (Horwitz et al., 1988), Pfr may exert posttranscriptional control on *cab* by regulating a step in chlorophyll synthesis (Briggs, 1987). Most plastid genes, on the other hand, appear to be constitutively transcribed and regulated only posttranscriptionally (see Gruissem, 1989).

INTERORGAN SIGNAL TRANSMISSION

Transduction of the phytochrome signal between cells adds yet another layer of complexity to the regulation of some genes. When Oelze-Karow transferred dark-grown mustard seedlings to WL, chlorophyll production began immediately, but there was a 15-min lag before the onset of photophosphorylation. A R pulse 12 h before the transfer abolished this lag, suggesting that phytochrome readies plastids to phosphorylate as soon as chlorophyll appears (Oelze-Karow and Mohr, 1978b).

Like phytochrome-mediated suppression of lipoxygenase activity (see Chapter 19), this induction of phosphorylative ability is a threshold response that occurs only if the hypocotyl hook is connected to the cotyledon (Oelze-Karow and Mohr, 1978a) for at least 2 min after an inductive R pulse. And photoreversibility is lost at 2 min (Oelze-Karow and Mohr, 1986). The response occurs when the hook alone—but not the cotyledon alone—is irradiated (Oelze-Karow and Mohr, 1988), and the required Pfr level is again 1.25% (Oelze-Karow and Mohr, 1986). Thus two responses that involve signal transmission between hook and cotyledon have the same threshold. Inhibition of hypocotyl growth by hook Pfr, which involves signal transmission in a different direction, operates above

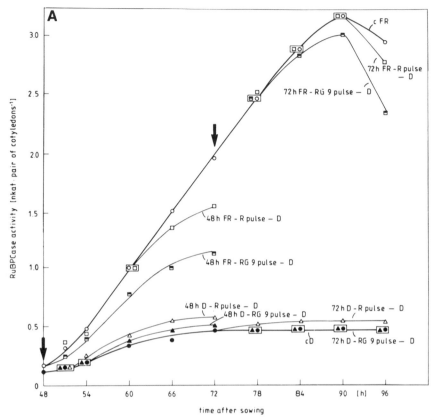

FIGURE 32-15   Time course of (A) ribulose-1,5-bisphosphate activity and (B) amount of *in vitro*–translatable *rbcS* mRNA isolated from mustard cotyledons between 48 and 96 h after sowing. A R pulse placed 80% of total phytochrome in the Pfr form, and an RG 9 pulse produced 1% Pfr. *Arrows* indicate transfer to darkness. (From Schuster et al., 1988. Copyright 1988, Springer-Verlag.)

much lower thresholds in etiolated seedlings (0.03%) and partly de-etiolated seedlings (possibly 0.0000003%) (Oelze-Karow and Mohr, 1989, 1990).

In the early 1970s, Jan A. de Greef induced greening of primary leaves by irradiating the primary axis of bean seedlings (de Greef and Caubergs, 1972b, 1973). Because the signal was rapidly transmitted (de Greef et al., 1976), he suggested that the "nature of the transfer can only be interpreted in biophysical terms" (de Greef and Caubergs, 1972a) and that "by means of a transmission system (primitive nervous system), phytochrome directs a flow of information of perceived light signals throughout the whole plant body" (de Greef et al., 1976).

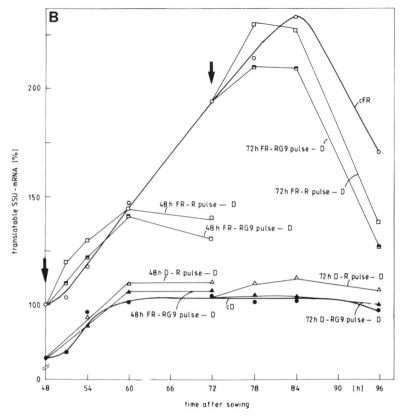

FIGURE 32-15   Continued

Mohr also favors a biophysical signal, "since the decisive facts (threshold regulation, precision and rapidity) are difficult to reconcile with a biochemical transmission process. . . . However, a physiological explanation of the kind of interorgan signal transmission we have observed in mustard requires a revision of our views about the nature of plants with regard to their ability to communicate between organs" (Oelze-Karow and Mohr, 1988).

REFERENCES

Adam, E., Fejes, E., Szell, M., Csizmadia, V., and Nagy, F. (1990). *Cis* and *trans* regulatory elements for phytochrome and circadian clock responsive expression of a wheat chlorophyll *a/b* gene. *In* "Phytochrome Properties and Biological Action." NATO Advanced Research Workshop, Chichester. Book of Abstracts, D3.

Apel, K. (1979). Phytochrome-induced appearance of mRNA activity for the apoprotein of the light-harvesting chlorophyll a/b protein of barley (*Hordeum vulgare*). *Eur. J. Biochem.* 97,183–188. Copyright 1979, Springer-Verlag.

Apel, K. (1981). The protochlorophyllide holochrome of barley (*Hordeum vulgare* L.). Phytochrome-induced decrease of translatable mRNA coding for the NADPH : protochlorophyllide oxidoreductase. *Eur. J. Biochem.* 120,89–93.

Apel, K, and Kloppstech, K. (1978). The plastid membranes of barley. *Eur. J. Biochem.* 85,581–588.

Apel, K., Santel, H.-J., Redlinger, T. E., and Falk, H. (1980). The protochlorophyllide holochrome of barley (*Hordeum vulgare* L.). *Eur. J. Biochem.* 111,251–258.

Batschauer, A., and Apel, K. (1984). An inverse control by phytochrome of the expression of two nuclear genes in barley (*Hordeum vulgare* L.). *Eur. J. Biochem.* 143,593–597.

Batschauer, A., Mösinger, E., Kreuz, K., Dörr, I., and Apel, K. (1986). The implication of a plastid-derived factor in the transcriptional control of nuclear genes encoding the light-harvesting chlorophyll a/b protein. *Eur. J. Biochem.* 154,625–634.

Bedbrook, J. R., Smith, S. M., and Ellis, R. J. (1980). Molecular cloning and sequencing of cDNA encoding the precursor to the small subunit of chloroplast ribulose-1,5-bisphosphate carboxylase. *Nature (Lond.)* 287,692–697.

Blair, G. E., and Ellis, R. J. (1973). Protein synthesis in chloroplasts. 1. Light-driven synthesis of the large subunit of fraction 1 protein by isolated pea chloroplasts. *Biochim. Biophys. Acta* 319,223–234.

Briggs, W. R. (1987). The molecular photophysiology of greening in etiolated pea seedlings. *In* "Plant Molecular Biology" (D. von Wettstein and N.-H. Chua, eds.), pp. 41–51. Plenum Press, New York, London.

Broglie, R., Coruzzi, G., Fraley, R. T., Rogers, S. G., Horsch, R. B., Niedermeyer, J. G., Fink, C. L., Flick, J. S., and Chua, N.-H. (1984a). Light-regulated expression of a pea ribulose-1,5-bisphosphate carboxylase small subunit gene in transformed plant cells. *Science* 224,838–843.

Broglie, R., Coruzzi, G., Keith, B., and Chua, N.-H. (1984b). Molecular biology of C4 photosynthesis in *Zea mays:* Differential localization of proteins and mRNAs in the two leaf cell types. *Plant Mol. Biol.* 3,431–444.

Brüning, K., Drumm, H., and Mohr, H. (1975). On the role of phytochrome in controlling enzyme levels in plastids. *Biochem. Physiol. Pflanz.* 168,141–156.

Buzby, J. S., Yamada, T., and Tobin, E. M. (1990). A light-regulated DNA-binding activity interacts with a conserved region of a *Lemna gibba rbcS* promoter. *Plant Cell* 2,805–814.

Castresana, C., Staneloni, R., Malik, V. S., and Cashmore, A. R. (1987). Molecular characterization of two clusters of genes encoding the Type I CAB polypeptides of PSII in *Nicotiana plumbaginifolia*. *Plant Mol. Biol.* 10,117–126.

Chilton, M.-D. (1983). A vector for introducing new genes into plants. *Sci. Am.* 248,51–59.

Chua, N.-H. and Schmidt, G. (1978). Post-translational transport into intact chloroplasts of a precursor to the small subunit of ribulose-1,5-bisphosphate carboxylase. *Proc. Natl. Acad. Sci. USA* 75,6110–6114.

Chua, N.-H., and Schmidt, G. W. (1979). Transport of proteins into mitochondria and chloroplasts. *J. Cell Biol.* 81,461–483.

Coen, D. M., Bedrook, J. R., Bogorad, L., and Rich, A. (1977). Maize chloroplast DNA encoding the large subunit of ribulose bisphosphate carboxylase. *Proc. Natl. Acad. Sci. USA*. 74,5487–5491.

Coruzzi, G., Broglie, R., Edwards, C., and Chua, N.-H. (1984). Tissue-specific and light-regulated expression of a pea nuclear gene encoding the small subunit of ribulose-1,5-bisphosphate carboxylase. *EMBO J.* 3,1671–1679.

Cuozzo-Davis, M. C., Yong, M.-H., Gilmartin, P. M., Goyvaerts, E., Kuhlemeier, C., Sarokin, L., and Chua, N.-H. (1990). Minimal sequence requirements for the regulated expression of *rbcS-3A* from *Pisum sativum* in transgenic tobacco plants. *Photochem. Photobiol.* 52,43–50.

Datta, N., and Cashmore, A. R. (1989). Binding of a pea nuclear protein to promoters of certain photoregulated genes is modulated by phosphorylation. *Plant Cell* 1,1069–1077.

deBlock M., Herrera-Estrella, L., van Montagu, M., Schell, J., and Zambryski, P. (1984). Expression of foreign genes in regenerated plants and in their progeny. *EMBO J.* 3,1681–1689.

de Greef, J. A., and Caubergs, R. (1972a). Morphogenic correlations between plant organs by transfer of light-induced stimuli. *Arch. Intern. Physiol. Biochim.* 80,390–391.

de Greef, J. A., and Caubergs, R. (1972b). Studies on greening of etiolated seedlings. I. Elimination of the lag phase of chlorophyll biosynthesis by a pre-illumination of the embryonic axis in intact plants. *Physiol. Plant.* 26,157–165.

de Greef, J. A., and Caubergs, R. (1973). Studies on greening of etiolated seedlings. II. Leaf greening by phytochrome action in the embryonic axis. *Physiol. Plant.* 28,71–76.

de Greef, J. A., Caubergs, R., Verbelen, J.-P., and Moereels, E. (1976). Phytochrome-mediated interorgan dependency and rapid transmission of the light stimulus. *In* "Light and Plant Development" (H. Smith, ed.), pp. 295–316. Butterworths, London.

Dobberstein, B., Blobel, G., and Chua, N.-H. (1977). *In vitro* synthesis and processing of a putative precursor for the small subunit of ribulose-1,5-bisphosphate carboxylase in *Chlamydomonas rheinhardtii. Proc. Natl. Acad. Sci. USA* 74,1082–1085.

Ellis, R. J. (1979). The most abundant protein in the world. *Trends Biochem. Sci.* 4,241–244.

Fejes, E., Pay, A., Kanevsky, I., Szell, M., Adam E., Kay, S., and Nagy, F. (1990). A 268 bp upstream sequence mediates the circadian clock regulated transcription of the wheat *cab-1* gene in transgenic plants. *Plant Mol. Biol.* 15,921–932.

Fluhr, R., and Chua, N.-H. (1986). Developmental regulation of two genes encoding ribulose-bisphosphate carboxylase small subunit in pea and transgenic petunia plants: Phytochrome response and blue-light induction. *Proc. Natl. Acad. Sci. USA* 83,2358–2362.

Fluhr, R., Kuhlemeier, C., Nagy, F., and Chua, N.-H. (1986a). Organ-specific and light-induced expression of plant genes. *Science* 232,1106–1112.

Fluhr, R., Moses, P., Morelli, G., Coruzzi, G., and Chua, N.-H. (1986b). Expression dynamics of the pea *rbcS* multigene family and organ distribution of the transcripts. *EMBO J.* 5,2063–2071.

Fluhr, R., Kuhlemier, C., Green, P. J., Kay, S. A., Strittmatter, G., Cuozzo, M., Nagy, F., and Chua, N.-H. (1987). Multiple positive and negative elements mediate the light-induced expression of a plant gene. *In* "Plant Molecular Biology" (D. von Wettstein and N.-H. Chua, eds.), pp. 583–588. Plenum Press, New York, London.

Fraley, R. T., Rogers, S. G., Horsch, R. B., Eichholtz, D. A., Flick, J. S., Fink, C. L., Hoffmann, N. L., and Sanders, P. R. (1985). The SEV system: A new disarmed T_i plasmid vector system for plant transformation. *Biotechnology* 3,629–635.

Fried, M., and Crothers, D. M. (1981). Equilibria and kinetics of *lac* repressor-operator interactions by polyacrylamide gel electrophoresis. *Nucleic Acids Res.* 9,6505–6525.

Gallagher, T. F., and Ellis, R. J. (1982). Light-stimulated transcription of genes for two chloroplast polypeptides in isolated pea leaf nuclei. *EMBO J.* 1,1493–1498.

Gallagher, T. F., Jenkins, G. I., and Ellis, R. J. (1985). Rapid modulation of transcription of nuclear genes encoding chloroplast proteins by light. *FEBS Lett.* 186,241–245.

Garner, M. M., and Revzin, A. (1981). A gel electrophoresis method for quantifying the binding of proteins to specific DNA regions: Application to components of the *Escherichia coli* lactase operon regulatory system. *Nucleic Acids Res.* 9,3047–3060.

Gidoni, D., Brosio, P., Bond-Nutter, D., Bedbrook, J., and Dunsmuir, P. (1989). Novel *cis*-acting elements in petunia *cab* gene promoters. *Mol. Gen. Genet.* 215,337–344.

Gilmartin, P. M., and Chua, N.-H. (1990a). Localization of a phytochrome-responsive element within the upstream region of pea *rbcS-3A. Mol. Cell. Biol.* 10,5565–5568.

Gilmartin, P. M., and Chua, N.-H., (1990b). Spacing between GT-1 binding sites within a light-responsive element is critical for transcriptional activity. *Plant Cell* 2,447–455.

Gilmartin, P. M., Sarokin, L., Memelink, J., and Chua, N.-H. (1990). Molecular light switches for plant genes. *Plant Cell* 2,369–378.

Giuliano, G., Pichersky, E., Malik, V. S., Timko, M. P., Scolnik, P. A., and Cashmore, A. R. (1988). An evolutionarily conserved protein binding sequence upstream of a plant light-regulated gene. *Proc. Natl. Acad. Sci. USA* 85,7089–7093.

Gollmer, I., and Apel, K. (1983). The phytochrome-controlled accumulation of mRNA sequences encoding the light-harvesting chlorophyll *a/b* protein of barley (*Hordeum vulgare* L.). *Eur. J. Biochem.* 133,309–313.

Graham, D., Grieve, A. M., and Smillie, R. M. (1971). Phytochrome-mediated plastid development in etiolated pea stem apices. *Phytochemistry* 10,2905–2914.

Green, P. J., Kay, S. A., and Chua, N.-H. (1987). Sequence-specific interactions of a pea nuclear factor with light-responsive elements upstream of the *rbcS-3A* gene. *EMBO J.* 6,2543–2549.

Green, P. J., Yong, M.-H., Cuozzo, M., Kano-Murakami, Y., Silverstein, P., and Chua, N.-H. (1988). Binding site requirements for pea nuclear protein factor GT-1 correlate with sequences required for light-dependent transcriptional activation of the *rbcS* gene. *EMBO J.* 7,4035–4044.

Grob, U., and Stuber, K. (1987). Discrimination of phytochrome dependent light inducible from non-light inducible plant genes. Prediction of a common light-responsive element (LRE) in phytochrome dependent light inducible plant genes. *Nucleic Acids Res.* 15,9957–9973.

Gruissem, W. (1989). Chloroplast gene expression: How plants turn their plastids on. *Cell* 56,161–170.

Harpster, M. H., Mayfield, S. P., and Taylor, W. C. (1984). Effects of pigment-deficient mutants on the accumulation of photosynthetic proteins in maize. *Plant Mol. Biol.* 3,59–71.

Herrera-Estrella, L., van den Broeck, G., Maenhaut, R., van Montagu, M., Schell, J., Timko, M., and Cashmore, A. (1984). Light-inducible and chloroplast-associated expression of a chimaeric gene introduced into *Nicotiana tabacum* using a Ti plasmid vector. *Nature (Lond.)* 310,115–120.

Highfield, P. E., and Ellis, R. J. (1978). Synthesis and transport of the small subunit of chloroplast ribulose bisphosphate carboxylase. *Nature (Lond.)* 271,420–424.

Hock, B., and Mohr, H. (1964). Die Regulation der O_2-Aufrahme von Senfkeimlingen (*Sinapis alba* L.) durch Licht. *Planta* 61,209–228.

Horsch, R. B., Fry, J. E., Hoffmann, N. L., Eichholtz, D., Rogers, S. G., and Fraley, R. T. (1985). A simple and general method for transferring genes into plants. *Science* 227, 1229–1231.

Horwitz, B. A., Thompson, W. F., and Briggs, W. R. (1988). Phytochrome regulation of greening in *Pisum. Plant Physiol.* 86,299–305.

Jenkins, G. I., Hartley, M. R., and Bennett, J. (1983). Photoregulation of chloroplast development: Transcriptional, translational and post-translational controls? *Philos. Trans. R. Soc. Lond. B* 303,419–431.

Kasemir, H., Oberdorfer, U., and Mohr, H. (1973). A twofold action of phytochrome in controlling chlorophyll *a* accumulation. *Photochem. Photobiol.* 18,481–486.

Kaufman, L. S., Thompson, W. F., and Briggs, W. R. (1984). Different red light requirements for phytochrome-induced accumulation of *cab* and *rbcS* RNA. *Science* 226,1447–1449.

Kaufman, L. S., Briggs, W. R., and Thompson, W. F. (1985). Phytochrome control of specific mRNA levels in developing pea buds. The presence of both very low fluence and low fluence responses. *Plant Physiol.* 78,388–393.

Kaufman, L. S., Roberts, L. L., Briggs, W. R., and Thompson, W. F. (1986). Phytochrome control of specific mRNA levels in developing pea buds. Kinetics of accumulation, reciprocity, and escape kinetics of the low fluence response. *Plant Physiol.* 81,1033–1038.

Kay, S. A., Nagatani, A., Keith, B., Deak, M., Furuya, M., and Chua, N.-H. (1989). Rice phytochrome is biologically active in transgenic tobacco. *Plant Cell* 1,775–782.

Keith, B., and Chua, N.-H. (1986). Monocot and dicot pre-mRNAs are processed with different efficiencies in transgenic tobacco. *EMBO J.* 5,2419–2425.

Klee, H. J., Yanofsky, M. F., and Nester, E. W. (1985). Vectors for transformation of higher plants. *Biotechnology* 3,637–642.

Kloppstech, K. (1985). Diurnal and circadian rhythmicity in the expression of light-induced plant nuclear messenger RNAs. *Planta* 165,502–506.

Kuhlemeier, C., Fluhr, R., Green, P. J., and Chua, N.-H. (1987). Sequences in the pea *rbcS-3A* gene have homology to constitutive mammalian enhancers but function as negative regulatory elements. *Genes Dev.* 1,247–255.

Kuhlemeier, C., Cuozzo, M., Green, P. J., Goyvaerts, E., Ward, K., and Chua, N.-H. (1988). Localization and conditional redundancy of regulatory elements in *rbcS-3A*, a pea gene encoding the small subunit of ribulose-bisphosphate carboxylase. *Proc. Natl. Acad. Sci. USA* 85,4662–4666.

Kung, S. D. (1976). Tobacco fraction 1 protein; A unique genetic marker. *Science* 191, 429–434.

Lam, E., and Chua, N.-H. (1990). GT-1 binding site confers light responsive expression in transgenic tobacco. *Science* 248,471–474.

Lam, E., Kano-Murakami, Y., Gilmartin, P., Niner, B., and Chua, N.-H. (1990). A metal-dependent DNA-binding protein interacts with a constitutive element of a light-responsive promoter. *Plant Cell* 2,857–866.

Lamppa, G., Morelli, G., and Chua, N.-H. (1985a). Structure and developmental regulation of a wheat gene encoding the major chlorophyll *a/b*-binding polypeptide. *Mol. Cell Biol.* 5,1370–1378.

Lamppa, G., Nagy, F., and Chua, N.-H. (1985b). Light-regulated and organ specific expression of a wheat *cab* gene in transgenic tobacco. *Nature (Lond.)* 316,750–752.

Luthe, D. S., and Quatrano, R. S. (1980a). Transcription in isolated wheat nuclei. I. Isolation of nuclei and elimination of endogenous ribonuclease activity. *Plant Physiol.* 65, 305–308.

Luthe, D. S., and Quatrano, R. S. (1980b). Transcription in isolated wheat nuclei. II. Characterization of RNA synthesized *in vitro*. *Plant Physiol.* 65,309–313.

Manzara, T., and Gruissem, W. (1988). Organization and expression of the genes encoding ribulose-1,5-bisphosphate carboxylase in higher plants. *Photosynth. Res.* 16,117–139.

Mapelston, R. E., and Griffiths, W. T. (1979). Effects of illumination of whole barley plants on the protochlorophyllide-activating system in the isolated plastids. *Biochem. Soc. Trans.* 5,319–321.

Mapelston, R. E., and Griffiths, W. T. (1980). Light-modulation of the activity of protochlorophyllide reductase. *Biochem. J.* 189,125–133.

Mayfield, S. P., and Taylor, W. C. (1984). Carotenoid-deficient maize seedlings fail to accumulate light-harvesting chlorophyll a/b binding protein (LHCP) mRNA. *Eur. J. Biochem.* 144,79–84.

Mohr, H. (1988). Control of plant development: Signals from without—signals from within. *Bot. Mag.* 101,79–101.

Morelli, G., Nagy, F., Fraley, R. T., Rogers, S. G., and Chua, N.-H. (1985). A short conserved sequence is involved in the light-inducibility of a gene encoding ribulose-1,5-bisphosphate carboxylase small subunit of pea. *Nature (Lond.)* 315,200–204.

Mösinger, E., Batschauer, A., Schäfer, E., and Apel, K. (1985). Phytochrome control of *in vitro* transcription of specific genes in isolated nuclei from barley (*Hordeum vulgare*). *Eur. J. Biochem.* 147,137–142.

Mösinger, E., Batschauer, A., Apel, K., Schäfer, E., and Briggs, W. R. (1988). Phytochrome regulation of greening in barley. Effects on mRNA abundance and on transcriptional activity of isolated nuclei. *Plant Physiol.* 86,706–710.

Nagy, F., Morelli, G., Fraley, R. T., Rogers, S. G., and Chua, N.-H. (1985). Photoregulated expression of a pea *rbcS* gene in leaves of transgenic plants. *EMBO J.* 4,3063–3068.

Nagy, F., Kay, S. A., Boutry, M., Hsu, M.-Y, and Chua, N.-H. (1986). Phytochrome-controlled expression of a wheat *cab* gene in transgenic tobacco seedlings. *EMBO J.* 5,1119–1124.

Nagy, F., Boutry, M., Hsu, M.-Y., Wong, M., and Chua, N.-H. (1987). The 5'-proximal region of the wheat *cab-1* gene contains a 268-bp enhancer-like sequence for phytochrome response. *EMBO J.* 6,2537–2542.

Nagy, F., Kay, S. A., and Chua, N.-H. (1988a). A circadian clock regulates transcription of the wheat *cab-1* gene. *Genes Dev.* 2,376–382.

Nagy, F., Kay, S. A., and Chua, N.-H. (1988b). Gene regulation by phytochrome. *Trends Genet.* 4,37–42.

Oelmüller, R., and Mohr, H. (1986). Photooxidative destruction of chloroplasts and its consequences for expression of nuclear genes. *Planta* 167,106–113.

Oelmüller, R., Levitan, I., Bergfeld, R., Rajasekhar, V. K., and Mohr, H. (1986b). Expression of nuclear genes as affected by treatments acting on the plastids. *Planta* 168, 484–492.

Oelmüller, R., Dietrich, G., Link, G., and Mohr, H. (1986a). Regulatory factors involved in gene expression (subunits of ribulose-1,5-bisphosphate carboxylase) in mustard (*Sinapis alba* L.) cotyledons. *Planta* 169,260–266.

Oelmüller R., Schuster, C., and Mohr, H. (1988). Physiological characterization of a plastidic signal required for nitrate-induced appearance of nitrate and nitrite reductases. *Planta* 174,75–83.

Oelze-Karow, H., and Mohr, H. (1978a). Die Bedeutung des Phytochroms für die Entwicklung der Kapazität für Photophosphorylierung. *Ber. Dtsch. Bot. Ges.* 91,603–610.

Oelze-Karow, H., and Mohr, H. (1978b). Phytochrome control of the development of photophosphorylation. *Photochem. Photobiol.* 27,255–258.

Oelze-Karow, H., and Mohr, H. (1986). Appearance of photophosphorylation capacity: Threshold *vs* graded control by phytochrome. *Photochem. Photobiol.* 44,221–230.

Oelze-Karow, H., and Mohr, H. (1988). Rapid transmission of a phytochrome signal from hypocotyl hook to cotyledons in mustard. *Photochem. Photobiol.* 47,447–450.

Oelze-Karow, H., and Mohr, H. (1989). An analysis of phytochrome-mediated threshold control of hypocotyl growth in mustard (*Sinapis alba* L.) seedlings. *Photochem. Photobiol.* 50,133–141.

Oelze-Karow, H., and Mohr, H. (1990). Phytochrome-mediated threshold control of hypocotyl growth in partly de-etiolated mustard (*Sinapis alba* L.) seedlings. *Photochem. Photobiol.* 52,115–121.

Price, L., and Klein, W. H. (1961). Red, far-red response and chlorophyll synthesis. *Plant Physiol.* 36,733–735.

Rajasekhar, V. K., and Mohr, H. (1986). Appearance of nitrite reductase in cotyledons of the mustard (*Sinapis alba* L.) seedlings as affected by nitrate, phytochrome and photooxidative damage of plastids. *Planta* 168,369–376.

Roberts, B. E., and Paterson, B. M. (1973). Efficient translation of tobacco mosaic virus RNA and rabbit globin 9s RNA in a cell-free system from commercial wheat germ. *Proc. Natl. Acad. Sci. USA* 70,2330–2334.

Rolfe, S. A., and Tobin, E. M. (1991). Deletion analysis of a phytochrome-regulated monocot *rbcS* promoter in a transient assay system. *Proc. Natl. Acad. Sci. USA* 88,2683–2686.

Santel, H.-J., and Apel, K. (1981). The protochlorophyllide holochrome of barley (*Hordeum vulgare* L.). The effect of light on the NADPH : protochlorophyllide oxidoreductase. *Eur. J. Biochem.* 120,95–103.

Schulze-Lefert, P., Becker-André, M., Schulz, W., Hahlbrock, K., and Dangl, J. L. (1989a). Functional architecture of the light-responsive chalcone synthase promoter from parsley. *Plant Cell* 1,707–714.

Schulze-Lefert, P., Dangl, J. L., Becker-André, M., Hahlbrock, K., and Schulz, W. (1989b). Inducible *in vivo* DNA footprints define sequences necessary for UV light activation of the parsley chalcone synthase gene. *EMBO J.* 8,651–656.

Schuster, C., and Mohr, H. (1990a). Photooxidative damage to plastids affects the abundance of nitrate-reductase mRNA in mustard cotyledons. *Planta* 181,125–128.

Schuster, C., and Mohr, H. (1990b). Appearance of nitrite reductase mRNA in mustard seedling cotyledons is regulated by phytochrome. *Planta* 181,327–334.

Schuster, C., Oelmüller, R., and Mohr, H. (1987). Signal storage in phytochrome action on nitrate-mediated induction of nitrate and nitrite reductases in mustard seedling cotyledons. *Planta* 171,136–143.

Schuster, C., Oelmüller, R., and Mohr, H. (1988). Control by phytochrome of the appearance of ribulose-1,5-bisphosphate carboxylase and the mRNA for its small subunit. *Planta* 174,426–432. Copyright 1988, Springer-Verlag.

Schuster, C., Schmidt, S., and Mohr, H. (1989). Effect of nitrate, ammonium, light and a plastidic factor on the appearance of multiple forms of nitrate reductase in mustard (*Sinapis alba* L.) cotyledons. *Planta* 177,74–83.

Sharma, A. K., and Sopory, S. K. (1984). Independent effects of phytochrome and nitrate on nitrate reductase and nitrite reductase activities in maize. *Photochem. Photobiol.* 39,491–493.

Silverthorne, J., and Tobin, E. M. (1984). Demonstration of transcriptional regulation of specific genes by phytochrome action. *Proc. Natl. Acad. Sci. USA* 81,1112–1116.

Silverthorne, J., and Tobin, E. M. (1990). Post-transcriptional regulation of organ-specific expression of individual *rbcS* mRNAs in *Lemna gibba*. *Plant Cell* 2,1181–1190.

Silverthorne, J., Wimpee, C. F., Yamada, T., Rolfe, S. A., and Tobin, E. M. (1990). Differential expression of individual genes encoding the small subunit of ribulose-1,5-bisphosphate carboxylase in *Lemna gibba*. *Plant Mol. Biol.* 15,49–58.

Simpson, J., von Montagu, M., and Herrera-Estrella, L. (1986). Photosynthesis-associated gene families: Differences in response to tissue-specific and environmental factors. *Science* 233,34–38.

Sisler, E. C., and Klein, W. H. (1963). The effect of age and various chemicals on the lag phase of chlorophyll synthesis in dark grown bean seedlings. *Physiol. Plant.* 16,315–322.

Slovin, J. P., and Tobin, E. M. (1982). Synthesis and turnover of the light-harvesting chlorophyll *a/b* protein in *Lemna gibba* grown with intermittent red light: Possible translational control. *Planta* 154,465–472.

Smith, S. E., and Ellis, R. J. (1981). Light-stimulated accumulation of transcripts of nuclear and chloroplast genes for ribulosephosphate carboxylase. *J. Mol. Appl. Genet.* 1, 127–137.

Staiger, D., Kaulen, H., and Schell, J. (1989). A CACGTG motif of the *Antirrhinum majus* chalcone synthase promoter is recognized by an evolutionarily conserved nuclear protein. *Proc. Natl. Acad. Sci. USA* 86,6930–6934.

Stiekema, W. J., Wimpee, C. F., Silverthorne, J., and Tobin, E. M. (1983). Phytochrome control of the expression of two nuclear genes encoding chloroplast proteins in *Lemna gibba* L. G-3. *Plant Physiol.* 72,717–724.

Thompson, W. F., Everett, M., Polans, N. O., Jorgensen, R. A., and Palmer, J. D. (1983). Phytochrome control of RNA levels in developing pea and mung-bean leaves. *Planta* 158,487–500.

Timko, M. P., Kausch, A. P., Castresana, C., Fassler, J., Herrera-Estrella, L., van den Broeck, G., van Montagu, M., Schell, J., and Cashmore, A. R. (1985). Light regulation of plant gene expression by an upstream enhancer-like element. *Nature (Lond.)* 318, 579–582.

Tobin, E. M. (1978). Light regulation of specific mRNA species in *Lemna gibba* L. G-3. *Proc. Natl. Acad. Sci. USA* 75,4749–4753.

Tobin, E. M. (1981a). Light-regulation of the synthesis of two major nuclear-coded chloroplast polypeptides in *Lemna gibba*. In "Photosynthesis V. Chloroplast Development" (G. Akoyunoglou, ed.), pp. 949–959. Balaban International Science Services, Philadelphia.

Tobin, E. M. (1981b). Phytochrome-mediated regulation of messenger RNAs for the small subunit of ribulose 1,5-bisphosphate carboxylase and the light-harvesting chlorophyll *a/b*-protein in *Lemna gibba*. *Plant Mol. Biol.* 1,35–51.

Tobin, E. M. (1981c). White light effects on the mRNA for the light harvesting chlorophyll *a/b*-protein in *Lemna gibba* L. G-3. *Plant Physiol.* 67,1078–1083.

Tobin, E. M., and Klein, A. O. (1975). Isolation and translation of plant messenger RNA. *Plant Physiol.* 56,88–92.

Tobin, E. M., Silverthorne, J., Stiekema, W. J., and Wimpee, C. F. (1984a). Phytochrome regulation of the synthesis of two nuclear-coded chloroplast proteins. In "Chloroplast Biogenesis" (R. J. Ellis, ed.), pp. 321–335. Cambridge University Press, Cambridge.

Tobin, E. M., Wimpee, C. F., Silverthorne, J., Stiekema, W. J., Neumann, G. A., and Thornber, J. P. (1984b). Phytochrome regulation of the expression of two nuclear-coded chloroplast proteins. In "Biosynthesis of the Photosynthetic Apparatus: Molecular Biology, Development and Regulation" (J. P. Thornber, L. A. Staehlin, and R. Hallick, eds.), pp. 325–334. Alan R. Liss, New York.

Tobin, E. M., Wimpee, C. F., Karlin-Neumann, G. A., Silverthorne, J., and Kohorn, B. D. (1985). Phytochrome regulation of nuclear gene expression. In "Molecular Biology of the Photosynthetic Apparatus" (C. J. Arntzen, L. Bogorad, S. Bonitz, and K. E. Steinback, eds.), pp. 373–380. Cold Spring Harber Laboratory, Cold Spring Harbor, New York.

Wimpee, C. F., Stiekema, W. J., and Tobin, E. M. (1983). Sequence heterogeneity in the RuBP carboxylase small subunit gene family of *Lemna gibba*. In "Plant Molecular Biology" (R. Goldberg, ed.), pp. 391–401. Alan R. Liss, New York.

Withrow, R. B., Wolff, J. B., and Price, L. (1956). Elimination of the lag phase of chlorophyll synthesis in dark-grown bean leaves by a pretreatment with low irradiances of monochromatic energy. *Plant Physiol.* 31 (suppl. xiii–xiv).

Autoregulation

With the advent of *in vitro* translation, Schäfer and Klaus Gottmann were able to compare phytochrome mRNA activities in light and darkness. When they isolated and translated poly(A) RNA from etiolated oat seedlings, a 125-kDa polypeptide that reacted with phytochrome antiserum became labeled. This protein was not among the translation products of poly(A) RNA from light-treated seedlings. But it reappeared when light-treated seedlings were placed back in darkness, suggesting that "the phytochrome system is possibly light dependent with respect to the synthetic potential for this protein" (Gottmann and Schäfer, 1982). Radioactive sulfur was incorporated into phytochrome in darkness but not in light, also indicating differential synthesis of the apoprotein (Gottmann and Schäfer, 1983).

In Madison, Quail and colleagues also failed to obtain phytochrome mRNA activity from light-treated oat (Quail et al., 1983). But Colbert was able to isolate poly(A) RNA from etiolated oat and immunoprecipitate phytochrome from the translation products. Fifteen minutes after a R pulse, the mRNA activity began to fall, halving within 1 h and becoming undetectable after 2 h. The amount of spectrally detectable phytochrome remained constant for the first hour and halved in $2\frac{1}{2}$ h. Phytochrome proved to be the photoreceptor for this response, which was R/FR-reversible to the level of the FR control. Just 1% Pfr decreased phytochrome mRNA activity by 60%, while 20% Pfr was saturating. Thus "conversion of Pfr not only enhances degradation of the chromoprotein but also decreases the rate of synthesis due to a concomitant reduction in the level of translatable phytochrome mRNA" (Fig. 33-1) (Colbert et al., 1983; Quail, 1984). Sixteen years earlier, Hillman had speculated "about phytochrome as a component of a feedback regulatory system" (Hillman, 1967).

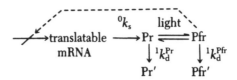

FIGURE 33-1   Autoregulation of phytochrome synthesis. (From Colbert et al., 1983.)

The phytochrome controlling this response appeared to be stable. Otto found that a R pulse at the end of a 24-h photoperiod was strongly inhibitory even many hours after the transition. Changing the photoequilibrium after 12 h darkness further affected phytochrome levels (Fig. 33-2) (Otto et al., 1983).

In 4-day-old etiolated pea seedlings, a R pulse decreased phytochrome mRNA activity to 5% of that in darkness within 30 min, and the response was partially R/FR-reversible (Otto et al., 1984). The effect was less marked in germinating peas, where the activity of phytochrome mRNA from embryonic axes began to increase 3 h after imbibition, before growth began. Fourteen hours later, axes had more than half the level of translatable mRNA in WL than in darkness (Sato and Furuya, 1985). Most dicots do not dramatically down-regulate phytochrome but synthesize similar amounts in light and darkness (Lissemore et al., 1987; Sharrock et al., 1988).

PHYTOCHROME mRNA ABUNDANCE

Colbert and Hershey, meanwhile, had prepared cDNAs for phytochrome (see Chapter 28), which enabled Colbert to measure directly the abundance of oat phytochrome mRNA (Hershey et al., 1984). Abundance declined within 30 min after a R pulse and halved within 1 h. By 5 h, less than 10% remained. In extended darkness, the level slowly rose, and half the initial amount was regained by 48 h (Colbert et al., 1985).

Using a cDNA probe and primer extension analysis, Naoki Sato discovered that pea $phyA^1$ was transcribed from different start sites in the 5' noncoding region to generate three transcripts, RNA 1, RNA 2, and RNA 3. These differed in the amount of RNA that preceded the coding sequence by 63, 285, or 465 nucleotides. Although total phytochrome mRNA decreased by about two-thirds during the first 3 h after a dark to WL transition, the three transcripts were not equally affected. The major transcript

1. That is, this gene is closer in sequence to *Arabidopsis phyA* than to *phyB* or *phyC*.

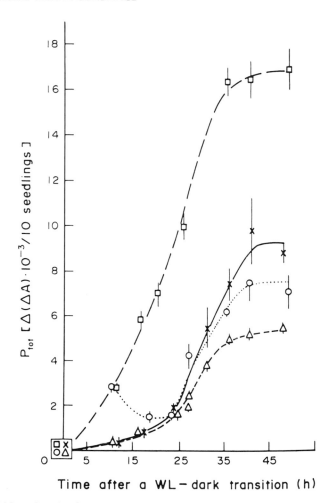

FIGURE 33-2   Levels of total spectroscopically detectable phytochrome in oat seedlings after 48 h dark/24 h WL followed by a 5-min light pulse + 12 h dark + a second 5-min light pulse. First light pulse: (□) FR, (○), FR, (△) R, (X) R. Second light pulse: (○) R, (X) FR. (From Otto et al., 1983.)

in etiolated seedlings, RNA 1, was strongly down-regulated, suggesting that the gene coded for phytochrome I. RNA 2, which initially was only about half as abundant as RNA 1, decreased its level less dramatically. The amount of RNA 3, the least abundant transcript, actually increased, providing the first example of light-induced *phy* expression (Sato, 1988). Phytochrome was the photoreceptor for the RNA 1 and RNA 2 responses, but the RNA 3 response was too small to be tested (Tomizawa et al., 1989).

TRANSCRIPTION RATES

Run-on transcription assays revealed that the phytochrome gene is regulated at the transcriptional level and that transcriptional rates change in less than 5 min (Quail et al., 1986). This rapid down-regulation suggested the "direct action of the phytochrome signal-transduction chain without the intervention of expression of other genes" (Quail et al., 1987). James L. Lissemore observed that transcription slowed within 5 min after a R pulse, dropped below detectable levels within 1 h, and recovered within 12 to 24 h (Fig. 33-3). A FR pulse also repressed *phy* transcription, but recovery began after only 3 h, as it did after a R/FR treatment. Thus recovery from repression was a photoreversible response, dependent on initial Pfr levels. Repression itself was an irreversible VLFR, saturating at less than 1% Pfr, because 0.03% Pfr or only about 30 molecules per cell decreased transcription to one-fifth the rate in etiolated tissue (Lissemore and Quail, 1988). "These are the principal data that enable us to say that this regulation is at the transcriptional level and that there is a reversible change in transcription in response to phytochrome," says Quail.

The data agreed well with the mRNA abundance measurements, but "Colbert's initial data," Quail says, "showed a discrepancy between the extent of the decrease in transcriptional regulation and the extent of the decrease in steady-state mRNA levels. We interpreted that difference as indicating that there was both transcriptional and post-transcriptional differences (Quail et al., 1986). Lissemore's data was obtained more rigorously, using internal hybridization standards, and he sees no indication of that. So the primary level of regulation may be transcriptional, although the possibility of a change in the turnover rate of the mRNA is not excluded."

SIGNAL TRANSDUCTION

When Lissemore placed wounded seedlings into cycloheximide and chloramphenicol solutions, no new proteins were made, but phytochrome regulation of *phy* transcription was unimpeded. This observation was "consistent with the notion that all of the signal transduction chain components required to regulate *phy* transcription are present prior to light treatment. . . . The data provide the first evidence for direct control of transcription of a gene by the phytochrome transduction chain"

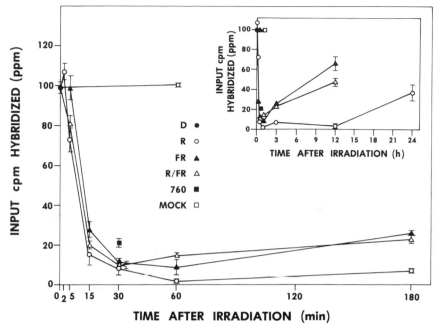

FIGURE 33-3 Effects of R, FR, R/FR, and 760-nm light pulses on transcription of phytochrome gene in dark-grown *Avena* seedlings. (From Lissemore and Quail, 1988.)

(Lissemore and Quail, 1988). Schäfer's group had reached a similar conclusion for the transduction chain between phytochrome and the *cab* gene, although they measured mRNA abundance rather than rates of transcription (Merkle and Schäfer, 1988).

Such rapid repression of *phy* resurrected the possibility that phytochrome itself is the repressor, even though its presence in nuclei had never been confirmed. But phytochrome failed to bind to any phytochrome promoter sequence tested in gel retardation assays (Quail et al., 1987), suggesting that a signal transduction pathway connects phytochrome with its gene.

In rice, which has a single phytochrome I gene, *phy* gene transcription is reduced to 15% of the dark level within 15 min and becomes undetectable within 30 min. The rice phytochrome gene has several familiar motifs in its 5'-flanking region, including a GC-rich sequence and two putative GT-1 core-binding sites (GGTTAA and GGTAAT). In gel retardation assays, a DNA fragment containing the −242/−220 set bound this *trans*-acting factor (Kay et al., 1989). During *in vitro* DNAse footprinting experiments at the Rockefeller University, three sites were bound by GT-1 in the rice *phy* promoter (Kay, 1991). All were competed by the box II core sequence but not by the GG → CC box II mutant. A second factor, ϕGC-1, was also identified; this binds the highly conserved GC-rich box I sequence that is present in rice, oat, and maize *phyA* promoters (Kay, 1991). In Quail's lab, Katayoon Dehesh identified another protein, GT-2, which binds exclusively to only one of the GT cores—GGTAATT. GT-2 has features in common with known transcriptional factors (Dehesh et al., 1990).

The USDA/U.C. Berkeley researchers had previously studied the oat *phyA* promoter, shooting particles coated with constructs (mutants of the oat *phyA* gene fused to a CAT reporter) into etiolated rice seedlings (Bruce et al., 1989). By progressively deleting the promoter, Wesley B. Bruce discovered a *cis*-acting element, PE1 (positive element 1), whose presence was required for full expression of *phyA* under low-Pfr conditions (Fig. 33-4). The 5′ boundary of PE1 lay between −381 and −348. Internal deletion analysis revealed a second major element, PE2, further upstream, between −635 and −489, which could substitute for PE1. PE3 (−110 to −76), which included a box I, was required in addition to PE1 or PE2 for

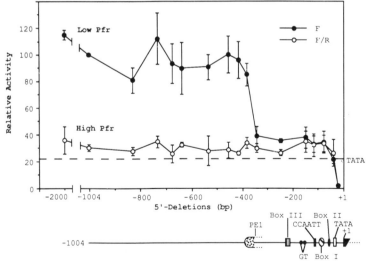

FIGURE 33-4   Deletion analysis of the oat *phyA3* promoter. After bombardment with oat *phyA3*/CAT constructs, etiolated rice seedlings were irradiated with a pulse of either FR (low Pfr) or FR/R (high Pfr) and returned to darkness for 24 h before extraction and measurement of CAT activity. Dashed line indicates basal activity driven by TATA alone (abnormally high—because cells contained multiple copies of a construct?). The 5′-flanking region is illustrated. CCAATT, like TATA, is a canonical sequence found in most eukaryotic promoters. (From Bruce and Quail, 1990. Reproduced by permission of the American Society of Plant Physiologists.)

maximal expression. Box I has been detected in all the known *phyA* genes (oat, rice, and maize).

PE1, PE2, and PE3 exhibited no sequence homology among themselves, and no regions of PE1 and PE2 corresponded to *cis*-acting elements in other genes or even nonoat *phyA* genes. The oat *phyA* promoter did contain several recognizable sequences, including a tandem pair of GT motifs (GGTTAAT) identical to the GT-1 binding motif in the rice *phyA* promoter. But removal or alteration of such sequences had only minor effects on CAT expression. Quail speculated that "*phyA* promoters may be modular with an array of different upstream *cis*-acting sequences having been recruited in different plants and in different locations within the promoters to serve the same transcriptional activation function, but with the down-stream element being strictly conserved and possibly exclusive to *phyA* promoters" (Bruce and Quail, 1990). The group recently detected a negatively acting sequence element, RE1 ($-80/-70$), CATGGGCGCGG. When this light switch was inactivated through sequence mutagenesis, *phyA3* was transcribed maximally in the presence of high Pfr levels (Bruce et al., 1991).

AUTOREGULATION?

Multiple phytochrome species were unknown when regulation of phytochrome synthesis was discovered. Therefore it was logical to propose that phytochrome was "a regulatory photoreceptor that controls the expression of its own gene" (Quail, 1984). But Otto's observations on phytochrome accumulation after a light–dark transition suggested the involvement of a stable photoreceptor. So did measurements of mRNA levels under similar conditions (Colbert et al., 1985).

After phytochrome II was discovered, Thomas began to question the concept of autoregulation. A visitor to his lab, Janet Hilton, was studying seed phytochrome, which resembled the pigment from light-grown plants in its stability and failure to react to an MAb specific for "seedling phytochrome" (Hilton and Thomas, 1985). This phytochrome became spectrophotometrically and immunochemically detectable in embryos 12–18 h after the beginning of imbibition. But a R pulse before this time reduced accumulation of immunodetectable—type I—phytochrome (Hilton and Thomas, 1987) and its mRNA (Thomas et al., 1989). Because this inhibition remained FR-reversible until 30 h after sowing (Hilton and Thomas, 1987), "the Pfr controlling the response was stable," Thomas says. "And phytochrome I did not appear to be in the seeds anyhow; it was difficult to see how it could control its own synthesis—although we couldn't eliminate the possibility that there was a low, undetectable level."

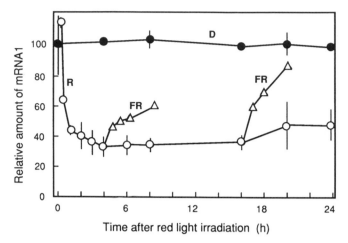

FIGURE 33-5   Effect of 3 min FR given after 4 or 16 h darkness on abundance of pea *phyA* mRNA1 in dark-grown pea seedling epicotyl hooks pretreated with 3 min R at zero time. D, dark control. (From Furuya et al., 1991. Copyright 1991, Springer-Verlag.)

Using a primer extension assay, Furuya's group monitored levels of RNA1 and RNA2 transcripts of the single-copy pea *phyA* gene after placing light-grown seedlings in darkness. A FR pulse immediately before darkness enhanced the accumulation of both mRNAs, and the effect was R/FR-reversible. Without the FR treatment, recovery was slow (Tomizawa et al., 1989). Moreover, FR also photoreversibly enhanced the accumulation of these transcripts when R-pretreated dark-grown pea seedlings received a FR pulse after 4 or even 16 h of darkness (Fig. 33-5). Thus a stable phytochrome appeared to control the levels of RNA1 and RNA2, because phytochrome I would be depleted by a R pretreatment and resynthesis would produce only Pr (Furuya et al., 1991). "But which molecular species results in stable phytochrome?" Furuya asks. "If type I accounts for the stable pool, transcription is autoregulated. But if type II forms the stable pool, it's no longer autoregulation. We have to answer this question with mutants."

REFERENCES

Bruce, W. B., and Quail, P. H. (1990). *Cis*-acting elements involved in photoregulation of an oat phytochrome promoter in rice. *Plant Cell* 2,1081–1089.

Bruce, W. B., Christensen, A. H., Klein, A. H., Fromm, M., and Quail, P. H. (1989). Photoregulation of a phytochrome gene promoter from oat transferred into rice by particle bombardment. *Proc. Natl. Acad. Sci. USA* 86,9692–9696.

Bruce, W. B., Deng, X.-W., and Quail, P. H. (1991). A negatively acting DNA sequence element mediates phytochrome-directed repression of *phyA* gene transcription. *EMBO J.* 10,3015–3024.

Colbert, J. T., Hershey, H. P., and Quail, P. H. (1983). Autoregulatory control of translatable phytochrome mRNA levels. *Proc. Natl. Acad. Sci. USA* 80,2248–2252.

Colbert, J. T., Hershey, H. P., and Quail, P. H. (1985). Phytochrome regulation of phytochrome mRNA abundance. *Plant Mol. Biol.* 5,91–101.

Dehesh, K., Bruce, W. B., and Quail, P. H. (1990). A *trans*-acting factor that binds to a GT-motif in a phytochrome gene promoter. *Science* 250,1397–1399.

Furuya, M., Ito, N., Tomizawa, K., and Schäfer, E. (1991). A stable phytochrome pool regulates the expression of the phytochrome I gene in pea seedlings. *Plants* 183,218–221.

Gottmann, K., and Schäfer, E. (1982). *In vitro* synthesis of phytochrome apoprotein directed by mRNA from light and dark grown *Avena* seedlings. *Photochem. Photobiol.* 35, 521–525.

Gottmann, K., and Schäfer, E. (1983). Analysis of phytochrome kinetics in light-grown *Avena sativa* L. seedlings. *Planta* 157,392–400.

Hershey, H. P., Colbert, J. T., Lissemore, J. L., Barker, R. F., and Quail, P. H. (1984). Molecular cloning of cDNA for *Avena* phytochrome. *Proc. Natl. Acad. Sci. USA* 81,2332–2336.

Hillman, W. S. (1967). The physiology of phytochrome. *Annu. Rev. Plant Physiol.* 18, 301–324.

Hilton, J. R., and Thomas, B. (1985). Comparison of seed and seedling phytochrome in *Avena sativa* L. using monoclonal antibodies. *J. Exp. Bot.* 36,1937–1946.

Hilton, J. R., and Thomas, B. (1987). Photoregulation of phytochrome synthesis in germinating embryos of *Avena sativa* L. *J. Exp. Bot.* 38,1704–1712.

Kay, S. A. (1991). *In vitro* protein–DNA interactions in the rice phytochrome promoter. *In* "Phytochrome Properties and Biological Action" (B. Thomas and C. B. Johnson, eds.), pp. 129–140. Springer, Berlin, Heidelberg, New York.

Kay, S. A., Keith, B., Shinozaki, K., Chye, M.-L., and Chua, N.-H. (1989). The rice phytochrome gene: Structure, autoregulated expression, and binding of GT-1 to a conserved site in the 5' upstream region. *Plant Cell* 1,351–360.

Lissemore, J. L., and Quail, P. H. (1988). Rapid transcriptional regulation by phytochrome of the genes for phytochrome and chlorophyll *a/b*-binding protein in *Avena sativa*. *Mol. Cell. Biol.* 8,4840–4850.

Lissemore, J. L., Colbert, J. T., and Quail, P. H. (1987). Cloning of cDNA for phytochrome from etiolated *Cucurbita* and coordinate photoregulation of the abundance of two distinct phytochrome transcripts. *Plant Mol. Biol.* 8,485–496.

Merkle, T., and Schäfer, E. (1988). Effect of cycloheximide on the inverse control by phytochrome of the expression of two nuclear genes in barley (*Hordeum vulgare* L.). *Plant Cell Physiol.* 29,1251–1254.

Otto, V., Mösinger, E., Sauter, M., and Schäfer, E. (1983). Phytochrome control of its own synthesis in *Sorghum vulgare* and *Avena sativa*. *Photochem. Photobiol.* 38,693–700.

Otto, V., Schäfer, E., Nagatani, A., Yamamoto, K. T., and Furuya, M. (1984). Phytochrome control of its own synthesis in *Pisum sativum*. *Plant Cell Physiol.* 25,1579–1584.

Quail, P. H. (1984). Phytochrome: A regulatory photoreceptor that controls the expression of its own gene. *Trends Biochem. Sci.* 9,450–453.

Quail, P. H., Bolton, G. W., Hershey, H. P., and Vierstra, R. D. (1983). Phytochrome: Molecular weight, *in vitro* translation and cDNA cloning. *In* "Current Topics in Plant Biochemistry and Physiology" (D. D. Randall, D. G. Blevins, and R. Larson, eds.), pp. 25–36. University of Missouri, Columbia.

Quail, P. H., Colbert, J. T., Peters, N. K., Christensen, A. H., Sharrock, R. A., and Lissemore, J. L. (1986). Phytochrome and the regulation of the expression of its genes. *Philos. Trans. R. Soc. Lond. B* 314,469–480.

Quail, P. H., Gatz, C., Hershey, H. P., Jones, A. M., Lissemore, J. L., Parks, B. M., Sharrock, R. A., Barker, R. F., Idler, K., Murray, M. G., Koornneef, M., and Kendrick, R. E. (1987). Molecular biology of phytochrome. *In* "Phytochrome and Photoregulation in Plants" (M. Furuya, ed.), pp. 23–37. Academic Press, Tokyo.

Sato, N. (1988). Nucleotide sequence and expression of the phytochrome gene in *Pisum sativum:* Differential regulation by light of multiple transcripts. *Plant Mol. Biol.* 11, 697–710.

Sato, N., and Furuya, M. (1985). Synthesis of translatable mRNA for phytochrome during imbibition in embryonic axes of *Pisum sativum* L. *Plant Cell Physiol.* 26,1511–1517.

Sharrock, R. A., Parks, B. M., Koornneef, M., and Quail, P. H. (1988). Molecular analysis of the phytochrome deficiency in an *aurea* tomato. *Mol. Gen. Genet.* 213,9–14.

Thomas, B., Penn, S. E., and Jordan, B. R. (1989). Factors affecting phytochrome transcripts and apoprotein synthesis in germinating embryos of *Avena sativa* L. *J. Exp. Bot.* 40,1299–1304.

Tomizawa, K.-I., Sato, N., and Furuya, M. (1989). Phytochrome control of multiple transcripts of the phytochrome gene in *Pisum sativum. Plant Mol. Biol.* 12,295–299.

CHAPTER **34** _____

Phytochrome and Flowering

"The recent literature abounds in hypotheses consistent with the data but, on the face of it, inconsistent with each other, or at least strikingly different," Hillman wrote in 1976. "Of course, aspects of the photoperiodic timing mechanism may differ between organisms. And some of the hypotheses may well reduce to the same basic structure" (Hillman, 1976).

It was accepted that photoperiodic timing involves endogenous rhythms, at least in SDP. But some workers interpreted their data with internal coincidence models, whereas others proposed that light must interact with the photoinducible phase of a single rhythm but were confused about how the rhythm was entrained. And although Pfr was known to promote flowering during the early hours of darkness and inhibit if formed by a light break later, the roles of phytochrome at dawn and dusk were unclear.

TIME MEASUREMENT IN *PHARBITIS*

The most detailed photoperiodic model has come from studies of *Pharbitis nil,* in which Takimoto and Hamner had linked the photoperiodic response to two separate rhythms, one entrained to dawn and one to dusk (Takimoto and Hamner, 1964, 1965).

In 1978, Vince-Prue (Fig. 34-1) moved to the Glasshouse Crops Research Institute in Littlehampton, where she at last had facilities for photoperiodism research. There she was joined by Thomas and doctoral student Peter Lumsden. Realizing that the growth of experimental plants in continuous light might have confounded many previous studies, the group chose to work with dark-grown *Pharbitis* seedlings, which can be induced if a single photoperiod precedes an inductive dark period.

The researchers discovered that a R light break during this dark period was most inhibitory (LB_{max}) at about 15 and 40 h after lights-on if the

FIGURE 34-1   Daphne Vince-Prue.

photoperiod was shorter than 6 h. Thus a circadian rhythm of inhibition was entrained to the beginning of the photoperiod. But when the photoperiod exceeded 6 h, the inhibitory peaks fell at 9 and 32.5 h after the *end* of the photoperiod, revealing that the rhythm was now entrained to the lights-off signal (Lumsden et al., 1982).

This suggested that a rhythm would begin when a seedling first emerged through the soil and saw light or, if it emerged at night, at dawn the next day. Six hours later, the rhythm would be "suspended"—at circadian time 6 (CT6)—and then released at dusk, still at CT6. Because daylength always exceeds 6 h during the growing season, the seedling would in practice time each night from the end of each photoperiod.

After prolonged (24-h) WL, the rhythm damped, almost erasing the second LB_{max} (Fig. 34-2). Thus a single point of inhibition punctuated the dark period, as it always did in *Xanthium*. The researchers therefore proposed that rapid damping, rather than the operation of a nonrhythmic timing mechanism, accounted for *Xanthium*'s failure to show a rhythmic light-break response (Lumsden et al., 1982).

Takimoto's student Hikaru Saji determined at Okazaki that critical night length, the major determinant of flowering in SDP, also relates to the beginning of a short photoperiod and the end of a longer one in *Pharbitis* (Saji et al., 1984). Lumsden made a similar observation (Lumsden, 1984). It therefore appeared that "the critical night-length depends on the same flowering rhythm as that which gives rise to the [light-break] response and is controlled by light in the same way" (Vince-Prue and Lumsden, 1987).

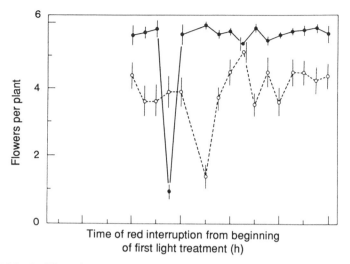

FIGURE 34-2   Flowering response of dark-grown seedlings to a 15-min R interruption of an inductive 48-h dark period after one continuous (●) or skeleton (○) 24-h photoperiod. (From Lumsden et al., 1982. Reproduced with permission from the American Society of Plant Physiologists.)

EXTERNAL COINCIDENCE

In 1967, Salisbury had suggested a similar scenario for light-grown *Xanthium*. But he attributed the effect of a light break to rhythm rephasing (see Chapter 14). To distinguish between this possibility and the direct action of a light break, Vince-Prue and co-workers substituted skeleton R photoperiods for a complete WL photoperiod, as Hillman had done. They also induced flowering with a single R pulse and simultaneous application of the cytokinin benzyladenine (Ogawa and King, 1979).

Plants exposed to the latter treatment were most sensitive to a light break 14–15 h after the pulse (R_1) and again after a further 24 h. Thus R_1 initiated a light-break rhythm. With R skeleton photoperiods, the point of maximal inhibition was delayed as the interval between the skeleton pulses increased. So under these conditions, the second R pulse (R_2) rephased the rhythm. Because the effect of a third R pulse (the light break) depended on its distance from the end of the skeleton photoperiod rather than from the end of the dark period (Fig. 34-3), the light break appeared to inhibit flowering by interacting with rather than rephasing the rhythm (Lumsden et al., 1982), indicating that an external coincidence model was applicable.

"The phase of a circadian timer is set at the transition to darkness (at dusk) by the action of the photoperiod," Vince-Prue concluded. "During

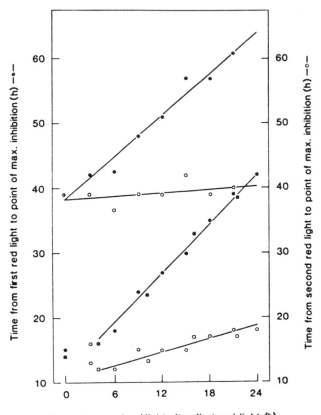

FIGURE 34-3 Time of maximum sensitivity to a R interruption of an inductive dark period as a function of the time interval between two R pulses beginning and ending a skeleton photoperiod. Time of maximum sensitivity in real-time (●,■); time of maximum sensitivity from the second R pulse (○,□). Plants were transferred to continuous light (to promote flower development) 72 h after the first R pulse (●,○) or 48 h after the second R pulse (■,□). (From Lumsden et al., 1982. Reproduced with permission from the American Society of Plant Physiologists.)

darkness, the circadian timer establishes phases of sensitivity to light and the photoperiodic response of the plant depends on whether the light-sensitive phase of the rhythm has been reached before the next exposure to light (dawn) occurs. This model for the photoperiodic mechanism envisages that there are two different actions of light. The first is to set the phase of the time-keeping rhythm and the second is to interact with a particular phase of the rhythm to inhibit flowering. Additionally, it has been found that removal of the Pfr form of phytochrome early in the night can, under some conditions, also inhibit induction [see Chapter 14]. Thus light has multiple actions in the photoperiodic control of flowering" (Vince Prue, 1985).

In seedlings de-etiolated with 24 h WL, hourly R pulses during the early part of the subsequent dark period constantly reset the rhythm to CT6, the circadian time at which the lights-off signal normally releases the rhythm (Lumsden and Vince-Prue, 1984; Lumsden et al., 1986). So the external coincidence model had to be tested further. Helen S. J. Lee detected different fluence responses for phase shifting and direct regulation of flowering, suggesting they were distinct processes (Lee et al., 1987).

In Okazaki, Lumsden was able to completely separate phase shifting and floral regulation in *Pharbitis* by scanning a 72-h dark period after a 24-h photoperiod. A 6-s R pulse 6 h into the dark period delayed the rhythm by 3 h but did not reduce the level of flowering. Two hours later, even a 200-s pulse failed to rephase the rhythm, but 6 s of R inhibited flowering (Lumsden and Furuya, 1986). So although a light break could rephase the rhythm, it appeared also to have a separate action, as the external coincidence model demanded.

This experiment implicated phytochrome in rephasing, because the required R exposure at 6 h was so brief. Lumsden's demonstration of FR reversibility confirmed this assumption (see Vince-Prue and Takimoto, 1987).

DAWN AND DUSK SIGNALS

Whether photochrome also sets the rhythm at dawn and dusk is still unclear. Because the rhythm begins in *Pharbitis* after a single R pulse (King et al., 1982), the pigment may convey a lights-on signal. FR reversibility has not been demonstrated, however (Lumsden, 1984), perhaps because of the direct requirement for Pfr and also because a FR pulse itself can start the rhythm (Lumsden, 1984).

The strongest evidence for the involvement of phytochrome in the onset of dark timing after the photoperiod comes from the observation that end-of-day FR advances the position of maximal sensitivity to a light break by 30–40 min compared with end-of-day R (Fig. 34-4) (Lumsden and Vince-Prue, 1984). Because the advance is so small, it can be observed only when the dark period is sampled at very frequent intervals. "This implies," Vince-Prue says, "that *this* Pfr is unstable or reverts rapidly, although we don't know for sure, since 'old' Pfr could become ineffective."

When Takimoto exposed light-grown *Pharbitis* seedlings to photoperiods of various colors, much lower intensities of R than of WL, B, or

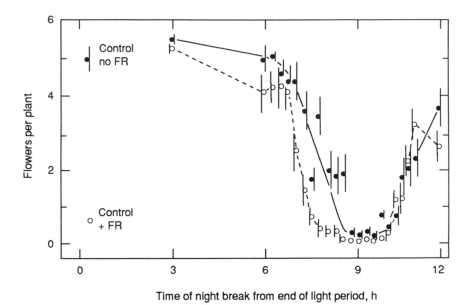

FIGURE 34-4   Effect of an end-of-day, 5-min FR exposure on the time of sensitivity of *Pharbitis* to a R light break. (From Lumsden and Vince-Prue, 1984.)

green produced a rhythmic response to light breaks. Moreover the effectiveness ratio for the various colors was exactly the same as that for the light break and reversal of FR light breaks, "suggesting that the same or very similar pigments are associated with all of these reactions." A FR photoperiod failed to produce a lights-off rhythm, but the requirement of Pfr during and after the photoperiod complicates the interpretation of this observation (Takimoto, 1967).

Perception of dusk and light breaks appears to involve different mechanisms, however, even if phytochrome is the photoreceptor in each case. Comparing the effects on floral induction in *Xanthium* of low levels of R and FR during the first 2 h and the middle of a 16-h dark period, Salisbury determined that higher irradiances were required to prevent lights-off timing than to give a light-break reaction. Initiation of timing was also somewhat temperature-sensitive. The results suggested that "the pigment system (phytochrome?) and/or responses to it may be significantly different as they function during twilight (initiation of dark measurement), and as they function during a light break several hours later" (Salisbury, 1981).

Theoretically, changes in light quality or light quantity could constitute the dusk signal. Shropshire speculated that perhaps "sudden rapid shifts in the FR/R ratio of solar radiation may provide the timing cues of those organisms which contain phytochrome" (Shropshire, 1973). But changes in light quality at dusk in diffuse light are less dramatic than those observed in the direct solar beam on the roof of the Smithsonian Institution.

Decreases in irradiance do occur rapidly at dusk, however, and do appear to affect the onset of photoperiodic timing. Takimoto observed that dark timing began after *Pharbitis* was transferred to a fluence of 0.04 W m^{-2} but not 1.0 W m^{-2}. Because the plants remained in fluorescent light, there was no change in light quality (Takimoto, 1967). When Lumsden lowered irradiance stepwise to more closely simulate dusk, time measurement began at 2.0 W m^{-2} (Vince-Prue, 1983; Lumsden and Vince-Prue, 1984). Morever, *Xanthium* sees darkness at fluence levels that are reached after only 5–12 min of twilight (Salisbury, 1981), suggesting that an irradiance response would allow plants to detect the onset of dusk with precision.

A drop in irradiance would halt cycling between Pfr and Pr, which might convey the dusk signal, Vince-Prue suggested "very speculatively" (Jose and Vince-Prue, 1978). But Salisbury discovered that low temperatures delayed the initiation of dark timing. Moreover, FR plus R was less effective than R in delaying the onset of dark measurement. Both observations argue against a cessation-of-cycling mechanism.

"The temperature-sensitive process that initiates dark measurement could well be conversion and/or destruction of Pfr," Salisbury suggested (Salisbury, 1981). Thus when irradiance fell, photoconversion of Pr to Pfr would at some point fail to compensate for removal of Pfr by dark reactions, and timing would begin. Alternatively, dissociation of a Pfr-receptor complex might initiate dark timing and account for the advance of the LB$_{max}$ position by FR (Lumsden and Vince-Prue, 1984).

Little is known about the behavior of Pfr at dusk in light-grown plants. But spectrophotometric measurements in *Pharbitis* cotyledons grown at 18°C (where almost no chlorophyll forms) suggested that Pfr reverts largely to Pr within an hour, with no loss of total phytochrome, when plants are transferred from fluorescent light to darkness (Vince-Prue, et al., 1978). In Wageningen, Johannes Rombach studied cotyledons of norflurazon-treated *Pharbitis* de-etiolated with three 12-h photoperiods. After a R pulse, Pfr levels decreased and Pr levels increased during the first 30 min of darkness (Fig. 34-5), indicating rapid reversion. This process was observed only in seedlings de-etiolated with at least 4 h WL (Rombach, 1986). Thus phytochrome reversion, triggered by a decrease in irradiance, might convey the dusk signal. But reversion has never been observed in Centrospermae and monocots.

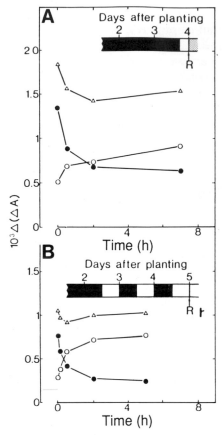

FIGURE 34-5   Pfr (●), Pr (○), and P_{total} (△) levels as a function of time in darkness at 22°C after (A) 6 h WL and then 10 min R, and (B) 3 × 12 h WL and then 10 min R. (From Rombach, 1986.)

TWO-PHYTOCHROME-POOL HYPOTHESIS

Although the role of phytochrome at dusk is unclear, it is apparent that Pfr acts during the photoperiod and early part of the darkness to directly promote flowering (see Chapter 13). Thus "if a rapidly decaying/reverting Pfr pool does function in dusk perception, a second pool must have different properties and a different function in the photoperiodic mechanism" (Vince-Prue, 1981). The simultaneous operation (see Fig. 13-7) of the Pfr-requiring reaction and the light-break effect also suggested that at least two pools of phytochrome might coexist (King et al., 1982). One pool was proposed to contain Pfr that persists in darkness and promotes floral induction without interacting with an endogenous rhythm. The Pfr in the other pool disappears rapidly in darkness and, if restored by a light break at a sensitive time in the circadian rhythm, inhibits flowering in SDP (Takimoto and Saji, 1984).

To determine whether two such phytochrome pools might coexist in darkness, Saji and Furuya exposed continuously light-grown *Lemna paucicostata* 441 to monochromatic light at various times in a single 14-h dark period, using an elaborate, computer-operated spectrograph that had recently been constructed at Okazaki (Watanabe et al., 1982). With very narrow wavebands at their disposal, the researchers could distinguish between phytochrome's two forms, because only Pfr absorbs pure 750-nm light.

Light breaks at 450, 550, and 650 nm inhibited flowering between hours 4 and 12 of darkness, but especially at hour 8. The most effective wavelength was 650 nm, suggesting the involvement of Pfr. FR at 750 nm did not reverse this effect, because it completely inhibited flowering at any time during the first 8 h. But inhibition by FR was R-reversible (although not at hour 8), implicating phytochrome. Because the FR-fluence response curves were identical at 0, 4, and 8 h, "removal of a constant fraction of Pfr is required to inhibit flowering regardless of time or irradiation (Saji et al., 1982). Thus at 8 h, Pfr was required for floral induction. But production of Pfr at this time with a R light break was inhibitory.

"There may exist," the researchers proposed, "two different phytochrome systems controlling photoperiodic floral induction, although other possibilities cannot be eliminated. Namely, one system must be in the Pfr form (responsible for the far-red effect) and the other phytochrome system must be in the Pr form (responsible for the red-effect) at the 8th hour of the dark period" (Saji et al., 1982).

Visiting Okazaki for two consecutive summers, Vince-Prue reached a similar conclusion with dark-grown *Pharbitis* seedlings. After a single R pulse plus benzyladenine, FR was inhibitory for the first 24 h of an inductive 48-h dark period. Because inhibition was reversed by FR and reinstated by subsequent R, Pfr was clearly present and active. A R light break was most inhibitory at hours 10 and 15 and most promotive at hours 5 and 24. At 15 h, where the effect was not FR-reversible, an action spectrum showed a sharp peak at 660 nm, a shoulder in the R, and a rapid cut-off in the FR (Fig. 34-6), confirming the photoconversion of Pr. Thus both forms of phytochrome appeared to coexist in *Pharbitis,* as in *Lemna.* Moreover, the Pfr had not reverted or been degraded 24 h after the initial R pulse (Saji et al., 1983). This Pfr did not have to act during the dark period if the photoperiod was sufficiently long, Lumsden discovered by extending

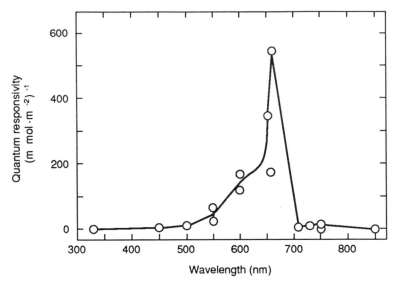

FIGURE 34-6   Action spectrum for inhibition of flowering in *Pharbitis* by a light break 15 h after an initial R pulse. (From Saji et al., 1983.)

the photoperiod to 32 h (Lumsden, 1984; see Vince Prue and Lumsden, 1987). Moreover, it acted more quickly in the light than in darkness, because FR was inhibitory for 55 h after a 12-h photoperiod but only for 20 h after a 24-h photoperiod (Thomas and Lumsden, 1984).

To test the two-pool hypothesis further, Lumsden and Saji irradiated *Lemna paucicostata* on the spectrograph at the beginning or middle of a 16-h dark period. At the beginning, the action spectrum for inhibition of floral induction strongly resembled that for Pfr (Fig. 34-7A). At 8 h, however, it peaked at 640 nm, the maximum for Pr in the presence of chlorophyll (Fig. 34-8A), but also extended beyond 720 nm, indicating the continued presence of Pfr. In fact, the FR portions of two curves could be superimposed (Fig. 34-8B). These results were "consistent with there being a stable pool of Pfr necessary for the induction of flowering and another pool of phytochrome in a different cellular environment which participates in the [light-break] reaction as Pr" (Lumsden et al., 1987).

MOLECULAR SPECIES

Noting that "interactions with the circadian rhythm are closely coupled to photochemical turnover of phytochrome whereas the end-of-day response is a persistent long-term reaction to a stable fraction of Pfr," Thomas and

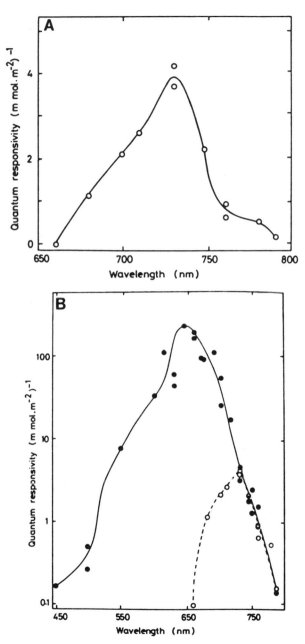

FIGURE 34-7   (A) Action spectrum for the inhibition of flowering in *Lemna paucicostata*
441 by light given during the first 20 min of a 16-h dark period. (B) The same but at 8 h. (From
Lumsden et al., 1987.)

FIGURE 34-8   Thomas (*left*) explaining mysteries of photoperiodism to British Prime Minister Margaret Thatcher at Institute of Horticultural Research, Wellesbourne, October 1988.

Lumsden were tempted'' to ascribe these actions respectively to labile and stable pools of phytochrome such as those which have been identified on the basis of their Pfr destruction kinetics'' (Thomas and Lumsden, 1984).

The discovery of multiple molecular species of phytochrome ''appeared to provide a potential let out for the conflicting actions of light,'' Thomas says, ''since the suggested different physiological pools of phytochrome could have been type I and type II phytochromes. More recently, I think that is probably not the case, mainly because in the system where we have the most evidence—*Pharbitis*—you need a rapidly lost lights-off signal but at the same time can show a Pfr-persistent response, which would seem to be due to type II phytochrome. If so, could the rapid response be type I phytochrome? But phytochrome synthesis is heavily inhibited in the light, so there is virtually no type I phytochrome at the end of the light period. So type I phytochrome is unlikely to be doing the timing.''

Thomas (Fig. 34-8) and Vince-Prue suggested that two—or more—pools might contain identical phytochrome molecules attached to different receptors. Thus Pfr would interact with one receptor to phase the circadian rhythm, another to effect the light-break response, and a third to directly promote flowering. Only the latter interaction would involve a long-lasting complex (Thomas and Vince-Prue, 1987). Such receptors could be cell-specific, although not organ-specific because the Pfr-requiring reaction in *Pharbitis,* like dark-period measurement, appeared to occur in the co-tyledon (Knapp et al., 1986).

Thomas's latest suggestion is that heterodimers may constitute a revert-ing pool of phytochrome and homodimers the stable pool. "There is some evidence," he says, "that heterodimers are less stable than homodimers (see Chapter 25). In WL, you will have significant proportions of each. So at the end of a WL period, you would have heterodimers that could revert very rapidly back to Pr and homodimers that were stable during the dark period." He also points out that, if phytochrome II encompasses two (or more) gene products, there might be light-stable phytochromes with differ-ing abilities to revert in darkness (Thomas, 1991).

LONG-DAY PLANTS

Vince-Prue observed that floral induction in some SDP, such as straw-berry (*Fragaria*), requires prolonged irradiation (Vince-Prue and Gut-tridge, 1973), while flowering in some LDP, such as *Fuchsia hybrida* (Vince-Prue, 1976), is controlled by a brief light break during the dark period, as in SDP. So in 1978, Hillman proposed the categories "light-dominant" and "dark-dominant" responses (see Vince-Prue, 1980). As well as needing prolonged exposures, the light-dominant response requires R/FR mixtures, and its action spectrum peaks in the FR (see Fig. 13-11). Moreover, the optimal R/FR ratio often shifts with time (see Fig. 13-10).

Light-dominant and dark-dominant responses may occur in the same plant. Floral induction in chrysanthemum requires prolonged exposures to light, but subsequent development of the inflorescence is under the control of night length (Cockshull and Vince-Prue, 1980).

ACTION OF FAR-RED LIGHT

The action of FR on light-dominant plants is far from understood, but it appears not to involve photosynthesis. Deitzer obtained very different fluence response curves for the two processes (Deitzer, 1983) and ob-served FR promotion of flowering in norflurazon-treated—chlorophyll free—*Arabidopsis* seedlings (Deitzer, 1985). Because R/FR ratios below 0.6 were most effective in barley and because the response showed strong irradiance dependence and reciprocity failure, Deitzer concluded that